T0135069

Studies in Computational Intelligence

Volume 898

Series Editor

Janusz Kacprzyk, Polish Academy of Sciences, Warsaw, Poland

The series "Studies in Computational Intelligence" (SCI) publishes new developments and advances in the various areas of computational intelligence—quickly and with a high quality. The intent is to cover the theory, applications, and design methods of computational intelligence, as embedded in the fields of engineering, computer science, physics and life sciences, as well as the methodologies behind them. The series contains monographs, lecture notes and edited volumes in computational intelligence spanning the areas of neural networks, connectionist systems, genetic algorithms, evolutionary computation, artificial intelligence, cellular automata, self-organizing systems, soft computing, fuzzy systems, and hybrid intelligent systems. Of particular value to both the contributors and the readership are the short publication timeframe and the world-wide distribution, which enable both wide and rapid dissemination of research output.

Indexed by SCOPUS, DBLP, WTI Frankfurt eG, zbMATH, SCImago.

More information about this series at http://www.springer.com/series/7092

Nguyen Ngoc Thach · Vladik Kreinovich ·
Nguyen Duc Trung
Editors

Data Science for Financial Econometrics

Springer

Editors
Nguyen Ngoc Thach
Institute for Research Science
and Banking Technology
Banking University Ho Chi Minh City
Ho Chi Minh, Vietnam

Vladik Kreinovich
Department of Computer Science
Institute for Research Science
and Banking Technology
El Paso, TX, USA

Nguyen Duc Trung
Banking University Ho Chi Minh City
Ho Chi Minh City, Vietnam

ISSN 1860-949X ISSN 1860-9503 (electronic)
Studies in Computational Intelligence
ISBN 978-3-030-48855-0 ISBN 978-3-030-48853-6 (eBook)
https://doi.org/10.1007/978-3-030-48853-6

This Springer imprint is published by the registered company Springer Nature Switzerland AG
The registered company address is: Gewerbestrasse 11, 6330 Cham, Switzerland

Preface

Researchers and practitioners have been analyzing data for centuries, by using techniques ranging from traditional statistical tools to more recent machine learning and decision-making methods. Until recently, however, limitations on computing abilities necessitates the use of simplifying assumptions and models, and processing data samples instead of all the available data. In the last decades, a steady progress both in computing power and in data processing algorithms has enabled us to directly process all the data—to the extent that sometimes (e.g., in applications of deep learning) our data collection lags behind our computing abilities. As a result, a new multi-disciplinary field has emerged: *data science*, a field that combines statistics, data analysis, machine learning, mathematics, computer science, information science, and their related methods in order to understand and analyze real-life phenomena.

Data science is largely in its infancy. In many application areas, it has already achieved great successes, but there are still many application areas in which these new techniques carry a great potential. One of such areas is econometrics—qualitative and numerical analysis of economic phenomena. A part of economics which is especially ripe for using data science is financial economics, in which all the data are numerical already.

This volume presents the first results and ideas of applying data science techniques to economic phenomena—and, in particular, financial phenomena. All this is still work in progress. Some papers build on the successes of the traditional methods and just hint on how a more in-depth application of new techniques can help, some use new methods more bravely, and some deal with theoretical foundations behind the new techniques—yet another area that still has many open questions. Overall, papers from this volume present a good picture of a working body of using data science to solve different aspects of economic problems.

This volume shows what can be achieved, but even larger is the future potential. We hope that this volume will inspire practitioners to learn how to apply various data science techniques to economic problems, and inspire researchers to further improve the existing techniques and to come up with new data science techniques for economics.

We want to thank all the authors for their contributions and all anonymous referees for their thorough analysis and helpful comments.

The publication of this volume is partly supported by the Banking University of Ho Chi Minh City, Vietnam. Our thanks to the leadership and staff of the Banking University, for providing crucial support. Our special thanks to Prof. Hung T. Nguyen for his valuable advice and constant support.

We would also like to thank Prof. Janusz Kacprzyk (Series Editor) and Dr. Thomas Ditzinger (Senior Editor, Engineering/Applied Sciences) for their support and cooperation in this publication.

Ho Chi Minh, Vietnam   Nguyen Ngoc Thach
El Paso, USA   Vladik Kreinovich
Ho Chi Minh, Vietnam   Nguyen Duc Trung
January 2020

Contents

Theoretical Research

A Theory-Based Lasso for Time-Series Data

Achim Ahrens, Christopher Aitken, Jan Ditzen, Erkal Ersoy, David Kohns, and Mark E. Schaffer

Abstract We present two new lasso estimators, the HAC-lasso and AC-lasso, that are suitable for time-series applications. The estimators are variations of the theory-based or 'rigorous' lasso of Bickel et al. (2009), Belloni et al. (2011), Belloni and Chernozhukov (2013), Belloni et al. (2016) and recently extended to the case of dependent data by Chernozhukov et al. (2019), where the lasso penalty level is derived on theoretical grounds. The rigorous lasso has appealing theoretical properties and is computationally very attractive compared to conventional cross-validation. The AC-lasso version of the rigorous lasso accommodates dependence in the disturbance term of arbitrary form, so long as the dependence is known to die out after q periods; the HAC-lasso also allows for heteroskedasticity of arbitrary form. The HAC- and AC-lasso are particularly well-suited to applications such as nowcasting, where the time series may be short and the dimensionality of the predictors is high. We present

Invited paper for the International Conference of Econometrics of Vietnam, 'Data Science for Financial Econometrics', Banking University of Ho-Chi-Minh City, Vietnam, 13–16 January 2020. Our exposition of the 'rigorous lasso' here draws in part on our paper Ahrens et al. (2020). All errors are our own.

A. Ahrens
ETH Zürich, Zürich, Switzerland
e-mail: achim.ahrens@gess.ethz.ch

C. Aitken · J. Ditzen · E. Ersoy · D. Kohns · M. E. Schaffer (✉)
Heriot-Watt University, Edinburgh, UK
e-mail: M.E.Schaffer@hw.ac.uk

C. Aitken
e-mail: christopher.f.aitken@gmail.com

J. Ditzen
e-mail: jan@ditzen.net

E. Ersoy
e-mail: e.ersoy@hw.ac.uk

D. Kohns
e-mail: david.kohns94@googlemail.com

N. Ngoc Thach et al. (eds.), *Data Science for Financial Econometrics*, Studies in Computational Intelligence 898, https://doi.org/10.1007/978-3-030-48853-6_1

some Monte Carlo comparisons of the performance of the HAC-lasso versus penalty selection by cross-validation approach. Finally, we use the HAC-lasso to estimate a nowcasting model of US GDP growth based on Google Trends data and compare its performance to the Bayesian methods employed by Kohns and Bhattacharjee (2019).

Keywords Lasso · Machine learning · Time-series · Dependence

1 Introduction

Machine learning methods are increasingly widely used in economic and econometric analysis (Varian 2014; Mullainathan and Spiess 2017; Athey 2017; Kleinberg et al. 2018). One of the most popular such methods is the lasso (Least Absolute Shrinkage and Selection Operator) of Tibshirani (1996). So far, however, most applications of the lasso in time-series analysis have focused on the problem of lag selection (see, e.g., Hsu et al. (2008) or Nardi and Rinaldo (2011)). In this paper, we present two new lasso estimators, the HAC-lasso and AC-lasso, that are suitable for time-series applications. The estimators are variations of the 'rigorous' or 'plug-in' lasso of Bickel et al. (2009), Belloni et al. (2011), Belloni and Chernozhukov (2013), Belloni et al. (2016) and recently extended to the case of dependent data by Chernozhukov et al. (2019), where the lasso penalty level is derived on theoretical grounds.

We present the theoretical foundations of the rigorous lasso, and then the HAC- and AC-lasso variations for the dependent data setting. We identify limitations and potential pitfalls in the way these lasso estimators can be used. Our proposed approach is appropriate for lag selection in time series models, including VARs, but is arguably even better suited to high-dimensional time series models.

The article is structured as follows. Section 2 introduces the concept of regularised regression and related notation, and Sect. 3 provides an overview of recent work on this theme in a time series context. In Sect. 4, we discuss sparsity and the rigorous lasso. Section 6 discusses the lasso with a focus on time series applications, and Sect. 7 presents Monte Carlo results demonstrating the predictive and model selection performance of the HAC-lasso. In Sect. 8 we use the HAC-lasso to estimate a nowcasting model of US GDP growth based on Google Trends data.

The statistical software used throughout is Stata except where noted. We use a version of the `rlasso` command in the software package lassopack by Ahrens, Hansen and Schaffer, modified to incorporate the HAC-lasso and AC-lasso estimators. The package without this modification is presented and discussed in Ahrens et al. (2020).

2 Basic Setup and Notation

We begin with the simple linear regression model in the cross-section setting with p independent variables.

$$y_i = \beta_0 + \beta_1 x_{1i} + \beta_2 x_{2i} + \cdots + \beta_p x_{pi} + \varepsilon_i \qquad (1)$$

In traditional least squares regression, estimated parameters are chosen to minimise the residual sum of squares (RSS):

$$RSS = \sum_{i=1}^{n} \left(y_i - \beta_0 - \sum_{j=1}^{p} \beta_j x_{ij} \right)^2 \qquad (2)$$

The problem arises when p is relatively large. If the model is too complex or flexible, the parameters estimated using the training dataset do not perform well with future datasets. This is where regularisation is key. By adding a shrinkage quantity to RSS, regularisation shrinks parameter estimates towards zero. A very popular regularised regression method is the lasso, introduced by Frank and Friedman (1993) and Tibshirani (1996). Instead of minimising the RSS, the lasso minimises

$$RSS + \lambda \sum_{j=1}^{p} |\beta_j| \qquad (3)$$

where λ is the tuning parameter that determines how much model complexity is penalised. At one extreme, if $\lambda = 0$, the penalty term disappears, and lasso estimates are the same as OLS. At the other extreme, as $\lambda \to \infty$, the penalty term grows and estimated coefficients approach zero.

Choosing the tuning parameter, λ, is critical. We discuss below both the most popular method of tuning parameter choice, cross-validation, and the theory-derived 'rigorous lasso' approach.

We note here that our paper is concerned primarily with prediction and model selection with dependent data rather than causal inference. Estimates from regularised regression cannot be readily interpreted as causal, and statistical inference on these coefficients is complicated and an active area of research.[1]

We are interested in applying the lasso in a single-equation time series framework, where the number of predictors may be large, either because the set of contemporaneous regressors is inherently large (as in a nowcasting application), and/or because the model has many lags.

We write a general time-series single-equation model with lags of the dependent variable and other predictors as follows. The dataset has $t = 1, \ldots, n$ observations. There are K contemporaneous regressors, x_{tj}, $j = 1, \ldots, K$, with coefficients β_j. The model may include lags of up to order R of the regressors, indexed by $r = 1, \ldots, R$, with coefficients β_{jr}. The model may also be autoregressive and include lags up to order R_y of the dependent variable with coefficients ρ_r. Denote the constant by β_0. The disturbance term is ε_t and may be serially dependent.

---

[1] See, for example, Buhlmann (2013), Meinshausen et al. (2009), Weilenmann et al. (2017), Wasserman and Roeder (2009), Lockhart et al. (2014).

We can write the general model parsimoniously using the lag operator applied to x_{tj} and y_t, respectively:

$$B_j(L) = 1 + \beta_{j1}L + \beta_{j2}L^2 + \cdots + \beta_{jR}L^R \qquad (4)$$

$$P(L) = 1 - \rho_1 L - \rho_2 L^2 - \cdots + \rho_{R_y}L^{R_y} \qquad (5)$$

The general model can therefore be written compactly by

$$P(L)y_t = \beta_0 + \sum_{j=1}^{K} B_j(L)x_{tj} + \varepsilon_t \qquad (6)$$

The model size (total number of predictors excluding the intercept) is $p := (K(R + 1) + R_y)$.

The model may be high-dimensional in either the number of lags (R and/or R_y), the dimension of the x predictors (K), or both. The general *HD-RK* model arises where the dimensionality of both the lags and predictors is large. The *HD-R* model describes the case with a large number of lags but the dimensionality of the predictors is small; a typical example would be a single equation in a VAR (vector autoregression). A special case of the HD-R model is the *high-dimensional univariate autoregressive* model (HD-UAR), where $K = 0$ and the model size is $p = R_y$; this is the model examined by e.g. Bergmeir et al. (2018).

The *HD-K* model refers to the case where the number of lags is small but the dimensionality K of the predictors is large. A typical example would be a nowcasting application, where many different predictors are used, but few or no lags of the predictors appear. A special case of the HD-K model is the *high-dimensional contemporaneous* model (HD-C), where no lags appear ($R = R_y = 0$) and the model size is $p = K$. We are particularly interested in this paper in applications of our proposed HAC-lasso and AC-lasso estimators to the HD-K and HD-C models.

3   Literature Review

The literature on lag selection in VAR and ARMA models is very rich. Lütkepohl (2005) notes that fitting a VAR(R) model to a VAR(R) process yields a better outcome in terms of mean square error than fitting a VAR($R + i$) model, because the latter results in inferior forecasts than the former, especially when the sample size is small. In practice, the order of data generating process (DGP) is, of course, unknown and we face a trade-off between out-of-sample prediction performance and model consistency. This suggests that it is advisable to avoid fitting VAR models with unnecessarily large orders. Hence, if an upper bound on the true order is known or suspected, the usual next step is to set up significance tests. In a causality context, Wald tests are useful. The likelihood ratio (LR) test can also be used to compare maximum log-likelihoods over the unrestricted and restricted parameter space.

If the focus is on forecasting, information criteria are typically favoured. In this vein, Akaike (1969, 1971) proposed using 1-step ahead forecast mean squared error (MSE) to select the VAR order, which led to the final prediction error (FPE) criterion. Akaike's information criterion (AIC), proposed by Akaike (1974), led to (almost) the same outcome through different reasoning. AIC, defined as $-2 \times$ log-likelihood $+ 2 \times$ no. of regressors, is approximately equal to FPE in moderate and large sample sizes (T).

Two further information criteria are popular in applied work: Hannan-Quinn criterion (Hannan and Quinn 1979) and Bayesian information criterion (Schwarz 1978). These criteria perform better than AIC and FPE in terms of order selection consistency under certain conditions. However, AIC and FPE have better small sample properties, and models based on these criteria produce superior forecasts despite not estimating the orders correctly (Lütkepohl 2005). Further, the popular information criteria (AIC, BIC, Hannan-Quinn) tend to underfit the model in terms of lag order selection in a small-t context (Lütkepohl 2005).

Although applications of lasso in a time series context are an active area of research, most analyses have focused solely on the use of lasso in lag selection. For example, Hsu et al. (2008) adopt the lasso for VAR subset selection. The authors compare predictive performance of two-dimensional VAR(5) models and US macroeconomic data based on information criteria (AIC, BIC), lasso, and combinations of the two. The findings indicate that the lasso performs better than conventional selection methods in terms of prediction mean squared errors in small samples. In a related application, Nardi and Rinaldo (2011) show that the lasso estimator is model selection consistent when fitting an autoregressive model, where the maximal lag is allowed to increase with sample size. The authors note that the advantage of the lasso with growing R in an AR(R) model is that the 'fitted model will be chosen among all possible AR models whose maximal lag is between 1 and [...] log(n)' (Nardi and Rinaldo 2011).

Medeiros and Mendes (2015) discuss the asymptotic properties of the adaptive lasso[2] (adaLASSO) of Zou (2006) in sparse, high-dimensional, linear time series models. The authors show that the adaLASSO is model selection consistent as the number of observations increases. They also demonstrate adaLASSO's oracle properties[3] even when the errors are non-Gaussian and conditionally heteroskedastic.

The fused lasso of Tibshirani et al. (2005) is designed to encourage sparse solutions while accounting for natural ordering of features. This is particularly useful when the number of features is much greater than the sample size. The main shortcoming of this approach is the speed of estimation, which could be ameliorated by hybridising it with the group lasso approach of Yuan and Lin (2006).

---

[2]Zou (2006) proposed adaLASSO with an ℓ_1-penalty but with the addition of weights that are data-dependent. The author noted that under some conditions, the weighted lasso could have oracle properties.

[3]Fan and Li (2001) note that an estimator has the oracle property if it both selects the correct subset of non-negligible variables and the estimates of non-zero parameters have the same asymptotic distribution as the ordinary least squares (OLS) estimator in a regression including only the relevant variables.

Regularisation methods, including the lasso, are closely related to information criteria. For example, AIC shares the same underlying principle as regularisation and is asymptotically equivalent to leave-one-out cross-validation. To see this, note that AIC can be interpreted as penalised likelihood which imposes a penalty on the number of predictors included in the model (Ahrens et al. 2020). The disadvantage of using this approach to find the model with the lowest AIC is two-fold. First, all different model specifications need to be estimated under the assumption that the true model is among those specifications. Second, even with a relatively small number of regressors, the number of different models quickly increases, making the approach intractable. In this regard, the clear advantage of regularisation is that model selection is reduced to a data-driven one-dimensional problem, which involves choosing λ.

Having said that, an attractive feature of information criteria is model selection consistency. For example, BIC is model selection consistent if the true model is among the candidate models, although AIC is not (Ahrens et al. 2020). This implies that the true model is selected with probability 1 as $n \to \infty$. By contrast, AIC is said to be loss efficient, because it minimises the average squared error among all candidate models. Therefore, we face a trade-off: if the objective is model selection, BIC is preferable; if prediction has higher priority, AIC performs better. Furthermore, Yang (2005) demonstrates that a loss efficient selection method, such as AIC, cannot be model selection consistent, and vice versa. Zhang et al. (2010) confirm this observation in penalised regressions. A key drawback to using information criteria in this context is the strong assumption that the true model is among those being tested. In fact, in some cases, the assumption that a true model exists is problematic. Furthermore, AIC is biased in small samples, and neither AIC nor BIC is appropriate in a large-p-small-n context, where they tend to select too many variables (Ahrens et al. 2020).

Although the discussion around high dimensionality in lags as well as the cross-section of regressors at each point in time is rather new in frequentist estimation of VARs, this problem has already received much attention in the Bayesian VAR literature. Starting with early contributions, such as Litterman (1986), and Doan et al. (1983), priors on the VAR coefficients incorporated the idea of different degrees of shrinkage for regressors and lags of the dependent variable, while also allowing for a global level of shrinkage. The success of global-local shrinkage priors has spawned a vast literature on large Bayesian VAR estimation (see for a survey Banbura et al. (2008)).

4   Sparsity and the Rigorous or Plug-In Lasso

4.1   High-Dimensional Data and Sparsity

The high-dimensional linear model is:

$$y_i = x'_i \beta + \varepsilon_i \tag{7}$$

Our initial exposition assumes independence, and to emphasise independence we index observations by i. Predictors are indexed by j. We have up to $p = \dim(\boldsymbol{\beta})$ potential predictors. p can be very large, potentially even larger than the number of observations n. For simplicity we assume that all variables have already been mean-centered and rescaled to have unit variance, i.e., $\sum_i y_i = 0$ and $\frac{1}{n} \sum_i y_i^2 = 1$, and similarly for the predictors x_{ij}.

If we simply use OLS to estimate the model and p is large, the result is very poor performance: we overfit badly and classical hypothesis testing leads to many false positives. If $p > n$, OLS is not even identified.

How to proceed depend on what we believe the 'true model' is. Does the true model (DGP) include a very large number of regressors? In other words, is the set of predictors that enter the model 'dense'? Or does the true model consist of a small number of regressors s, and all the other $p - s$ regressors do not enter (or equivalently, have zero coefficients)? In other words, is the set of predictors that enter the model 'sparse'?

In this paper, we focus primarily on the 'sparse' case and in particular an estimator that is particularly well-suited to the sparse setting, namely the *lasso* introduced by Tibshirani (1996).

In the **exact sparsity** case of the p potential regressors, **only s regressors belong in the model**, where

$$s := \sum_{j=1}^{p} \mathbb{1}\{\beta_j \neq 0\} \ll n. \tag{8}$$

In other words, most of the true coefficients β_j are actually zero. The problem facing the researcher is that which are zeros and which are not is unknown.

We can also use the weaker assumption of **approximate sparsity**: some of the β_j coefficients are well-approximated by zero, and the approximation error is sufficiently 'small'. The discussion and methods we present in this paper typically carry over to the approximately sparse case, and for the most part we will use the term 'sparse' to refer to either setting.

The sparse high-dimensional model accommodates situations that are very familiar to researchers and that typically presented them with difficult problems where traditional statistical methods would perform badly. These include both settings where the number p of observed potential predictors is very large and the researcher does not know which ones to use, and settings where the number of observed variables is small but the number of potential predictors in the model is large because of interactions and other non-linearities, model uncertainty, temporal and spatial effects, etc.

4.2   The Penalisation Approach and the Lasso

The basic idea behind the lasso and its high-dimensional-friendly relatives is *penalisation*: put a penalty or 'price' on the use of predictors in the objective function that the estimator minimizes.

The lasso estimator minimizes the mean squared error subject to a penalty on the *absolute size* of coefficient estimates (i.e., using the ℓ_1 norm):

$$\hat{\boldsymbol{\beta}}_{\text{lasso}}(\lambda) = \arg\min \frac{1}{n} \sum_{i=1}^{n} \left(y_i - \boldsymbol{x}_i'\boldsymbol{\beta}\right)^2 + \frac{\lambda}{n} \sum_{j=1}^{p} \psi_j |\beta_j|. \tag{9}$$

The tuning parameter λ controls the overall penalty level and ψ_j are predictor-specific penalty loadings.

The intuition behind the lasso is straightforward: there is a cost to including predictors, the unit 'price' per regressor is λ, and we can reduce the value of the objective function by removing the ones that contribute little to the fit. The bigger the λ, the higher the 'price', and the more predictors are removed. The penalty loadings ψ_j introduce the additional flexibility of putting different prices on the different predictors, x_{ij}. The natural base case for standardised predictors is to price them all equally, i.e., the individual penalty loadings $\psi_j = 1$ and they drop out of the problem. We will see shortly that separate pricing for individual predictors turns out to be important for our proposed estimators.

We can say 'remove' because in fact the effect of the penalisation with the ℓ_1 norm is that *the lasso sets the $\hat{\beta}_j$s for some variables to zero*. This is what makes the lasso so suitable to sparse problems: the estimator itself has a sparse solution. The lasso is also computationally feasible: the path-wise coordinate descent ('shooting') algorithm allows fast estimation.

The lasso, like other penalized regression methods, is subject to an attenuation bias. This bias can be addressed by post-estimation using OLS, i.e., re-estimate the model using the variables selected by the first-stage lasso (Belloni and Chernozhukov 2013):

$$\hat{\boldsymbol{\beta}}_{\text{post}} = \arg\min \frac{1}{n} \sum_{i=1}^{n} \left(y_i - \boldsymbol{x}_i'\boldsymbol{\beta}\right)^2 \qquad \text{subject to} \qquad \beta_j = 0 \text{ if } \tilde{\beta}_j = 0, \tag{10}$$

where $\tilde{\beta}_j$ is the first-step lasso estimator. In other words, the first-step lasso is used exclusively as a model selection technique, and OLS is used to estimate the selected model. This estimator is sometimes referred to as the 'Post-lasso' (Belloni and Chernozhukov 2013).

4.3   The Lasso: Choice of Penalty Level

The penalisation approach allows us to simplify the model selection problem to a one-dimensional problem, namely the choice of the penalty level λ. In this section we discuss two approaches: (1) *cross-validation* and (2) '*rigorous*' or *plug-in* penalisation. Cross-validation is widely-used for choosing the lasso penalty level, whereas the rigorous lasso is less well-known. The contribution of this paper is to propose versions of the rigorous lasso, the HAC-lasso and AC-lasso, that are suitable for time-series applications. We compare the performance of the HAC-lasso to the more commonly-used cross-validation approach to selecting the penalty. We therefore summarise the theory of the rigorous lasso in some detail, but first we briefly discuss cross-validation for independent and dependent data.

4.4 Cross-Validation

The objective in *cross-validation* is to choose the lasso penalty parameter based on predictive performance. Typically, the dataset is repeatedly divided into a portion which is used to fit the model (the 'training' sample) and the remaining portion which is used to assess predictive performance (the 'validation' or 'holdout' sample), usually with mean squared prediction error (MSPE) as the criterion. Arlot and Celisse (2010) survey the theory and practice of cross-validation.

In the case of independent data, common approaches are 'leave-one-out' (LOO) cross-validation and the more general 'K-fold' cross-validation.

In 'K-fold' cross-validation, the dataset is split into K portions or 'folds'; each fold is used once as the validation sample and the remainder are used to fit the model for some value of λ. For example, in 10-fold cross-validation (a common choice of K) the MSPE for the chosen λ is the MSPE across the 10 different folds when used for validation. LOO cross-validation is a special case where $K = 1$, i.e., every observation is used once as the validation sample while the remaining $n - 1$ observations are used to fit the model (Fig. 1).

Cross-validation is computationally intensive because of the need to repeatedly estimate the model and check its performance across different folds and across a grid of values for λ. Standardisation of data adds to the computational cost because it needs to be done afresh for each training sample; standardising the entire dataset once up-front would violate a key principle of cross-validation, which is that a training

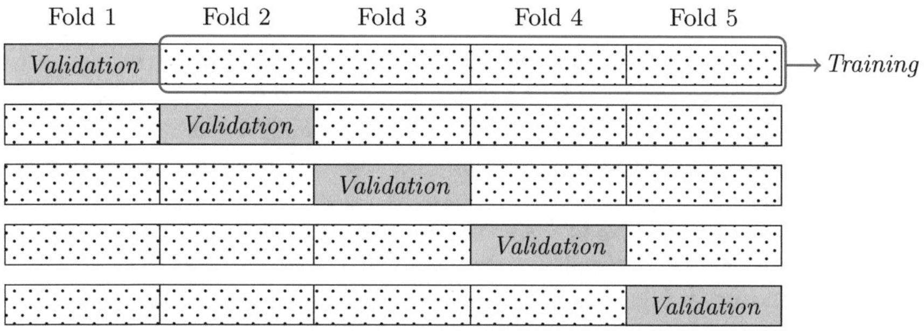

Fig. 1 This is K-fold cross-validation

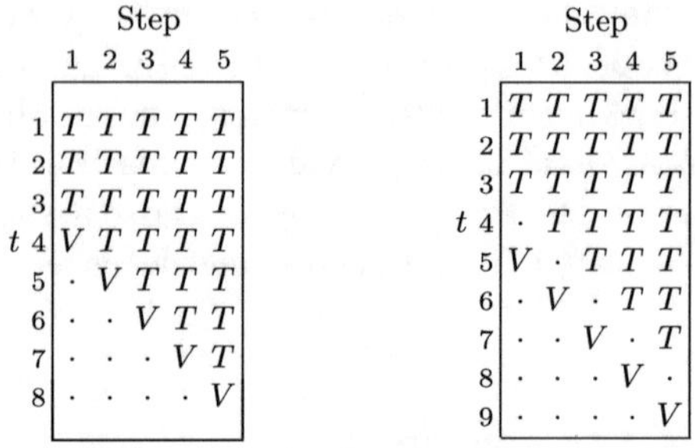

Fig. 2 Rolling h-step ahead cross-validation with expanding training window. 'T' and 'V' denote that the observation is included in the training and validation sample, respectively. A dot ('.') indicates that an observation is excluded from both training and validation data

(a) $h = 1$, expanding window

(b) $h = 2$, expanding window

dataset cannot contain any information from the corresponding validation dataset. LOO is a partial exception because the MSPE has a closed-form solution for a chosen λ, but a grid search across λ and repeated standardisation are still needed.

4.5 Cross-Validation for Time Series

Cross-validation with dependent data adds further complications because we need to be careful that the validation data are independent of the training data. It is possible that some settings, standard K-fold cross-validation is appropriate. Bergmeir et al. (2018) show that standard cross-validation that ignores the time dimension is valid in the pure auto-regressive model if one is willing to assume that the errors are uncorrelated. This implies, for example, that K-fold cross-validation can be used with auto-regressive models that include a sufficient number of lags, since the errors will be uncorrelated (if the model is not otherwise misspecified).

In general, however, researchers typically use a version of 'non-dependent cross validation' (Bergmeir et al. 2018), whereby prior information about the nature of the dependence is incorporated into the structure of the cross-validation and possibly dependent observations are omitted from the validation data. For example, one approach used with time-series data is 1-step-ahead cross-validation (Hyndman and Athanasopoulos 2018), where the predictive performance is based on a training sample with observations through time t and the forecast for time $t + 1$.

Rolling h-step ahead CV is an intuitively appealing approach that directly incorporates the ordered nature of time series-data (Hyndman and Athanasopoulos 2018).[4] The procedure builds on repeated h-step ahead forecasts. The procedure is illustrated in Figs. 2 and 3.

Figure 2a displays the case of 1-step ahead cross-validation. 'T' and 'V' refer to the training and validation samples, respectively. In the first step, observations 1 to 3 constitute the training data set and observation 4 used for validation; the remaining observations are unused as indicated by a dot ('.'). Figure 2b illustrates 2-step ahead

---

[4]Another approach is a variation of LOO cross-validation known as h-block cross-validation (Burman et al. 1994), which omits h observations between training and validation data.

Fig. 3 Rolling h-step ahead cross-validation with fixed training window

Step

	1	2	3	4	5
1	T	·	·	·	·
2	T	T	·	·	·
3	T	T	T	·	·
t 4	V	T	T	T	·
5	·	V	T	T	T
6	·	·	V	T	T
7	·	·	·	V	T
8	·	·	·	·	V

(a) $h = 1$, fixed window

Step

	1	2	3	4	5
1	T	·	·	·	·
2	T	T	·	·	·
3	T	T	T	·	·
t 4	·	T	T	T	·
5	V	·	T	T	T
6	·	V	·	T	T
7	·	·	V	·	T
8	·	·	·	V	·
9	·	·	·	·	V

(b) $h = 2$, fixed window

cross-validation. Figures 2a and 3b both illustrate cross-validation where the training window expands incrementally. Figure 3 displays a variation of rolling CV where the training window is fixed in length.

We use 1-step-ahead rolling CV with a fixed window for the comparisons in this paper.

5 The 'Rigorous' or 'Plug-in' Lasso

Bickel et al. (2009) present a theoretically-derived penalisation method for the lasso. Belloni, Chernozhukov, Hansen, and coauthors in a series of papers (e.g., Belloni et al. (2011), Belloni and Chernozhukov (2013), Belloni et al. (2016), Chernozhukov et al. (2015) and, most recently, Chernozhukov et al. (2019)) proposed feasible algorithms for implementing the 'rigorous' or 'plug-in' lasso and extended it to accommodate heteroskedasticity, non-Gaussian disturbances, and clustered data. The estimator has two attractive features for our purposes. First, it is theoretically and intuitively appealing, with well-established properties. Second, it is computationally attractive compared to cross-validation. We first present the main results for the rigorous lasso for the independent data, and then briefly summarise the 'cluster-lasso' of Belloni et al. (2016) before turning to the more general time-series setting analysed in Chernozhukov et al. (2019).

The rigorous lasso is consistent in terms of prediction and parameter estimation under three main conditions:

- **Sparsity**
- **Restricted sparse eigenvalue condition**
- **The 'regularisation event'.**

We consider each of these in turn.

Exact sparsity we have already discussed: there is a large set of potentially relevant predictors, but the true model has only a small number of regressors. Exact sparsity is a strong assumption, and in fact it is stronger than is needed for the rigorous lasso.

Instead, we assume *approximate sparsity*. Intuitively, some true coefficients may be non-zero but small enough in absolute size that the lasso performs well even if the corresponding predictors are not selected.

Belloni et al. (2012) define the *approximate sparse model (ASM)*,

$$y_i = f(\mathbf{w}_i) + \varepsilon_i = \mathbf{x}_i'\boldsymbol{\beta}_0 + r_i + \varepsilon_i. \tag{11}$$

where ε_i are independently distributed, but possibly heteroskedastic and non-Gaussian errors. The elementary predictors $\mathbf{w}_i$ are linked to the dependent variable through the unknown and possibly non-linear function $f(\cdot)$. The objective is to approximate $f(\mathbf{w}_i)$ using the target parameter vector $\boldsymbol{\beta}_0$ and the transformations $\mathbf{x}_i := P(\mathbf{w}_i)$, where $P(\cdot)$ is a set of transformations. The vector of predictors $\mathbf{x}_i$ may be large relative to the sample size. In particular, the setup accommodates the case where a large number of transformations (polynomials, dummies, etc.) approximate $f(\mathbf{w}_i)$.

Approximate sparsity requires that $f(\mathbf{w}_i)$ can be approximated sufficiently well using only a small number of non-zero coefficients. Specifically, the target vector $\boldsymbol{\beta}_0$ and the sparsity index s need to satisfy

$$\left\|\boldsymbol{\beta}_0\right\|_0 := s \ll n \quad \text{with} \quad \frac{s^2 \log^2(p \vee n)}{n} \to 0 \tag{12}$$

and the resulting approximation error $r_i = f(\mathbf{w}_i) - \mathbf{x}_i'\boldsymbol{\beta}_0$ satisfied the bound

$$\sqrt{\frac{1}{n}\sum_{i=1}^{n} r_i^2} \le C\sqrt{\frac{s}{n}}, \tag{13}$$

where C is a positive constant.

For example, consider the case where $f(\mathbf{w}_i)$ is linear with $f(\mathbf{w}_i) = \mathbf{x}_i'\boldsymbol{\beta}^\star$, but the true parameter vector $\boldsymbol{\beta}^\star$ is high-dimensional: $\left\|\boldsymbol{\beta}^\star\right\|_0 > n$. Approximate sparsity means we can still approximate $\boldsymbol{\beta}^\star$ using the sparse target vector $\boldsymbol{\beta}_0$ as long as $r_i = \mathbf{x}_i'(\boldsymbol{\beta}^\star - \boldsymbol{\beta}_0)$ is sufficiently small as specified in (13).

The *Restricted sparse eigenvalue condition* (RSEC) relates to the Gram matrix, $n^{-1}X'X$. The RSEC condition specifies that sub-matrices of the Gram matrix of size m are well-behaved (Belloni et al. 2012). Formally, the RSEC requires that the minimum sparse eigenvalues

$$\phi_{\min}(m) = \min_{1 \le \|\boldsymbol{\delta}\|_0 \le m} \frac{\boldsymbol{\delta}'X'X\boldsymbol{\delta}}{\|\boldsymbol{\delta}\|_2^2} \quad \text{and} \quad \phi_{\max}(m) = \max_{1 \le \|\boldsymbol{\delta}\|_0 \le m} \frac{\boldsymbol{\delta}'X'X\boldsymbol{\delta}}{\|\boldsymbol{\delta}\|_2^2}$$

are bounded away from zero and from above. The requirement that $\phi_{\min}(m)$ is positive means that all sub-matrices of size m have to be positive definite.[5]

---

[5]Bickel et al. (2009) use instead the weaker *restricted eigenvalue condition (REC)*. The RSEC implies the REC and has the advantage of being sufficient for both the lasso and the post-lasso.

The *regularisation event* is the third central condition required for the consistency of the rigorous lasso.

Denote by $S = \nabla \hat{Q}(\boldsymbol{\beta})$, the gradient of the objective function $\hat{Q}$ at the true value $\boldsymbol{\beta}$. $S_j = \frac{2}{n} \sum_{i=1}^{n} x_{ij} \varepsilon_i$ is the jth element of the score vector.

The idea is to select the lasso penalty level(s) to control the scaled estimation noise as summarised by the score vector. Specifically, the overall penalty level λ and the predictor-specific penalty loadings ψ_j are chosen so that the 'regularisation event'

$$\frac{\lambda}{n} \geq c \max_{1 \leq j \leq p} \left| \psi_j^{-1} S_j \right| \tag{14}$$

occurs with high probability, where $c > 1$ is a constant slack parameter.

Denote by $\Lambda = \max_j |\psi_j^{-1} S_j|$ the maximal element of the score vector scaled by the predictor-specific penalty loadings ψ_j, and denote by $q_\Lambda(\cdot)$ the quantile function for Λ, i.e., the probability that Λ is at most a is $q_\Lambda(a)$. In the rigorous lasso, we choose the penalty parameters λ and ψ_j and confidence level γ so that

$$\frac{\lambda}{n} \geq c q_\Lambda(1 - \gamma) \tag{15}$$

The intuition behind this approach is clear from a simple example. Say that no predictors appear in the true model ($\beta_j = 0 \; \forall \; j = 1, \ldots, p$). For the lasso to select no variables, the penalty parameters λ and ψ_j need to satisfy $\lambda \geq 2 \max_j | \sum_i \psi_j^{-1} x_{ij} y_i |$.[6] Because none of the regressors appear in the true model, $y_i = \varepsilon_i$, and the requirement for the lasso to correctly identify the model without regressors is therefore $\lambda \geq 2 \max_j | \sum_i \psi_j^{-1} x_{ij} \varepsilon_i |$. Since $x_{ij} \varepsilon_i$ is the score for observation i and predictor j, this is equivalent to requiring $\lambda \geq n \max_j |\psi_j^{-1} S_j|$, which is the regularisation event in (14). We want this event to occur with high probability of at least $(1 - \gamma)$. We therefore choose values for λ and ψ_j such that $\frac{\lambda}{n} \geq q_\Lambda(1 - \gamma)$. Since $q_\Lambda(\cdot)$ is a quantile function, by definition we will choose the correct model (no predictors) with probability of at least $(1 - \gamma)$. This yields (15), the rule for choosing penalty parameters.[7]

The procedure for choosing λ is not yet feasible, because the quantile function $q_\Lambda(\cdot)$ for the maximal element of the score vector is unknown, as is the predictor-specific penalty loadings ψ_j. We discuss how these issues are addressed in practice in the next subsection.

If the sparsity and restricted sparse eigenvalue assumptions ASM and RSEC are satisfied, if certain other technical conditions are satisfied,[8] and if λ and ψ_j are estimated as described below, then Belloni et al. (2012) show the lasso and post-lasso obey:

---

[6] See, for example, Hastie et al. (2015, Chap. 2).

[7] In this special case, the requirement of the slack is loosened and $c = 1$.

[8] These conditions relate to the use of the moderate deviation theory of self-normalized sums (Jing et al. 2003) that allows the extension of the theory to cover non-Gaussianity. See Belloni et al. (2012).

$$\sqrt{\frac{1}{n}\sum_{i=1}^{n}\left(x_i'\hat{\boldsymbol{\beta}} - x_i'\boldsymbol{\beta}\right)^2} = O\left(\sqrt{\frac{s\log(p \vee n)}{n}}\right), \tag{16}$$

$$\|\hat{\boldsymbol{\beta}} - \boldsymbol{\beta}\|_1 = O\left(\sqrt{\frac{s^2\log(p \vee n)}{n}}\right), \tag{17}$$

$$\|\hat{\boldsymbol{\beta}}\|_0 = O(s) \tag{18}$$

Equation (16) provides an asymptotic bound for the prediction error. Equation (17) provides an asymptotic bound for the bias in the estimated $\hat{\boldsymbol{\beta}}$. Equation (18) provides a sparsity bound; the number of selected predictors in the estimated model does not diverge relative to the true model.

The 'oracle' estimator is the estimator obtained if the s predictors in the model were actually known. This provides a useful theoretical benchmark for comparison. Here, if the s predictors in the model were known, the prediction error would converge at the oracle rate $\sqrt{s/n}$. Thus, the logarithmic term $\log(p \vee n)$ can be interpreted as the cost of not knowing the true model. For this reason, Belloni et al. (2012) describe these rates of convergence as *near-oracle* rates.

For the case of the lasso with theory-driven regularisation, Belloni and Chernozhukov (2013) have shown that post-estimation OLS, also referred to as post-lasso, achieves the same convergence rates as the lasso and can outperform the lasso in situations where consistent model selection is feasible (see also Belloni et al. 2012)).

The rigorous lasso has recently been shown to have certain appealing properties vis-a-vis the K-fold cross-validated lasso. The rates of convergence of the rigorous lasso are faster than those for the K-fold cross-validated lasso derived in Chetverikov et al. (forthcoming). Moreover, the sparsity bound for the K-fold cross-validated lasso derived in Chetverikov et al. (forthcoming) does not exclude situations where rule out situations where (18) fails badly, in the sense that the number of predictors selected via cross-validation is much larger than s. One of the implications is that cross-validation will select a penalty level λ that is 'too small' in the sense that the regularisation event (14) will no longer be guaranteed to occur with high probability.

5.1 Implementing the Rigorous Lasso

The quantile function $q_\Lambda(\cdot)$ for the maximal element of the score vector is unknown. The most common approach to addressing this is to use a theoretically-derived upper bound that guarantees that the regularisation event (14) holds asymptotically.[9] Specifically, Belloni et al. (2012) show that

---

[9]The alternative is to simulate the distribution of the score vector. This is known as the 'exact' or *X-dependent* approach. See Belloni and Chernozhukov (2011) for details and Ahrens et al. (2020) for a summary discussion and an implementation in Stata.

$$P\left(\max_{1\le j\le p} c\,|S_j| \le \frac{\lambda\psi_j}{n}\right) \to 1 \text{ as } n \to \infty,\, \gamma \to 0 \tag{19}$$

if the penalty levels and loadings are set to

$$\lambda = 2c\sqrt{n}\Phi^{-1}(1-\gamma/(2p)) \qquad \psi_j = \sqrt{\frac{1}{n}\sum_i x_{ij}^2 \varepsilon_i^2} \tag{20}$$

c is the slack parameter from above and $\gamma \to 0$ means the probability of the regularisation event converges towards 1. Common settings for c and γ, based on Monte Carlo studies are $c = 1.1$ and $\gamma = 0.1/\log(n)$, respectively.

The only remaining element is estimation of the ideal penalty loadings ψ_j. Belloni et al. (2012, 2014) recommend an iterative procedure based on some initial set of residuals $\hat{\varepsilon}_{0,i}$. One choice is to use the d predictors that have the highest correlation with y_i and regress y_i on these using OLS; $d = 5$ is their suggestion. The residuals from this OLS regression can be used to obtain an initial set of penalty loadings $\hat{\psi}_j$ according to (20). These initial penalty loadings and the penalty level from (20) are used to obtain the lasso or post-lasso estimator $\hat{\boldsymbol{\beta}}$. This estimator is then used to obtain a updated set of residuals and penalty loadings according to (20), and then an updated lasso estimator. The procedure can be iterated further if desired.

The framework set out above requires only independence across observations; heteroskedasticity, a common issue facing empirical researchers, is automatically accommodated. For this reason we refer to it as the 'heteroskedastic-consistent rigorous lasso' or HC-lasso. The reason is that heteroskedasticity is captured in the penalty loadings for the score vector.[10] Intuitively, heteroskedasticity affects the probability that the term $\max_j |\sum_i x_{ij}\varepsilon_i|$ takes on extreme values, and this needs to be captured via the penalty loadings. In the special case of homoskedasticity, the ideal penalisation in (20) simplifies:

$$\lambda = 2c\sigma\sqrt{n}\Phi^{-1}(1-\gamma/(2p)), \qquad \psi_j = 1. \tag{21}$$

This follows from the fact that we have standardised the predictors to have unit variance and hence homoskedasticity implies $E(x_{ij}^2 \varepsilon_i^2) = \sigma^2 E(x_{ij}^2) = \sigma^2$. The iterative procedure above is used to obtain residuals to form an estimate $\hat{\sigma}^2$ of the error variance σ^2. We refer to the rigorous lasso with this simplification as the 'standard' or 'basic' rigorous lasso.

5.2 The Rigorous Lasso for Panel Data

The rigorous lasso has been extended to cover a special case of dependent data, namely panel data. The 'cluster-lasso' proposed by Belloni et al. (2016) allows

---

[10]The formula in (20) for the penalty loading is familiar from the standard Eicker-Huber-White heteroskedasticty-robust covariance estimator.

for arbitrary within-panel correlation. The theoretical justification for the cluster-lasso also supports the use of the rigorous lasso in a pure time series setting, and specifically the HAC-lasso and AC-lasso proposed in this paper. Belloni et al. (2016) prove consistency of the rigorous cluster-lasso for both the large n, fixed T and large n, large T settings. The large n-fixed T results apply also to the specific forms of the HAC-lasso and AC-lasso proposed here. We first outline the Belloni et al. (2016) cluster-lasso and then our proposed estimators.

Belloni et al. (2016) present the approach in the context of a fixed-effects panel data model with balanced panels, but the fixed effects and balanced structure are not essential and the approach applies to any setups with clustered data. For presentation purposes we simplify and write the model as a balanced panel:

$$y_{it} = x_{it}'\beta + \varepsilon_{it} \qquad i = 1, \ldots, n, \ t = 1, \ldots, T \qquad (22)$$

The general intuition behind the rigorous lasso is to control the noise in the score vector $S = (S_1, \ldots, S_j, \ldots, S_p)$ where $S_j = \frac{2}{n}\sum_{i=1}^{n} x_{ij}\varepsilon_i$. Specifically, we choose the overall penalty λ and the predictor-specific penalty loading ψ_j so that $\frac{\lambda}{n}$ exceeds the maximal element of the scaled score vector $|\psi_j^{-1}S_j|$ with high probability. In effect, the ideal penalty loading ψ_j scales the jth element of the score by its standard deviation. In the benchmark heteroskedastic case, the ideal penalty loading is $\psi_j = \sqrt{\frac{1}{n}\sum_i x_{ij}^2\varepsilon_i^2}$; under homoskedasticity, the ideal penalty loading is simply $\psi_j = \sigma \ \forall \ j$ and hence can be absorbed into the overall penalty λ.

The cluster-lasso of Belloni et al. (2016) extends this to accommodate the case where the score is independent across but not within panels i. In this case, the ideal penalty loading is just an application of the standard cluster-robust covariance estimator, which provides a consistent estimate of the variance of the jth element of the score vector. The ideal penalty loadings for the cluster-lasso are simply

$$\psi_j = \sqrt{\frac{1}{nT}\sum_{i=1}^{n} u_{ij}^2} \qquad \text{where } u_{ij} := \sum_t x_{ijt}\varepsilon_{it} \qquad (23)$$

Belloni et al. (2016) show that this ideal penalty can be implemented in the same way as the previous cases, i.e., by using an initial set of residuals and then iterating. They recommend that the overall penalty level is the same as in the heteroskedastic case, $\lambda = 2c\sqrt{n}\Phi^{-1}(1 - \gamma/(2p))$, except that γ is $0.1/\log(n)$, i.e., it uses the number of clusters rather than the number of observations.

6 The Rigorous Lasso for Time-Series Data

We propose two estimators, the HAC-lasso and AC-lasso, that extend the rigorous lasso to the pure time-series setting. These estimators are, in effect, special cases of the rigorous lasso for dependent data presented in Chernozhukov et al. (2019).

We first present the HAC-lasso and then AC-lasso as a special case. For simplicity we consider the contemporaneous high-dimensional model, using t to denote observations numbered $1, \ldots, n$ but not including lags:

$$y_t = x_t'\beta + \varepsilon_t \tag{24}$$

The HAC-lasso uses the HAC (heteroskedastic- and autocorrelation-consistent) covariance estimator to estimate the variance of the jth element of the score vector. The implementation we propose is a simplified version of the estimator in Chernozhukov et al. (2019). The simplification follows from the additional assumption that the score is autocorrelated up to order q where q is finite, fixed and known a priori. The form of autocorrelation of this $MA(q)$ process can be arbitrary. Denote the HAC sample autocovariance s of the score for predictor j by Γ_{js}^{HAC}:

$$\Gamma_{js}^{HAC} := \frac{1}{n} \sum_{t=s+1}^{n} (x_{tj}\varepsilon_t)(x_{t-s,j}\varepsilon_{t-s}) \tag{25}$$

The sample variance of the score for predictor j is

$$\Gamma_{j0}^{HAC} := \frac{1}{n} \sum_{i=1}^{n} (x_{tj}\varepsilon_t)^2 \tag{26}$$

The variance of the jth element of the score vector can be consistently estimated using the truncated kernel with bandwidth q (Hayashi 2000, p. 408), and hence the HAC ideal penalty loading is

$$\psi_j^{HAC} = \sqrt{\Gamma_{j0}^{HAC} + 2 \sum_{s=1}^{q} \Gamma_{js}^{HAC}} \tag{27}$$

Feasible penalty loadings can be estimated using the same procedure used with the other forms of the rigorous lasso, i.e., starting with some initial residuals and iterating.

The assumption of a fixed and known q means that the HAC-lasso is closely related to the cluster-lasso. The cluster-robust covariance estimator also uses the truncated kernel with a fixed bandwidth (in the latter case, the bandwidth is set to $T - 1$ where T is the length of the panel). In the cluster-lasso, observations on the score are allowed to be correlated within clusters of length T but are assumed independent across clusters; in the HAC-lasso, observations are allowed to be correlated within a window of $q + 1$ observations but are assumed to be independent vs observations outside that window. This also motivates our suggested choice of the rigorous lasso parameter γ as $0.1/\log(\frac{n}{(q+1)})$; the term $\frac{n}{(q+1)}$ can be interpreted as the number of independent windows in the time-series setting (vs. the number of clusters in the panel setting).

The AC-lasso is a special case of the HAC lasso where we assume homoskedasticity but still allow for arbitrary autocorrelation of up to order q in the construction of the ideal penalty loadings. (See Hayashi (2000, pp. 413–414) for a general discussion of the AC covariance estimator.) Under homoskedasticity the sample autocovariance s of the score for predictor j is:

$$\Gamma_{js}^{AC} := \left(\frac{1}{n} \sum_{t=s+1}^{n} (x_{tj} x_{t-s,j}) \right) \left(\frac{1}{n} \sum_{t=s+1}^{n} (\varepsilon_t \varepsilon_{t-s}) \right) \tag{28}$$

Since predictors have been standardised to have unit variance, the variance of the score for predictor j is simply the variance of the disturbance ε

$$\Gamma_{j0}^{AC} := \sigma^2 \tag{29}$$

The AC ideal penalty loading for the $MA(q)$ case using the truncated kernel and q known a priori is

$$\psi_j^{AC} = \sqrt{\Gamma_{j0}^{AC} + 2 \sum_{s=1}^{1} \Gamma_{js}^{AC}} \tag{30}$$

Again, feasible penalty loadings can be estimated by starting with some initial residuals and iterating.

The extension to the high-dimensional model with lags,

$$y_t = X'B + \varepsilon_t \tag{31}$$

is straightforward. For the HAC-lasso, the only difference is that we need to construct penalty loadings for the full set of current and lagged predictors, ψ_{jl}^{HAC}, $j = 1, \ldots, p$, $l = 1, \ldots, L$. The AC-lasso case is similarly straightforward.

7 Monte Carlo

In this section, we present results of Monte Carlo simulations to assess the performance of the HAC-lasso estimator. We focus attention on the HD-C model with only contemporaneous predictors and $p = K$; our motivation is that this resembles the nowcasting application we discuss in the next section. The underlying data generation process for the dependent variable with p explanatory variables is:

$$y_t = \beta_0 + \sum_{j=1}^{p} \beta_j x_{tj} + \varepsilon_t.$$

A total of $p = 100$ predictors are generated, but only the first s predictors are non-zero. Therefore, in all specifications, the coefficients on the predictors β_j are defined as:

$$\beta_j = \mathbb{1}\{j \leq s\} \ \forall j = 1, \ldots, p$$

where we set the number of non-zero predictors to $s = 5$. β_0 is a constant and set to 1 in all simulations.

The error component ε_t for the dependent variable is an MA(q) process:

$$\varepsilon_t = \sum_{r=0}^{q} \theta_r \eta_{t-r}$$

$$\eta_t \sim N(0, \sigma_\eta^2).$$

We use three DGPs with $q = 0$, $q = 4$, and $q = 8$. For all DGPs, the MA coefficient θ_r is fixed such that $\theta_r = \theta = 1$, $\forall \ l = 1, \ldots, q$. The standard deviation varies across $\sigma_\eta = [0.5; 1; 2; 4; 5]$.

The predictors x_{tj} follow an $AR(1)$ process:

$$x_{tj} = \pi_j x_{t-1,j} + \xi_{tj}, \ \forall \ j = 1, \ldots, p$$

The AR coefficients across all predictors are the same with $\pi_j = \pi = 0.8$.

The random component $\boldsymbol{\xi}_t = (\xi_{t1}, \ldots, \xi_{tp})'$ is multivariate normal, generated as:

$$\boldsymbol{\xi}_t = MVN(0, \Sigma_\xi),$$

where Σ_ξ is a $p \times p$ covariance matrix. In this approach, we specify error components that are independent over time, and that are either also contemporaneously independent or correlated across p. In a first step the Monte Carlo specifies uncorrelated error components for the predictors x and Σ_ξ is diagonal with elements $\sigma_{\xi(1)}^2 = \cdots = \sigma_{\xi(p)}^2 = 1$.

The sample size is set to $T = 100$, with the same number of periods as burn-in when generating the data. In addition we generate a further sample of size T from $t = T + 1, \ldots, 2T$ for out-of-sample prediction. Each Monte Carlo run is repeated 1000 times for the rigorous lasso, the oracle estimator and 100 times for the cross-validated lasso.

In the results, we report the number of correctly and incorrectly selected predictors, the bias of the lasso and post-lasso coefficients, the root mean squared error (RMSE), and, to analyse out-of-sample performance, the root mean squared prediction error (RMSPE):

$$RMSE = \sqrt{\frac{1}{T} \sum_{t=1}^{T} \left(y_t - \hat{y}_{t,T}\right)^2} \text{ and } RMSPE = \sqrt{\frac{1}{T} \sum_{t=T+1}^{2T} \left(y_t - \hat{y}_{t,T}\right)^2}. \quad (32)$$

Table 1 Specifications for Monte Carlo simulations. For all specifications $p = 100$, $s = 5$ and $\sigma_\eta = [0.5; 1; 2; 4; 5]$

Specification	MA(q)	Table
1	0	Table 2
2	4	Table 3
3	8	Table 4

For each DGP, we compare the performance of six estimation methods:

- The 'basic' or 'standard' rigorous lasso estimator that assumes independent data but assumes homoskedasticity.
- The HC-lasso, i.e., the rigorous lasso estimator that assumes independent data but allows for heteroskedastic-consistent penalty loadings.
- The HAC-lasso using the truncated kernel and bandwidth = 4.
- The HAC-lasso using the truncated kernel and bandwidth = 8.
- The CV-lasso, i.e., the cross-validated lasso using one-step-ahead CV, a fixed window size and a holdout sample of 10 observations.
- The Oracle estimator, which is OLS on the full set of true regressors (i.e. variables 1 to s).

The HAC-lasso uses modified versions of the `rlasso` estimator for Stata, described in Ahrens et al. (2020). The default settings for the modified estimator are the same as those for `rlasso`. The HAC penalty loadings are calculated as described in Sect. 6.[11] The CV-lasso uses the `cvlasso` estimator for Stata, also described in Ahrens et al. (2020).

In all cases, we report both the lasso and post-lasso OLS performance. Table 1 summarizes the parameters for our 3 specifications.

We are particularly interested in performance in terms of model selection. We expect that if ε_t is independent ($q = 0$) the basic rigorous lasso and HC-lasso should perform somewhat better than the HAC-lasso estimators, since in effect the HAC lasso is estimating covariances that are zero and the basic and HC-lasso incorporate this assumption. If ε_t is $MA(q)$, $q > 0$, the HAC-lasso is expected to select a smaller set of non-zero predictors than the basic rigorous and HC-lasso. The CV-lasso is likely to perform better than the rigorous lasso in terms of pure prediction but worse in terms of false positives, i.e., including irrelevant predictors. In terms of bias, the oracle estimator should perform best, but of course is infeasible because the true model is not known.

---

[11] The sole exception addresses what happens if, during the iteration process, an estimated penalty loading is negative. This is possible in finite samples with the truncated kernel if $\hat{\Gamma}_{j0}^{HAC} + 2\sum_{s=1}^{q} \hat{\Gamma}_{js}^{HAC} < 0$. In this case, we replace this in the expression for the penalty loading with $\hat{\Gamma}_{j0}^{HAC} + \sum_{s=1}^{q} \hat{\Gamma}_{js}^{HAC}$. A similar change is made if a negative penalty loading is encountered when iterating to obtain the AC-lasso estimator.

Table 2 presents the results for the case where ε_t is independent, i.e., it is $MA(0)$. As expected, the basic and the HC-lasso perform similarly in terms of model selection. In comparison to the HAC-lasso estimators they select more variables and in particular more falsely included predictors. The HAC-lasso estimators perform worse in terms of missing out relevant predictors (false negatives). It is interesting to note that while there is almost no difference between the HAC-lasso with bandwidth 4 and 8, the HC-lasso misses three times as many relevant variables as the simple one.

Table 2 Monte Carlo simulation for $p = 100$, $s = 5$ and $q = 0$

	σ	Rigorous lasso				CV lasso	Oracle
	Bandwidth	Basic	HC	HAC(4)	HAC(8)	-	-
Selected $\hat{s}$	0.5	7.64	7.64	4.53	5.44	27.27	-
	1.0	7.36	7.09	4.49	5.28	32.76	-
	2.0	5.81	5.40	4.23	5.08	27.35	-
	3.0	3.92	3.78	3.81	4.62	28.32	-
	5.0	1.64	1.73	2.93	3.93	28.30	-
False pos.	0.5	2.74	3.06	1.96	2.96	22.27	-
	1.0	2.63	2.83	1.92	2.85	27.76	-
	2.0	2.04	2.09	1.93	2.88	22.47	-
	3.0	1.36	1.45	1.91	2.78	23.52	-
	5.0	0.59	0.69	1.74	2.71	24.05	-
False neg.	0.5	0.10	0.42	2.43	2.52	0.00	-
	1.0	0.27	0.74	2.43	2.57	0.00	-
	2.0	1.23	1.69	2.70	2.79	0.12	-
	3.0	2.44	2.67	3.10	3.16	0.20	-
	5.0	3.95	3.96	3.81	3.77	0.75	-
Bias	0.5	1.484	2.207	4.120	4.333	0.795	0.091
		(0.473)	(1.086)	(3.549)	(3.784)	(1.064)	-
	1.0	2.335	2.838	4.120	4.275	1.941	0.182
		(1.005)	(1.747)	(3.572)	(3.838)	(2.647)	-
	2.0	3.625	3.839	4.353	4.514	3.386	0.365
		(2.497)	(3.029)	(3.912)	(4.190)	(4.408)	-
	3.0	4.307	4.376	4.585	4.785	4.923	0.542
		(3.695)	(3.948)	(4.398)	(4.689)	(6.945)	-
	5.0	4.798	4.809	4.945	5.183	9.198	0.905
		(4.885)	(4.973)	(5.360)	(5.727)	(12.222)	-
RMSE	0.5	0.912	1.226	2.389	2.313	0.475	0.491
		(0.529)	(0.730)	(1.863)	(1.878)	(0.438)	-
	1.0	1.565	1.770	2.545	2.482	0.925	0.983
		(1.051)	(1.257)	(2.038)	(2.063)	(0.858)	-
	2.0	2.788	2.868	3.133	3.063	1.958	1.964
		(2.179)	(2.316)	(2.667)	(2.676)	(1.805)	-
	3.0	3.870	3.876	3.889	3.820	2.869	2.936
		(3.272)	(3.333)	(3.470)	(3.464)	(2.687)	-
	5.0	5.833	5.809	5.636	5.532	4.751	4.919
		(5.375)	(5.367)	(5.263)	(5.211)	(4.477)	-
RMSFE	0.5	1.265	1.780	3.182	3.216	0.476	0.492
		(0.674)	(1.088)	(2.894)	(2.968)	(0.438)	-
	1.0	2.069	2.402	3.290	3.320	0.919	0.983
		(1.356)	(1.801)	(3.030)	(3.120)	(0.849)	-
	2.0	3.438	3.554	3.837	3.854	1.926	1.970
		(2.921)	(3.176)	(3.651)	(3.713)	(1.784)	-
	3.0	4.470	4.495	4.565	4.592	2.848	2.952
		(4.239)	(4.326)	(4.487)	(4.532)	(2.661)	-
	5.0	6.220	6.220	6.208	6.220	4.748	4.914
		(6.270)	(6.289)	(6.346)	(6.364)	(4.498)	-

Notes: $\hat{s}$ denotes the number of selected variables excluding the constant. 'False pos.' and 'False neg.' denote the number of falsely included and falsely excluded variables, respectively. 'Bias' is the ℓ_1-norm bias defined as $\sum_j |\hat{\beta}_j - \beta_j|$ for $j = 1, \ldots, p$. 'RMSE' is the root mean squared error (a measure of in-sample fit) and 'RMSFE' is the root mean squared forecast error (a measure of out-of-sample prediction performance); see equation (1.32). Post-estimation OLS results are shown in parentheses if applicable. `cvlasso` results are using one-step-ahead CV, a fixed window size and a holdout sample of 10 observations. The oracle estimator applies OLS to all predictors in the true model (i.e., variables 1 to s). Thus, the false positive and false negative frequency is zero by design for the oracle. The number of replications is 1,000 for `rlasso`, the oracle estimator and 100 for `cvlasso`.

The bias, RMSE and RMSPE are similar within the two groups, whereas it is larger for the HAC-lasso estimators. Due to the better selection performance of the simple and HC-lasso, this behaviour was expected.

In the next set of results, we consider the case where ε_t is dependent. We set the order of the MA process to $q = 4$ and $q = 8$, as shown in Tables 3 and 4. In this setting, the HAC-lasso estimators should perform better than their counterparts that assume independence. Indeed, the HAC-lasso estimator outperforms the basic rigorous lasso and HC-lasso estimator in terms of selecting wrongly predictors (false positives), but is still more likely to falsely exclude predictors. The bias for all the

Table 3 Monte Carlo simulation for $p = 100$, $s = 5$ and $q = 4$

σ	Rigorous lasso				CV lasso	Oracle
Bandwidth	Basic	HC	HAC(4)	HAC(8)	-	-
Selected s 0.5	9.44	8.62	4.28	5.20	55.92	-
1.0	8.37	7.61	3.68	4.65	57.36	-
2.0	6.57	6.12	2.54	4.00	53.25	-
3.0	5.84	5.55	2.11	3.70	53.65	-
5.0	5.12	4.94	1.66	3.73	47.43	-
False pos. 0.5	4.71	4.34	1.93	2.94	50.92	-
1.0	4.72	4.35	1.90	2.93	52.38	-
2.0	4.70	4.43	1.74	3.08	48.59	-
3.0	4.71	4.50	1.66	3.14	49.57	-
5.0	4.55	4.41	1.48	3.40	44.17	-
False neg. 0.5	0.27	0.71	2.65	2.74	0.00	-
1.0	1.34	1.74	3.23	3.27	0.02	-
2.0	3.13	3.31	4.20	4.08	0.34	-
3.0	3.87	3.95	4.56	4.44	0.92	-
5.0	4.43	4.47	4.82	4.67	1.74	-
Bias 0.5	2.597	3.059	4.294	4.495	3.368	0.369
	(1.686)	(2.227)	(3.850)	(4.139)	(4.427)	-
1.0	4.018	4.189	4.678	4.850	7.159	0.743
	(3.819)	(4.120)	(4.629)	(4.933)	(9.058)	-
2.0	5.337	5.395	5.188	5.721	13.090	1.484
	(6.851)	(6.864)	(6.185)	(6.645)	(17.156)	-
3.0	6.067	6.113	5.449	6.319	19.797	2.215
	(9.021)	(8.897)	(7.122)	(7.976)	(25.684)	-
5.0	7.073	7.218	5.773	10.918	28.561	3.732
	(12.458)	(12.233)	(8.411)	(10.071)	(38.677)	-
RMSE 0.5	1.468	1.684	2.630	2.536	0.650	1.039
	(0.912)	(1.123)	(2.066)	(2.075)	(0.567)	-
1.0	2.499	2.589	3.288	3.145	1.296	2.079
	(1.821)	(1.983)	(2.696)	(2.677)	(1.143)	-
2.0	4.267	4.295	4.955	4.699	2.644	4.171
	(3.388)	(3.508)	(4.339)	(4.192)	(2.363)	-
3.0	5.994	6.012	6.784	6.446	3.985	6.258
	(4.857)	(4.987)	(6.092)	(5.828)	(3.613)	-
5.0	9.604	9.577	10.685	10.082	6.942	10.447
	(7.925)	(8.072)	(9.784)	(9.265)	(6.365)	-
RMSFE 0.5	2.232	2.551	3.445	3.477	0.635	1.051
	(1.648)	(2.009)	(3.214)	(3.287)	(0.563)	-
1.0	3.634	3.747	4.125	4.133	1.280	2.104
	(3.422)	(3.598)	(4.092)	(4.146)	(1.135)	-
2.0	5.762	5.797	5.886	5.961	2.630	4.200
	(6.162)	(6.183)	(6.222)	(6.282)	(2.375)	-
3.0	7.867	7.899	7.876	8.010	3.879	6.319
	(8.654)	(8.648)	(8.467)	(8.572)	(3.597)	-
5.0	12.232	12.283	12.160	13.246	6.855	10.539
	(13.635)	(13.663)	(13.143)	(13.384)	(6.273)	-

Notes: See also notes in Table 1.2.

Table 4 Monte Carlo simulation for $p = 100$, $s = 5$ and $q = 8$

σ Bandwidth	Rigorous lasso Basic	HC	HAC(4)	HAC(8)	CV lasso -	Oracle -
Selected s						
0.5	10.13	9.21	4.19	5.07	62.59	-
1.0	9.52	8.61	3.76	4.45	62.49	-
2.0	8.80	8.11	3.11	4.17	60.20	-
3.0	8.82	8.21	2.91	3.87	60.07	-
5.0	8.89	8.19	2.89	3.82	52.42	-
False pos.						
0.5	5.60	5.14	2.05	3.02	57.59	-
1.0	6.31	5.78	2.32	3.06	57.62	-
2.0	7.21	6.70	2.53	3.53	55.69	-
3.0	7.79	7.27	2.55	3.45	56.00	-
5.0	8.20	7.57	2.67	3.55	49.35	-
False neg.						
0.5	0.48	0.93	2.86	2.95	0.00	-
1.0	1.79	2.17	3.56	3.61	0.13	-
2.0	3.40	3.59	4.42	4.36	0.49	-
3.0	3.97	4.06	4.64	4.59	0.93	-
5.0	4.31	4.38	4.78	4.73	1.93	-
Bias						
0.5	3.137	3.503	4.474	4.661	4.525	0.571
	(2.442)	(2.912)	(4.163)	(4.466)	(5.507)	-
1.0	4.688	4.808	4.958	5.180	9.110	1.130
	(5.095)	(5.295)	(5.387)	(5.650)	(11.130)	-
2.0	6.561	6.616	5.703	6.292	17.494	2.289
	(9.165)	(9.109)	(7.564)	(7.972)	(21.755)	-
3.0	8.085	8.109	6.200	6.886	26.120	3.438
	(12.662)	(12.387)	(9.150)	(9.816)	(32.657)	-
5.0	10.679	10.626	7.112	8.258	38.232	5.657
	(18.600)	(17.974)	(12.046)	(13.072)	(49.349)	-
RMSE						
0.5	1.661	1.848	2.781	2.693	0.668	1.352
	(1.043)	(1.254)	(2.196)	(2.204)	(0.587)	-
1.0	2.687	2.778	3.623	3.505	1.378	2.694
	(1.932)	(2.096)	(2.958)	(2.964)	(1.222)	-
2.0	4.526	4.590	5.706	5.486	2.824	5.413
	(3.411)	(3.580)	(4.862)	(4.769)	(2.527)	-
3.0	6.401	6.466	8.080	7.765	4.169	8.203
	(4.851)	(5.067)	(6.933)	(6.770)	(3.790)	-
5.0	10.097	10.214	12.775	12.279	7.720	13.533
	(7.662)	(8.007)	(10.960)	(10.746)	(6.963)	-
RMSFE						
0.5	2.716	2.963	3.668	3.694	0.643	1.366
	(2.250)	(2.550)	(3.498)	(3.576)	(0.566)	-
1.0	4.355	4.431	4.698	4.737	1.305	2.705
	(4.342)	(4.469)	(4.829)	(4.879)	(1.187)	-
2.0	7.284	7.324	7.304	7.381	2.659	5.443
	(7.863)	(7.895)	(7.901)	(7.931)	(2.434)	-
3.0	10.276	10.306	10.150	10.284	4.017	8.170
	(11.260)	(11.287)	(11.122)	(11.193)	(3.653)	-
5.0	16.418	16.475	16.124	16.328	7.340	13.546
	(18.143)	(18.146)	(17.755)	(17.861)	(6.660)	-

Notes: See also notes in Table 1.2.

rigorous lasso estimators is close, while the RMSE and RMSPE are lower for the non-HAC-lasso estimators. The cross-validated lasso selects more predictors than the rigorous lasso. This comes with the benefit of almost always correctly selecting the predictors that are in the model, but at the cost of incorrectly selecting many additional irrelevant predictors. This unconservative selection leads to a lower bias, RMSE and RMSPE. Thus is terms of prediction the cross-validated lasso performs better than the basic, HC- and HAC-lasso, but worse in terms of model selection.

The Monte Carlo simulation also allows to assess the computational costs of each estimation method. The average run time and the standard deviation for each of the

Table 5 Average run times with $p = 100$

Method	Stata call	Seconds	
		Average	SD
Rigorous lasso			
Basic	`rlasso y*`	0.032	0.004
HC	`rlasso y x*` `robust`	0.033	0.005
HAC(4)	`rlasso y x*` `robust bw(4)`	0.035	0.004
HAC(8)	`rlasso y x*` `robust bw(8)`	0.042	0.004

Cross-validated	`cvlasso y x*`
	`rolling`
	`fixedwindow`
		17.358	1.38

PC specification: Intel Core i5-7500 with 8GB RAM, Windows 10 and Stata 15.1

estimators are shown in Table 5. There are small differences in computation time as we move from the simple rigorous lasso to the HAC-lasso for with a bandwidth of 8; the HAC(8) lasso is about 30% slower than the simple rigorous lasso that assumes homoskedasticity. Unsurprisingly, the cross-validated lasso comes with a far higher computational cost: in these examples it is over 400 times slower than the slowest rigorous lasso estimator, the HAC(8) lasso. The standard deviation for `rlasso` is small and almost negligible, implying that the run time of the estimator is identical across the Monte Carlo draws. The standard deviation for the cross-validated lasso is much larger implying a variation of more than a second across draws.

Monte Carlo Tables

8   Application to Nowcasting

In this section, we illustrate how the properties of the HAC-lasso and AC-lasso estimators are particularly useful for model consistency for fore- and nowcasting and that it produces competitive nowcasts at low computational cost.

The objective of nowcast models is to produce 'early' forecasts of the target variable which exploits the real time data publication schedule of the explanatory data set. Such real time data sets are usually in higher frequency and are published with a considerably shorter lag than the target variable of interest. Nowcasting is particularly relevant for central banks and other policy environments where key economic indices such as GDP or inflation are published with a lag of up to 7 weeks with respect to their reference period.[12] In order to conduct informed forward-looking policy decisions, policy makers require accurate nowcasts where it is now common to combine, next to traditional macroeconomic data, ever more information from Big Data sources such as internet search terms, satellite data, scanner data, etc. (Buono et al. 2018).

A data source which has garnered much attention in the recent nowcast literature is Google Trends (GT), Google's search term indices. GT provides on a scale of 1–100, for a given time frame and location, the popularity of certain search terms entered into the Google search engine. Due to their timeliness as compared to conventional macro data and ability to function as an index of sentiment of demand and supply (Scott and Varian 2014), they have celebrated wide spread use in nowcasting applications in many disparate fields of economics (see Choi and Varian (2012), and Li (2016) for surveys). They have proven especially useful in applications where searches are directly related to the variable of interest, such as unemployment data where internet search engines provide the dominant funnel through which job seekers find jobs (Smith 2016). Only recently has Google Trends been applied to nowcasting such aggregate economic variables as GDP (Kohns and Bhattacharjee 2019).

Since the policymaker, in addition to the accuracy of the nowcast, is also interested in the interpretation of why the release of new information contributes to the fit in the nowcast, the nowcast methodology should ideally be model consistent and offer a high degree of interpretability. The HAC-lasso estimator is well-suited to the nowcasting setting, because of its features of model consistency, robustness to heteroskedasticity and serial correlation, and computational speed. In the nowcasting applications, we are interested in exploiting short-term fluctuations (monthly or quarterly) of our regressor set for forecasting, and we assume that the unmodelled serial correlation in the disturbance term can be ignored after a relatively short period. We select the appropriate MA order in the HAC-lasso on this basis below.

Besides the problems of high dimensionality and model consistency under sparsity, from the methodological standpoint in nowcasting there is the issue of mixed frequency and asynchronous data publications (ragged edges). Here, we take the approach of Kohns and Bhattacharjee (2019), who use U-MIDAS sampling. The MIDAS approach, as originally popularised by Ghysels et al. (2004), utilises additive distributed lag (ADL) functions as kernels such as the Almon or Beta function, to allow for a parsimonious way in bridging high frequency observations to lower frequency target data. Foroni and Marcellino (2015) show in simulations as well as a nowcasting exercise on U.S. GDP growth that if the frequency mismatch between the lower frequency and higher frequency variable is not too large, such as in a quarterly to monthly mismatch, leaving the intra-period coefficients unconstrained can increase forecasting fit. To illustrate loosely how higher frequency explanatory data is treated in the U-MIDAS approach, we represent this for a single regressor in monthly frequency and a quarterly dependent variable below:

---

[12]The exact lag in publications of GDP and inflation depends as well on which vintage of data the econometrician wishes to forecast. Since early vintages of aggregate quantities such as GDP can display substantial variation between vintages, this is not a trivial issue.

$$\begin{pmatrix} y_{1stquarter} & | & x_{Mar} & x_{Feb} & x_{Jan} \\ y_{2ndquarter} & | & x_{Jun} & x_{May} & x_{Apr} \\ . & | & . & . & . \\ . & | & . & . & . \\ . & | & . & . & . \end{pmatrix} \tag{33}$$

Since the objective of this study—to recover the consistent model under sparsity in high dimensional nowcast settings—is closest in spirit to the paper by Kohns and Bhattacharjee (2019), we base our empirical exercise on the same data. The dataset is composed of a monthly macro dataset and a monthly Google Trends dataset, which together are used to nowcast real quarterly US GDP growth.

A few methodological differences should be stressed here: (1) While the HAC-lasso assumes sparsity or approximate sparsity in the DGP, the spike-and-slab estimator, as used by Kohns and Bhattacharjee (2019), is agnostic about the degree of sparsity. This is achieved by separating variable selection from shrinkage of the parameter covariance through two different hyperparameters. Since, arguably, macro data sets are traditionally modeled as dense (e.g., Stock et al. (2002); Mol et al. (2009)) and Big Data as sparse (e.g. Giannone et al. (2018), Buono et al. (2018)), resulting sparse models could be an artefact of the prior degree of shrinkage imposed (Giannone et al. 2018). Theoretically, as well as empirically, it is uncertain which representation sparse estimators recover when the DGP is dense or mixed. (2) Spike-and-slab, as opposed to lasso-style estimators, offers shrinkage with positive probability. This follows from the fact that in Bayesian estimation lasso regularisation is imposed through a double exponential Laplace prior. Hence, spike-and-slab allows for exact shrinkage. (3) The spike-and-slab estimator is able to yield credible intervals around the regression parameter means. This results from it being able to obtain fully tractable posterior coefficient and variance distributions which allow also for predictive inference. With the HAC-lasso, we only obtain point estimates of the coefficients.

8.1   Data

As early data vintages of U.S. GDP can exhibit substantial variation compared to final vintages (Croushore 2006; Romer and Romer 2000; Sims 2002), it is not trivial which data to use in evaluating nowcast models on historical data. Further complications can arise through changing definitions or methods of measurements of data (Carriero et al. 2015). However, as in our application, only a few explanatory variables have recorded real time vintages (see Giannone et al. (2016)) and our training window is restricted to begin with 2004 only, since this is the earliest data point for the Google Trends data base, we decided to use final vintages of our data. We therefore consider a pseudo-real time data set: we use the latest vintage of data, but, at each point of the forecast horizon, we use only the data published up to that point in time.

The target variable for this application is deseasonalised U.S. real GDP growth (GDP growth) of as downloaded from the FRED website. The deseasonalisation pertains here to the X-13-ARIMA method and was performed prior to download from the FRED-MD website. As Google Trends are only available from 01/01/2004– 01/06/2019, at the time of download, the period under investigation pertains to the same period in quarters (61 quarters). We have split the data set into a training sample of 45 quarters (2004q2–2015q2) and a forecast sample of 15 quarters (2015q3– 2019q1).

The macro data set pertains to an updated version of the data base of the seminal paper by Giannone et al. (2016) (henceforth, 'macro data'). It contains 13 time series which are closely watched by professional and institutional forecasters (Giannone et al. 2016) such as real indicators (industrial production, house starts, total construction expenditure etc.), price data (CPI, PPI, PCE inflation), financial market data (BAA–AAA spread) and credit, labour and economic uncertainty measures (volume of commercial loans, civilian unemployment, economic uncertainty index etc.) (Giannone et al. 2016). Table 6 gives an overview over all data along with FRED codes.

Our sample of search terms comprises 27 Google Trends which have been chosen based on the root term methodology as in Koop and Onorante (2016), and Bock (2018). In general, there is no consensus on how to optimally pre-screen search terms for the final estimation. Methods which have been proposed by the previous literature fall into: (i) pre-screening through correlation with the target variable (e.g.: Scott and Varian (2014), Choi and Varian (2012) and references therein) or through cross-validation (Ferrara and Simoni 2019), (ii) use of prior economic intuition where search terms are selected through backward induction (e.g.: Smith (2016), Ettredge et al. (2005), Askitas and Zimmermann (2009)), and (iii) root terms, which similarly specify a list of search terms through backward induction, but additionally download "suggested" search terms from the Google interface. This serves to broaden the semantic variety of search terms in a semi-automatic way. From the authors' perspective, the root term methodology currently provides the best guarantee of finding economically relevant Google Trends, as methodologies based on pure correlation do not preclude spurious relationships (Scott and Varian 2014).

Since search terms can display seasonality, we deseasonalise all Google Trends by the Loess filter which is implemented with the 'stl' command in R.[13] Finally, our pseudo-real time calendar can be found in Table 6 and has been constructed after the data's real publication schedule. It comprises in total 30 vintages which are estimated until the final GDP observation is released.

---

[13]To mitigate any inaccuracy stemming from sampling error, we downloaded the set of Google Trends seven times between 01/07–08/07/2019 and took the cross-sectional average. Since we used the same IP address and googlemail account, there might still be some unaccounted measurement error which could be further mitigated by using web-crawling.

Table 6 Pseudo real time calendar based on actual publication dates. Transformation: 1 = monthly change, 2 = monthly growth rate, 3 = no change, 4 = LOESS decomposition. Pub. lag: m = refers to data for the given month within the reference period, m-1 = refers to data with a months' lag to publication in the reference period, m-2 = refers to data with 2 months' lag to publication in the reference period

Vintage	Timing	Release	Variable name	Pub. lag	Transformation	FRED Code
1	Last day of month 1	Fed. funds rate and credit spread	fedfunds and baa	m	3	FEDFUNDS and BAAY10
2	Last day of month 1	Google Trends		m	4	–
3	1st bus. day of month 2	Economic policy uncertainty index	uncertainty	m-1	1	USEPUINDXM
4	1st Friday of month 2	Employment situation	hours and unrate	m-1	2	AWHNONAG and UNRATE
5	Middle of month 2	CPI	cpi	m-1	2	CPI
6	15th-17th of month 2	Industrial production	indpro	m-1	2	INDPRO
7	3rd week of month 2	Credit and M2	loans and m2	m-1	2	LOANS and M2
8	Later part of month 2	Housing starts	housst	m-1	1	HOUST
9	Last week of month 2	PCE and PCEPI	pce and pce2	m-1	2	PCE and PCEPI
10	Last day of month 2	Fed. funds rate and credit spread	fedfunds and baa	m	3	FEDFUNDS and BAAY10
11	Last day of month 2	Google Trends		m	4	–
12	1st bus. day of month 3	Economic policy uncertainty Index	uncertainty	m-1	1	USEPUINDXM
13	1st bus. day of month 3	Construction starts	construction	m-2	1	TTLCONS
14	1st Friday of month 3	Employment situation	hours and unrate	m-1	2	AWHNONAG and UNRATE

(continued)

Table 6 (continued)

Vintage	Timing	Release	Variable name	Pub. lag	Transformation	FRED Code
15	Middle of month 3	CPI	cpi	m-1	2	CPI
16	15th–17th of month 3	Industrial production	indpro	m-1	2	INDPRO
17	3rd week of month 3	Credit and M2	loans and m2	m-1	2	LOANS and M2
18	Later part of month 3	Housing starts	housst	m-1	1	HOUST
19	Last week of month 3	PCE and PCEPI	pce and pce2	m-1	2	PCE and PCEPI
20	Last day of month 3	Fed. funds rate and credit spread	fedfunds and baa	m	3	FEDFUNDS and BAAY10
21	Last day of month 3	Google Trends		m	4	–
22	1st bus. day of month 4	Economic policy uncertainty Index	uncertainty	m-1	1	USEPUINDXM
23	1st bus. day of month 4	Construction starts	construction	m-2	1	TTLCONS
24	1st Friday of month 4	Employment situation	hours and unrate	m-1	2	AWHNONAG and UNRATE
25	Middle of month 4	CPI	cpi	m-1	2	CPI
26	15th–17th of month 4	Industrial production	indpro	m-1	2	INDPRO
27	3rd week of month 4	Credit and M2	loans and m2	m-1	2	LOANS and M2
28	Later part of month 4	Housing starts	housst	m-1	1	HOUST
29	Last week of month 4	PCE and PCEPI	pce and pce2	m-1	2	PCE and PCEPI
30	Later part of month 5	Housing starts	housst	m-2	1	HOUST

Table 7 Estimated coefficients of HAC-lasso and AC-lasso models. All estimations refer to post-lasso OLS. The number appended to AC and HAC refers to the MA order of the disturbance ε_t assumed for consistent estimation of the penalty loadings. The numbers appended to the variable names refer to the month within the reference period as per Table 6

Variable	AC-4	HAC-4	AC-8	HAC-8
Constant	2.167	2.4922	1.9097	2.2093
Baa-1	−0.2323	−0.282	−0.2208	−0.2452
GT-foreclosure-2	−0.0443		−0.0486	
GT-jobless claims-1		−0.0078		
GT-real gdp growth-0	−0.0233	−0.0272	−0.0212	−0.031
M2-0	−0.0073	−0.0082	−0.0066	−0.0075
PCE-2			0.3971	
PCE2-1		−0.0655		0.1263
Unemployment rate-0		−0.0189		−0.0341
Unemployment rate-2	−0.0441		−0.0373	

8.2 Results

Results for the in-sample HAC-lasso and AC-lasso estimations can be found in Table 7. The number appended to model name (H)AC refers to the MA order of the disturbance ε_t assumed for consistent estimation of the penalty loadings. As noted above, we assume that the autocorrelation in the scores does not persist for long periods. We select the appropriate order in the HAC-lasso to be 4 and 8. Hence, we robustify the penalty loadings for the scores with respect to MA terms of quarterly growth for an entire year back and up to 2 years, including the current quarter.

We find that the models select a very similar set of variables, where in particular the credit spread variable BAA-1, the monthly M2-0 base money growth, and the Google search term 'gdp-growth-0' feature consistently. Model sizes are also very similar with on average 6.5 regressors included. Considering that the original regressor set includes 120 variables, the resulting model is indeed sparse.

One of the main advantages over competing non-parametric machine learning methods, is that the analytic lasso allows for a high degree of interpretability of the included regressors. And indeed, most of the included variables have signs which correspond to economic intuition. For instance, an increase in the credit spread, which in macro models is often used as a gauge of recessionary fears, features with a negative sign in all the models. Similarly, unemployment rates are also predicted to have a negative impact on GDP, as intuition suggests. Similarly to Kohns and Bhattacharjee (2019), we also find that a particular subset of Google Trends improve fit, and more importantly that these Google Trends cluster on topics of financial distress or recessionary fears. As economic intuition would suggest an increase in recessionary fears has a negative impact on GDP growth. This finding is supported in

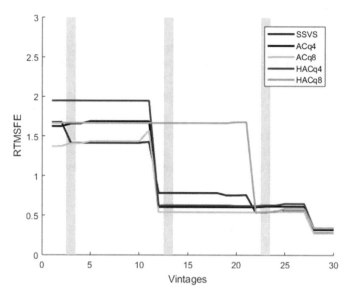

Fig. 4 RTMSFE of the models of table over 30 data vintages. Publication vintages of Google Trends are highlighted with grey vertical bars

the broader Google Trends literature which finds that search terms are a good gauge of future expectations of the variable under investigation.

Figure 4 shows the real-time root-mean-squared-forecast-error (RTMSFE) of the lasso models and of the SSVS estimator. To make the estimation procedures as comparable as possible, we use the SSVS estimator of Kohns and Bhattacharjee (2019) with a prior model size of 12 in order to enforce parsimony.

The (H)AC-lasso models provide very similar and competitive nowcasts compared to the SSVS estimator. An exception is provided by the HAC-lasso with 8 MA terms in early vintages. Since it mostly selects variables which are published in the last month of a given reference period, its forecasting performance in early vintages is worse than the other models. The fact that both methods select in a very similar subset of regressors supports a clear sparse representation given that one is willing to assume sparsity. In contrast to the SSVS estimator, which improves its nowcast accuracy by 16% with the first publication of Google Trends (as indicated by the first grey vertical bar), lasso models don't improve their nowcasts with any Google Trends publication. Hence, most of those models' performance seems to be driven by the macro variables.

In general, these results are impressive keeping in mind that the sampling and re-sampling approach involved in Bayesian computation requires much longer time. Running the lasso models takes 0.1–0.2 s, compared to 1–2 min with the SSVS, a speed advantage on the order of 500 times. A caveat that must be kept in mind, is that in contrast to the SSVS approach, one does not receive any credible intervals around the parameter estimates.

To summarise, the results show that:

- The HAC-lasso recovers, consistent with competing high dimensional estimators designed for time series contexts, sparse models which offer very competitive forecasts at a much lower computational time.
- Once all relevant macro data is out, all models offer similar nowcasting performance.

These results suggest that the HAC-lasso is a very efficient first regularisation tool for model exploration in dependent time series contexts, which paired with its low computational time make it a valuable extension for policy environments.

References

Ahrens, A., Hansen, C. B., & Schaffer, M. E. (2020). lassopack: Model selection and prediction with regularized regression in Stata. *The Stata Journal, 20*, 176–235.

Akaike, H. (1969). Fitting autoregressive models for prediction. *Annals of the Institute of Statistical Mathematics, 21*, 243–247.

Akaike, H. (1971). Autoregressive model fitting for control.

Akaike, H. (1974). A new look at the statistical model identification. *IEEE Transactions on Automatic Control, 19*, 716–723.

Arlot, S., & Celisse, A. (2010). A survey of cross-validation procedures for model selection. *Statistics Surveys, 4*, 40–79.

Askitas, N., & Zimmermann, K. F. (2009). Google econometrics and unemployment forecasting. *Applied Economics Quarterly*.

Athey, S. (2017). The impact of machine learning on economics.

Banbura, M., Giannone, D., & Reichlin, L. (2008). Large Bayesian VARs. *ECB Working Paper Series, 966*.

Belloni, A., Chen, D., Chernozhukov, V., & Hansen, C. (2012). Sparse models and methods for optimal instruments with an application to Eminent domain. *Econometrica, 80*, 2369–2429.

Belloni, A., & Chernozhukov, V. (2011). High dimensional sparse econometric models: An introduction. In P. Alquier, E. Gautier, & G. Stoltz (Eds.), *Inverse problems and high-dimensional estimation SE-3* (pp. 121–156). Lecture Notes in Statistics. Berlin, Heidelberg: Springer.

Belloni, A., & Chernozhukov, V. (2013). Least squares after model selection in high-dimensional sparse models. *Bernoulli, 19*, 521–547.

Belloni, A., Chernozhukov, V., & Hansen, C. (2011). Inference for high-dimensional sparse econometric models.

Belloni, A., Chernozhukov, V., & Hansen, C. (2014). Inference on treatment effects after selection among high-dimensional controls. *Review of Economic Studies, 81*, 608–650.

Belloni, A., Chernozhukov, V., Hansen, C., & Kozbur, D. (2016). Inference in high dimensional panel models with an application to gun control. *Journal of Business & Economic Statistics, 34*, 590–605.

Bergmeir, C., Hyndman, R. J., & Koo, B. (2018). A note on the validity of cross-validation for evaluating autoregressive time series prediction. *Computational Statistics & Data Analysis, 120*, 70–83.

Bickel, P. J., Ritov, Y., & Tsybakov, A. B. (2009). Simultaneous analysis of Lasso and Dantzig selector. *The Annals of Statistics, 37*, 1705–1732.

Bock, J. (2018). Quantifying macroeconomic expectations in stock markets using Google trends. *SSRN Electronic Journal*.

Buhlmann, P. (2013). Statistical significance in high-dimensional linear models. *Bernoulli, 19,* 1212–1242.

Buono, D., Kapetanios, G., Marcellino, M., Mazzi, G., & Papailias, F. (2018). Big data econometrics: Now casting and early estimates. *Bocconi Working Paper Series.*

Burman, P., Chow, E., & Nolan, D. (1994). A cross-validatory method for dependent data. *Biometrika, 81,* 351–358.

Carriero, A., Clark, T. E., & Marcellino, M. (2015). Realtime nowcasting with a Bayesian mixed frequency model with stochastic volatility. *Journal of the Royal Statistical Society. Series A: Statistics in Society.*

Chernozhukov, V., Hansen, C., & Spindler, M. (2015). Post-selection and post-regularization inference in linear models with many controls and instruments. *American Economic Review, 105,* 486–490.

Chernozhukov, V., Härdle, W., Huang, C., & Wang, W. (2019). LASSO-driven inference in time and space. arXiv:1806.05081v3.

Chetverikov, D., Liao, Z., & Chernozhukov, V. (forthcoming). On cross-validated lasso in high dimensions. Annals of Statistics.

Choi, H., & Varian, H. (2012). Predicting the present with google trends. *Economic Record, 88,* 2–9.

Croushore, D. (2006). Forecasting with real-time macroeconomic data. In *Handbook of economic forecasting.* Elsevier.

Doan, T., Litterman, R. B., & Sims, C. A. (1983). Forecasting and conditional projection using realistic prior distributions. *NBER Working Paper Series.*

Ettredge, M., Gerdes, J., & Karuga, G. (2005). Using web-based search data to predict macroeconomic statistics.

Fan, J., & Li, R. (2001). Variable selection via nonconcave penalized likelihood and its Oracle properties. *Journal of the American Statistical Association, 96,* 1348–1360.

Ferrara, L., & Simoni, A. (2019). When are Google data useful to nowcast GDP? An approach via pre-selection and shrinkage. *Banque de France Working Paper.*

Foroni, C., & Marcellino, M. (2015). A comparison of mixed frequency approaches for nowcasting Euro area macroeconomic aggregates. *International Journal of Forecasting, 30,* 554–568.

Frank, l. E., & Friedman, J. H. (1993). A statistical view of some chemometrics regression tools. *Technometrics, 35,* 109–135.

Ghysels, E., Santa-Clara, P., & Valkanov, R. (2004). *The MIDAS touch: Mixed data sampling regression models.* Discussion Paper, University of California and University of North Carolina.

Giannone, D., Lenza, M., & Primiceri, G. E. (2018). Economic predictions with big data: The illusion of sparsity. *SSRN Electronic Journal.*

Giannone, P., Monti, F., & Reichlin, L. (2016). Exploiting monthly data flow in structural forecasting. *Journal of Monetary Economics, 88,* 201–216.

Hannan, E. J., & Quinn, B. G. (1979). The determination of the order of an autoregression. *Journal of the Royal Statistical Society: Series B (Methodological), 41,* 190–195.

Hastie, T., Tibshirani, R., & Wainwright, M. J. (2015). *Statistical learning with sparsity: The Lasso and generalizations, monographs on statistics and applied probability.* Boca Raton: CRC Press, Taylor & Francis.

Hayashi, F. (2000). *Econometrics.* Princeton: Princeton University Press.

Hsu, N. J., Hung, H. L., & Chang, Y. M. (2008). Subset selection for vector autoregressive processes using Lasso. *Computational Statistics and Data Analysis, 52,* 3645–3657.

Hyndman, R. J., & Athanasopoulos, G. (2018). *Forecasting: Principles and practice,* 2nd ed.

Jing, B.-Y., Shao, Q.-M., & Wang, Q. (2003). Self-normalized Cramér-type large deviations for independent random variables. *The Annals of Probability, 31,* 2167–2215.

Kleinberg, J., Lakkaraju, H., Leskovec, J., Ludwig, J., & Mullainathan, S. (2018). Human decisions and machine predictions. *The Quarterly Journal of Economics, 133,* 237–293.

Kohns, D., & Bhattacharjee, A. (2019). Interpreting big data in the macro economy: A Bayesian mixed frequency estimator. In *CEERP Working Paper Series.* Heriot-Watt University.

Koop, G., & Onorante, L. (2016). Macroeconomic nowcasting using Google probabilities.

Li, X. (2016). Nowcasting with big data: Is Google useful in the presence of other information?

Litterman, R. B. (1986). Forecasting with Bayesian vector autoregressions—five years of experience. *Journal of Business & Economic Statistics, 4,* 25–38.

Lockhart, R., Taylor, J., Tibshirani, R. J., & Tibshirani, R. (2014). A significance test for the Lasso. *Annals of Statistics, 42,* 413–468.

Lütkepohl, H. (2005). *New introduction to multiple time series analysis.*

Medeiros, M. C., & Mendes, E. F. (2015). L1-regularization of high-dimensional time-series models with non-Gaussian and heteroskedastic errors. *Journal of Econometrics, 191,* 255–271.

Meinshausen, N., Meier, L., & Bühlmann, P. (2009). p-values for high-dimensional regression. *Journal of the American Statistical Association, 104,* 1671–1681.

Mol, C. D., Vito, E. D., & Rosasco, L. (2009). Elastic-net regularization in learning theory. *Journal of Complexity, 25,* 201–230.

Mullainathan, S., & Spiess, J. (2017). Machine learning: An applied econometric approach. *Journal of Economic Perspectives, 31,* 87–106.

Nardi, Y., & Rinaldo, A. (2011). Autoregressive process modeling via the Lasso procedure. *Journal of Multivariate Analysis, 102,* 528–549.

Romer, C. D., & Romer, D. H. (2000). Federal reserve information and the behavior of interest rates. *American Economic Review, 90,* 429–457.

Schwarz, G. (1978). Estimating the dimension of a model. *The Annals of Statistics, 6,* 461–464.

Scott, S., & Varian, H. (2014). Predicting the present with Bayesian structural time series. *International Journal of Mathematical Modelling and Numerical Optimisation, 5.*

Sims, C. A. (2002). The role of models and probabilities in the monetary policy process. In *Brookings Papers on Economic Activity.*

Smith, P. (2016). Google's MIDAS touch: Predicting UK unemployment with internet search data. *Journal of Forecasting, 35,* 263–284.

Stock, J. H., Wright, J. H., & Yogo, M. (2002). A survey of weak instruments and weak identification in generalized method of moments.

Tibshirani, R. (1996). Regression shrinkage and selection via the Lasso. *Journal of the Royal Statistical Society. Series B (Methodological), 58,* 267–288.

Tibshirani, R., Saunders, M., Rosset, S., Zhu, J., & Knight, K. (2005). Sparsity and smoothness via the fused lasso. *Journal of the Royal Statistical Society. Series B: Statistical Methodology, 67,* 91–108.

Varian, H. R. (2014). Big data: New tricks for econometrics. *The Journal of Economic Perspectives, 28,* 3–27.

Wasserman, L., & Roeder, K. (2009). High-dimensional variable selection. *Annals of Statistics, 37,* 2178–2201.

Weilenmann, B., Seidl, I., & Schulz, T. (2017). The socio-economic determinants of urban sprawl between 1980 and 2010 in Switzerland. *Landscape and Urban Planning, 157,* 468–482.

Yang, Y. (2005). Can the strengths of AIC and BIC be shared? A conflict between model indentification and regression estimation. *Biometrika, 92,* 937–950.

Yuan, M., & Lin, Y. (2006). Model selection and estimation in additive regression models. *Journal of the Royal Statistical Society. Series B (Methodological), 68,* 49–67.

Zhang, Y., Li, R., & Tsai, C.-L. (2010). Regularization parameter selections via generalized information criterion. *Journal of the American Statistical Association, 105,* 312–323.

Zou, H. (2006). The adaptive Lasso and its Oracle properties. *Journal of the American Statistical Association, 101,* 1418–1429.

Why LASSO, EN, and CLOT: Invariance-Based Explanation

Hamza Alkhatib, Ingo Neumann, Vladik Kreinovich, and Chon Van Le

Abstract In many practical situations, observations and measurement results are consistent with many different models—i.e., the corresponding problem is ill-posed. In such situations, a reasonable idea is to take into account that the values of the corresponding parameters should not be too large; this idea is known as regularization. Several different regularization techniques have been proposed; empirically the most successful are LASSO method, when we bound the sum of absolute values of the parameters, and EN and CLOT methods in which this sum is combined with the sum of the squares. In this paper, we explain the empirical success of these methods by showing that they are the only ones which are invariant with respect to natural transformations—like scaling which corresponds to selecting a different measuring unit.

1 Formulation of the Problem

Need for solving the inverse problem. Once we have a model of a system, we can use this model to predict the system's behavior, in particular, to predict the results of future measurements and observations of this system. The problem of estimating future measurement results based on the model is known as the *forward problem*.

H. Alkhatib · I. Neumann
Geodetic Institute, Leibniz University of Hannover,
Nienburger Str. 1, 30167 Hannover, Germany
e-mail: alkhatib@gih.uni-hannover.de

I. Neumann
e-mail: neumann@gih.uni-hannover.de

V. Kreinovich (✉)
University of Texas at El Paso, El Paso, TX 79968, USA
e-mail: vladik@utep.edu

C. Van Le
International University – VNU HCMC, Ho Chi Minh City, Vietnam
e-mail: lvchon@hcmiu.edu.vn

N. Ngoc Thach et al. (eds.), *Data Science for Financial Econometrics*, Studies in Computational Intelligence 898, https://doi.org/10.1007/978-3-030-48853-6_2

In many practical situations, we do not know the exact model. To be more precise, we know the general form of a dependence between physical quantities, but the parameters of this dependence need to be determined from the observations and from the results of the experiment. For example, often, we have a linear model $y = a_0 + \sum_{i=1}^{n} a_i \cdot x_i$, in which the parameters a_i need to be experimentally determined. The problem of determining the parameters of the model based on the measurement results is known as the *inverse problem*.

To actually find the parameters, we can use, e.g., the Maximum Likelihood method. For example, when the errors are normally distributed, then the Maximum Likelihood procedure results in the usual Least Squares estimates; see, e.g., Sheskin (2011). For example, for a general linear model with parameters a_i, once we know several tuples of corresponding values $(x_1^{(k)}, \ldots, x_n^{(k)}, y^{(k)})$, $1 \le k \le K$, then we can find the parameters from the condition that

$$\sum_{k=1}^{K} \left(y^{(k)} - \left(a_0 + \sum_{i=1}^{n} a_i \cdot x_i^{(k)} \right) \right)^2 \rightarrow \min_{a_0, \ldots, a_n} . \tag{1}$$

Need for regularization. In some practical situations, based on the measurement results, we can determine all the model's parameters with reasonably accuracy. However, in many other situations, the inverse problem is *ill-defined* in the sense that several very different combinations of parameters are consistent with all the measurement results.

This happens, e.g., in dynamical systems, when the observations provide a smoothed picture of the system's dynamics. For example, if we are tracing the motion of a mechanical system caused by an external force, then a strong but short-time force in one direction followed by a similar strong and short-time force in the opposite direction will (almost) cancel each other, so the same almost-unchanging behavior is consistent both with the absence of forces and with the above wildly-oscillating force. A similar phenomenon occurs when, based on the observed economic behavior, we try to reconstruct the external forces affecting the economic system.

In such situations, the only way to narrow down the set of possible solutions is to take into account some general a priori information. For example, for forces, we may know—e.g., from experts—the upper bound on each individual force, or the upper bound on the overall force. The use of such a priori information is known as *regularization*; see, e.g., Tikhonov and Arsenin (1977).

Which regularizations are currently used. There are many possible regularizations. Many of them have been tried, and, based on the results of these tries, a few techniques turned out to be empirically successful.

The most widely used technique of this type is known as LASSO technique (short of Least Absolute Shrinkage and Selection Operator), where we look for solutions for which the sum of the absolute values $\|a\|_1 \overset{\text{def}}{=} \sum_{i=0}^{n} |a_i|$ is bounded by a given value; see, e.g., Tibshirani (1996). Another widely used method is a *ridge regression* method, in which we limit the sum of the squares $S \overset{\text{def}}{=} \sum_{i=0}^{n} a_i^2$ or, equivalently, its square root $\|a\|_2 \overset{\text{def}}{=} \sqrt{S}$; see, e.g., Hoerl and Kennard (1970), Tikhonov (1943). Very promising are also:

- the *Elastic Net* (EN) method, in which we limit a linear combination $\|a\|_1 + c \cdot S$ (see, e.g., Kargoll et al. (2018), Zou and Hastie (2005)), and
- the *Combined L-One and Two* (CLOT) method in which we limit a linear combination $\|a\|_1 + c \cdot \|a\|_2$; see, e.g., Ahsen et al. (2017).

Why: remaining question and what we do in this paper. The above empirical facts prompt a natural question: why the above regularization techniques work the best? In this paper, we show that the efficiency of these methods can be explained by the natural invariance requirements.

2 General and Probabilistic Regularizations

General idea of regularization and its possible probabilistic background. In general, regularization means that we dismiss values a_i which are too large or too small. In some cases, this dismissal is based on subjective estimations of what is large and what is small. In other cases, the conclusion about what is large and what is not large is based on past experience of solving similar problem—i.e., on our estimate of the frequencies (= probabilities) with which different values have been observed in the past. In this paper, we consider both types of regularization.

Probabilistic regularization: towards a precise definition. There is no a priori reason to believe that different parameters have different distributions. So, in the first approximation, it makes sense to assume that they have the same probability distribution. Let us denote the probability density function of this common distribution by $\rho(a)$.

In more precise terms, the original information is invariant with respect to all possible permutations of the parameters; thus, it makes sense to conclude that the resulting joint distribution is also invariant with respect to all these permutations—which implies, in particular, that all the marginal distributions are the same.

Similarly, in general, we do not have a priori reasons to prefer positive or negative values of each the coefficients, i.e., the a priori information is invariant with respect to changing the sign of each of the variables: $a_i \rightarrow -a_i$. It is therefore reasonable to conclude that the marginal distribution should also be invariant, i.e., that we should have $\rho(-a) = \rho(a)$, and thus, $\rho(a) = \rho(|a|)$.

Also, there is no reason to believe that different parameters are positively or negatively correlated, so it makes sense to assume that their distributions are statistically independent. This is in line with the general Maximum Entropy (= Laplace Indeterminacy Principle) ideas Jaynes and Bretthorst (2003), according to which we should not pretend to be certain—to be more precise, if several different probability distributions are consistent with our knowledge:

- we should *not* select distributions with small entropy (measure of uncertainty),
- we *should* select the one for which the entropy is the largest.

If all we know are marginal distributions, then this principle leads to the conclusion that the corresponding variables are independent; see, e.g., Jaynes and Bretthorst (2003).

Due to the independence assumption, the joint distribution of n variables a_i take the form $\rho(a_0, a_1, \ldots, a_n) = \prod_{i=0}^{n} \rho(|a_i|)$. In applications of probability and statistics, it is usually assumed, crudely speaking, that events with very small probability are not expected to happen. This is the basis for all statistical tests—e.g., if we assume that the distribution is normal with given mean and standard deviation, and the probability that this distribution will lead to the observed data is very small (e.g., if we observe a 5-sigma deviation from the mean), then we can conclude, with high confidence, that experiments disprove our assumption. In other words, we take some threshold t_0, and we consider only the tuples $a = (a_0, a_1, \ldots, a_n)$ for which $\rho(a_0, a_1, \ldots, a_n) = \prod_{i=0}^{n} \rho(|a_i|) \geq t_0$. By taking logarithms of both sides and changing signs, we get an equivalent inequality

$$\sum_{i=0}^{n} \psi(|a_i|) \leq p_0, \tag{2}$$

where we denoted $\psi(z) \stackrel{\text{def}}{=} -\ln(\rho(z))$ and $p_0 \stackrel{\text{def}}{=} -\ln(t_0)$. (The sign is changed for convenience, since for small $t_0 \ll 1$, logarithm is negative, and it is more convenient to deal with positive numbers.)

Our goal is to avoid coefficients a_i whose absolute values are too large. Thus, if the absolute values $(|a_0|, |a_1| \ldots, |a_n|)$ satisfy the inequality (2), and we decrease one of the absolute values, the result should also satisfy the same inequality. So, the function $\psi(z)$ must be increasing.

We want to find the minimum of the usual least squares (or similar) criterion under the constraint (2). The minimum is attained:

- either when in (2), we have strict inequality
- or when we have equality.

If we have a strict inequality, then we get a local minimum, and for convex criteria like least squares (where there is only one local minimum which is also global), this means that we have the solution of the original constraint-free problem—and we started this whole discussion by considering situations in which this straightforward approach does not work. Thus, we conclude that the minimum under constraint (2) is attained when we have the equality, i.e., when we have

$$\sum_{i=0}^{n} \psi(|a_i|) = p_0 \tag{3}$$

for some function $\psi(z)$ and for some value p_0.

In practice, most probability distributions are continuous—step-wise and point-wise distributions are more typically found in textbooks than in practice. Thus, it is reasonable to assume that the probability density $\rho(x)$ is continuous. Then, its logarithm $\psi(z) = \ln(\rho(z))$ is continuous as well. Thus, we arrive at the following definition.

Definition 1 By a *probabilistic constraint*, we mean a constraint of the type (3) corresponding to some continuous increasing function $\psi(z)$ and to some number p_0.

General regularization. In the general case, we do not get any probabilistic justification of our approach, we just deal with the values $|a_i|$ themselves, without assigning probability to different possible values. In general, similarly to the probabilistic case, there is no reason to conclude that large positive values of a_i are better or worse than negative values with similar absolute value. Thus, we can say that a very large value a and its opposite $-a$ are equally impossible. The absolute value of each coefficient can be thus used as its "degree of impossibility": the larger the number, the less possible it is that this number will appear as the absolute value of a coefficient a_i.

Based on the degrees of impossibility of a_0 and a_1, we need to estimate the degree of impossibility of the pair (a_0, a_1). Let us denote the corresponding estimate by $|a_0| * |a_1|$. If the second coefficient a_1 is 0, it is reasonable to say that the degree of impossibility of the pair $(a_0, 0)$ is the same as the degree of impossibility of a_0, i.e., equal to $|a_0|$: $|a_0| * 0 = |a_0|$. If the second coefficient is not 0, the situation becomes slightly worse that when it was 0, so: if $a_1 \neq 0$, then $|a_0| * |a_1| > |a_0| * 0 = |a_0|$. In general, if the absolute value of one of the coefficients increases, the overall degree of impossibility should increase.

Once we know the degree of impossibility $|a_0| * |a_1|$ of a pair, we can combine it with the degree of impossibility $|a_2|$ of the third coefficient a_2, and get the estimated degree of impossibility $(|a_0| * |a_1|) * |a_2|$ of a triple (a_0, a_1, a_2), etc., until we get the degree of impossibility of the whole tuple.

The result of applying this procedure should not depend on the order in which we consider the coefficients, i.e., we should have $a * b = b * a$ (commutativity) and $(a * b) * c = a * (b * c)$ (associativity).

We should consider only the tuples for which the degree of impossibility does not exceed a certain threshold t_0:

$$|a_0| * |a_1| * \cdots * |a_n| \leq t_0 \tag{4}$$

for some t_0. Thus, we arrive at the following definitions.

Definition 2 By a *combination operation*, we mean a function $*$ that maps two non-negative numbers into a new non-negative number which is:

- commutative,
- associative,
- has the property $a * 0 = a$ and
- *monotonic* in the sense that if $a < a'$, then $a * b < a' * b$.

Definition 3 By a *general constraint*, we means a constraint of the type (4) for some combination operation $*$ and for some number $t_0 > 0$.

3 Natural Invariances

Scale-invariance: general idea. The numerical values of physical quantities depend on the selection of a measuring unit. For example, if we previously used meters and now start using centimeters, all the physical quantities will remain the same, but the numerical values will change—they will all get multiplied by 100.

In general, if we replace the original measuring unit with a new measuring unit which is λ times smaller, then all the numerical values get multiplied by λ:

$$x \to x' = \lambda \cdot x.$$

Similarly, if we change the original measuring units for the quantity y to a new unit which is λ times smaller, then all the coefficients a_i in the corresponding dependence $y = a_0 + \cdots + a_i \cdot x_i + \cdots$ will also be multiplied by the same factor: $a_i \to \lambda \cdot a_i$.

Scale-invariance: case of probabilistic constraints. It is reasonable to require that the corresponding constraints should not depend on the choice of a measuring unit. Of course, if we change a_i to $\lambda \cdot a_i$, then the value p_0 may also need to be accordingly changed, but overall, the constraint should remain the same. Thus, we arrive at the following definition.

Definition 4 We say that probability constraints corresponding to the function $\psi(z)$ are *scale-invariant* if for every p_0 and for every $\lambda > 0$, there exists a value p'_0 such that

$$\sum_{i=0}^{n} \psi(|a_i|) = p_0 \Leftrightarrow \sum_{i=0}^{n} \psi(\lambda \cdot |a_i|) = p'_0. \tag{5}$$

Scale-invariance: case of general constraints. In general, the degree of impossibility is described in the same units as the coefficients themselves. Thus, invariance would mean that if replace a and b with $\lambda \cdot a$ and $\lambda \cdot b$, then the combined value $a * b$ will be replaced by a similarly re-scaled value $\lambda \cdot (a * b)$. Thus, we arrive at the following definition:

Definition 5 We say that a general constraint corresponding to a combination operation $*$ is *scale-invariance* if for every a, b, and λ, we have

$$(\lambda \cdot a) * (\lambda \cdot b) = \lambda \cdot (a * b). \tag{6}$$

In this case, the corresponding constraint is naturally scale-invariant: if $*$ is scale-invariant operation, then, for all a_i and for all λ, we have

$$|\lambda \cdot a_0| * |\lambda \cdot a_1| * \cdots * |\lambda \cdot a_n| = \lambda * (|a_0| * |a_1| * \cdots * |a_n|)$$

and thus,

$$|a_0| * |a_1| * \cdots * |a_n| = t_0 \Leftrightarrow |\lambda \cdot a_0| * |\lambda \cdot a_1| * \cdots * |\lambda \cdot a_n| = t_0' \stackrel{\text{def}}{=} \lambda \cdot t_0.$$

Shift-invariance: general idea. Our goal is to minimize the deviations of the coefficients a_i from 0. In the ideal case, when the model is exact and when measurement errors are negligible, in situations when there is no signal at all (i.e., when $a_i = 0$ for all i), we will measure exactly 0s and reconstruct exactly 0 values of a_i. In this case, even if we do not measure some of the quantities, we should also return all 0s. In this ideal case, any deviation of the coefficients from 0 is an indication that something is not right.

In practice, however, all the models are approximate. Because of the model's imperfection and measurement noise, even if we start with a case when $a_i = 0$ for all i, we will still get some non-zero values of y and thus, some non-zero values of a_i (hopefully, small, but still non-zero). In such situations, small deviations from 0 are OK, they do not necessarily indicate that something is wrong.

We can deal with this phenomenon in two different ways:

- we can simply have this phenomenon in mind when dealing with the original values of the coefficients a_i—and do not change any formulas,
- or we can explicitly subtract an appropriate small tolerance level $\varepsilon > 0$ from the absolute values of all the coefficients, i.e., replace the original values $|a_i|$ with the new values $|a_i| - \varepsilon$ thus explicitly taking into account that deviations smaller than this tolerance level are OK, and only values above this level are problematic.

It is reasonable to require that the corresponding constraints do not change under this shift $|a| \to |a| - \varepsilon$.

Shift-invariance: case of probabilistic constraints. If we change $|a_i|$ to $|a_i| - \varepsilon$, then the coefficient p_0 may also need to be accordingly changed, but overall, the constraint should remain the same. Thus, we arrive at the following definition.

Definition 6 We say that probability constraints corresponding to the function $\psi(z)$ are *shift-invariant* if for every p_0 and for every sufficiently small $\varepsilon > 0$, there exists a value p_0' such that

$$\sum_{i=0}^{n} \psi(|a_i|) = p_0 \Leftrightarrow \sum_{i=0}^{n} \psi(|a_i| - \varepsilon) = p_0'. \tag{7}$$

Shift-invariance: case of general constraints. In general, the degree of impossibility is described in the same units as the coefficients themselves. Thus, invariance would mean that if replace a and b with $a - \varepsilon$ and $b - \varepsilon$, then the combined value $a * b$

will be replaced by a similarly re-scaled value $(a * b) - \varepsilon'$. Here, ε' may be different from ε, since it represents deleting two small values, not just one. A similar value should exist for all n. Thus, we arrive at the following definition:

Definition 7 We say that a general constraint corresponding to a combination operation $*$ is *shift-invariance* if for every n and for all sufficiently small $\varepsilon > 0$ there exists a value $\varepsilon' > 0$ such that for every $a_0, a_1, \ldots, a_n > 0$, we have

$$(a_0 - \varepsilon) * (a_1 - \varepsilon) * \cdots * (a_n - \varepsilon) = (a_0 * a_1 * \cdots * a_n) - \varepsilon'. \tag{8}$$

In this case, the corresponding constraint is naturally shift-invariant: if $*$ is a shift-invariant operation, then, for all a_i and for all sufficiently small $\varepsilon > 0$, we have

$$|a_0| * |a_1| * \cdots * |a_n| =$$

$$t_0 \Leftrightarrow (|a_0| - \varepsilon) * (|a_1| - \varepsilon) * \cdots * |(|a_n| - \varepsilon) = t_0' \stackrel{\text{def}}{=} t_0 - \varepsilon.$$

4 Why LASSO: First Result

Let us show that for both types of constraints, natural invariance requirements lead to LASSO formulas.

Proposition 1 *Probabilistic constraints corresponding to a function $\psi(z)$ are shift- and scale-invariant if and only if $\psi(z)$ is a linear function $\psi(z) = k \cdot z + \ell$.*

Discussion. For a linear function, the corresponding constraint $\sum_{i=0}^{n} \psi(|a_i|) = p_0$ is equivalent to the LASSO constraint $\sum_{i=1}^{n} |a_i| = t_0'$, with $t_0' \stackrel{\text{def}}{=} (t_0 - \ell)/k$. Thus, Proposition 1 explains why probabilistic constraints should be LASSO constraints: LASSO constraints are the only probabilistic constraints that satisfy natural invariance requirements.

Proposition 2 *General constraints corresponding to a combination function $*$ are shift- and scale-invariant if and only if the operation $*$ is addition $a * b = a + b$.*

Discussion. For addition, the corresponding constraint $\sum_{i=0}^{n} |a_i| = t_0$ is exactly the LASSO constraints. Thus, Proposition 2 explains why general constraints should be LASSO constraints: LASSO constraints are the only general constraints that satisfy natural invariance requirements.

Proof of Proposition 1. Scale-invariance implies that if $\psi(a) + \psi(b) = \psi(c) + \psi(0)$, then, for every $\lambda > 0$, we should have $\psi(\lambda \cdot a) + \psi(\lambda \cdot b) = \psi(\lambda \cdot c) + \psi(0)$. If we subtract $2\psi(0)$ from both sides of each of these equalities, then we can conclude that for the auxiliary function $\Psi(z) \overset{\text{def}}{=} \psi(a) - \psi(0)$, if $\Psi(a) + \Psi(b) = \Psi(c)$, then $\Psi(\lambda \cdot a) + \Psi(\lambda \cdot b) = \Psi(\lambda \cdot c)$. So, for the mapping $f(z)$ that transforms $z = \Psi(a)$ into $f(z) = \Psi(\lambda \cdot a)$—i.e., for $f(z) \overset{\text{def}}{=} \Psi(\lambda \cdot \Psi^{-1}(z))$, where $\Psi^{-1}(z)$ denotes the inverse function—we conclude that if $z + z' = z''$ then $f(z) + f(z') = f(z'')$. In other words, $f(z + z') = f(z) + f(z')$. It is known that the only monotonic functions with this property are linear functions $f(z) = c \cdot z$; see, e.g., Aczel and Dhombres (1989).

Since $z = \Psi(a)$ and $f(z) = \Psi(\lambda \cdot a)$, we thus conclude that for every λ, there exists a value c (which, in general, depends on λ) for which $\Psi(\lambda \cdot a) = c(\lambda) \cdot \Psi(a)$. Every monotonic solution to this functional equation has the form $\Psi(a) = A \cdot a^\alpha$ for some A and α, so $\psi(a) = \Psi(a) + \psi(0) = A \cdot a^\alpha + B$, where $B \overset{\text{def}}{=} \psi(0)$.

Similarly, shift-invariance implies that if $\psi(a) + \psi(b) = \psi(c) + \psi(d)$, then, for each sufficiently small $\varepsilon > 0$, we should have

$$\psi(a - \varepsilon) + \psi(b - \varepsilon) = \psi(c - \varepsilon) + \psi(d - \varepsilon).$$

The inverse is also true, so the same property holds for $\varepsilon = -\delta$, i.e., if $\psi(a) + \psi(b) = \psi(c) + \psi(d)$, then, for each sufficiently small $\delta > 0$, we should have

$$\psi(a + \delta) + \psi(b + \delta) = \psi(c + \delta) + \psi(d + \delta).$$

Substituting the expression $\psi(a) = A \cdot a^\alpha + B$, subtracting $2B$ from both sides of each equality and dividing both equalities by A, we conclude that if $a^\alpha + b^\alpha = c^\alpha + d^\alpha$, then $(a + \delta)^\alpha + (b + \delta)^\alpha = (c + \delta)^\alpha + (d + \delta)^\alpha$. In particular, the first equality is satisfied if we have $a = b = 1$, $c = 2^{1/\alpha}$, and $d = 0$. Thus, for all sufficiently small δ, we have $2 \cdot (1 + \delta)^\alpha = (2^{1/\alpha} + \delta)^\alpha + \delta^\alpha$.

On both sides, we have analytical expressions. When $\alpha < 1$, then for small δ, the left-hand side term and the first term in the right-hand side start with linear term δ, and the terms $\delta^\alpha \gg \delta$ is not compensated. If $\alpha > 1$, then by equating terms linear in δ in the corresponding expansions, we get $2\alpha \cdot \delta$ in the left-hand side and $\alpha \cdot (2^{1/\alpha})^{\alpha-1} \cdot \delta = 2^{1-1/\alpha} \cdot \alpha \cdot \delta$ in the right-hand side—the coefficients are different, since the corresponding powers of two are different: $1 \ne 1 - 1/\alpha$. Thus, the only possibility is $\alpha = 1$. The proposition is proven.

Proof of Proposition 2. It is known (see, e.g., Autchariyapanitkul (2018)) that every scale-invariant combination operation has the form $a * b = (a^\alpha + b^\alpha)^{1/\alpha}$ or $a * b = \max(a, b)$. The second case contradicts the requirement that $a * b$ be strictly increasing in both variables. For the first case, similarly to the proof of Proposition 1, we conclude that $\alpha = 1$. The proposition is proven.

5   Why EN and CLOT

Need to go beyond LASSO. In the previous section, we showed that if we need to select a single method for all the problems, then natural invariance requirements lead to LASSO, i.e., to bounds on the sum of the absolute values of the parameters. In some practical situations, this works, while in others, it does not lead to good results. To deal with such situations, instead of fixing a *single* method for all the problems, a natural idea is to select a *family* of methods, so that in each practical situation, we should select an appropriate method from this family. Let us analyze how we can do it both for probabilistic and for general constraints.

Probabilistic case. Constraints in the probabilistic case are described by the corresponding function $\psi(z)$. The LASSO case corresponds to a 2-parametric family $\psi(z) = c_0 + c_1 \cdot z$. In terms of the corresponding constraints, all these functions from this family are equivalent to $\psi(z) = z$.

To get a more general method, a natural idea is to consider a 3-parametric family, i.e., a family of the type $\psi(z) = c_0 + c_1 \cdot z + c_2 \cdot f(z)$ for some function $f(z)$. Constraints related to this family are equivalent to using the functions $\psi(z) = z + c \cdot f(z)$ for some function $f(z)$. Which family—i.e., which function $f(z)$—should we choose? A natural idea is to again use scale-invariance and shift-invariance.

Definition 8 We say that functions $\psi_1(z)$ and $\psi_2(z)$ are *constraint-equivalent* if:

- for each n and for each c_1, there exists a value c_2 such that the condition $\sum_{i=0}^{n} \psi_1(a_i) = c_1$ is equivalent to $\sum_{i=0}^{n} \psi_2(a_i) = c_2$, and
- for each n and for each c_2, there exists a value c_1 such that the condition $\sum_{i=0}^{n} \psi_2(a_i) = c_2$ is equivalent to $\sum_{i=0}^{n} \psi_1(a_i) = c_1$.

Definition 9

- We say that a family $\{z + c \cdot f(z)\}_c$ is *scale-invariant* if for each c and λ, there exists a value c' for which the re-scaled function $\lambda \cdot z + c \cdot f(\lambda \cdot z)$ is constraint-equivalent to $z + c' \cdot f(z)$.
- We say that a family $\{z + c \cdot f(z)\}_c$ is *shift-invariant* if for each c and for each sufficiently small number ε, there exists a value c' for which the shifted function $z - \varepsilon + c \cdot f(z - \varepsilon)$ is constraint-equivalent to $z + c' \cdot f(z)$.

Proposition 3 *A family $\{z + c \cdot f(z)\}_c$ corresponding to a smooth function $f(z)$ is scale- and shift-invariant if and only if the function $f(z)$ is quadratic.*

Discussion. Thus, it is sufficient to consider functions $\psi(z) = z + c \cdot z^2$. This is exactly the EN approach—which is thus justified by the invariance requirements.

Proof Similarly to the proof of Proposition 1, from the shift-invariance, for $c = 1$, we conclude that $z - \varepsilon + f(z - \varepsilon) = A + B \cdot (z + c' \cdot f(z))$ for some values A, B, and c' which are, in general, depending on ε. Thus,

$$f(z - \varepsilon) = A_0(\varepsilon) + A_1(\varepsilon) \cdot z + A_2(\varepsilon) \cdot f(z), \tag{9}$$

where $A_0(\varepsilon) \overset{\text{def}}{=} A + \varepsilon$, $A_1(\varepsilon) \overset{\text{def}}{=} B - 1$, and $A_2(\varepsilon) \overset{\text{def}}{=} B \cdot c'$.

By considering three different values x_k ($k = 1, 2, 3$), we get a system of three linear equations for three unknowns $A_i(\varepsilon)$. Thus, by using Cramer's rule, we get an explicit formula for each A_i in terms of the values x_k, $f(x_k)$, and $f(x_k - \varepsilon)$. Since the function $f(z)$ is smooth (differentiable), these expressions are differentiable too. Thus, we can differentiate both sides of the formula (9) with respect to ε. After taking $\varepsilon = 0$, we get the following differential equation $f'(z) = B_0 + B_1 \cdot z + B_2 \cdot f(z)$, where we denoted $B_i \overset{\text{def}}{=} A_i'(0)$. For $B_2 = 0$, we get $f'(z) = B_0 + B_1 \cdot z$, so $f(z)$ is a quadratic function.

Let us show that the case $B_2 \neq 0$ is not possible. Indeed, in this case, by moving all the terms containing f to the left-hand side, we get $f'(z) - B_2 \cdot f(z) = B_0 + B_1 \cdot z$. Thus, for the auxiliary function $F(z) \overset{\text{def}}{=} \exp(-B_2 \cdot z) \cdot f(z)$, we get

$$F'(z) = \exp(-B_2 \cdot z) \cdot f'(z) - B_2 \cdot \exp(-B_2 \cdot z) \cdot f(z) =$$

$$\exp(-B_2 \cdot z) \cdot (f'(z) - B_2 \cdot f(z)) = \exp(-B_2 \cdot z) \cdot (B_0 + B_1 \cdot z).$$

Integrating both sides, we conclude that

$$F(z) = f(z) \cdot \exp(-B_2 \cdot z) = (c_0 + c_1 \cdot z) \cdot \exp(-B_2 \cdot z) + c_2$$

for some constants c_i, thus

$$f(z) = c_0 + c_1 \cdot z + c_2 \cdot \exp(B_2 \cdot z). \tag{10}$$

From scale-invariance for $c = 1$, we similarly get

$$\lambda \cdot z + f(\lambda \cdot z) = D + E \cdot (z + c' \cdot f(z))$$

for some values D, E, and c' which are, in general, depending on λ. Thus,

$$f(\lambda \cdot z) = D_0(\lambda) + D_1(\lambda) \cdot z + D_2(\lambda) \cdot f(z) \tag{11}$$

for appropriate $D_i(\lambda)$. Similarly to the case of shift-invariance, we can conclude that the functions D_i are differentiable. Thus, we can differentiate both sides of the formula (11) with respect to λ. After taking $\lambda = 1$, we get the following differential equation $z \cdot f'(z) = D_0 + D_1 \cdot z + D_2 \cdot f(z)$ for appropriate values D_i. Substituting the expression (10) with $B_2 \neq 0$ into this formula, we can see that this equation is not satisfied. Thus, the case $B_2 \neq 0$ is indeed not possible, so the only possible case is $B_2 = 0$ which leads to a quadratic function $f(z)$. The proposition is proven.

Comment. The general expression $\psi(z) = g_0 + g_1 \cdot z + g_2 \cdot z^2$ is very natural for a different reason as well: it can be viewed as keeping the first terms in the Taylor expansion of a general function $\psi(z)$.

Case of general constraints. For the case of probabilistic constraints, we used a linear combination of different functions $\psi(z)$. For the case of general constraints, it is natural to use a linear combination of combination operations. As we have mentioned in the proof of Proposition 2, scale-invariant combination operations have the form $\|a\|_p \stackrel{\text{def}}{=} \left(\sum_{i=0}^n |a_i|^p\right)^{1/p}$. According to Proposition 3, it makes sense to use quadratic terms, i.e., $\|a\|_2$. Thus, it makes sense to consider the combination $\|a\|_1 + c \cdot \|a\|_2$—which is exactly CLOT.

Another interpretation of CLOT is that we combine $\|a\|_1$ and $c \cdot \|a\|_2$ by using shift- and scaling-invariant combination rule—which is, according to Proposition 2, simply addition.

Comments.

- An interesting feature of CLOT—as opposed to EN—is that it is scale-invariant.
- Not only we got a justification of EN and CLOT, we also got an understanding of when we should use EN and when CLOT: for probabilistic constraints, it is more appropriate to use EN, while for general constraints, it is more appropriate to use CLOT.

6 Beyond EN and CLOT?

Discussion. What if 1-parametric families like EN and CLOT are not sufficient? In this case, we need to consider families

$$F = \{z + c_1 \cdot f_1(z) + \cdots + c_n \cdot f_m(z)\}_{c_1,\ldots,c_m}$$

with more parameters.

Definition 10

- We say that a family $\{z + c_1 \cdot f_1(z) + \cdots + c_m \cdot f_m(z)\}_{c_1,\ldots,c_m}$ is *scale-invariant* if for each $c = (c_1,\ldots,c_m)$ and λ, there exists a tuple $c' = (c'_1,\ldots,c'_m)$ for which the re-scaled function

$$\lambda \cdot z + c_1 \cdot f_1(\lambda \cdot z) + \cdots + c_m \cdot f_m(\lambda \cdot z)$$

is constraint-equivalent to $z + c'_1 \cdot f_1(z) + \cdots + c'_m \cdot f_m(z)$.
- We say that a family $\{z + c_1 \cdot f_1(z) + \cdots + c_m \cdot f_m(z)\}_{c_1,\ldots,c_m}$ is *shift-invariant* if for each tuple c and for each sufficiently small number ε, there exists a tuple c' for which the shifted function

$$z - \varepsilon + c_1 \cdot f_1(z - \varepsilon) + \cdots + c_m \cdot f_m(z - \varepsilon)$$

is constraint-equivalent to $z + c_1' \cdot f_1(z) + \cdots + c_m' \cdot f_m(z)$.

Proposition 4 *A family* $\{z + c_1 \cdot f_1(z) + \cdots + c_m \cdot f_m(z)\}_{c_1,\ldots,c_m}$ *corresponding to a smooth functions* $f_i(z)$ *is scale- and shift-invariant if and only if all the functions* $f_i(z)$ *are polynomials of order* $\leq m + 1$.

Discussion. So, if EN and CLOT are not sufficient, our recommendation is to use a constraint $\sum_{i=0}^{n} \psi(|a_i|) = c$ for some higher order polynomial $\psi(z)$.

Proof of Proposition 4 is similar to the s of Proposition 3, the only difference is that instead of a single differential equation, we will have a system of linear differential equations.

Comment. Similarly to the quadratic case, the resulting general expression $\psi(z) = g_0 + g_1 \cdot z + \cdots + a_{m+1} \cdot z^{m+1}$ can be viewed as keeping the first few terms in the Taylor expansion of a general function $\psi(z)$.

Acknowledgements This work was supported by the Institute of Geodesy, Leibniz University of Hannover. It was also supported in part by the US National Science Foundation grants 1623190 (A Model of Change for Preparing a New Generation for Professional Practice in Computer Science) and HRD-1242122 (Cyber-ShARE Center of Excellence).
This paper was written when V. Kreinovich was visiting Leibniz University of Hannover.

References

Aczel, J., & Dhombres, J. (1989). *Functional equations in several variables*. Cambridge, UK: Cambridge University Press.

Ahsen, M. E., Challapalli, N., & Vidyasagar, M. (2017). Two new approaches to compressed sensing exhibiting both robust sparse recovery and the grouping effect. *Journal of Machine Learning Research, 18*, 1–24.

Autchariyapanitkul, K., Kosheleva, O., Kreinovich, V., & Sriboonchitta, S. (2018). Quantum econometrics: How to explain its quantitative successes and how the resulting formulas are related to scale invariance, entropy, and fuzziness. In: V.-N. Huynh, M. Inuiguchi, D.-H. Tran, & T. Denoeux (Eds.), *Proceedings of the International Symposium on Integrated Uncertainty in Knowledge Modelling and Decision Making IUKM'2018*, Hanoi, Vietnam, March 13–15, 2018.

Hoerl, A. E., & Kennard, R. W. (1970). Ridge regression: Biased estimation for nonorthogonal problems. *Technometrics, 12*(1), 55–67.

Jaynes, E. T., & Bretthorst, G. L. (2003). *Probability theory: The logic of science*. Cambridge, UK: Cambridge University Press.

Kargoll, B., Omidalizarandi, M., Loth, I., Paffenholz, J.-A., & Alkhatib, H. (2018). An iteratively reweighted least-squares approach to adaptive robust adjustment of parameters in linear regression models with autoregressive and t-distributed deviations. *Journal of Geodesy, 92*(3), 271–297.

Sheskin, D. J. (2011). *Handbook of parametric and nonparametric statistical procedures*. Boca Raton, Florida: Chapman and Hall/CRC.

Tibshirani, R. (1996). Regression shrinkage and selection via the lasso. *Journal of the Royal Statistical Society, 58*(1), 267–288.

Tikhonov, A. N. (1943). On the stability of inverse problems. *Doklady Akademii Nauk SSSR, 39*(5), 195–198.

Tikhonov, A. N., & Arsenin, V. Y. (1977). *Solutions of ill-posed problems*. Washington, DC: Winston and Sons.

Zou, H., & Hastie, T. (2005). Regularization and variable selection via the elastic net. *Journal of the Royal Statistical Society B, 67*, 301–320.

Composition of Quantum Operations and Their Fixed Points

Umar Batsari Yusuf, Parin Chaipunya, Poom Kumam, and Sikarin Yoo-Kong

Abstract Fixed point sets of quantum operations are very important in correcting quantum errors. In this work, using single qubit two-state quantum system; Bloch sphere, we consider the situation whereby the quantum system undergoes evolution through multiple channels (composition of quantum operations). The outputs of both Bloch sphere deformation and amplitude damping were analyzed using the density matrix representation and the Bloch vector representation respectively. Also, the common fixed point sets were characterized.

Keywords Bloch sphere · Quantum operation · Fixed point · Quantum system · Quantum state · Composition of operations

2010 Mathematics Subject Classification 47H09 · 47H10 · 81P45

U. B. Yusuf
Department of Mathematics and Statistics, College of Science and Technology,
Hassan Usman Katsina Polytechnic, Katsina, Katsina State, Nigeria
e-mail: umar.batsari@mail.kmutt.ac.th; ubyusuf@hukpoly.edu.ng

U. B. Yusuf · P. Chaipunya · P. Kumam
Department of Mathematics, Faculty of Science, King Mongkut's, University of Technology
Thonburi (KMUTT), 126 Pracha-Uthit Road, Bang Mod, Thung Khru, Bangkok
10140, Thailand
e-mail: chaipunya.p@gmail.com

U. B. Yusuf · P. Chaipunya · P. Kumam (✉) · S. Yoo-Kong
Center of Excellence in Theoretical and Computational Science (TaCS-CoE),
Science Laboratory Building, King Mongkut's University of Technology Thonburi (KMUTT),
126 Pracha-Uthit Road, Bang Mod, Thrung Khru, Bangkok 10140, Thailand
e-mail: poom.kumam@mail.kmutt.ac.th

S. Yoo-Kong
e-mail: syookong@googlemail.com

S. Yoo-Kong
The Institute for Fundamental Study (IF), Naresuan University, Phitsanulok 65000, Thailand

N. Ngoc Thach et al. (eds.), *Data Science for Financial Econometrics*,
Studies in Computational Intelligence 898,
https://doi.org/10.1007/978-3-030-48853-6_3

1   Introduction

Researchers have been utilizing the quantum theory in their field of expertise. Fixed point theory area is not left behind (Arias et al. 2002; Busch and Singh 1998; Theodore et al. 2014; Smolin 1982).

Let H be a Hilbert space, and $\mathscr{B}(H)$ the set of bounded linear operators on H. An *observable* A in a quantum system is a bounded, linear, and self adjoint operator on H (Nguyen and Dong 2018). Each observable A represents a physical quantity (Nguyen and Dong 2018). To know more information of a particle/observable in a quantum state, we need quantum measurements that give the probability of the observable to be in a given interval (neighborhood) (Robert 2017; Stephan 2015). Nevertheless, quantum measurements comes with error that we call quantum effects (Arias et al. 2002; Davies 1976; Busch and Singh 1998), these effects sometimes can be seen as superposition, interference etc. (Davies 1976). In fact, in quantum mechanics, at a given time, repeated measurements under the same state of the quantum system give different values of physical quantity (Nguyen and Dong 2018). General quantum measurements that have more than two values are described by effect-valued measures (Arias et al. 2002). Let $\mathscr{E}(H) = \{A \in \mathscr{B}(H) : 0 \leq A \leq I\}$ be the collection of *quantum effects*. Consider the discrete effect-valued measures $E_i \in \mathscr{E}(H), i = 1, 2, \ldots$, whose infinite sum is an identity operator I. So, the probability for an outcome i occurring in the state ρ can be seen as $P_\rho(E_i)$ and the post-measurement state given occurrence of the outcome i is $\frac{E_i^{\frac{1}{2}} \rho E_i^{\frac{1}{2}}}{tr(\rho E_i)}$ (Arias et al. 2002). Furthermore, the resulting state after the execution of measurement without making any observation is given by

$$\phi(\rho) = \sum E_i^{\frac{1}{2}} \rho E_i^{\frac{1}{2}}. \tag{1}$$

If the measurement does not disturb the state ρ, then we have $\phi(\rho) = \rho$ **(fixed point equation)**. It was established in Busch and Singh (1998) that, fixed point equation exist if and only if the fixed/invariant state commutes with every E_i, $i = 1, 2, \ldots$. One can say there exist compatibility between $E_i, i = 1, 2, \ldots$ and ρ, this is well known as the *generalized Lüders theorem*. Also, the likelihood that an effect A took place in the state ρ given that, the measurement was performed can be seen as

$$P_{\phi(\rho)}(A) = tr\left[A \sum E_i^{\frac{1}{2}} \rho E_i^{\frac{1}{2}}\right] = tr\left(\sum E_i^{\frac{1}{2}} A E_i^{\frac{1}{2}} \rho\right). \tag{2}$$

If A is not disturbed by the measurement in any state we have

$$\sum E_i^{\frac{1}{2}} A E_i^{\frac{1}{2}} = A,$$

and by defining $\phi(A) = E_i^{\frac{1}{2}} A E_i^{\frac{1}{2}}$, we end up with $\phi(A) = A$.

Other areas such as quantum information theory, quantum computation, and quantum dynamics utilizes the concept of measurements too (see Busch et al. (1996), Davies (1976), Nielsen and Chuang (2010)).

Let $\mathscr{A} = \{A_i, A_i^* : A_i \in \mathscr{B}(H) \, i = 1, 2, 3, \ldots\}$ be a collection of operators such that, $\sum A_i A_i^* \leq I$. A mapping $\phi : \mathscr{B}(H) \rightarrow \mathscr{B}(H)$ defined as $\phi_{\mathscr{A}}(B) = \sum A_i B A_i^*$ is known as **quantum operation** (see Arias et al. (2002)). Whenever A_i's are self adjoint so is the mapping $\phi_{\mathscr{A}}$ too. Define the **commutant** $\mathscr{A}' = \{S \in \mathscr{B}(H) : AS = SA \, \forall A \in \mathscr{A}\}$.

Let ρ, σ be two quantum states. Then, the trace distance between the two quantum states is defined as

$$D(\rho, \sigma) \equiv \frac{1}{2} tr |\rho - \sigma|,$$

where here the modulus is seen as $|\beta| \equiv \sqrt{\beta^{\dagger} \beta}$, see Nielsen and Chuang (2010).

Denote the fixed point set of $\phi_{\mathscr{A}}$ by $\mathscr{B}(H)^{\phi_{\mathscr{A}}}$. The fixed points of $\phi_{\mathscr{A}}$ are sometimes called $\phi_{\mathscr{A}}$-fixed points. It is true that, $\mathscr{A}' \subseteq \mathscr{B}(H)^{\phi_{\mathscr{A}}}$ (Arias et al. 2002).

The below **Lüders operation** is a clear example for a self-adjoint quantum operation

$$L_{\mathscr{A}}(B) = \sum A_i^{\frac{1}{2}} B A_i^{\frac{1}{2}}, \tag{3}$$

for A_i's being positive and whose sum is identity ($\sum A_i = I$) (Arias et al. 2002). A **faithful** quantum operation $\phi_{\mathscr{A}}$ is that with the property $\phi_{\mathscr{A}}(B^*B) = 0$ implies B as a zero operator, and it is **unital** if the identity operator is invariant ($\sum A_i A_i^* = I$). Also, if $\sum A_i^* A_i = I$ then $\phi_{\mathscr{A}}$ is **trace preserving** (Arias et al. 2002).

Suppose the trace class operator on the Hilbert space H is denoted by $\mathscr{T}(H)$ while $\mathscr{D}(H) = \{\rho \in \mathscr{T}(H)^+ : tr(\rho) = 1\}$ denote the set of **density operators** representing quantum states in a quantum system (Arias et al. 2002).

In 1950, Lüders (1950) asserted that, the operator assigned to the quantity R possesses the spectral representation $R = \sum r_k P_k$ where the r_k represent the eigenvalues (measured values) and the corresponding projection operators satisfy $P_j P_k$, $\sum P_k = I$.

In 1998, Busch and Singh (1998) proved a proposition that generalized Lüders theorem as follows: *Let E_i, $i = 1, 2, 3, \ldots, N \leq \infty$ be a complete family of effects (quantum effects). Let $B = \sum_k b_k P_k$ be an effect with discrete spectrum given by the strictly decreasing sequence of eigenvalues $b_k \in [0, 1]$ and spectral projection P_k, $k = 1, 2, 3, \ldots, k \leq \infty$. Then, $tr[\mathscr{I}_L(\rho) \cdot B] = tr[\rho \cdot B]$ holds for all states ρ, exactly when all E_i commute with B.* Note that, $tr[\rho \cdot B] = P_\rho(B)$; Probability that an effect B occurs at state ρ. Also, they proved that, *the measurement $\mathscr{I}_L$ will not disturb the state ρ if ρ commutes with every effect E_i.*

In 2002, Arias et al. (2002) proved that, *any quantum operation (ϕ) is a quantum operation is weakly continuous and completely positive map.* Furthermore, they proved the below three theorems.

Theorem 1.1 (Arias et al. 2002) *Suppose $\phi_{\mathscr{A}}$ is a quantum operation. Then,*

1. $\phi_{\mathscr{A}}$ is a weakly continuous and completely positive mapping.

2. *Any trace preserving $\phi_{\mathscr{A}}$ is faithful and $tr(\phi_{\mathscr{A}}(B)) = tr(B)$, $B \in \mathscr{T}(H)$.*

Theorem 1.2 (Arias et al. 2002) *For a trace preserving and unital quantum operation $\phi_{\mathscr{A}}$, $\mathscr{B}(H)^{\phi_{\mathscr{A}}} = \mathscr{A}'$ whenever $dim(H) < \infty$.*

Theorem 1.3 (Arias et al. 2002) *For any unital quantum operation $\phi_{\mathscr{A}}$, (a) there exists a completely positive, unital and idempotent mapping $\psi : \mathscr{B}(H) \to \mathscr{B}(H)$ such that $ran(\psi) = \mathscr{B}(H)^{\phi_{\mathscr{A}}}$. (b) $\mathscr{B}(H)^{\phi_{\mathscr{A}}} = \mathscr{A}'$ if and only if ψ is a conditional expectation. (c) $\mathscr{A}'$ is an injective von Neumann algebra whenever $\mathscr{B}(H)^{\phi_{\mathscr{A}}} = \mathscr{A}'$.*

Along the applications direction, in 2008, Blume-Kohou et al. (2008) proved that, *fixed point set of quantum operations are isometric to some information preserving structures (IPS)*. Furthermore, they showed that, *fixed point sets of quantum operations can be used in forming quantum error correction codes (QECC)*. While in 2010 (Blume-Kohou et al. 2010), one of their result was *an efficient algorithm of finding channels information preserving structures (IPS), using the fixed points of the channel (quantum operation)*. Note that, formulating quantum error correction codes (QECC) is more easier through fixed point sets via information preserving structures (IPS) (Blume-Kohou et al. 2010).

In 2009, Felloni et al. (2010) studied the process involved in the evolution of general quantum system using diagrams of states with the aid of quantum circuits. Although, the study was limited to a single qubit decoherence and errors. The method prove to be most useful whenever the quantum operations to be analyzed are described by very sparse matrices. Also, the diagrams revealed the significant pattern along which quantum information is processed from input to output, clearly and immediately.

In 2011, Long and Zhang (2011) gave some necessary and sufficient conditions for finding a non-trivial fixed point set of quantum operations.

In 2012, Zhang and Ji (2012) studied the fixed point sets of generalized quantum operation and discussed some equivalent and sufficient conditions for the set to be non-trivial. Below is the definition of the generalized quantum operation.

Let H be a separable complex Hilbert space, and $\mathscr{B}(H)$ be the set of all bounded linear operators on H. Let $\mathscr{A} = \{A_k\}_{k=1}^n$ (n is a positive integer or ∞) be a finite or countable subset of $\mathscr{B}(H)$. If

$$\sum_{k=1}^n A_k A_k^* \leq I,$$

then $\mathscr{A}$ is called a row contraction (Zhang and Ji 2012). Now, let $\mathscr{A}$ and $\mathscr{B}$ be row contractions. Then, a function $\phi_{\mathscr{A},\mathscr{B}}$ is called a generalized quantum operation if

$$\phi_{\mathscr{A},\mathscr{B}}(X) = \sum_{k=1}^n A_k X B_k^*, \quad \forall X \in \mathscr{B}(H),$$

where the series converges in the strong operator topology (Zhang and Ji 2012).

In 2015, Zhang and Mingzhi (2015) gave an equivalent condition for the power of a quantum operation to converge to a projection in the strong operator topology. Let $\mathcal{Q}_{\mathscr{A}} = s.o. - \lim_{j \to \infty} \phi_{\mathscr{A}}^j (I)$ *(strong operator topology limit) where*

$$\phi_{\mathscr{A},\mathscr{A}}(X) = \phi_{\mathscr{A}}(X) = \sum_{k=1}^{n} A_k X A_k^*, \quad \forall X \in \mathscr{B}(H),$$

H is a separable complex Hilbert space, $\mathscr{B}(H)$ is the algebra of all bounded linear operators on H, $\mathscr{A}$ is a row contraction. Then, they gave the below theorems.

Theorem 1.4 (Zhang and Mingzhi 2015) *Let $\mathscr{A} = \{A_k\}_{k=1}^{n}$ be a row contraction. Then, $\mathcal{Q}_{\mathscr{A}}$ is a projection iff $\mathcal{Q}_{\mathscr{A}} A_k = A_k \mathcal{Q}_{\mathscr{A}}$ for all $k = 1, 2, 3, \ldots, n$.*

Theorem 1.5 (Zhang and Mingzhi 2015) *Let $\mathscr{A} = \{A_k\}_{k=1}^{n}$ be a trace preserving and commuting operator sequence. Then, the following statements hold:*

1. *$\mathscr{A}$ is a row contraction.*
2. *$\mathcal{Q}_{\mathscr{A}}$ is a projection.*
3. *$\mathscr{B}(H)^{\phi_{\mathscr{A}}} = \mathcal{Q}_{\mathscr{A}} \mathscr{A}' \mathcal{Q}_{\mathscr{A}}$ for all $k = 1, 2, 3, \ldots, n$.*

In 2016, Zhang and Si (2016) investigated the conditions for which the fixed point set of a quantum operation $(\phi_{\mathscr{A}})$ with respect to a row contraction $\mathscr{A}$ equals to the fixed point set of the power of the quantum operation $\phi_{\mathscr{A}}^j$ for some $1 \leq j < \infty$.

Theorem 1.6 (Zhang and Si 2016) *Let $A = \{E_k\}_{k=1}^{n}$ be a unital and commutable. Suppose there is k_0 such that, $E_{k_0} \in \mathscr{A}$ is positive and invertible. Then, $\mathscr{B}(H)^{\phi_{\mathscr{A}}^j} = \mathscr{B}(H)^{\phi_{\mathscr{A}}}$, $1 \leq j < \infty$.*

Applications of quantum operations are found in discrete dynamics (Bruzda et al. 2009), quantum measurement theory, quantum probability, quantum computation, and quantum information (Arias et al. 2002).

Motivated by the application of fixed points in Quantum error correction codes, work of Felloni et al. (2010), and work of Batsari et al. (2019), we study the fixed points of quantum operations associated with the deformation of Bloch sphere (Bit flip, Phase flip and Bit-Phase flip) and amplitude damping (displacement of origin) and the fixed points of their respective compositions.

2 Main Results

Consider the Bloch sphere below (Nielsen and Chuang 2010). The states of a single bit two-level ($|0\rangle$, $|1\rangle$) quantum bit (qubit) are described by the Bloch sphere above with $0 \leq \theta \leq \pi$, $0 \leq \varphi \leq 2\pi$; qubit is just a quantum system.

Fig. 1 Cross section of a
Bloch sphere

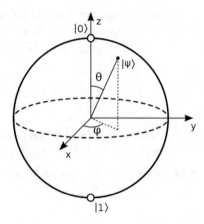

A single qubit quantum state ρ can be represented with below density matrix,

$$\rho = \frac{1}{2} \begin{pmatrix} 1 + \eta \cos\theta & \eta e^{-i\varphi} \sin\theta \\ \eta e^{i\varphi} \sin\theta & 1 - \eta \cos\theta \end{pmatrix}, \eta \in [0, 1], \quad 0 \leq \theta \leq \pi, \text{ and } 0 \leq \varphi \leq 2\pi. \tag{4}$$

Also, the density matrix can take below representation (Nielsen and Chuang 2010),

$$\rho = \frac{1}{2}[I + \bar{r} \cdot \bar{\sigma}] = \frac{1}{2}\begin{bmatrix} 1 + r_z & r_x - i r_y \\ r_x + i r_y & 1 - r_z \end{bmatrix}, \tag{5}$$

where $\bar{r} = [r_x, r_y, r_z]$ is the Bloch vector with $\|\bar{r}\| \leq 1$, and $\bar{\sigma} = [\sigma_x, \sigma_y, \sigma_z]$ for $\sigma_x, \sigma_y, \sigma_z$ being the Pauli matrices.

$$\sigma_x = \begin{pmatrix} 0 & 1 \\ 1 & 0 \end{pmatrix}, \quad \sigma_y = \begin{pmatrix} 0 & -i \\ i & 0 \end{pmatrix}, \quad \sigma_z = \begin{pmatrix} 1 & 0 \\ 0 & -1 \end{pmatrix}$$

Let T_1, T_2, T_3 and $T_{\pm h}$ denote the Bit flip operation, Phase flip operation, Bit-Phase flip operation and Displacements operation on a Bloch sphere respectively, for $h \in \{x, y, z\}$. Denote the Bloch sphere by $\mathcal{D}$ and $\mathcal{D}_T$ be the deformation of the Bloch sphere after an operation T. Let $F(T)$ denote the fixed point set of the operation T.

Proposition 2.1 *Suppose $p \in [0, 1]$ is the same for T_1, T_2 and T_3; $p_{T_1} = p_{T_2} = p_{T_3} = p$. Then, the six different compositions obtained from the permutation of T_i, $i = 1, 2, 3$ gives the same out put,*

Proof Let ρ be a qubit (quantum bit) state in/on the Bloch sphere. Suppose the general representation of ρ using the density matrix is

$$\rho = \begin{pmatrix} a & b \\ c & d \end{pmatrix}$$

where $a, b, c, d \in \mathbb{C}$. Then,

1. Bit flip is given by $T_1(\rho) = E_0 \rho E_0^\dagger + E_1 \rho E_1^\dagger$, where

$$E_0 = \sqrt{p_{T_1}} \begin{pmatrix} 1 & 0 \\ 0 & 1 \end{pmatrix}, \quad E_1 = \sqrt{1 - p_{T_1}} \begin{pmatrix} 0 & 1 \\ 1 & 0 \end{pmatrix}, \quad p_{T_1} \in [0, 1].$$

2. Phase flip is given by $T_2(\rho) = G_0 \rho G_0^\dagger + G_1 \rho G_1^\dagger$, where

$$G_0 = \sqrt{p_{T_2}} \begin{pmatrix} 1 & 0 \\ 0 & 1 \end{pmatrix}, \quad G_1 = \sqrt{1 - p_{T_2}} \begin{pmatrix} 1 & 0 \\ 0 & -1 \end{pmatrix}, \quad p_{T_2} \in [0, 1].$$

3. Bit-Phase flip is given by $T_3(\rho) = R_0 \rho R_0^\dagger + R_1 \rho R_1^\dagger$, where

$$R_0 = \sqrt{p_{T_3}} \begin{pmatrix} 1 & 0 \\ 0 & 1 \end{pmatrix}, \quad R_1 = \sqrt{1 - p_{T_3}} \begin{pmatrix} 0 & -i \\ i & 0 \end{pmatrix}, \quad p_{T_3} \in [0, 1].$$

Note that, $E_i^\dagger, G_i^\dagger, R_i^\dagger$ for $i \in \{0, 1\}$ are the conjugate transpose of E_i, G_i, R_i for $i \in \{0, 1\}$ respectively.

Furthermore, we now have

$$T_1\left(\begin{pmatrix} a & b \\ c & d \end{pmatrix}\right) = \begin{pmatrix} pa + (1-p)d & pb + (1-p)c \\ pc + (1-p)b & pd + (1-p)a \end{pmatrix},$$

$$T_2\left(\begin{pmatrix} a & b \\ c & d \end{pmatrix}\right) = \begin{pmatrix} a & (2p-1)b \\ (2p-1)c & d \end{pmatrix},$$

$$T_3\left(\begin{pmatrix} a & b \\ c & d \end{pmatrix}\right) = \begin{pmatrix} pa + (1-p)d & pb - (1-p)c \\ pc - (1-p)b & pd + (1-p)a \end{pmatrix}.$$

Moreover, we have the following result in six cases:

Case I:

$$T_1\left(T_2\left(T_3\left(\begin{pmatrix} a & b \\ c & d \end{pmatrix}\right)\right)\right)$$

$$= T_1\left(T_2\left(\begin{pmatrix} pa + (1-p)d & pb - (1-p)c \\ pc - (1-p)b & pd + (1-p)a \end{pmatrix}\right)\right)$$

$$= T_1\left(\begin{pmatrix} pa + (1-p)d & (2p-1)(pb - (1-p)c) \\ (2p-1)(pc - (1-p)b) & pd + (1-p)a \end{pmatrix}\right)$$

$$= \begin{pmatrix} 2p(1-p)(d-a) + a & 4pb(p-1) + b \\ 4pc(p-1) + c & 2p(1-p)(a-d) + d \end{pmatrix}.$$

Case II:

$$T_2\left(T_3\left(T_1\left(\begin{pmatrix} a & b \\ c & d \end{pmatrix}\right)\right)\right)$$

$$= T_2\left(T_3\left(\begin{pmatrix} pa + (1-p)d & pb + (1-p)c \\ pc + (1-p)b & pd + (1-p)a \end{pmatrix}\right)\right)$$

$$= T_2\left(\begin{pmatrix} p^2a + (1-p)(2pd + (1-p)a) & p^2b - (1-p)b \\ p^2c - (1-p)^2c & p^2d + (1-p)(2pa + (1-p)d) \end{pmatrix}\right)$$

$$= \begin{pmatrix} 2p(1-p)(d-a) + a & 4pb(p-1) + b \\ 4pc(p-1) + c & 2p(1-p)(a-d) + d \end{pmatrix}.$$

Case III:

$$T_3\left(T_1\left(T_2\left(\begin{pmatrix} a & b \\ c & d \end{pmatrix}\right)\right)\right)$$

$$= T_3\left(T_1\left(\begin{pmatrix} a & (2p-1)b \\ (2p-1)c & d \end{pmatrix}\right)\right)$$

$$= T_3\left(\begin{pmatrix} pa + (1-p)d & (2p-1)(pb + (1-p)c) \\ (2p-1)(pc + (1-p)b) & pd + (1-p)a \end{pmatrix}\right)$$

$$= \begin{pmatrix} 2p(1-p)(d-a) + a & 4pb(p-1) + b \\ 4pc(p-1) + c & 2p(1-p)(a-d) + d \end{pmatrix}.$$

Case IV:

$$T_1\left(T_3\left(T_2\left(\begin{pmatrix} a & b \\ c & d \end{pmatrix}\right)\right)\right)$$

$$= T_1\left(T_3\left(\begin{pmatrix} a & (2p-1)b \\ (2p-1)c & d \end{pmatrix}\right)\right)$$

$$= T_1\left(\begin{pmatrix} pa + (1-p)d & (2p-1)(pb - (1-p)c) \\ (2p-1)(pc - (1-p)b) & pd + (1-p)a \end{pmatrix}\right)$$

$$= \begin{pmatrix} 2p(1-p)(d-a) + a & 4pb(p-1) + b \\ 4pc(p-1) + c & 2p(1-p)(a-d) + d \end{pmatrix}.$$

Case V:

$$T_2\left(T_1\left(T_3\left(\begin{pmatrix} a & b \\ c & d \end{pmatrix}\right)\right)\right)$$

$$= T_2\left(T_1\left(\begin{pmatrix} pa + (1-p)d & pb - (1-p)c \\ pc - (1-p)b & pd + (1-p)a \end{pmatrix}\right)\right)$$

$$= T_2\left(\begin{pmatrix} p^2a + (1-p)(2pd + (1-p)a) & p^2b - (1-p)^2b \\ p^2c - (1-p)^2c & p^2d + (1-p)(2pa + (1-p)d) \end{pmatrix}\right)$$

$$= \begin{pmatrix} 2p(1-p)(d-a) + a & 4pb(p-1) + b \\ 4pc(p-1) + c & 2p(1-p)(a-d) + d \end{pmatrix}.$$

Case VI:

$$T_3\left(T_2\left(T_1\left(\begin{pmatrix} a & b \\ c & d \end{pmatrix}\right)\right)\right)$$

$$= T_3\left(T_2\left(\begin{pmatrix} pa+(1-p)d & pb+(1-p)c \\ pc+(1-p)b & pd+(1-p)a \end{pmatrix}\right)\right)$$

$$= T_3\left(\begin{pmatrix} pa+(1-p)d & (2p-1)(pb+(1-p)c) \\ (2p-1)(pc+(1-p)b) & pd+(1-p)a \end{pmatrix}\right)$$

$$= \begin{pmatrix} 2p(1-p)(d-a)+a & 4pb(p-1)+b \\ 4pc(p-1)+c & 2p(1-p)(a-d)+d \end{pmatrix}. \qquad \square$$

Below is a simple example using real valued functions to illustrate the above proposition.

Example 1 Let $T_i : \mathbb{R}_+ \to \mathbb{R}_+$, $i = 1, 2, 3$ be defined by $T_1(x) = x$, $T(x) = x^2$, $T_3(x) = \sqrt{x}$.

Using the definition of T_1, T_2 and T_3 in the above example, we can check and see that, all the six different compositions gives same output as x. Also, the common fixed point set $\bigcap_{i=1}^{3} F(T_i) = \{1, 0\}$.

Corollary 2.2 *The compositions obtained from the permutation of any two different T_i ($i = 1, 2, 3$) with uniform $p \in [0, 1]$ for T_1, T_2 and T_3, gives the same out put,*

Proposition 2.3 *The output obtained from the composition of Bit flip (T_1), phase flip (T_2) and Bit-Phase flip (T_3) with uniform deformation parameter $p \in [0, 1]$ yields a depolarizing channel (T_4).*

Proof In Proposition 2.1, we have seen that, the order of composition does not matters. Also, the depolarizing channel (T_4) is defined as $T_4(\rho) = \frac{1}{2}p + (1-p)\rho$ for any density matrix ρ.

Now, in reference to Eq. (4) and Proposition 2.1 we have

$$\rho = \begin{pmatrix} a & b \\ c & d \end{pmatrix} = \frac{1}{2}\begin{pmatrix} 1+\eta\cos\theta & \eta e^{-i\phi}\sin\theta \\ \eta e^{i\phi}\sin\theta & 1-\eta\cos\theta \end{pmatrix}, \text{ for } \eta \in [0, 1].$$

So we have,

$$a = \frac{1}{2} + \frac{1}{2}\eta\cos\theta$$

$$b = \frac{\eta e^{-i\phi}\sin\theta}{2}$$

$$c = \frac{\eta e^{i\phi}\sin\theta}{2}$$

$$d = \frac{1}{2} - \frac{\eta\cos\theta}{2}$$

$$d - a = -\eta\cos\theta$$

$$a - d = \eta\cos\theta.$$

Therefore we have

$$T_1\left(T_2\left(T_3\left(\begin{pmatrix} a & b \\ c & d \end{pmatrix}\right)\right)\right)$$

$$= \begin{pmatrix} 2p(1-p)(d-a)+a & 4pb(p-1)+b \\ 4pc(p-1)+c & 2p(1-p)(a-d)+d \end{pmatrix}$$

$$= \begin{pmatrix} 2p(1-p)(-\eta\cos\theta)+\frac{1}{2}+\frac{1}{2}\eta\cos\theta & 4p\frac{1}{2}\eta e^{-i\phi}\sin\theta(p-1)+\frac{1}{2}\eta e^{-i\phi}\sin\theta \\ 4p\frac{1}{2}\eta e^{i\phi}\sin\theta(p-1)+\frac{1}{2}\eta e^{i\phi}\sin\theta & 2p(1-p)\eta\cos\theta+\frac{1}{2}-\frac{1}{2}\eta\cos\theta \end{pmatrix}$$

$$= \frac{1}{2}\begin{pmatrix} 4p(1-p)(-\eta\cos\theta)+1+\eta\cos\theta & 4p\eta e^{-i\phi}\sin\theta(p-1)+\eta e^{-i\phi}\sin\theta \\ 4p\eta e^{i\phi}\sin\theta(p-1)+\eta e^{i\phi}\sin\theta & 4p(1-p)\eta\cos\theta+1-\eta\cos\theta \end{pmatrix}$$

$$= \frac{1}{2}\begin{pmatrix} 1+(1-4p(1-p))\eta\cos\theta & (1-4p(1-p))\eta e^{-i\phi}\sin\theta \\ (1-4p(1-p))\eta e^{i\phi}\sin\theta & 1-(1-4p(1-p))\eta\cos\theta \end{pmatrix}.$$

Clearly, for $p \in [0, 1]$ then $4p(1-p) \in [0, 1]$, take $4p(1-p) = \zeta$. Therefore,

$$T_1\left(T_2\left(T_3\left((\rho)\right)\right)\right) = \frac{1}{2}\begin{pmatrix} 1+(1-\zeta)\eta\cos\theta & (1-\zeta)\eta e^{-i\phi}\sin\theta \\ (1-\zeta)\eta e^{i\phi}\sin\theta & 1-(1-\zeta)\eta\cos\theta \end{pmatrix}.$$

From the definition of T_4, we conclude that, $T_1(T_2(T_3(\rho)))$ is a depolarizing channel (see Yusuf et al. (2020)). □

Proposition 2.4 *The composition under different set of displacement operations* $\{T_x, T_y, T_z\}$ *or* $\{T_{-x}, T_{-y}, T_{-z}\}$ *of a Bloch sphere along the coordinate axes yields different outputs, and has non empty common fixed point set, for some* $0 \le \theta \le \pi$. *But, for any coordinate axis say h, there is no common fixed point between* T_{-h} *and* T_{+h}, *for* $h \in \{x, y, z\}$ *and* $0 < \theta < \pi$.

Proof Consider the Kraus operators of amplitude damping say F_0 and F_1 below. Let the Bloch vector transformations $([r_x, r_y, r_z])$ be as follows (Felloni et al. 2010):

1. The operation T_x has

$$F_0 = \frac{1}{2}\begin{bmatrix} 1+\cos\frac{\theta}{2} & 1-\cos\frac{\theta}{2} \\ 1-\cos\frac{\theta}{2} & 1+\cos\frac{\theta}{2} \end{bmatrix}, \quad F_1 = \frac{1}{2}\begin{bmatrix} -\sin\frac{\theta}{2} & \sin\frac{\theta}{2} \\ -\sin\frac{\theta}{2} & \sin\frac{\theta}{2} \end{bmatrix},$$

$$T_x([r_x, r_y, r_z]) = [-\sin^2\theta + \cos^2\theta r_x, \cos\theta r_y, \cos\theta r_z].$$

and the operator T_{-x} has

$$F_0 = \frac{1}{2}\begin{bmatrix} 1+\cos\frac{\theta}{2} & -(1-\cos\frac{\theta}{2}) \\ -(1-\cos\frac{\theta}{2}) & 1+\cos\frac{\theta}{2} \end{bmatrix}, \quad F_1 = \frac{1}{2}\begin{bmatrix} \sin\frac{\theta}{2} & \sin\frac{\theta}{2} \\ -\sin\frac{\theta}{2} & -\sin\frac{\theta}{2} \end{bmatrix},$$

$$T_{-x}([r_x, r_y, r_z]) = [\sin^2\theta - \cos^2\theta r_x, \cos\theta r_y, \cos\theta r_z].$$

2. The operation T_y has

$$F_0 = \frac{1}{2}\begin{bmatrix} 1+\cos\frac{\theta}{2} & i(1-\cos\frac{\theta}{2}) \\ -i(1-\cos\frac{\theta}{2}) & 1+\cos\frac{\theta}{2} \end{bmatrix}, \quad F_1 = \frac{1}{2}\begin{bmatrix} i\sin\frac{\theta}{2} & \sin\frac{\theta}{2} \\ \sin\frac{\theta}{2} & -i\sin\frac{\theta}{2} \end{bmatrix},$$

$T_y([r_x, r_y, r_z]) = [\cos\theta r_x, -\sin^2\theta + \cos^2\theta r_y, \cos\theta r_z].$

and the operator T_{-y} has

$$F_0 = \frac{1}{2}\begin{bmatrix} 1+\cos\frac{\theta}{2} & -i(1-\cos\frac{\theta}{2}) \\ i(1-\cos\frac{\theta}{2}) & 1+\cos\frac{\theta}{2} \end{bmatrix}, \quad F_1 = \frac{1}{2}\begin{bmatrix} -i\sin\frac{\theta}{2} & \sin\frac{\theta}{2} \\ \sin\frac{\theta}{2} & i\sin\frac{\theta}{2} \end{bmatrix},$$

$T_{-y}([r_x, r_y, r_z]) = [\cos\theta r_x, \sin^2\theta - \cos^2\theta r_y, \cos\theta r_z].$

3. The operation T_z has

$$F_0 = \begin{bmatrix} 1 & 0 \\ 0 & \cos\frac{\theta}{2} \end{bmatrix}, \quad F_1 = \begin{bmatrix} 0 & \sin\frac{\theta}{2} \\ 0 & 0 \end{bmatrix},$$

$T_z([r_x, r_y, r_z]) = [\cos\theta r_x, \cos\theta r_y, -\sin^2\theta + \cos^2\theta r_z].$

and the operator T_{-z} has

$$F_0 = \frac{1}{2}\begin{bmatrix} \cos\frac{\theta}{2} & 0 \\ 0 & 1 \end{bmatrix}, \quad F_1 = \frac{1}{2}\begin{bmatrix} 0 & 0 \\ \sin\frac{\theta}{2} & 0 \end{bmatrix},$$

$T_{-z}([r_x, r_y, r_z]) = [\cos\theta r_x, \cos\theta r_y, \sin^2\theta - \cos^2\theta r_z].$

So we proceed with the following cases, starting with the positive axes:
medskip **Case I**:

$T_z(T_x(T_y([r_x, r_y, r_z])))$
$= T_z(T_x([\cos\theta r_x, -\sin^2\theta + \cos^2\theta r_y, \cos\theta r_z]))$
$= T_z([-\sin^2\theta + \cos^3\theta r_x, -\cos\theta\sin^2\theta + \cos^3\theta r_y, \cos^2\theta r_z])$
$= [-\cos\theta\sin^2\theta + \cos^4\theta r_x, -\cos^2\theta\sin^2\theta + \cos^4\theta r_y, -\sin^2\theta + \cos^4\theta r_z].$

Case II:

$T_z(T_y(T_x([r_x, r_y, r_z])))$
$= T_z(T_y([-\sin^2\theta + \cos^2\theta r_x, \cos\theta r_y, \cos\theta r_z]))$
$= T_z([-\cos\theta\sin^2\theta + \cos^3\theta r_x, -\sin^2\theta + \cos^3\theta r_y, \cos^2\theta r_z])$
$= [-\cos^2\theta\sin^2\theta + \cos^4\theta r_x, -\cos\theta\sin^2\theta + \cos^4\theta r_y, -\sin^2\theta + \cos^4\theta r_z].$

Case III:

$$T_y(T_x(T_z([r_x, r_y, r_z])))$$
$$= T_y(T_x([\cos\theta r_x, \cos\theta r_y, -\sin^2\theta + \cos^2\theta r_z]))$$
$$= T_y([-\sin^2\theta + \cos^3\theta r_x, \cos^2\theta r_y, -\cos\theta\sin^2\theta + \cos^3\theta r_z])$$
$$= [-\cos\theta\sin^2\theta + \cos^4\theta r_x, -\sin^2\theta + \cos^4\theta r_y, -\cos^2\theta\sin^2\theta + \cos^4\theta r_z].$$

Case IV:

$$T_y(T_z(T_x([r_x, r_y, r_z])))$$
$$= T_y(T_z([-\sin^2\theta + \cos^2\theta r_x, \cos\theta r_y, \cos\theta r_z]))$$
$$= T_y([-\cos\theta\sin^2\theta + \cos^3\theta r_x, \cos^2\theta r_y, -\sin^2\theta + \cos^3\theta r_z])$$
$$= [-\cos^2\theta\sin^2\theta + \cos^4\theta r_x, -\sin^2\theta + \cos^4\theta r_y, -\cos\theta\sin^2\theta + \cos^4\theta r_z].$$

Case V:

$$T_x(T_z(T_y([r_x, r_y, r_z])))$$
$$= T_x(T_z([\cos\theta r_x, -\sin^2\theta + \cos^2\theta r_y, \cos\theta r_z]))$$
$$= T_x([\cos^2\theta r_x, -\cos\theta\sin^2\theta + \cos^3\theta r_y, -\sin^2\theta + \cos^3\theta r_z])$$
$$= [-\sin^2\theta + \cos^4\theta r_x, -\cos^2\theta\sin^2\theta + \cos^4\theta r_y, -\cos\theta\sin^2\theta + \cos^4\theta r_z].$$

Case VI:

$$T_x(T_y(T_z([r_x, r_y, r_z])))$$
$$= T_x(T_y([\cos\theta r_x, \cos\theta r_y, -\sin^2\theta + \cos^2\theta r_z]))$$
$$= T_x([\cos^2\theta r_x, -\sin^2\theta + \cos^3\theta r_y, -\cos\theta\sin^2\theta + \cos^3\theta r_z])$$
$$= [-\sin^2\theta + \cos^4\theta r_x, -\cos\theta\sin^2\theta + \cos^4\theta r_y, -\cos^2\theta\sin^2\theta + \cos^4\theta r_z].$$

Similarly, for the negative axes we have

Case I:

$$T_{-z}(T_{-x}(T_{-y}([r_x, r_y, r_z])))$$
$$= T_{-z}(T_{-x}([\cos\theta r_x, \sin^2\theta - \cos^2\theta r_y, \cos\theta r_z]))$$
$$= T_{-z}([\sin^2\theta - \cos^3\theta r_x, \cos\theta\sin^2\theta - \cos^3\theta r_y, \cos^2\theta r_z])$$
$$= [\cos\theta\sin^2\theta - \cos^4\theta r_x, \cos^2\theta\sin^2\theta - \cos^4\theta r_y, \sin^2\theta - \cos^4\theta r_z].$$

Case II:

$$T_{-z}(T_{-y}(T_{-x}([r_x, r_y, r_z])))$$
$$= T_{-z}(T_{-y}([\sin^2\theta - \cos^2\theta r_x, \cos\theta r_y, \cos\theta r_z]))$$
$$= T_{-z}([\cos\theta \sin^2\theta - \cos^3\theta r_x, \sin^2\theta - \cos^3\theta r_y, \cos^2\theta r_z])$$
$$= [\cos^2\theta \sin^2\theta - \cos^4\theta r_x, \cos\theta \sin^2\theta - \cos^4\theta r_y, \sin^2\theta - \cos^4\theta r_z].$$

Case III:

$$T_{-y}(T_{-x}(T_{-z}([r_x, r_y, r_z])))$$
$$= T_{-y}(T_{-x}([\cos\theta r_x, \cos\theta r_y, \sin^2\theta - \cos^2\theta r_z]))$$
$$= T_{-y}([\sin^2\theta - \cos^3\theta r_x, \cos^2\theta r_y, \cos\theta \sin^2\theta - \cos^3\theta r_z])$$
$$= [\cos\theta \sin^2\theta - \cos^4\theta r_x, \sin^2\theta - \cos^4\theta r_y, \cos^2\theta \sin^2\theta - \cos^4\theta r_z].$$

Case IV:

$$T_{-y}(T_{-z}(T_{-x}([r_x, r_y, r_z])))$$
$$= T_{-y}(T_{-z}([\sin^2\theta - \cos^2\theta r_x, \cos\theta r_y, \cos\theta r_z]))$$
$$= T_{-y}([\cos\theta \sin^2\theta - \cos^3\theta r_x, \cos^2\theta r_y, \sin^2\theta - \cos^3\theta r_z])$$
$$= [\cos^2\theta \sin^2\theta - \cos^4\theta r_x, \sin^2\theta - \cos^4\theta r_y, \cos\theta \sin^2\theta - \cos^4\theta r_z].$$

Case V:

$$T_{-x}(T_{-z}(T_{-y}([r_x, r_y, r_z])))$$
$$= T_{-x}(T_{-z}([\cos\theta r_x, \sin^2\theta - \cos^2\theta r_y, \cos\theta r_z]))$$
$$= T_{-x}([\cos^2\theta r_x, \cos\theta \sin^2\theta - \cos^3\theta r_y, \sin^2\theta - \cos^3\theta r_z])$$
$$= [\sin^2\theta - \cos^4\theta r_x, \cos^2\theta \sin^2\theta - \cos^4\theta r_y, \cos\theta \sin^2\theta - \cos^4\theta r_z].$$

Case VI:

$$T_{-x}(T_{-y}(T_{-z}([r_x, r_y, r_z])))$$
$$= T_{-x}(T_{-y}([\cos\theta r_x, \cos\theta r_y, \sin^2\theta - \cos^2\theta r_z]))$$
$$= T_{-x}([\cos^2\theta r_x, \sin^2\theta - \cos^3\theta r_y, \cos\theta \sin^2\theta - \cos^3\theta r_z])$$
$$= [\sin^2\theta - \cos^4\theta r_x, \cos\theta \sin^2\theta - \cos^4\theta r_y, \cos^2\theta \sin^2\theta - \cos^4\theta r_z].$$

Clearly, for $\theta = 180$, $\{[r_x, r_y, r_z] : \|\bar{r}\| \leq 1\}$ is a common fixed point set for all of the six different composition formed by the displacement operations T_x, T_y, and T_z. Also, for $\theta = 180$, $[0, 0, 0]$ is a common fixed point for all of the six different compositions formed by the displacement operations T_{-x}, T_{-y}, and T_{-z}.

Furthermore, from the fact that, the origin/center of the Bloch sphere is shifted to the positive direction by T_h, and to the negative direction by T_{-h}. The two transformations can never have a common point for $0 < \theta < \pi$. Thus, T_{-h} and T_h have no common fixed point. □

Theorem 2.5 *Let $\mathscr{Q}$ represent a Bloch Sphare and T_4 be a depolarizing quantum operation on $\mathscr{Q}$. Then, T_4 has a fixed point.*

Proof Let ρ be a pure quantum state on $\mathscr{Q}$. Suppose γ is any mixed quantum state of $\mathscr{Q}$ and on the straight line joining the center of $\mathscr{Q}$ say η.

Define a relation $\preceq$ on $\mathscr{Q}$ by $x \preceq y$ if and only if the straight line joining x and η passes through y, for $x, y \in \mathscr{Q}$. Then, the followings can be deduced:

1. For a trace distance function D, if $x \preceq y$, then $D(T_4^n x, T_4^n y) \to 0$, for $x, y \in \mathscr{Q}$.
2. $\eta \preceq T_4 \eta$.
3. $T_4^n \gamma \preceq \eta$, $\forall n \in \mathbb{N}$.
4. T_4 is order preserving.
5. For a trace distance function D, if $x \preceq y$ and $y \preceq z$, then $D(x, z) \geq \max\{D(x, y), D(y, z)\}$ where $x, y \in \mathscr{Q}$.
6. $T_4^n \gamma \preceq \eta \preceq T_4^n \eta$, $\forall n \in \mathbb{N}$. So, we have

$$D(\eta, T_4\eta) \leq D(\eta, T_4^n \eta), \forall n \in \mathbb{N}.$$
$$\leq D(T_4^n \gamma, T_4^n \eta) \to 0.$$

Thus, $\eta \in \mathscr{Q}$ is a fixed point of T_4. □

Corollary 2.6 *Suppose $p \in [0, 1]$ is the same for T_1, T_2 and T_3; $p_{T_1} = p_{T_2} = p_{T_3} = p$. Then, the compositions obtained from the permutation of T_i, $i = 1, 2, 3$ has a fixed point.*

3 Conclusion

Whenever a quantum system undergoes an evolution (deformation) under multiple channels, by knowing the common fixed point set of the different channels involved, with the help of IPS, a substantial amount of error can be corrected through the formation of QECC. Also, the order of arranging quantum channels that involves deformation of a Bloch sphere does not matters, as all arrangements give the same output. While the order of arranging multiple quantum channels involving amplitude damping matters.

Acknowledgements The authors acknowledge the financial support provided by King Mongkut's University of Technology Thonburi through the "KMUTT 55th Anniversary Commemorative Fund". Umar Batsari Yusuf was supported by the Petchra Pra Jom Klao Doctoral Academic Scholarship for Ph.D. Program at KMUTT. Moreover, the second author was supported by Theoretical and Computational Science (TaCS) Center, under Computational and Applied Science for Smart Innovation Cluster (CLASSIC), Faculty of Science, KMUTT.

References

Arias, A., Gheondea, A., & Gudder, S. (2002). Fixed points of quantum operations. *Journal of Mathematics and Physics*, *43*(12), 5872–5881.

Batsari, U. Y., Kumam, P., & Dhompongsa, S. (2019). Fixed points of terminating mappings in partial metric spaces. *The Journal of Fixed Point Theory and Applications*, *21*(39), 20 pages. https://doi.org/10.1007/s11784-019-0672-4.

Blume-Kohout, R., Ng, H. K., Paulin, D., & Viola, L. (2008). Characterizing the structure of preserved information in quantum processes. *Physical Review Letters*, *100*(3), 4 pages. https://doi.org/10.1103/PhysRevLett.100.030501.

Blume-Kohout, R., Ng, H. K., Paulin, D., & Viola, L. (2010). Information preserving structures; A general framework for quantum zero-error information. *Physical Review Letters*, *82*, 29 pages. https://doi.org/10.1103/PhysRevA.82.062306.

Bruzda, W., Cappellini, V., Sommers, H.-J., & Życzkowski, K. (2009). Random quantum operations. *Physics Letters A*, *373*, 320–324. https://doi.org/10.1016/j.physleta.2008.11.043.

Busch, P., Lahti, P. J., & Mittelstaedt, P. (1996). *The quantum theory of measurements*. Heidelberg, Berlin, Germany: Springer-Verlag.

Busch, P., & Singh, J. (1998). Lüders theorem for unsharp quantum measurements. *Physics Letters A*, *249*, 10–12. https://doi.org/10.1016/S0375-9601(98)00704-X.

Davies, E. B. (1976). *Quantum theory of open systems*. London, England: Academic Press.

Felloni, S., Leporati, A., & Strini, G. (2010). Evolution of quantum systems by diagrams of states; An illustrative tutorial. *International Journal of Unconventional Computing*, *6*(3–4). arXiv:0912.0026v1 [quant-ph], 1 Dec 2009.

Long, L., & Zhang, S. (2011). Fixed points of commutative super-operators. *Journal of Physics A: Mathematical and Theoretical*, *44*, 10 pages.

Lüders, G. (1950). Über die Zustandsänderung durch den Meßprozeß, *443*(8), 322–328. Translated by K. A. Kirkpatrick as Lüders, G. (2006). Concerning the state-change due to the measurement process. *Annalen der Physik*, *15*, 663–670. arXiv:quant-ph/0403007.

Nguyen, H. T., & Dong, L. S. (2018). An invitation to quantum econometrics. In L. Anh, L. Dong, V. Kreinovich, & N. Thach (Eds.), *Econometrics for financial applications*. ECONVN; Studies in Computational Intelligence (Vol. 760, pp. 44–62). Cham: Springer. https://doi.org/10.1007/978-3-319-73150-6_3.

Nielsen, M. A., & Chuang, I. L. (2010). Quantum computation and quantum information. In *10th Anniversary Edition*. Cambridge, United Kingdom: Cambridge University Press.

Robert, B. G. (2017). What quantum measurements measure. *Physical Review A*, *96*, 032110; 1–20. https://doi.org/10.1103/PhysRevA.96.032110.

Smolin, L. (1982). A fixed point for quantum gravity. *Nuclear Physics B*, *208*(3), 439–466.

Stephan, F. (2015). *Mathematical foundations of quantum mechanics*. Lecture Note, version (July 17 2015). Baden-Württemberg, Germany: ULM University. https://www.uni-ulm.de/fileadmin/website_uni_ulm/mawi.inst.020/fackler/SS15/qm/lnotes_mathematical_found_qm_temp.pdf.

Theodore, J. Y., Guang, H. L., & Isaac, L. S. (2014). Fixed point quantum search with an optimal number of queries. *Physical Review Letters*, *113*(210501), 1–5.

Yusuf, U. B., Kumam, P., & Yoo-Kong, S. (2020). Some Generalised Fixed Point Theorems Applied to Quantum Operations. *12*(5), 759.

Zhang, H., & Ji, G. (2012). A note on fixed point of general quantum operations. *Reports on Mathematical Physics*, *70*(1), 111–117.

Zhang, H., & Mingzhi, X. (2015). Fixed points of trace preserving completely positive maps. *Linear Multilinear Algebra*, *64*(3), 404–411. https://doi.org/10.1080/03081087.2015.1043718.

Zhang, H., & Si, H. (2016). Fixed points associated to power of normal completely positive maps*. *Journal of Applied Mathematics and Physics*, *4*(5), 925–929. https://doi.org/10.4236/jamp.2016.45101.

Information Quality: The Contribution of Fuzzy Methods

Bernadette Bouchon-Meunier

Abstract Data and information quality have been pointed out as key issues in data science. We detail the parts played by the trustworthiness of the source, the intrinsic quality of data, including accuracy and completeness, the qualities of information content such as relevance, trust and understandability, as well as the explainable character of the data mining tool extracting information from data. We focus on fuzzy-set based contributions to these aspects of information quality.

Keywords Information quality · Fuzzy methods · Data quality · Information veracity · Trustworthiness · Source quality · Explainable information

1 Introduction

Information quality is a major problem in data science and the management of digital information, for various reasons occurring at all stages of the process. The sources of information, be they witnesses, experts, journalists, newspapers or social networks, are more or less relevant and qualified, depending on the case, and they can express information in a natural language tainted by uncertainty or imprecision. The data themselves, available on the web or stored in huge databases, can be imperfect, complex, incomplete, with missing or imprecise data, possibly containing errors. The artificial intelligence engine, achieving the automatic extraction of information from data, and in particular the model used by the engine, is more or less efficient in providing information corresponding to the user's expectations. The last stage of the data mining process is the final information proposed to the user, that can contain errors or ambiguities, and that is not always associated with explanations about the reasons why they have been obtained.

Several major players in economy have pointed out data and information quality as key issues in data science. Already in 2012, IBM (http://www.ibmbigdatahub.com/

B. Bouchon-Meunier (✉)
Sorbonne Université, CNRS, LIP6, F-75005 Paris, France
e-mail: Bernadette.Bouchon-Meunier@lip6.fr

© The Editor(s) (if applicable) and The Author(s), under exclusive license
to Springer Nature Switzerland AG 2021
N. Ngoc Thach et al. (eds.), *Data Science for Financial Econometrics*,
Studies in Computational Intelligence 898,
https://doi.org/10.1007/978-3-030-48853-6_4

infographic/four-vs-big-data) estimates to $3.1 the yearly cost of poor quality data in the US. DARPA writes (https://www.darpa.mil/program/explainable-artificial-int elligence) in 2016 in its description of Explainable Artificial Intelligence that «the effectiveness of Artificial Intelligence systems is limited by the machine's current inability to explain their decisions and actions to human users». SAS claims in 2017 that «Interpretability is crucial for trusting AI and machine learning» (https://blogs. sas.com/content/subconsciousmusings/2017/12/18/).

It is interesting to distinguish objective quality evaluation from subjective quality evaluation of information. Objective quality corresponds to the accuracy, validity and completeness of data, independently of any opinion or personal appreciation expressed by the user or the experts of these data. Imprecisions coming from errors in measurement or inaccurate sensors provide examples of objective uncertainties, the exact value of the observed variable remaining uncertain. Interval computation, probabilistic methods and fuzzy models can be used to deal with such defects to improve the data quality. Incomplete or partial information provide other objective uncertainty. For instance, it is impossible to identify authors of digital documents or online users with a full certainty, the only available information being the pseudonym or the IP address. In the case of e-commerce, such incompleteness is not critical. In the case of e-reputation analysis or strategic intelligence, it is important to take it into account.

The second kind of flaw in information quality is subjective. The confidence of experts in information sources and the reliability of these sources are examples of factors of subjectivity involved in the evaluation of data veracity.

It is worth noting that all forms of natural language descriptions of real-world phenomena entail subjective uncertainty as well because of the imprecision of expressed information.

Among the most complex concepts involved in big data, we can mention opinions and emotions. Opinions are currently managed through very elementary models by social medias, considering scoring on simple scales or written comments. A more sophisticated analysis of emotions and opinions requires subtle natural language processing. The irony used by a customer to indicate his/her appreciation or exasper-ation is an example of a very difficult expression mode we have to handle if we want to determine the positive or negative opinion of this customer. It is clear that statis-tical methods are not sufficient to manage the subtlety of emotions and opinions and fuzzy models can help by treating the graduality between neighboring emotions and avoiding simplistic like/do not like or positive/negative appreciations (https://www. darpa.mil/program/explainable-artificial-intelligence). Biological sensors in elder-care or sometimes in online education provide another environment where emotions and psychological states, such as stress or depression, are present in big data. In this case again, fuzzy models can collaborate with machine learning methods to avoid strict categories and to take into account the natural graduality of such emotions or states.

In the sequel, we review various aspects of data and information quality in Sect. 2. We then present a sample of solutions based on fuzzy approaches to solve both data quality problems in Sect. 3 and information quality problems in Sect. 4 along the main dimensions of relevance, trust or veracity and understandability. We eventually conclude in Sect. 5.

2 Aspects of Information Quality

The view of information quality differs, depending on the domain of application and the kind of business generated by using data science (Batini and Scannapieco 2016; Capet and Revault d'Allonnes 2013; Revault d'Allonnes 2013; Lesot and Revault d'Allonnes 2017). The concept is complex and based on many components. Data quality has always been a major issue in databases, information systems and risk forecasting. The difficulties are accentuated in the environment of big data.

There exists a jungle of definitions of information quality and proposals of solutions in the framework of data mining and data science. In this jungle, you can recognize different aspects of information quality. Relevance, appropriateness, accessibility, compatibility correspond to the good matching between the retrieved information and the expectation of the user. Understandability, expressiveness describe the capacity of the system to speak a language familiar to the user. Accuracy may be necessary in specific domains. Comprehensibility, consistency, coherence, completeness represent the quality of the set of information as a whole. Timeliness, operationality, security are technical qualities. Veracity, validity, trust, reliability, plausibility, credibility represent various means to define the confidence the user can have in the obtained information.

The factors of information quality are dependent on the nature of sources, that can be news streams, databases, open source data, sensor records or social networks. They are also dependent on the form of data, for instance text, images, videos, temporal series, graphs or database records. The evaluation of information quality is also related to the expectations of the end user and purpose of the information extraction, the requirement of accuracy, for instance, not being the same in finance or for a student preparing a report.

To describe the chaining in the information quality components, we can consider that the source quality has an influence on data quality, which is one of the factors of information quality, as well as the artificial intelligence-based model quality. Finally, the user satisfaction rests on both information quality and model quality. To illustrate the diversity of views of information quality according to the considered domain of application, we can roughly consider that specialists of business intelligence and intelligence services pay a great attention to source quality, while financial engineering and econometrics are focused on data quality. In medical diagnosis, the quality of the model is very important to explain the diagnosis, whereas information retrieval is centered on global information quality. The user satisfaction is a priority for domains such as social networks or targeted advertising.

Each of these constituents of what we can call global information quality requires appropriate solutions to the best possible. In the following, we focus on fuzzy set-based solutions, among all those provided in computational intelligence, thanks to the capacity of fuzzy systems to handle uncertainties, imprecisions, incompleteness and reliability degrees in a common environment. The diversity of aggregation methods available for the fusion of elements and the richness of measures of similarity are additional reasons to choose fuzzy methods.

3 Fuzzy Solutions to Data Quality Problems

Defaults in data quality are the most primary among the problems occurring in information quality, and they have been studied for years, in particular in databases (Ananthakrishna et al. 2002; Janta-Polczynski and Roventa 1999) and more generally in products and services (Loshin 2011). Accuracy, completeness and consistency of data (Huh et al. 1990) have always been major concerns in industrial products.

In the modern environments, it is necessary to have a more general approach of data quality and to consider both intrinsic and extrinsic factors. The former include defaults such as imprecision, measurement errors, vague linguistic descriptions, incompleteness, inaccuracies, inconsistencies and discrepancies in data elements. The latter mainly refers to insufficient trustworthiness of sources and inconsistency between various sources.

A general analysis of data quality is proposed in Pipino et al. (2002) by means of objective and subjective assessments of data quality. The question of measuring the quality of data is addressed in Bronselaer et al. (2018a, b) through the presentation of a measure-theoretic foundation for data quality and the proposal of an operational measurement.

A fuzzy set-based knowledge representation is of course an interesting solution to the existence of inaccuracies, vagueness and incompleteness, as well as the necessary bridge between linguistic and numerical values.

The issue of incomplete data is very frequent in all environments, for various reasons such as the absence of answer to a specific request, the impossibility to obtain a measurement, a loss of pieces of information or the necessity to hide some elements to protect privacy. Various fuzzy methods have been proposed for the imputation of missing values, based on very different techniques. For instance, a fuzzy K-means clustering algorithm is used in Liao et al. (2009) and Li et al. (2004), a neuro-fuzzy classifier is proposed in Gabrys (2002), evolutionary fuzzy solutions are investigated in Carmona et al. (2012). Rough fuzzy sets are incorporated in a neuro-fuzzy structure to cope with missing data in Nowicki (2009). Various types of fuzzy rule-based classification systems are studied in Luengo et al. (2012) to overcome the problem of missing data. Missing pixels in images are also managed by means of fuzzy methods, for instance with the help of an intuitionistic fuzzy C-means clustering algorithm in Balasubramaniam and Ananthi (2016). All these works exemplify the variety of solutions available in a fuzzy setting.

4 Fuzzy Approaches to Other Information Quality Dimensions

Information is extracted from data by means of an artificial intelligence engine and it is supposed to fulfill users' needs. The expected level of quality of information is different, depending on the purpose of information extraction. In information retrieval, the challenge is to find images, texts or videos corresponding to a user's query and the information quality reflects the adequacy and completeness of the obtained information. In domains such as prevision, prediction, risk or trend forecasting, the quality of information is evaluated on the basis of the comparison between the forecasting and the real world. In real time decision making, where a diagnosis or a solution to a problem must be provided rapidly, the completeness and accuracy of information is crucial. In cases where data must be analyzed instantaneously to act on a system, for instance in targeted advertising, adaptive interfaces or online sales, the timeliness and operationality are more important than the accuracy. In business intelligence or e-reputation analysis, where opinions, blogs or customer's evaluations are analyzed, the trustworthiness of information is crucial.

We propose to structure the analysis of information quality along three main dimensions, namely the relevance of information, its trust or veracity, and its understandability.

4.1 Relevance of Information

This aspect corresponds to the adequacy of the retrieved information provided to the user with his/her query, or more generally his/her needs or expectations. This issue has extensively been analyzed and many kinds of solutions have been proposed in a fuzzy setting, as described in Zadeh's paper (2006), which pointed out the various aspects of relevance in the semantic web, in particular topic relevance, question relevance and the consideration of perception-based information. Fuzzy compatibility measures are investigated in Cross (1994) to evaluate the relevance in information retrieval.

Fuzzy formal concept analysis brings efficient solutions to the representation of information in order to retrieve relevant information (Medina et al. 2009; Lai and Zhang 2009; De Maio et al. 2012). Fuzzy ontologies are also used to improve relevance, as described in Calegari and Sanchez (2007), Akinribido et al. (2011) and their automatic generation is studied in Tho et al. (2006). Fuzzy or possibilistic description logic can also be used (Straccia 1998; Straccia 2006; Qi et al. 2007) to facilitate the identification of significant elements of information to answer a query and to avoid inconsistencies (Couchariere et al. 2008; Lesot et al. 2008). Fuzzy clustering was also used to take into account relevance in text categorization (Lee and Jiang 2014).

The pertinence of images retrieved to satisfy a user query has particularly attracted the attention of researchers. Similarity measures may be properly chosen to achieve a satisfying relevance (Zhao et al. 2003; Omhover and Detyniecki 2004). Fuzzy graph models are presented in Krishnapuram et al. (2004) to enable a matching algorithm to compare the model of the image to the model of the query. Machine learning methods are often used to improve the relevance, for instance active learning requesting the participation of the user in Chowdhury et al. (2012) or semi-supervised fuzzy clustering performing a meaningful categorization that will help image retrieval to be more relevant (Grira et al. 2005). Concept learning is performed thanks to fuzzy clustering in order to take advantage of past experiences (Bhanu and Dong 2002). The concept of fuzzy relevance is also addressed (Yap and Wu 2003) and linguistic relevance is explored (Yager and Petry 2005) to take into account perceptual subjectivity and human-like approach of relevance.

All these methods are representative of attempts to improve the relevance of the information obtained by a retrieval system to satisfy the user's needs, based on a fuzzy set-based representation.

4.2   Trust or Veracity of Information

The trustworthiness of information is crucial for all domains where users look for information. A solution to take this factor of quality into account lies in the definition of a degree of confidence attached to a piece of information (Lesot and Revault d'Allonnes 2017) to evaluate the uncertainty it carries and the confidence the user can have in it. We focus here on fuzzy set-based or possibilistic approaches.

First of all, the sources of information have a clear influence on the user's trust of information (Revault d'Allonnes 2014), because of their own reliability mainly based on their importance and their reputation. Their competence on the subject of the piece of information is another element involved in the trust of information, for instance the *Financial Times* is more renowned and expert in economics than the *Daily Mirror*. The relevance of the source with respect to the event is an additional component of the user's trust in a piece of information, be the relevance geographical or relating to the topic of the event. For instance, a local website such as wildfiretoday.com may be more relevant to obtain precise and updated information on bushfires than well-known international media such as *BBC News*. Moreover, a subjective uncertainty expressed by the source, such as *"We believe"* or *"it seems"*, is an element of the trustworthiness of the source.

The content of the piece of information about an event also bears a part of uncertainty environ is inherent in the formulation itself through numerical imprecisions (*"around* 150 persons died" or *"between* 1000 and 1200 cases of infection") or symbolic ones (*"many* participants"). Linguistic descriptions of uncertainty can also be present (*"probably"*, *"almost certainly"*, "69 homes *believed* destroyed", "2 will *probably* survive"). Uncertain information can also be the consequence of an insufficient compatibility between several pieces of information on the same event. A fuzzy set-based knowledge representation contributes to taking into account imprecisions and to evaluating the compatibility between several descriptions such as «more than

70» and «*approximately* 75». The large range of aggregation methods in a fuzzy setting helps to achieve the fusion of pieces of information on a given event in order to confirm or invalidate each of them through a comparison with others, and to therefore overcome compatibility problems.

The trust in a piece of information results from the aggregation of the trustworthiness of the source and the confidence in the content of this piece of information. In the case of the analysis of reputation in social networks, where sources are complex, fuzzy models have for instance been chosen to manage social relations in peer-to-peer communities (Aringhieri et al. 2006). The case of geospatial information based on volunteered geographic data and crowdsourced information is remarkable because it involves social media content as well as information provided by volunteers about geographical or citizen issues such as crisis events. The community has extensively studied data quality in such environments by for several years and fuzzy set-based methods are promoted to cope with this problem (Bordogna et al. 2014).

In the case of open data, an example of semi-automatic information scoring process is proposed in Lesot et al. (2011) in a possibilistic setting. A query of the form "Did event e occur?" is presented by a user to a database of structured linguistic information extracted from various media. For example, the query "Did the event *"Around 70 houses were destroyed by bushfires in New South Wales in March 2018"* occur" is asked to several sources including wildfiretoday.com, The Guardian or BBC News. The result is a confidence degree expressed as a possibility distribution $(\Pi(e), \Pi(\neg e))$ resulting from the fusion of individual confidence degrees assigned to pieces of information provided by each of the sources. In addition to the above-mentioned components of the trust, contextual elements are taken into account to calculate these confidence degrees.

Relations between sources are considered to weight differences which may exist in pieces of information provides by different sources on a given event. Networks of sources show friendly (or hostile) relations between them, that entail a natural similarity (or dissimilarity) in opinions and presentation of events. For instance *Al Jazeera* and *Amaq* are expected to show similar views of events in the region they cover, these views being often different from those proposed by *i24News*.

The evolution of a situation over time is considered to compare pieces of information provided at different moments and a criterion of obsolescence reflects the necessity to attach less and less importance to a piece of information when time passes and eventually to forget it.

Various other approaches of the scoring of information have been proposed in the literature. In Revault d'Allonnes and Lesot (2014), Revault d'Allonnes and Lesot (2015), a multivalued logic-based model considers four components of trust: the source reliability, the source competence, the information plausibility for the user and the information credibility. Evidence theory is used in Pichon et al. (2012), Cholvy (2010) to evaluate degrees of confidence in a piece of information, on the basis of uncertain metaknowledge on the source relevance and truthfulness. Modal logic (Cholvy 2012) is a solution to describe the belief in a piece of information reported by a source citing another one. Choquet integral has been called on in a multidimensional approach (Pichon et al. 2014) to aggregate various dimensions of the reliability of a source of information.

4.3 Understandability of Information

The last component of the quality of information content we consider is its understandability or expressiveness. It is a complex notion (Marsala and Bouchon-Meunier 2015; Hüllermeier 2015), dealing with the understandability of the process leading to the presented information, as well as the easiness for the end user to interpret the piece of information he receives. This component has been widely investigated since the introduction of Explainable Artificial Intelligence (XAI) by DARPA in 2016 (https://www.darpa.mil/program/explainable-artificial-intelligence), that requires an explainable model and an explanation interface.

Fuzzy models are recognized for their capability to be understood. In particular, fuzzy rule-based systems are considered to be easily understandable because rules of the form "*If the current return is Low or the current return is High, then low or high future returns are rather likely*" (Van den Berg et al. 2004) contain symbolic descriptions similar to what specialists express. Fuzzy decision trees (Laurent et al. 2003; Bouchon-Meunier and Marsala 1999) are also very efficient models of the reasons why a conclusion is presented to the user. Nevertheless, a balance between complexity, accuracy and understandability of such fuzzy models is necessary (Casillas et al. 2003). The capacity of the user to understand the system represented by the fuzzy model depends not only on the semantic interpretability induced by natural-language like descriptions, but also on the number of attributes involved in premises and the number of rules (Gacto et al. 2011). The interpretability of a fuzzy model is a subjective appreciation and it is possible to distinguish high-level criteria such as compactness, completeness, consistency and transparency of fuzzy rules, from low-level criteria such as coverage, normality or distinguishability of fuzzy modalities (Zhou and Gan 2008).

The interpretability of information presented to the user depends on the expertise of the user. Linguistic descriptions are not the only expected form of information extracted from time series or large databases, for instance. Formal logic, statistics or graphs may look appealing to experts. However, we focus here on fuzzy methods that provide linguistic information by means of fuzzy modalities such as "*rapid increase*" or "*low cost*" and fuzzy quantifiers like "*a large majority*" or "*very few*". The wide range of works on linguistic summarization of big databases and time series by means of fuzzy descriptions and so-called protoforms (Zadeh 2002) shows the importance of the topic, starting from seminal definitions of linguistic summaries (Yager 1982; Kacprzyk and Yager 2001). Their interpretability can be questioned (Lesot et al. 2016) and improved by means of automatic methods, such as mathematical morphology (Moyse et al. 2013) or evolutionary computation (Altintop et al. 2017), among other methods. Local or global periodicity of time series, including regularity modeling or detection of periodic events, for instance "*The first two months, the series is highly periodic (0.89) with a period of approximately one week"* (Moyse and Lesot 2016; Almeida et al. 2013), analysis of trends with expressions of the form *"Most slowly decreasing trends are of a very low variability"* (Kacprzyk et al. 2006; Kacprzyk et al. 2008) and analysis of their dynamical characteristics, are among the most elaborated aspects of linguistic summarization. Various domains of application have been investigated, for instance marketing data analysis (Pilarski 2010), shares quotations from the Warsaw stock exchange (Kacprzyk and Wilbik 2009) or energy consumption (Van der Heide and Triviño 2009).

5   Conclusion

Data quality problems, in particular insufficient trustworthiness, are drawbacks inducing some uncertainty on available data. They play a part in the quality of information extracted from them, especially in the case of big data. Another component of information quality is inherent in methods to deal with data, pertaining to their capability to be explainable to the final user, to enable him to understand the way information has been obtained as well as to grasp the content of the presented information.

We have reviewed the main aspects of information quality and we have pointed out the importance of non-statistical models to cope with it. Our purpose was to present alternative solutions which can supplement statistics and statistical machine learning in data science, to provide examples of fuzzy set-based models rather than an exhaustive study, and to point out the interest of opening new possibilities to solve the difficult problem of quality in big data analysis.

References

https://blogs.sas.com/content/subconsciousmusings/2017/12/18/.
http://www.ibmbigdatahub.com/infographic/four-vs-big-data.
https://www.darpa.mil/program/explainable-artificial-intelligence.
Akinribido, C. T., Afolabi, B. S., Akhigbe, B. I., & Udo, I. J. (2011). A fuzzy-ontology based information retrieval system for relevant feedback. *International Journal of Computer Science Issues, 8*(1), 382–389.
Almeida, R. J., Lesot, B., Bouchon-Meunier, M.-J., Kaymak, U., & Moyse, G. (2013). Linguistic summaries of categorical time series patient data. In *Proceedings of the IEEE International Conference on Fuzzy Systems* (pp. 1–8). FUZZ-IEEE 2013.
Altintop, T., Yager, R., Akay, D., Boran, E., & Ünal, M. (2017). Fuzzy linguistic summarization with genetic algorithm: An application with operational and financial healthcare data. *International Journal of Uncertainty, Fuzziness and Knowledge-Based Systems, IJUFKS, 25*(04), 599–620.
Ananthakrishna, R., Chaudhuri, S., & Ganti, V. (2002). Eliminating fuzzy duplicates in data warehouses. In *Proceedings of the 28th VLDB Conference*, Hong Kong, China.

Aringhieri, R., Damiani, E., Di Vimercati, S. D. C., Paraboschi, S., & Samarati, P. (2006). Fuzzy techniques for trust and reputation management in anonymous peer-to-peer systems. *Journal of the Association for Information Science and Technology, 57,* 528–537.

Balasubramaniam, P., & Ananthi, V. P. (2016). Segmentation of nutrient deficiency in incomplete crop images using intuitionistic fuzzy C-means clustering algorithm. *Nonlinear Dynamics, 83*(1–2), 849–866.

Batini, C., & Scannapieco, M. (2016). *Data and information quality.* Springer International Publishing.

Bhanu, B., & Dong, A. (2002). Concepts learning with fuzzy clustering and relevance feedback. *Engineering Applications of Artificial Intelligence, 15*(2), 123–138.

Bordogna, G., Carrara, P., Criscuolo, L., Pepe, M., & Rampini, A. (2014). A linguistic decision making approach to assess the quality of volunteer geographic information for citizen science. *Information Sciences, 258*(10), 312–327.

Bouchon-Meunier, B., & Marsala, C. (1999). Learning fuzzy decision rules. In J.C. Bezdek, D. Dubois & H. Prade (Eds.), *Fuzzy sets in approximate reasoning and information systems.* The Handbooks of Fuzzy Sets Series (Vol. 5). Boston, MA: Springer.

Bronselaer, A., Nielandt, J., Boeckling, T., & De Tré, G. (2018a). A measure-theoretic foundation for data quality. *IEEE Transactions on Fuzzy Systems, 26*(2), 627–639.

Bronselaer, A., Nielandt, J., Boeckling, T., & De Tré, G. (2018b). Operational measurement of data quality. In J. Medina, M. Ojeda-Aciego, J. Verdegay, I. Perfilieva, B. Bouchon-Meunier & R. Yager (Eds.), *Information processing and management of uncertainty in knowledge-based systems. Applications. IPMU 2018.* Communications in Computer and Information Science (Vol. 855, pp. 517–528). Springer International Publishing.

Calegari, S., & Sanchez, E. (2007). A fuzzy ontology-approach to improve semantic information retrieval. In *Proceedings of the Third ISWC Workshop on Uncertainty Reasoning for the Semantic Web*, Busan, Korea.

Capet, P., & Revault d'Allonnes, A. (2013). Information evaluation in the military domain: Doctrines, practices and shortcomings. In P. Capet & T. Delavallade (Eds.), *Information evaluation* (pp 103–128). Wiley.

Carmona, C. J., Luengo, J., González, P., & del Jesus, M. J. (2012). A preliminary study on missing data imputation in evolutionary fuzzy systems of subgroup discovery. In *2012 IEEE International Conference on Fuzzy Systems*, Brisbane (pp. 1–7).

Casillas, J., Cordón, O., Herrera, F., & Magdalena, L. (2003). Interpretability improvements to find the balance interpretability-accuracy in fuzzy modeling: An overview. In J. Casillas, O. Cordón, F. Herrera & L. Magdalena (Eds.), *Interpretability issues in fuzzy modeling.* Studies in Fuzziness and Soft Computing (Vol. 128). Berlin, Heidelberg: Springer.

Cholvy, L. (2010). Evaluation of information reported: A model in the theory of evidence. In E. Hüllermeier, R. Kruse & F. Hoffmann (Eds.), *Information processing and management of uncertainty in knowledge-based systems. Theory and methods. IPMU 2010.* Communications in Computer and Information Science (Vol. 80). Springer.

Cholvy, L. (2012). Collecting information reported by imperfect information sources. In S. Greco, B. Bouchon-Meunier, G. Coletti, M. Fedrizzi, B. Matarazzo & R. R. Yager (Eds.), *Advances in Computational Intelligence. IPMU 2012.* Communications in Computer and Information Science (Vol. 299). Springer.

Chowdhury, M., Das, S., & Kundu, M. K. (2012). Interactive content based image retrieval using ripplet transform and fuzzy relevance feedback. In *Perception and Machine Intelligence—First Indo-Japan Conference.* LNCS (Vol. 7143, pp. 243–251).

Couchariere, O., Lesot, M.-J., & Bouchon-Meunier, B. (2008). Consistency checking for extended description logics. In *International Workshop on Description Logics* (DL 2008) (Vol. 353), Dresden, Germany, CEUR.

Cross, V. (1994). Fuzzy information retrieval. *Journal of Intelligent Information Systems, 3*(1), 29–56.

De Maio, C., Fenza, G., Loia, V., & Senatore, S. (2012). Hierarchical web resources retrieval by exploiting fuzzy formal concept analysis. *Information Processing and Management, 48*(3), 399–418.

Gabrys, B. (2002). Neuro-fuzzy approach to processing inputs with missing values in pattern recognition problems. *International Journal of Approximate Reasoning, 35,* 149–179.

Gacto, M. J., Alcalá, R., & Herrera, F. (2011). Interpretability of linguistic fuzzy rule-based systems: An overview of interpretability measures. *Information Sciences, 181*(20), 4340–4360.

Grira, N., Crucianu, M., & Boujemaa, N. (2005). Semi-supervised fuzzy clustering with pairwise-constrained competitive agglomeration. In *The 14th IEEE International Conference on Fuzzy Systems, 2005. FUZZ-IEEE 2005*, Reno, USA (pp. 867–872).

Huh, Y. U., Keller, F. R., Redman, T. C., & Watkins, A. R. (1990). Data quality. *Information and Software Technology, 32*(8), 559–565.

Hüllermeier, E. (2015). Does machine learning need fuzzy logic? *Fuzzy Sets and Systems, 281,* 292–299.

Janta-Polczynski, M., & Roventa, E. (1999). Fuzzy measures for data quality. In *18th International Conference of the North American Fuzzy Information Processing Society—NAFIPS*, New York, NY, USA (pp. 398–402).

Kacprzyk, J., & Wilbik, A. (2009). Using fuzzy linguistic summaries for the comparison of time series: An application to the analysis of investment fund quotations. In *Conference Proceedings of the Joint 2009 International Fuzzy Systems Association World Congress and 2009 European Society of Fuzzy Logic and Technology Conference*, Lisbon, Portugal (pp. 1321–1326).

Kacprzyk, J., & Yager, R. R. (2001). Linguistic summaries of data using fuzzy logic. *International Journal of General Systems, 30,* 33–154.

Kacprzyk, J., Wilbik, A., & Zadrozny, S. (2006). On some types of linguistic summaries of time series. In *Proceedings of the Third International IEEE Conference on Intelligent Systems* (pp. 373–378). New York, London, UK: IEEE Press.

Kacprzyk, J., Wilbik, A., & Zadrozny, S. (2008). Linguistic summarization of time series using a fuzzy quantifier driven aggregation. *Fuzzy Sets and Systems, 159,* 1485–1499.

Krishnapuram, R., Medasani, S., Jung, S. K., Choi, Y. S., & Balasubramaniam, R. (2004). Content-based image retrieval based on a fuzzy approach. *IEEE Transactions on Knowledge and Data Engineering, 16–10,* 1185–1199.

Lai, H., & Zhang, D. (2009). Concept lattices of fuzzy contexts: Formal concept analysis vs. rough set theory. *International Journal of Approximate Reasoning, 50*(5), 695–707.

Laurent, A., Marsala, C., & Bouchon-Meunier, B. (2003). Improvement of the interpretability of fuzzy rule based systems: Quantifiers, similarities and aggregators, In: J. Lawry, J. Shanahan & A. L. Ralescu (Eds.), *Modelling with words*. Lecture Notes in Computer Science (Vol. 2873, pp. 102–123). Berlin, Heidelberg: Springer.

Lee, S. J., & Jiang, J. Y. (2014). Multilabel text categorization based on fuzzy relevance clustering. *IEEE Transactions on Fuzzy Systems, 22*(6), 1457–1471.

Lesot, M.-J., & Revault d'Allonnes, A. (2017). Information quality and uncertainty. In V. Kreinovich (Ed.), *Uncertainty modeling* (pp. 135–146). Springer.

Lesot, M.-J., & Revault d'Allonnes, A. (2017). Information quality and uncertainty. In: V. Kreinovich (Ed.), *Uncertainty modeling* (pp. 135–146). Springer.

Lesot, M.-J., Couchariere, O., Bouchon-Meunier, B., & Rogier, J.-L. (2008). Inconsistency degree computation for possibilistic description logic: An extension of the tableau algorithm. In *NAFIPS 2008*, New York.

Lesot, M.-J., Delavallade, T., Pichon F., Akdag, H., Bouchon-Meunier, B., & Capet, P. (2011). Proposition of a semi-automatic possibilistic information scoring process. In *Proceedings of the 7th Conference of the European Society for Fuzzy Logic and Technology (EUSFLAT-2011) and LFA-2011* (pp. 949–956). Atlantis Press.

Lesot, M.-J., Moyse, G., & Bouchon-Meunier, B. (2016). Interpretability of fuzzy linguistic summaries. *Fuzzy Sets and Systems, 292*(1), 307–317.

Li, D., Deogun, J., Spaulding, W., & Shuart, B. (2004). Towards missing data imputation: A study of fuzzy K-means clustering method. In S. Tsumoto, R. Słowinski, J. Komorowski & J. W. Grzymała-Busse (Eds.), *Rough sets and current trends in computing* (pp. 573–579). Berlin, Heidelberg: Springer.

Liao, Z., Lu, X., Yang, T., & Wang, H. (2009). Missing data imputation: A fuzzy K-means clustering algorithm over sliding window. In Y. Chen & D. Zhang (Eds.), *Proceedings of the 6th International Conference on Fuzzy Systems and Knowledge Discovery* (FSKD'09) (Vol. 3, pp. 133–137). Piscataway, NJ, USA: IEEE Press.

Loshin, D. (2011). Dimensions of data quality. In D. Loshin (Ed.), *The practitioner's guide to data quality improvement*. MK Series on Business Intelligence (pp. 129–146). Morgan Kaufmann.

Luengo, J., Sáez, J. A., & Herrera, F. (2012). Missing data imputation for fuzzy rule-based classification systems. *Soft Computing, 16*(5), 863–881.

Marsala, C., & Bouchon-Meunier, B. (2015). Fuzzy data mining and management of interpretable and subjective information. *Fuzzy Sets and Systems* (Vol. 281, pp. 252–259). Elsevier.

Medina, J., Ojeda-Aciego, M., & Ruiz-Calvino, J. (2009). Formal concept analysis via multi-adjoint concept lattices. *Fuzzy Sets and Systems, 160*(2), 30–144.

Moyse, G., & Lesot, M.-J. (2016). Linguistic summaries of locally periodic time series. *Fuzzy Sets and Systems, 285,* 94–117.

Moyse, G., Lesot, M.-J., & Bouchon-Meunier, B. (2013). Linguistic summaries for periodicity detection based on mathematical morphology. In *Proceedings of IEEE Symposium on Foundations of Computational Intelligence, FOCI 2013* (pp. 106–113). Singapore.

Nowicki, R. (2009). Rough neuro-fuzzy structures for classification with missing data. *IEEE Transactions on Systems, Man, and Cybernetics. Part B, Cybernetics, 39*(6), 1334–1347.

Omhover, J.-F., & Detyniecki, M. (2004). STRICT: An image retrieval platform for queries based on regional content. In *International Conference on Image and Video Retrieval CIVR 2004.*

Pichon, F., Dubois, D., & Denoeux, T. (2012). Relevance and truthfulness in information correction and fusion. *International Journal of Approximate Reasoning, 53,* 159–175.

Pichon, F., Labreuche, C., Duqueroie, B. & Delavallade, T. (2014) Multidimensional approach to reliability evaluation of information sources. In Capet, P., Delavallade, T. (eds.), *Information Evaluation*. Wiley, pp. 129–160.

Pilarski, D. (2010). Linguistic summarization of databases with quantirius: A reduction algorithm for generated summaries. *International Journal of Uncertainty, Fuzziness and Knowledge-Based Systems (IJUFKS), 18*(3), 305–331.

Pipino, L. L., Lee, Y. W., & Wang, R. Y. (2002, April). Data quality assessment. *Communications ACM, 45*(4), 211–218.

Qi, G., Pan, J. Z., & Ji, Q. (2007). Possibilistic extension of description logics. In *Proceedings of the 2007 International Workshop on Description Logics* (DL2007), Brixen-Bressanone, near Bozen-Bolzano, Italy.

Revault d'Allonnes, A. (2013). An architecture for the evolution of trust: Definition and impact of the necessary dimensions of opinion making. In P. Capet & T. Delavallade (Eds.), *Information evaluation* (pp. 261–294). Wiley.

Revault d'Allonnes, A. (2014). An architecture for the evolution of trust: Definition and impact of the necessary dimensions of opinion making. In P. Capet & T. Delavallade (Eds.), *Information evaluation* (pp. 261–294). Wiley.

Revault d'Allonnes, A., & Lesot, M. J. (2014). Formalising information scoring in a multivalued logic framework. In A. Laurent, O. Strauss, B. Bouchon-Meunier, & R.R. Yager (Eds.), *Information processing and management of uncertainty in knowledge-based systems. IPMU 2014.* Communications in Computer and Information Science (Vol. 442, pp. 314–323). Springer.

Revault d'Allonnes, A., & Lesot, M.-J. (2015). Dynamics of trust building: Models of information cross-checking in a multivalued logic framework. In 2015 *IEEE International Conference on Fuzzy Systems*, FUZZ-IEEE 2015.

Straccia, U. (1998). A fuzzy description logic. In *Proceedings of AAAI-98, 15th National Conference on Artificial Intelligence*, Madison, Wisconsin.

Straccia, U. (2006). A fuzzy description logic for the semantic web. In E. Sanchez (Ed.), *Fuzzy logic and the semantic web* (pp. 73–90). Amsterdam: Elsevier.

Tho, Q. T., Hui, S. C., Fong, A. C. M., & Cao, T. H. (2006). Automatic fuzzy ontology generation for semantic Web. *IEEE Transactions on Knowledge and Data Engineering, 18*(6), 842–856.

Van den Berg, J., Kaymak, U., & van den Bergh, W.-M. (2004). Financial markets analysis by using a probabilistic fuzzy modelling approach. *International Journal of Approximate Reasoning, 35*(3), 291–305.

Van der Heide, A., & Triviño, G. (2009). Automatically generated linguistic summaries of energy consumption data. In *Proceedings of ISDA'09* (pp. 553–559).

Yager, R. R. (1982). A new approach to the summarization of data. *Information Sciences, 28*(1), 69–86.

Yager, R. R., & Petry, F. E. (2005). A framework for linguistic relevance feedback in content-based image retrieval using fuzzy logic. *Information Sciences, 173*(4), 337–352.

Yap, K. H., & Wu, K. (2003). Fuzzy relevance feedback in content-based image retrieval. In *Proceedings of the 2003 Joint Fourth International Conference on Information, Communications and Signal Processing, 2003 and the Fourth Pacific Rim Conference on Multimedia* (Vol. 3, pp. 1595–1599).

Zadeh, L. A. (2002). A prototype-centered approach to adding deduction capabilities to search engines—The concept of a protoform. In *Proceedings of the Annual Meeting of the North American Fuzzy Information Processing Society (NAFIPS 2002)* (pp. 523–525).

Zadeh, L. A. (2006). From search engines to question answering systems—The problems of world knowledge, relevance, deduction and precipitation. In E. Elie Sanchez (Ed.), *Fuzzy logic and the semantic web* (pp. 163–210). Elsevier.

Zhao, T., Tang, L. H., Ip, H. H. S., & Qi, F. (2003). On relevance feedback and similarity measure for image retrieval with synergetic neural nets. *Neurocomputing, 51*.

Zhou, S., & Gan, J. Q. (2008). Low-level interpretability and high-level interpretability: A unified view of data-driven interpretable fuzzy system modeling. *Fuzzy Sets and Systems, 159*, 3091–3131.

Parameter-Centric Analysis Grossly Exaggerates Certainty

William M. Briggs

Abstract The reason probability models are used is to characterize uncertainty in observables. Typically, certainty in the parameters of fitted models based on their parametric posterior distributions is much greater than the predictive uncertainty of new (unknown) observables. Consequently, when model results are reported, uncertainty in the observable should be reported and not uncertainty in the parameters of these models. If someone mistook the uncertainty in parameters for uncertainty in the observable itself, a large mistake would be made. This mistake is exceedingly common, and almost exclusive in some fields. Reported here are some possible measures of the over-certainty mistake made when parametric uncertainty is swapped with observable uncertainty.

Keywords Model reporting · Posterior distributions · Predictive probability · Uncertainty

1 Over-Certainty

Suppose an observable $y \sim \text{Normal}(0, 1)$; i.e., we characterize the uncertainty in an observable y with a normal distribution with known parameters (never mind how we know them). Obviously, we do not know with exactness what any future value of y will be, but we can state probabilities (of intervals) for future observables using this model.

It might seem an odd way of stating it, but in a very real sense we are infinitely more certain about the value of the model parameters than we are about values of the observable. We are *certain* of the parameters' values, but we have uncertainty in the observable. In other words, we know what the parameters are, but we don't know what values the observable will take. If the amount of uncertainty has any kind of measure, it would be 0 for the value of the parameters in this model, and something

W. M. Briggs (✉)
New York, NY, USA
e-mail: matt@wmbriggs.com

© The Editor(s) (if applicable) and The Author(s), under exclusive license to Springer Nature Switzerland AG 2021
N. Ngoc Thach et al. (eds.), *Data Science for Financial Econometrics*, Studies in Computational Intelligence 898, https://doi.org/10.1007/978-3-030-48853-6_5

positive for the value of the observable. The ratio of these uncertainties, observable to parameters, would be infinite.

That trivial deduction is the proof that, at least for this model, certainty in model parameters is not equivalent to certainty in values of the observable. It would be an obvious gaff, not even worth mentioning, were somebody to report uncertainty in the parameters *as if* it were the same as the uncertainty in the observable.

Alas, this is what is routinely done in probability models, see Chap. 10 of Briggs (2016). Open the journal of almost any sociology or economics journal, and you will find the mistake being made everywhere. If predictive analysis were used instead of parameteric or testing-based analysis, this mistake would disappear; see e.g. Ando (2007), Arjas and Andreev (2000), Berkhof and van Mechelen (2000), Clarke and Clarke (2018). And then some measure of sanity would return to those fields which are used to broadcasting "novel" results based on statistical model parameters.

The techniques to be described do not work for all probability models; only those models where the parameters are "like" the observables in a sense to be described.

2   Theory

There are several candidates for a measure of total uncertainty in a proposition. Since all probability is conditional, this measure will be, too. A common measure is variance; another is the length of the highest (credible) density interval. And there are more, such as entropy, which although attractive has a limitation described in the final section. I prefer here the length of credible intervals because they are stated in predictive terms in many models in units of the observable, made using plain-language probability statements. Example: "There is a 90% chance y is in (a, b)."

In the $y \sim \text{Normal}(0, 1)$ example, the variance of the uncertainty of either parameter is 0, as is the length of any kind of probability interval around them. The variance of the observable is 1, and the length of the $1 - \alpha$ density interval around the observable y is well known to be $2z_{\alpha/2}$, where $z_{\alpha/2} \approx 2$. The ratio of variances, parameter to observable, is $0/1 = 0$. The ratio of the length of confidence intervals, here observable to parameter, is $4/0 = \infty$.

We pick the ratio of the length of the $1 - \alpha$ credible intervals as observable to parameter to indicate the amount of *over-certainty*. If not otherwise indicated, I let α equal the magic number.

In the simple Normal example, as said in the beginning, if somebody were to make the mistake of claiming the uncertainty in the observable was identical to the uncertainty of the parameters, he would be making the worst possible mistake. Naturally, in situations like this, few or none would this blunder.

Things change, though, and for no good reason, when there exists or enters uncertainty in the parameter. In these cases, the mistake of confusing kinds of uncertainty happens frequently, almost to the point of exclusively.

The simplest models with parameter uncertainty follow this schema:

$$p(y|\text{DB}) = \int_{\theta} p(y|\theta, \text{DB}) p(\theta|\text{DB}) d\theta, \tag{1}$$

where $D = y_1, \ldots, y_n$ represents old measured or assumed values of the observable, and B represents the background information that insisted on the model formulation used. D need not be present. B must *always* be; it will contain the reasoning for the model form $p(y|\theta\text{DB})$, the form of the model of the uncertainty in the parameters $p(\theta|\text{DB})$, and the values of hyperparameters, if any. Obviously, if there are two (or more) contenders i and j for priors on the parameters, then in general $p(y|\text{DB}_k) \neq p(y|\text{DB}_l)$. And if there are two (or more) sets of D, k and l, then in general $p(y|D_iB) \neq p(y|D_jB)$. Both D and B may differ simultaneously, too.

It is worth repeating that unless one can *deduce* from B the form of the model (from the first principles of B), observables do not "have" probabilities. All probability is conditional: change the conditions, change the probability. All probability models are conditional on some D (even if null) and B. Change either, change the probability. Thus all measures of over-certainty are also conditional on D and B.

If D is not null, i.e. past observations exist, then of course

$$p(\theta|\text{DB}) = \frac{p(y|\theta\text{DB}) p(\theta|\text{DB})}{\int_{\theta} p(y|\theta\text{DB}) p(\theta|\text{DB}) d\theta} \tag{2}$$

The variances of $p(y|\text{DB})$ or $p(\theta|\text{DB})$ can be looked up if the model forms are common, or estimated if not.

Computing the highest density regions or intervals (HDI) of a probability distribution is only slightly more difficult, because multi-modal distributions may not have contiguous regions. We adopt the definition of Hyndman (2012). The $1 - \alpha$ highest-density region R is the subset $R(p_\alpha)$ of y such that $R(p_\alpha) = \{y : p(y) \geq p_\alpha\}$ where p_α is the largest constant such that $\Pr\left(y \in R(p_\alpha)|\text{DB}\right) \geq 1 - \alpha$. For unimodal distributions, this boils down to taking the shortest continuous interval containing $1 - \alpha$ probability. These, too, are computed for many packaged distributions. For the sake of brevity, all HDI will be called here "credible intervals."

It will turn out that comparing parameters to observables cannot always be done. This is when the parameters is not "like" the observable; when they are not measured in the same units, for example. This limitation will be detailed in the final section.

3 Analytic Examples

The analytic results here are all derived from well known results in Bayesian analysis. See especially Bernardo and Smith (2000) for the form of many predictive posterior distributions.

3.1 Poisson

Let $y \sim$ Poisson(λ), with conjugate prior $\lambda \sim$ Gamma(α, β). The posterior on λ is distributed Gamma($\sum y + \alpha, n + \beta$) (shape and scale parameters). The predictive posterior distribution is Negative Binomial, with parameters ($\sum y + \alpha, \frac{1}{n+\beta+1}$). The mean of both the parameter posterior and predictive posterior are $\frac{\sum y + \alpha}{n+\beta}$. The variance of the parameter posterior is $\frac{\sum y + \alpha}{(n+\beta)^2}$, while the variance of the predictive posterior is $\frac{\sum y + \alpha}{(n+\beta)^2}(n + \beta + 1)$. The ratio of the means, independent of both α and β, is 1. The ratio of the parameter to predictive variance, independent of α, is $1/(n + \beta + 1)$.

It is obvious, for finite β, that this ratio tends to 0 at the limit. This recapitulates the point that eventually the value of the parameter becomes certain, i.e. with a variance tending toward 0, while the uncertainty in the observable y remains at some finite level. One quantification of the exaggeration of certainty is thus equal to $(n + \beta + 1)$.

Although credible intervals for both parameter and predictive posteriors can be computed easily in this case, it is sometimes an advantage to use normal approximations. Both the Gamma and Negative Binomial admit normal approximations for large n. The normal approximation for a Gamma($\sum y + \alpha, n + \beta$) is Normal($(\sum y + \alpha)/(n + \beta), (\sum y + \alpha)/(n + \beta)^2$). The normal approximation for a Negative Binomial($\sum y + \alpha, \frac{1}{n+\beta+1}$) is Normal($(\sum y + \alpha)/(n + \beta), (n + \beta + 1) * (\sum y + \alpha)/(n + \beta)^2$).

The length of the $1 - \tau$ credible interval, equivalently the $z_{\tau/2}$ interval, for any normal distribution is $2z_{\tau/2}\sigma$. Thus the ratio of predictive to parameter posterior interval lengths is independent of τ and to first approximation equal to $\sqrt{n + \beta + 1}$. Stated another way, the predictive posterior interval will be about $\sqrt{n + \beta + 1}$ times higher than the parameter posterior interval. Most pick a β of around or equal to 1, thus for large n the over-certainty grows as $\sqrt{n}$. That is large over-certainty by any definition.

Also to a first approximation, the ratio of length of credible intervals also tends to 0 with n. Stated another way, the length of the credible interval for the parameter tends to 0, while the length of the credible interval for the observable tends to a fixed finite number.

3.2 Normal, σ^2 Known

Let $y \sim$ Normal(μ, σ^2), with σ^2 known, and with conjugate prior $\mu \sim$ Normal (θ, τ^2). The parameter posterior on μ is distributed Normal with parameters $\sigma_n^2 \left(\frac{\theta}{\tau^2} + \frac{n\bar{y}}{\sigma^2} \right)$ and $\sigma_n^2 = \sigma^2\tau^2/(n\tau^2 + \sigma^2)$. The posterior predictive is distributed as Normal, too, with the same central parameter and with the spread parameter $\sigma_n^2 + \sigma^2$.

The ratio of parameter to predictive posterior variances is $\sigma_n^2/(\sigma_n^2 + \sigma^2)$, which equals $\tau^2/((n + 1)\tau^2 + \sigma^2)$. This again goes to 1 in the limit, as expected. The ratio of credible interval lengths, predictive to the posterior, is the square root of the inverse of that, or $\sqrt{n+1+\tau^2/\sigma^2}$. As with Poisson distributions, this gives over-certainty which also increases proportionally to $\sqrt{n}$.

3.3 Normal, σ^2 Unknown

Let $y \sim \text{Normal}(\mu, \sigma^2)$, with a Jeffrey's prior over both parameters, which is proportional to $1/\sigma^2$. Then the marginal parameter posterior for μ (considering σ^2 to be of no direct interest) is a scaled T distribution with parameters $(\bar{y}, s^2/n)$ and with $n-1$ degrees of freedom. The predictive posterior is also a scaled T distribution also with $n-1$ degrees of freedom, and with parameters $(\bar{y}, s^2(n-1)/n)$.

For modest n, a normal approximation to the scaled T is sufficient. Thus the ratio of parameter to predictive posterior variances is equal to $1/(n-1)$. As before, this tends to 0 with increasing n. The ratio of the length of credible intervals is obvious, which again shows over-certainty rises proportionally to about $\sqrt{n}$.

Consider conjugate priors instead. Conditional on σ^2, the distribution of μ is a Normal with parameters $(\theta, \sigma^2/\tau)$. And the distribution of σ^2 is and Inverse Gamma with parameters $(\alpha/2, \beta/2)$. Then the conditional parameter posterior of μ is distributed as a scaled T with $\alpha + n$ degrees of freedom and with parameters $((\tau\theta + n\bar{y})/(\tau + n), (n-1)^2 s^2/\theta)$. The predictive posterior is also a scaled T with $\alpha + n$ degrees of freedom and with the same central parameter, but with a spread parameter equal to parametric posterior but multiplied by $\tau + n$.

Obviously, the ratio of parametric to predictive posterior variances is $1/(\tau + n)$, which again tends to 0 with n. Using the same normal approximation shows the credible interval ratio gives an over-certainty multiplier of $\sqrt{\tau + n}$.

The choice of Jeffrey's improper or the conjugate prior makes almost no difference to amount of over-certainty, as expected.

3.4 Regression, σ^2 Known

A regression model for observable y with predictor measures x is $y = x\beta + \varepsilon$, where the uncertainty in ε is characterized by a Normal distribution with parameters $(0, (\lambda I)^{-1})$, where λ is a scalar and I the identity matrix. The parameter posterior for β is a Normal distribution with parameters $(x'x + \lambda\sigma^2 I)^{-1} x'y$ and $\sigma^2(x'x + \lambda\sigma^2 I)^{-1}$. The predictive posterior for a new or assumed x, which we can write as w (a single vector), is also a Normal distribution, with parameters $w'(x'x + \lambda\sigma^2 I)^{-1} x'y$ and $\sigma^2(1 + w'(x'x + \lambda\sigma^2 I)^{-1} w)$.

Now as $\lambda \to 0$ the prior more resembles a Jeffrey's prior. Then the "ratio" of parametric to predictive variances is $\sigma^2(x'x)^{-1}(\sigma^2)^{-1}(1 + w'(x'x)^{-1}w)^{-1} = (x'x)^{-1}(1 + w'(x'x)^{-1}w)^{-1}$. The quantity $(1 + w'(x'x)^{-1}w)$ will be some scalar

a, thus the ratio becomes $(x'x)^{-1}/a$. The ratio therefore depends on the measures x and their inherent variability. This will become clearer in the numerical examples.

3.5 Other Models

There are a host of models where the calculation pickings are easy and analytic. However, with complex forms, analytics solutions are not readily available. In these cases, we use standard simulation approaches. This will be demonstrated below for a general regression example, which introduces an extension of over-certainty computation. The same limitation mentioned at the beginning about these techniques only working for situation where the parameters are "like" the observable still applies.

4 Numeric Examples

4.1 Poisson

The length of years (rounded up) served by each Pope of the Catholic Church was collected by the author. These ran from 1 year, shared by many Popes, to 36 years, which was the longest and was for St Peter, the first Pope. The mean was 8.3 years. There were 263 past Popes (and one current one). A natural first model to try is the Poisson, as above. The hyperparameters chosen were the commonly picked $\alpha = \beta = 1$. The model is shown in Fig. 1.

The spikes are the frequency of observations (a modified histogram), the dashed red line the Gamma parameter posterior, and the black solid the Negative Binomial predictive posterior. The peak of the parameter posterior is not shown, though the center of the distribution is clear.

It is clear the certainty in the λ parameter is not anywhere close to the certainty in the future values of length of reign. That is, the certainty in the parameter is vastly stronger than the certainty in the observable.

The length of the 95% credible interval for the parameter posterior is 0.69 years, while it is 11 years for the predictive posterior, a ratio of 15.9. This is very close to the normal approximation value of $\sqrt{n + \beta + 1} = 16.3$.

It is clear from the predictive posterior that the model does not fit especially well. It gives too little weight to shorter reigns, and not enough to the longest. Clearly better alternatives are available, but they were not attempted. I kept this poor model because the graph makes an important point.

It would be wrong to say "every time", but very often those who use probability models never check them against observables in this predictive manner. An estimate of the parameter is made, along with a credible (or confidence) interval of that parameter, and it is that which is reported, as stated above. Predictive probabilities of

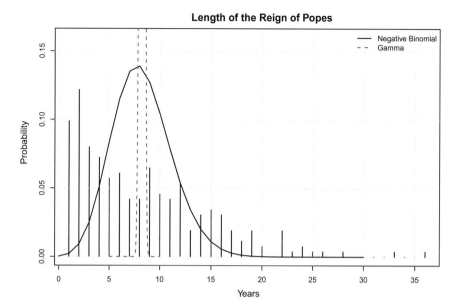

Fig. 1 The spikes are the frequency of observations (a modified histogram), the dashed red line the Gamma parameter posterior, and the black solid the Negative Binomial predictive posterior. The peak of the parameter posterior, scaled to the frequency, is not shown because it is so far off the scale, though the center of the distribution is clear

observables, the very point of modeling, is usually forgotten. Because of this, many poor models are released into the wild.

4.2 Normal, σ Unknown

We skip over the normal example with σ known and move to where σ is unknown. The rape rate per 100,000 for each of the 50 United States in 1973 was gathered from the built-in R dataset USArrests. A normal model was fit to it using Jeffrey's prior. The result is in Fig. 2.

The histogram is superimposed by the parameter posterior (red, dashed), and predictive posterior (black, solid). The models fit is in the ballpark, but not wonderful. It is clear uncertainty in the parameters is much greater than in the observables.

The length of the parameter posterior credible interval is 5.3 (rapes per 100,000), while for the predictive posterior it was 37.3. The ratio of predictive to parameter is 7. In other words, reporting only the parameter uncertainty results is seven times too certain.

Why one would do this model is also a question. We have the rates, presumably measured without error, available at each state. There is no reason to model unless one

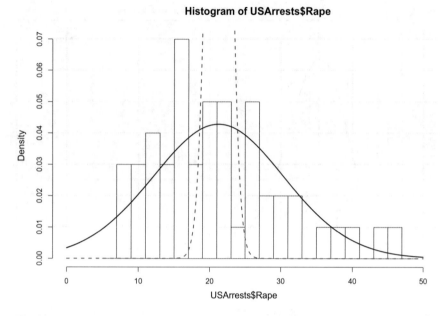

Fig. 2 The histogram is superimposed by the parameter posterior (red, dashed), and predictive posterior (black, solid). The models fit is in the ballpark, but not wonderful. It is clear uncertainty in the parameters is much greater than in the observables

wanted to make predictive statements of future (or past unknown) rates, conditional on assuming this model's adequacy.

There is also the small matter of "probability leakage", Briggs (2013), shown a the far left of the predictive distribution, which we leave until the regression examples. Probability leakage is probability given to observations known to be impossible, but which is information not told to B. This will be clear in the examples.

4.3   Regression 1, σ^2 Known

We use here the beloved cars dataset from R, which measured the speed of 50 cars (in mph) and how long it took them to stop (in feet). Speeds ranged from 4 to 25 mph, and distances from 2 to 120 feet. In order for σ to be known, we cheated, and used the estimate from the ordinary linear regression of speed on distance. Figure 3 shows the results.

Now this data set has been used innumerable times to illustrate regression techniques, but I believe it is the first time it has been demonstrated how truly awful regression is here.

In each panel, the predictive posterior is given in black, and the parameter posterior is given in dashed red. In order to highlight the comparisons, the parameter

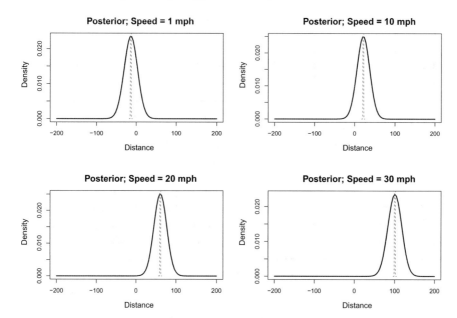

Fig. 3 In each panel, the predictive posterior is given in black, and the parameter posterior is given in dashed red. In order to highlight the comparisons, the parameter posterior, which is fixed regardless of the distance, was shifted to the peak of the predictive posterior distributions

posterior was shifted to the peak of the predictive posterior distributions. The parameter posterior is of course fixed—and at the "effect" size for speed. Here it has a mean of 3.9, with credible interval of (3.1, 4.8).

It is immediately clear just reporting the parameter posterior implies vastly more certainty than the predictive posteriors. We do not have just one predictive posterior, but one for every possible level of speed. Hence we also have varying levels of over-certainty. The ratio of predictive to parameter credible intervals was, 40.1 (at 1 mph), 37.8 (10 mph), 37.6 (20 mph), and 40.1 (30 mph),

The over-certainty is immense at any speed. But what is even more interesting is the enormous probability leakage at low speeds. Here we have most probability for predictive stopping distances of less than 0, a physical impossibility. The background information B did not account for this impossibility, and merely said, as most do say, that a regression is a fine approximation. It is not. It stinks for low speeds.

But this astonishingly failure of the model would have gone forever unnoticed had the model not been cast in its predictive form. The parameter-centric analysis would be have missed, and almost always does miss, this glaring error.

If instead of σ^2 being known, and faked, we let it be unknown, the results for this model are much the same, only with greater uncertainties in the posteriors. We skip that and move to a more general example.

4.4   Regression 2, σ^2 Unknown

It is well to take an example which in most respects passes muster, even in a predictive sense, but which nevertheless still exaggerates certainty. We use the `oats` data in the *MASS* package. This was an experiment with three oat varieties planted at 6 blocks, and with four amounts of added nitrogen via manure: none added, 0.2 cwt, 0.4 cwt, and 0.6 cwt per acre. The outcome is yield of oats in quarter pounds. The intent and supposition was that greater amounts of nitrogen would lead to greater yields.

The first step is computing an ordinary linear regression; here using Gibbs sampling via the `rstanarm` package version 2.18.2 in R version 3.5.2, using default priors, but here using 6 chains of 20,000 iterations each for greater resolution. Diagnostics (not shown) suggest the model converged. The results are in Fig. 4.

The predictive posteriors are presented as histograms, using posterior observable draws, and the parameter posteriors as red dashed lines. We chose the first block and the Golden rain oat variety to create the predictions. There is some difference between blocks, but little between varieties. A full analysis would, of course, consider all these arrangements, but here these choices will be sufficient. The parameter posteriors are cut off as before so as not to lose resolution on the observables. The parameter posteriors are also re-centered for easy visual comparison by adding the mean intercept value.

The over-certainty as measured by length of the predictive to parameteric credible intervals is about 3 times for each amount of nitrogen. However, since the purpose of this was to demonstrate increasing nitrogen boosts yields, this measure may seem out of place. Rather, the mistake of assuming the uncertainties are the same is not likely to be made.

There is still lurking over-certainty, though.

The posterior probability the parameter for 0.2 cwt added nitrogen is greater than 0 is very close to 1 (0.9998). Any researcher would go away with the idea that adding nitrogen guarantees boosting yield. But the posterior predictive chance that a new plot with 0.2 cwt will have a greater yield than a plot with no added nitrogen is only 0.8. This implies an over-certainty of $\approx 1/0.8 = 1.25$.

Likewise, although all the parameters in the model for nitrogen have extremely high probabilities of differing from each other, the predictive probabilities of greater yields do not differ as much. The posterior probability the parameter for 0.4 cwt is greater than 0.2 cwt is 0.997, and the parameter for 0.6 cwt is greater than 0.4 cwt is 0.96.

The predictive probability of greater yields in plots with 0.4 cwt over 0.2 cwt is 0.75, with over-certainty of $0.997/0.75 = 1.33$. And the predictive probability of greater yields in plots with 0.6 cwt over 0.4 cwt is 0.66, with over-certainty of $0.96/0.66 = 1.45$.

Notice that we were able to compute over-certainty in this case because we were comparing probabilities for "like" things, here oats yields. We cannot always make comparisons, however, as the last section details.

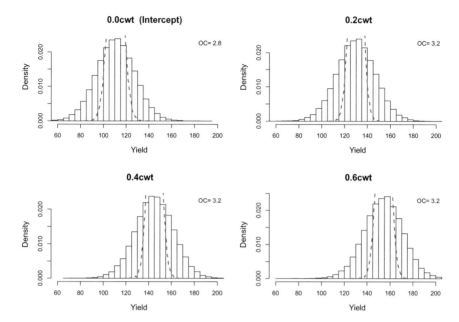

Fig. 4 The predictive posteriors are presented as histograms, using posterior observable draws, and the parameter posteriors as red dashed lines. These are cut off as before so as not to lose resolution on the observables. The parameter posteriors are also re-centered for easy visual comparison by adding the mean intercept value. The length of the predictive to parameteric credible intervals are given in the top corner (as OC = Over-certainty)

5 Comparing Apples and Kiwis

There are many more probability models beside regression and the like, but the techniques described in this paper only work where the parameters represent the same kind of thing as the observable. A normal model for the observable, say, weight, and the central parameter for that model are both in pounds (or kilograms), making comparison natural. But in other models, no such comparison can be made.

Consider the simplest logistic model with $p = \Pr(y|3DB)$ and

$$\log\left(\frac{p}{1-p}\right) = \beta, \tag{3}$$

with B specifying β and the model form, and with D (here) null.

Obviously, the odds of y are equal to e^β. If β is known, there is still uncertainty in the observable, which has probability $p = e^\beta/(1 + e^\beta)$. Any uncertainty interval around β would be 0, but what does it mean to have an "interval" around p? The probability p is also certain: it is equal to p! It is the observable which is uncertain. It will happen or not, with probability p. There is no interval.

One might guess entropy might rescue the notion of uncertainty, but it is not so. The entropy for our uncertainty of β is 0, while for the observable it is $p \log(p) + (1 - p) \log(1 - p) > 0$. At first glance it thus appears entropy will help. But consider that B instead specifies we have β or $\beta + \varepsilon$ where $\varepsilon \ll 1$, both with a probability of 0.5. The entropy of the probability distribution of the parameter is $\log(2)$. But the entropy of the observable is based on the distribution $p = e^{\beta}/(1 + e^{\beta}) \times 0.5 + e^{\beta+\varepsilon}/(1 + e^{\beta+\varepsilon}) \times 0.5$. Since ε is small, $p \approx e^{\beta}/(1 + e^{\beta})$ and the entropy is also approximately the same: $p \log(p) + (1 - p) \log(1 - p)$.

If β is at all large, the entropy of the observable will be near 0. Thus, according to entropy with "large" β, there is more uncertainty in the parameter than in the observable! There is now the idea of an interval around the parameter, but not around y.

The reason entropy doesn't work, nor credible intervals, nor variance which is here similar to entropy, is because the parameter is of a different kind than the observable, of a different nature. In logistic models the "βs" are usually multipliers of odds, and multipliers of odds aren't anything "like" observables, which are "successes" or "failures".

All that one can do in these instances, then, is to emphasize that uncertainty of parameters just isn't anything like uncertainty in the observable. The two do not translate. It's like comparing apples to kumquats. When we see authors make the mistake of swapping uncertainties, all we can do is tell them to cut it out.

References

Ando, T. (2007). *Biometrika*, *94*, 443.

Arjas, E., & Andreev, A. (2000). *Lifetime Data Analysis*, *6*, 187.

Berkhof, J., & van Mechelen, I. (2000). *Computational Statistics*, *15*, 337.

Bernardo, J. M., & Smith, A. F. M. (2000). *Bayesian theory*. New York: Wiley.

Briggs, W. M. (2013). arxiv.org/abs/1201.3611.

Briggs, W. M. (2016). *Uncertainty: The soul of probability, modeling and statistics*. New York: Springer.

Clarke, B. S., & Clarke, J. L. (2018). *Predictive Statistics*. Cambridge: Cambridge University Press.

Hyndman, R. J. (2012). *The American Statistician*, *50*, 120.

Three Approaches to the Comparison of Random Variables

Bernard De Baets

Abstract Decision making inevitably involves the comparison of real variables. In the presence of uncertainty, this entails the comparison of real-valued random variables. In this short contribution, we briefly review three approaches to such comparison: (i) *stochastic dominance*: an approach based on a pointwise comparison of cumulative distribution functions; (ii) *statistical preference*: an approach based on a pairwise comparison in terms of winning probabilities; (iii) *probabilistic preference*: an approach based on multivariate winning probabilities. Whereas the first and third approaches are intrinsically transitive, the second approach requires considerable mathematical effort to unveil the underlying transitivity properties. Moreover, the first approach ignores the possible dependence between the random variables and is based on univariate distribution functions, the second approach is by definition built on bivariate joint distribution functions, while the third approach is based on the overall joint distribution function.

Keywords Stochastic dominance · Statistical preference · Probabilistic preference · Winning probabilities · Transitivity

1 Introduction

In decision making under risk, the alternatives are usually modelled in terms of random variables, whence the need for tools that allow to establish an ordering between random variables. Two of the most prominent such tools are stochastic dominance (Levy 1998; Müller and Stoyan 2002; Shaked and Shanthikumar 2007) and statistical preference (De Schuymer et al. 2003a, b). Stochastic dominance is undoubtedly the most popular one and has been applied in a variety of disciplines.

B. De Baets (✉)
KERMIT, Faculty of Bioscience Engineering, Department of Data Analysis and Mathematical Modelling, Ghent University, Ghent, Belgium
e-mail: bernard.debaets@ugent.be

© The Editor(s) (if applicable) and The Author(s), under exclusive license to Springer Nature Switzerland AG 2021
N. Ngoc Thach et al. (eds.), *Data Science for Financial Econometrics*, Studies in Computational Intelligence 898, https://doi.org/10.1007/978-3-030-48853-6_6

Unfortunately, as stochastic dominance is based on marginal distributions only, it does not take into account the possible dependence between the random variables. To cope with this problem, various generalizations have been proposed, such as the stochastic precedence order (Arcones et al. 2002; Li and Hu 2008) and probability dominance (Wrather and Yu 1982). From the same point of view, statistical preference has been introduced (De Schuymer et al. 2003a, b) as a graded version of stochastic dominance, in the sense that it is based on a reciprocal relation expressing the winning probabilities between the random variables.

Stochastic dominance between two random variables can be characterized in terms of the comparison of the expectations of increasing transformations of these random variables. In the same vein, statistical preference is closer to the median, another location parameter (Montes et al. 2015). Under common conditions such as independence, stochastic dominance implies statistical preference (De Schuymer et al. 2005; Montes et al. 2010; Montes and Montes 2016). Both concepts are even equivalent in specific cases, such as normally distributed random variables with the same variance.

Pairwise methods like stochastic dominance and statistical preference have a number of drawbacks, whence the need arises to develop alternative comparison tools. One such approach consists of computing multivariate winning probabilities based on the joint distribution of all the random variables considered and thus uses all the available information. The multivariate winning probabilities allow to rank (with ties) the random variables, and thus naturally lead to a preference relation called probabilistic preference, which is an extension of the notion of statistical preference to a multivariate setting (Montes et al. 2019).

Throughout this contribution, we consider a probability space (Ω, Σ, P). A random variable $X : \Omega \to \mathbb{R}$ is a Σ-measurable function. Its cumulative distribution function $F_X : \mathbb{R} \to [0, 1]$ is given by $F_X(t) = P(X \le t)$, where $\{X \le t\}$ denotes the set $\{\omega \in \Omega \mid X(\omega) \le t\}$.

2 Stochastic Dominance

The pairwise comparison of random variables is a widely studied topic and several approaches have been proposed over the last decades. The simplest approach is the expected utility model of von Neumann and Morgenstern (1953): given a utility function $u : \mathbb{R} \to \mathbb{R}$, a random variable X is preferred to a random variable Y if $E[u(X)] \ge E[u(Y)]$ (assuming that both expectations exist). An obvious multi-utility generalization (Durba et al. 2004) is given by: given a set of utility functions $\mathcal{U}$, X is preferred to Y if $E[u(X)] \ge E[u(Y)]$, for every $u \in \mathcal{U}$. On the other hand, (first degree) stochastic dominance is based on the direct comparison of the associated cumulative distribution functions (Levy 1998; Müller and Stoyan 2002; Shaked and Shanthikumar 2007): for two random variables X and Y with cumulative distribution functions F_X and F_Y,

(i) X is said to stochastically dominate Y, denoted by $X \succeq_{\text{FSD}} Y$, if $F_X(t) \leq F_Y(t)$, for any $t \in \mathbb{R}$;

(ii) X is said to strictly stochastically dominate Y, denoted by $X \succ_{\text{FSD}} Y$ if $X \succeq_{\text{FSD}} Y$ while not $Y \succeq_{\text{FSD}} X$.

Obviously, stochastic dominance constitutes a pre-order relation on the set of random variables considered.

Remarkably, stochastic dominance between two random variables can be characterized in terms of the comparison of the expectations of some appropriate transformations of these random variables (see, e.g., Levy (1998)): $X \succeq_{\text{FSD}} Y \Leftrightarrow E[u(X)] \geq E[u(Y)]$, for any increasing function u (whenever both expectations exist). Obviously, stochastic dominance can be seen as a special case of the above multi-utility approach. For more information on the notion of stochastic dominance, such as second and third degree stochastic dominance, we refer to Levy (1998), Müller and Stoyan (2002), Shaked and Shanthikumar (2007).

3 Statistical Preference

The notion of statistical preference is based on that of a reciprocal relation. A mapping $Q : \mathscr{A} \times \mathscr{A} \to [0, 1]$ is called a reciprocal relation on $\mathscr{A}$ if it satisfies the reciprocity property: $Q(a, b) + Q(b, a) = 1$, for any $a, b \in \mathscr{A}$. Let $\mathscr{A}$ be the set of distinct random variables of interest, then the mapping Q defined by $Q(X, Y) = P(X > Y) + \frac{1}{2}P(X = Y)$ is a reciprocal relation on $\mathscr{A}$. The number $Q(X, Y)$ is called the winning probability of X over Y; the greater the winning probability $Q(X, Y)$, the stronger the preference of X over Y. Hence, the closer the value $Q(X, Y)$ to 1, the more X is preferred to Y; the closer $Q(X, Y)$ to 0, the more Y is preferred to X; and if $Q(X, Y)$ equals 0.5, then both random variables are considered indifferent. Formally,

(i) X is said to be statistically preferred to Y, denoted by $X \succeq_{\text{SP}} Y$, if $Q(X, Y) \geq \frac{1}{2}$;

(ii) X is said to be strictly statistically preferred to Y, denoted by $X \succ_{\text{SP}} Y$, if $Q(X, Y) > \frac{1}{2}$;

(iii) X and Y are said to be statistically indifferent if $Q(X, Y) = \frac{1}{2}$.

Note that statistical preference can be characterized in terms of the cumulative distributions of $X - Y$ and $Y - X$ (Montes et al. 2015): $X \succeq_{\text{SP}} Y \Leftrightarrow F_{X-Y}(0) \leq F_{Y-X}(0)$.

One obvious advantage of statistical preference compared to stochastic dominance is the possibility of establishing winning probabilities between the alternatives. Another advantage is the fact that it takes into account the possible dependence between the random variables since it is based on the joint distribution, while stochastic dominance only uses the marginal distributions. Note that in case of independent random variables, statistical preference is less restrictive than stochastic dominance, meaning that the latter implies the former. Moreover, statistical preference establishes a complete relation, while one can find pairs of random variables that are incomparable under first degree stochastic dominance. While statistical preference rules out incomparability, it has its own important drawback: its lack of transitivity in general. Even worse, it is possible that statistical preference results in cycles (De Schuymer et al. 2003b, 2005). There exists various frameworks for studying the transitivity of reciprocal relations (De Baets and De Meyer 2005; De Baets et al. 2006; Switalski 2003), and there is abundant literature on the positioning of winning probability relations therein (De Baets and De Meyer 2008, 2019; De Baets et al. 2010, 2015; De Meyer et al. 2007; De Schuymer et al. 2003b, 2005, 2007).

4 Probabilistic Preference

In an attempt to overcome the limitations of statistical preference, a genuinely multivariate approach has been proposed recently (Montes et al. 2019). Let $\mathcal{A}$ be the set of distinct random variables of interest, then we consider the mapping $\Pi_{\mathcal{A}} : \mathcal{A} \to [0, 1]$ defined by

$$\Pi_{\mathcal{A}}(X) = \sum_{\mathcal{Y} \subseteq \mathcal{A} \setminus \{X\}} \frac{1}{1 + |\mathcal{Y}|} P\Big((\forall Z \in \mathcal{Y})(\forall W \in \mathcal{A} \setminus \{X\} \setminus \mathcal{Y})(X = Z > W)\Big).$$

In case $\mathcal{A} = \{X, Y\}$, we retrieve the winning probability: $\Pi_{\mathcal{A}}(X) = Q(X, Y)$. The value $\Pi_{\mathcal{A}}(X)$ is called the multivariate winning probability of X in $\mathcal{A}$. The greater $\Pi_{\mathcal{A}}(X)$, the stronger the preference for X in $\mathcal{A}$. Formally,

   (i) X is said to be probabilistically preferred to Y in $\mathcal{A}$, denoted by $X \succeq_{\text{PP}} Y$, if $\Pi_{\mathcal{A}}(X) \geq \Pi_{\mathcal{A}}(Y)$;
   (ii) X is said to be strictly probabilistically preferred to Y in $\mathcal{A}$, denoted by $X \succ_{\text{PP}} Y$, if $\Pi_{\mathcal{A}}(X) > \Pi_{\mathcal{A}}(Y)$;
   (iii) X is said to be probabilistically indifferent to Y in $\mathcal{A}$, denoted by $X \equiv_{\text{PP}} Y$, if $\Pi_{\mathcal{A}}(X) = \Pi_{\mathcal{A}}(Y)$.

Similarly, given $X \in \mathcal{A}$, X is probabilistically preferred to the random variables in $\mathcal{A} \setminus \{X\}$, denoted by $X \succeq_{\text{PP}} \mathcal{A} \setminus \{X\}$, if $\Pi_{\mathcal{A}}(X) \geq \Pi_{\mathcal{A}}(Y)$ for any $Y \in \mathcal{A} \setminus \{X\}$. The mapping $\Pi_{\mathcal{A}}$ can be seen as a probability distribution on $\mathcal{A}$.

Note that in contrast to statistical preference, probabilistic preference is transitive, because it is based on the comparison of the multivariate winning probabilities, whence $\succeq_{\text{PP}}$ is a weak order ranking the random variables in decreasing order of multivariate winning probability. Probabilistic preference is based on the joint distribution of all random variables in $\mathcal{A}$, and not only bivariate distributions, and therefore possibly points out a different preferred random variable than statistical preference.

As mentioned above, for independent random variables, stochastic dominance implies statistical preference. Also in the multivariate setting, one can establish an interesting correspondence between stochastic dominance and probabilistic preference when the random variables are independent (Montes et al. 2019): if the random variables in $\mathscr{A}$ are mutually independent, and if $X \succeq_{\mathrm{FSD}} Y$ for any $Y \in \mathscr{A} \setminus \{X\}$, then $X \succeq_{\mathrm{PP}} \mathscr{A} \setminus \{X\}$. The independence clearly compensates for the fact that stochastic dominance only considers the marginal distributions.

5 Conclusion

In this short contribution, we have reviewed three approaches to the comparison of a given set of random variables, have discussed their pros and cons, and have pointed out some basic links between them, mainly in the case of mutually independent random variables.

References

Arcones, M. A., Kvam, P. H., & Samaniego, F. J. (2002). Nonparametric estimation of a distribution subject to a stochastic preference constraint. *Journal of the American Statistical Association, 97*, 170–182.

De Baets, B., & De Meyer, H. (2005). Transitivity frameworks for reciprocal relations: Cycle-transitivity versus *F G*-transitivity. *Fuzzy Sets and Systems, 152*, 249–270.

De Baets, B., & De Meyer, H. (2008). On the cycle-transitive comparison of artificially coupled random variables. *International Journal of Approximate Reasoning, 47*, 306–322.

De Baets, B., & De Meyer, H. (2019). Cutting levels of the winning probability relation of random variables pairwisely coupled by a same Frank copula. *International Journal of Approximate Reasoning, 112*, 22–36.

De Baets, B., De Loof, K., & De Meyer, H. (2015). A frequentist view on cycle-transitivity of reciprocal relations. *Fuzzy Sets and Systems, 281*, 198–218.

De Baets, B., De Meyer, H., & De Loof, K. (2010). On the cycle-transitivity of the mutual rank probability relation of a poset. *Fuzzy Sets and Systems, 161*, 2695–2708.

De Baets, B., De Meyer, H., De Schuymer, B., & Jenei, S. (2006). Cyclic evaluation of transitivity of reciprocal relations. *Social Choice and Welfare, 26*, 217–238.

De Meyer, H., De Baets, B., & De Schuymer, B. (2007). On the transitivity of the comonotonic and countermonotonic comparison of random variables. *Journal of Multivariate Analysis, 98*, 177–193.

De Schuymer, B., De Meyer, H., & De Baets, B. (2003a). *A fuzzy approach to stochastic dominance of random variables*. Lecture Notes in Artificial Intelligence (Vol. 2715, pp. 253–260).

De Schuymer, B., De Meyer, H., De Baets, B., & Jenei, S. (2003b). On the cycle-transitivity of the dice model. *Theory and Decision, 54*, 261–285.

De Schuymer, B., De Meyer, H., & De Baets, B. (2005). Cycle-transitive comparison of independent random variables. *Journal of Multivariate Analysis, 96*, 352–373.

De Schuymer, B., De Meyer, H., & De Baets, B. (2007). Extreme copulas and the comparison of ordered lists. *Theory and Decision, 62*, 195–217.

Durba, J., Maccheroni, F., & Ok, E. A. (2004). Expected utility theory without the completeness axiom. *Journal of Economic Theory, 115*, 118–133.

Levy, H. (1998). *Stochastic dominance*. Kluwer Academic Publishers.

Li, X., & Hu, X. (2008). Some new stochastic comparisons for redundancy allocations in series and parallel systems. *Statistics and Probability Letters, 78*, 3388–3394.

Montes, I., Martinetti, D., Díaz, S., & Montes, S. (2010). Comparison of random variables coupled by Archimedean copulas. In *Combining soft computing and statistical methods in data analysis*. Advances in Intelligent and Soft Computing (Vol. 77, pp. 467–474). Berlin/Heidelberg: Springer.

Montes, I., Martinetti, D., Díaz, S., & Montes, S. (2015). Interpretation of statistical preference in terms of location parameters. *Information Systems and Operational Research, 53*, 1–12.

Montes, I., & Montes, S. (2016). Stochastic dominance and statistical preference for random variables coupled by an Archimedean copula or by the Fréchet-Hoeffding upper bound. *Journal of Multivariate Analysis, 143*, 275–298.

Montes, I., Montes, S., & De Baets, B. (2019). Multivariate winning probabilities. *Fuzzy Sets and Systems, 362*, 129–143.

Müller, A., & Stoyan, D. (2002). *Comparison methods for stochastic models and risks*. Wiley.

Shaked, M., & Shanthikumar, J. G. (2007). *Stochastic orders*. New York: Springer-Verlag.

Switalski, Z. (2003). General transitivity conditions for fuzzy reciprocal preference matrices. *Fuzzy Sets and Systems, 137*, 85–100.

von Neumann, J., & Morgenstern, O. (1953). *Theory of games and economic behavior*. Princeton University Press.

Wrather, C., & Yu, P. L. (1982). Probability dominance in random outcomes. *Journal of Optimization Theory and Applications, 36*, 315–334.

A QP Framework: A Contextual Representation of Agents' Preferences in Investment Choice

Polina Khrennikova and Emmanuel Haven

Abstract Contextual decisions and beliefs and their impact upon market outcomes are at the core of research in behavioural finance. We describe some of the notable probabilistic fallacies that underpin investor behaviour and the consequent deviation of asset prices from the rational expectations equilibrium. In real financial markets, the complexity of financial products and the surrounding ambiguity calls for a more general formalization of agents belief formation than offered by the standard probability theory and dynamic models based on classical stochastic processes. The main advantage of quantum probability (QP) is that it can capture contextuality of beliefs through the notion of *non-commuting* prospect observables. QP has the potential to model myopia in asset return evaluation, as well as inter-asset valuation. Moreover, the interference term of the agents' comparison state can provide a quantitative description of their vacillating ambiguity perception characterized by non-additive beliefs of agents. Some of the implications of non-classicality in beliefs for the composite market outcomes can also be modelled with the aid of QP. As a final step we also discuss the contributions of the growing body of psychological studies that reveal a true (quantum type) contextuality in human preference statistics showing that the classical probability theory is too restrictive to capture the very strong non-classical correlations between preference outcomes and beliefs.

Keywords Quantum probability · Contextuality · Belief state · Interference effects · Complementarity of observables · Behavioural finance · Narrow framing · Ambiguous beliefs

P. Khrennikova (✉)
School of Business, University of Leicester, Leicester, United Kingdom
e-mail: pk228@le.ac.uk

E. Haven
Faculty of Business Administration, Memorial University, St. John's, Canada
e-mail: ehaven@mun.ca

© The Editor(s) (if applicable) and The Author(s), under exclusive license
to Springer Nature Switzerland AG 2021
N. Ngoc Thach et al. (eds.), *Data Science for Financial Econometrics*,
Studies in Computational Intelligence 898,
https://doi.org/10.1007/978-3-030-48853-6_7

1   Introduction

Starting with the seminal paradoxes revealed in thought experiments by Allais (1953) and Ellsberg (1961) classical economic theory was preoccupied with modelling of the impact of ambiguity and risk upon agent's probabilistic belief and preference formation. Broadly speaking, the main causes of contextual or state dependent behaviour were attributed to cognitive and psychological influences coupled with environmental conditions elaborated (among others) in the works by, Kahneman and Tversky (1972), Kahneman (2003). Deviations from classical probability based information processing (using the state dependence of economic agents' valuation of payoffs), has far-reaching implications for their trading in financial markets, by fuelling disequilibrium prices of the traded risky assets, Shiller (2014) and Thaler and Johnson (1990).

In classical decision theories due to Von Neumann and Morgenstern (1944) and Savage (1954) there are two core components of a decision making process. Firstly, the agents form beliefs about subjective and objective risks via classical probability measures. Agents also make inference by employing a Bayesian scheme, as new information about fundamentals or other important macroeconomic events is accumulated. Secondly, preference formation is derived from a context independent utility ranking of monetary values. These two building blocks of rational decision making serve as the core pillars behind asset trading frameworks in finance, starting with modern portfolio theory that is based on mean-variance optimization and the Capital Asset Pricing model.[1] The core premise of the frameworks is that beliefs about the returns suppose a similar historical pattern in the absence of new information, and are homogeneous across economic agents. The predictions of asset allocation and asset trading are grounded in the assumption that all agents are being Bayesian rational in their wealth maximization. Agents are not assumed to possess a contextual perception of sequences of gains and losses. Agents do form joint probability distributions of all asset class returns for the whole investment period in order to assess their mean returns, standard deviations and degree of price co-movement. Non-linearity in beliefs, as well as their dependence on the negative, or positive changes in wealth was well captured through an inflected probability weighting function, devised in the works by Kahneman and Tversky (1979), Tversky and Kahneman (1992).

The notion of ambiguity that surrounds future events, and its possible implications for agents' beliefs about returns of risky assets also attracted heightened attention in the finance literature. Most of these frameworks are aimed at modelling ambiguity in asset prices. Most well known models are targeting Ellsberg-type ambiguity aversion that results in more pessimistic beliefs and in shunning of complex risks. For instance, agents can be ambiguous with respect to the prior likelihoods, as well as being affected by ambiguous information that produces deviations of asset prices from the rational equilibrium, Epstein and Schneider (2008). We also refer to work on ambiguity markets for risky assets and managerial decisions related to project evaluation, detected in experimental studies by Mukerji and Tallan (2001), Sarin and

---

[1]The interested reader is referred to core texts in finance, e.g. Bodie et al. (2014).

Weber (1993), Ho et al. (2002) and Roca et al. (2006). Paradoxical findings on the differences between objective risk and ambiguity perception, attracted a vast body of research in both behavioural economics and finance fields. Furthermore, more recent works revealed that preference for ambiguous alternatives is also contextual and can be shaped by the framing of the ambiguous and risky prospects, as well as previously experienced returns. The celebrated "Max-min expected utility" due to Gilboa (1989) and Schemeidler (1989) provides a good account for the representation of the pessimistic beliefs that can explain an additional 'ambiguity premium' on assets with complex and unknown risks.[2] Consequentialism is also not satisfied in many decision making tasks that involve "what if" reasoning by agents, as they are confronted with different states of the world. The studies that exposed a "disjunction effect" in risky and ambiguous choice show that agents often prefer not to think through a decision tree and remain at "uncertain nodes" when making a choice, see the original findings in Tversky and Kahneman (1992), Shafir (1994).

In the search for a different theory of probability that could serve as a more general descriptive framework to accommodate context sensitivity of beliefs and preferences, quantum probability (QP) that is based on operational representation of beliefs in a linear vector space, started to be widely explored in decision theory and its applications to psychology, politics, economics, game theory and finance, see some contributions by Aerts et al. (2018), Asano et al. (2017), Bagarello (2007), Busemeyer et al. (2006), Pothos and Busemeyer (2009), Busemeyer and Bruza (2012), Haven (2002), Haven et al. (2018), Haven and Khrennikova (2018), Khrennikova and Haven (2016) and Khrennikova and Patra (2018). Please also see surveys in Khrennikov (2010), Busemeyer and Bruza (2012), Haven and Khrennikov (2013), Haven and Sozzo (2016) and Pothos and Busemeyer (2013).

While quantum probability has proven to provide for a good descriptive account for, (i) ambiguity perception and its vacillating nature; (ii) Non-Bayesian updating of information, the ultimate goal of quantum probability based decision theory is to develop a descriptive and predictive theoretical tool for capturing the context-sensitivity of the agents' information processing, as well as to formalize the resolution of ambiguity in the process of their preference formation. In this vein, the contributions in economics and finance addressed well the Ellsberg and Machina type ambiguity, see works by Aerts et al. (2018), Asano et al. (2017), Basieva et al. (2019). State dependence and contextuality of human judgments have been extensively explored in questionnaires and experiments, where framing of choices was manipulated to trigger the contextual perception of outcomes. These studies showed violations of the joint distribution rule, known in psychology as "order effects", see Trueblood and Busemeyer (2011), Wang and Busemeyer (2013), Khrennikov et al. (2014). Proper contextuality in quantum physics has a more precise meaning than context dependence of preferences and judgments (as shown under order effects and violations of joint probability distributions of outcomes of some random variables). A certain condition on the correlations between the statistics of some random variables

---

[2]A comprehensive account of theoretical frameworks of ambiguity aversion in human preference formation can be found in a monograph by Gilboa (2009).

has to be satisfied, in order for a system to obey the laws of classical probability and statistical physics. Many experiments were already implemented in various fields of psychology to test for quantum (Bell-type) contextuality, and ascertain, which QP model is appropriate to describe measurements on psychological random variables. Such experiments were first performed by Asano et al. (2014), Conte et al. (2008), followed by more complex setups in Cervantes and Dzhafarov (2019), Cervantes and Dzhafarov (2018) and Basieva et al. (2019).

In this review we will focus on depicting the most notable deviations from rational information processing and its impacts on asset trading. We will also provide a brief tutorial on distinctions between CP and QP and finish off with a review of recent experimental work that is aimed at quantifying contextuality in psychological data. The remainder of this survey is structured as follows: in the next Sect. 2, we present a non-technical introduction to review the basics of CP and QP calculus. In Sect. 3 we look at the core information processing biases explored in behavioural finance studies and analyse more deeply their probabilistic underpinnings. We look at the mechanism behind the emergence of "myopia" in return evaluations as well as evaluate contextuality of agents' attitude to ambiguous events in investment choice and managerial decision making. In Sect. 3.3 we review some of the recent experimental studies on different combinations of psychological random variables aimed at testing the existence of quantum contextuality in human reasoning. Finally, in Sect. 4, we conclude and consider some possible future venues of research in the domain of applications of QP to information processing in behavioural finance.

2   Classical Versus Quantum Probability

In this section we provide a short non-technical introduction to classical probability (CP) and quantum probability (QP) calculus.

CP was mathematically formalized by Kolmogorov (1933, 1956). CP is a calculus of probability measures where a non-negative weight $p(A)$ is assigned to any event A. Probability measures are additive to unity. Given two disjoint events A_1, A_2, the probability of a disjunction of these events equals to the sum of their probabilities:

$$P(A_1 \vee A_2) = P(A_1) + P(A_2).$$

Quantum probability (QP) is a calculus of complex amplitudes. The main distinction is that instead of operations on probability measures, one operates with vectors in a complex (Hilbert) space. Hence, one can allude to QP as a *vector model of probabilistic reasoning*. Each complex amplitude ψ can be equated to classical probability through application of the Born's rule: *Probability is obtained as the square of the absolute value of the complex amplitude.*

$$p = |\psi|^2.$$

This distinction in computation of probability allows, amongst other, to relax the additivity constraint on probability measures from CP.

2.1 Interference of Probabilities

By operating with complex probability amplitudes, instead of direct operations, such as addition and multiplication of probabilities, QP is relaxing the constraints imposed by the rules of classical CP. For instance, for two disjoint events, the probability of their disjunction can become strictly smaller, or larger than the sum of their probabilities. For CP calculus, this inequality leads to non-satisfaction of the "Formula of total probability" (FTP).

$$P(A_1 \vee A_2) < P(A_1) + P(A_2)$$

or

$$P(A_1 \vee A_2) > P(A_1) + P(A_2),$$

QP calculus leads to a more general formula aimed at quantifying sub- or super-additivity of probabilities.

$$P(A_1 \vee A_2) = P(A_1) + P(A_2) + 2\cos\theta\sqrt{P(A_1)P(A_2)}. \tag{1}$$

The additional term on the right-hand side is known as the *interference term*. We recall that interference is a basic feature of waves, hence we can speak about constructive or destructive interference of probability waves.

2.2 Non-Bayesian Probability Inference

We remark that the Bayesian approach to conditional probability is central in the CP modelling of belief formation and consequential reasoning. FTP is a manifestation of the agents' ability to employ consequential reasoning across future states of the world. The additivity of CP and the Bayes formula in the definition of conditional probability are key for FTP satisfaction, namely,

$$P(B|A) = \frac{P(B \cap A)}{P(A)}, \quad P(A) > 0. \tag{2}$$

Consider the pair, a and b, of discrete classical random variables. Then

$$P(b = \beta) = \sum_\alpha P(a = \alpha)P(b = \beta|a = \alpha). \tag{3}$$

Thus b-probability distribution can be calculated from the a-probability distribution and the conditional probabilities $P(b = \beta | a = \alpha)$. From Eq. 3 we can see that non-satisfaction of FTP stems from the non-additivity and non-commutativity of probabilistic beliefs and preferences.

In this vein, QP-modeling can be considered as an extension of the boundaries of Bayesian probability inference. In QP, formula (1) leads to the perturbation of classical FTP, (3) as FTP with the interference term:

$$P(b = \beta) = \sum_{\alpha} P(a = \alpha) P(b = \beta | a = \alpha) \tag{4}$$

$$+2 \sum_{\alpha_1 < \alpha_2} \cos \theta_{\alpha_1 \alpha_2} \sqrt{P(a = \alpha_1) P(b = \beta | a = \alpha_1) P(a = \alpha_2) P(b = \beta | a = \alpha_2)}$$

The interference term can serve as a parameter aimed at quantifying the deviation of beliefs from objective probabilities (prior and posterior) from CP.

Due to its more general axiomatics, QP has gained recognition in the modelling of contextual beliefs and preferences in various fields of decision making.

3 Agents' (contextual) Investment Behaviour and Deviations from Rational Expectations

In this section we will review the most prominent information processing fallacies (from the viewpoint of CP and Bayesian updating) and the resulting impact upon the agents' investment behaviour. We will discuss a fundamental deviation from rational information processing, known as myopia that gives raise to the violation of the conjunction rule. This rule is central to the axiomatics of the Expected Utility representation of preferences. We also discuss how non- additivity of beliefs can emerge under informational ambiguity that is often present in financial markets, but also in managerial decision making. Preferences for ambiguous stocks as well as financial projects can be also of a contextual nature. Another important aspect of beliefs is that ambiguity may be present over time and no predetermined probability distributions exist in an agent's decision-making state. The same concept of "indeterminacy" can apply to choices for asset allocation, investment projects, voting, see monographs, Busemeyer and Bruza (2012), Haven and Khrennikov (2013) and reviews in Pothos and Busemeyer (2013), Haven et al. (2018), Haven and Sozzo (2016).

3.1 Myopia and Investment Behaviour

As shown in an array of behavioural finance studies, investors have a tendency to evaluate market outcomes frequently rather than thinking about the long-term horizon of returns on a risky asset. Under standard EUT (Expected Utility Theory) the length of evaluation period would not affect market prices since the expected utility would be constant for gains and losses. In the pioneering works by Merton and Samuelson (see review in Benartzi and Thaler 1995) it was shown that under certain assumptions about the behaviour of the financial market (random walk, power utility and other technical assumptions) the evaluation horizon, and consequently investment horizon, would make no difference in the allocation to risky funds for a risk averse investor. Yet in real world investment choices as well as in market experiments, the allocation of funds to risky assets is fluctuating among investors. Some of the main drivers of these fluctuations are contextual beliefs and contextual preferences that are further affected by informational ambiguity. The deviations from rational information processing can be considered as a manifestation of state dependence in agents' belief formation that affects their trading. Shiller (2014) notes that 'overreaction' does not necessarily mean that new information about fundamentals is released, and can manifest itself in over-optimism and deviation from Bayesian updates, affected by past trends in asset prices. At the same time, it was documented in Benartzi and Thaler (1995) that an investor holding contextual (myopic) beliefs about the price behaviour of a risky asset coupled with fluctuations in her risk attitude, will be unwilling to invest in the risky asset for a shorter investment period. More specifically the agent would treat future investment periods as *complementary to each other* in the process of her belief formation, see experimental findings on myopia coupled with loss aversion and a detailed exploration in Gneezy et al. (2003) and Langer and Weber (2001).

3.1.1 Probabilistic Account of Myopia

When a decision maker acts in a myopic manner the evaluation of each single investment horizon takes place in an incompatible mode. The operators $A_1...A_n$ are non-commuting and hence, the sequence of risky investment payoffs is not evaluated jointly by following the reduction axiom that lies at the core of expected utility theory, Von Neumann and Morgenstern (1944). The reduction axiom applies to compound lotteries, which are lotteries consisting of lotteries rather than a probabilistic distribution of outcomes. As such, these lotteries contain a "double layer of risk" for the decision maker. A sequence of risky or uncertain investments can be treated as such lottery. In the simplest set-up with two dichotomous lotteries, $L_a = (x_1, p_a; x_2, (1 - p_a))$ and $L_b = (y_1, p_b; y_2(1 - p_b))$ a compound lottery is: $L = (L_a, p; L_2, 1 - p)$. It can be reduced via rules of CP through a formation of a joint distribution of these nested lotteries, given the overall probabilities of these lotteries $(p, (1 - p))$. An equivalent single (reduced) lottery will have the form: $L = ((x_1, p_a; x_2, (1 - p_a), p; (y_1, p_b; y_2(1 - p_b), (1 - p))$.

Myopia, or narrow bracketing is a clear manifestation of the incompatibility of different events in the belief state of an agent. As noted by Busemeyer and Bruza, Busemeyer and Bruza (2012), (2012:141) *"Compatible representations may only become available after a person gains sufficient experience with joint observations of the variables to form a good estimate of the joint probabilities."* Hence, one can draw here a clear parallel with related empirical findings on the human tendency to make sequential judgements of events, as well as to exhibit order effects in a wide range of preferences, which are not jointly predetermined in a decision maker's cognitive state, but can be well modelled through non-commuting operators in a QP framework, cf. Kahneman (2003), Trueblood and Busemeyer (2011), Wang and Busemeyer (2013) and Khrennikov et al. (2014).

3.1.2   Complementarity of Beliefs About Prices of Financial Assets

Myopia can exist for some events that take place across a time horizon, but also for some concurrent events. When a decision making task attains additional complexity, for instance, when the assets are opaque and the investor is not able to form a joint evaluation of their return distribution, an evaluation of these assets' performance in a portfolio setting can become problematic. A joint evaluation of assets' performance within portfolios, as well as their return correlation lies at the heart of the mean-variance approach of standard portfolio allocation, due to Modern Portfolio Theory by Markowitz, see e.g., introduction in Bodie et al. (2014).

Given that the agents hold contextual beliefs about portfolio returns, QP can account for the complementarity of beliefs about these assets. In an example of two risky assets, A and B, an agent is uncertain about the price dynamics of these assets and does not possess a *joint probability evaluation* over their returns. The agent is in a *superposition state* and interference effects exist with respect to the beliefs about the price realization of these assets. If an agent evaluates the future returns of the assets sequentially, the contextuality of the evaluation order affects her final beliefs about the return distribution.

An investor needs to form beliefs about price dynamics $\alpha = \pm 1$ of the first asset A. Mathematically, such a belief elicitation is given as a projection of agent's belief states ψ onto the eigenvector $|\alpha_i\rangle$ that corresponds to an eigenstate for a particular price realization for the considered asset.[3] After the agent has formed a belief about the price realization of the asset A, she forms a belief about the possible price behaviour of the complementary asset B. Now the agent is in a different (updated) belief state $|+_i\rangle$ and her state transition with respect to the price behaviour of that asset B with eigenvalues $\beta = \pm 1$ is given by a Born rule, with the transition probabilities:

$$p_{A \to B}(\alpha \to \beta) = |\langle \alpha_A | \beta_B \rangle|^2. \tag{5}$$

---

[3]For simplicity we devise only two eigenvectors $|\alpha_+\rangle$ and $|\alpha_-\rangle$, corresponding to eigenvalues $a = \pm 1$. These projectors can be for instance beliefs about a price increase, or price decrease in the next trading period.

The above exposition of state transition allows to obtain quantum transition probabilities that denote agents' beliefs with respect to the asset B price distribution, when she firstly observes her beliefs on the price realization of an asset A. Hence, the computed probabilities via the Born rule can be interpreted as subjective probabilities, or an agent's degree of belief about the distribution of asset prices. Transition probabilities also have an objective (frequentist) interpretation. Considering an ensemble of agents in the same state ψ, who form a firm belief α, with respect to the price behaviour of the asset A, we could observe the frequency of those agents who are fully confident that the price of A will go up, or down. As a next step, the agents form preferences about asset's B price distribution.[4] It is possible to find the frequency (approximated by probability) $p_{A \to B}(\alpha \to \beta)$. To contrast with CP models, we can consider quantum probabilities as analogues of the conditional probabilities obtained in the Bayesian update, $p_{A \to B}(\alpha \to \beta) \equiv p_{B|A}(\beta|\alpha)$. We remark that belief formation about asset prices in this setup, takes place under informational ambiguity (subjective risk) that motivates the modelling of beliefs as superposition states.

Given the probabilities, in (5) we can define a quantum joint probability distribution for forming beliefs about both of the two assets A and B.

$$p_{AB}(\alpha, \beta) = p_A(\alpha) p_{B|A}(\beta|\alpha). \tag{6}$$

This joint probability does not obey commutativity of CP and respects the order of observations, where:

$$p_{AB}(\alpha, \beta) \neq p_{BA}(\beta, \alpha), \tag{7}$$

We stress that sequential information processing is a manifestation of order effects, or contextuality in belief formation that is not in accord with the classical Bayesian probability update, see e.g., Pothos and Busemeyer (2013), Wang and Busemeyer (2013), Khrennikov et al. (2014). The order effect implies a non-satisfaction of the joint probability distribution and brings a violation of the commutativity principle that is central to classical probability theory, Kolmogorov (1956).

To sum up, in the setting of narrow framing and sequential information processing, when risks and ambiguity are present, the QP framework aids to depict the agents' non-definite opinions about the price behaviour for the *complementary assets* that she holds. Non-classical information processing can take place in the presence of a vague probabilistic composition of the future price state realizations of the set of traded assets. In the case of such assets, an agent forms her beliefs *sequentially*, and not jointly as is the case in the standard finance frameworks. QP allows to describe subjective belief formation of a representative agent, by exploring the 'bets', or price observations of an ensemble of agents and approximate these frequencies by probabilities, see some studies that combine experimental statistics in decision making with a QP calculus, Aerts et al. (2018), Haven and Sozzo (2016), Haven et al. (2018) and Haven and Khrennikova (2018).

---

[4]Before the agents have elicited their beliefs that inform their trading behaviour, they are in a superposition state and their beliefs' distribution is not classical.

3.2 Ambiguous Preference States: Non-neutral Ambiguity Attitude Of investors

In everyday investment decisions, agents are confronted with situations where known lottery-type risks are not given to them and subjective beliefs have to be formed. We note that standard finance frameworks operate under the assumptions that agents are forming homogeneous subjective beliefs following the rules of CP theory. This is the essence of preference formation under Subjective Expected Utility Theory (SEUT) conceived by Savage (1954). The subjective beliefs of agents are not observable per se, yet these can be elicited from their preferences, or trading behaviour. However, empirical evidence in behavioural finance studies and psychological experiments showed that agents usually dislike ambiguity and prefer known risks. This behaviour is known as "Ellsberg-type" ambiguity aversion, Ellsberg (1961). There is a stream of asset trading models that utilize the notion or ambiguity aversion. Most known frameworks build upon "Maxmin Expected Utility" (MEU) that models agents prior beliefs as a set of priors, whereby expected utility with respect to the most "pessimistic forecast" is maximized, see for instance Epstein and Schneider (2008). MEU due to Gilboa and Schmeidler, Gilboa (1989), is resting upon a convex transformation of probability measures to capture non-additive probabilities. Implications of non-neutral ambiguity attitudes for market outcomes were also extensively studied in the finance literature. The work by Mukerji and Tallan (2001) shows that agents' ambiguity aversion with respect to asset specific risks brings forth under-pricing and results in an existence of a negative welfare effect for investors. The ambiguity attitude is also widely researched in experimental studies, see e.g., Sarin and Weber (1993), Ho et al. (2002) and some studies on non-consequential preferences under ambiguous risks, Shafir (1994), Aerts et al. (2018). At the same time, investors can also possess, (i) a heterogeneous attitude towards ambiguity, (ii) state dependent shifts in their attitude towards some kinds of uncertainties. The study by Roca et al. (2006) detects a more rare phenomenon of 'ambiguity seeking', as a result of agents shifting reference points due to previously experienced gains or losses. Finally, the study by Ho et al. (2002) shows that ambiguity preferences can be enhanced when compared with high probability risks of a loss or low probability risks of a gain. There are also other contextual factors that can come into play, showing that the agents' ambiguity attitude is not stable and can deviate from the CP rules of information processing.

3.2.1 QP Contributions to Ambiguity Sensitive Behaviour in Economic and Finance Decisions

In a framework elaborated in Khrennikova (2016), the investors' ambiguous beliefs can affect their trading preferences, giving raise to long term deviations of asset prices from their fundamental values. The dynamics of assets prices from agents' beliefs are modelled with the aid of a quantum dynamical framework (quantum Markovian dynamics). Prices stabilize to (QP) induced equilibrium that captures the agents' resolution of ambiguity with respect to asset values.[5] Ambiguity attitudes of agents are affecting over-time the price realization of financial assets. Given the findings of behavioural finance, the QP dynamical framework aims to show the mechanism of bubble formation from the contextual beliefs of agents. The idea of investors' non-classical beliefs under ambiguity is further developed in Khrennikova and Patra (2018) whereby the agents can be categorized according to the type of their ambiguity attitude. In line with some earlier works on asset trading under heterogeneous beliefs, agents can behave over-optimistically under informational ambiguity, or show ambiguity aversion through pessimistic beliefs about asset prices.[6] In a QP based framework of asset trading devised in Khrennikova and Patra (2018), agents hold non classical prior beliefs, whereby heterogeneity in their ambiguous beliefs is captured through different ψ vectors (they are given with respect to different bases, where the parameter, $\theta_1 - \theta_2$, provides a measure of divergence in ambiguous beliefs). Asset trading takes place, as agents get information about some fundamentals, such as dividends. Upcoming information interferes with agents' prior ambiguous beliefs, advancing contextual trading behaviour that produces over-pricing and in other periods leads to under-pricing of financial assets.[7]

In a similar vein, the Ellsberg and Machina paradox-type behaviour from context dependence and ambiguous beliefs is explained in Haven and Sozzo (2016) through positive and negative interference effects. An ambiguity sensitive probability weighting function is derived in Asano et al. (2017) with an special parameter capturing the attitude towards risk from an interference term λ.

We can witness that QP provides an advantage in considering ambiguity sensitive beliefs as well as provide a unified framework to quantify the deviations from rational Savage-type, Savage (1954) subjective beliefs of economic agents.

3.3 Quantum-Type Contextuality of Human Preferences: Entanglement and Non-separable States

There are two degrees of contextuality in the distributions of random variables that are measured in decision making, judgements and secondary statistics. As noted by Cervantes and Dzhafarov (2019) one form of context impact is *information carrying*, characterized by a disturbance of the random variable realization by the preceding context(s).[8]

---

[5] We note that the prices reflect agents' beliefs about fundamentals and not the objective information that the prices would contain. Given that agents' beliefs are contextual, these prices can deviate from the fundamental values.

[6] See for instance the framework by Scheinkman and Xiong (2003) and references herein.

[7] Under some assumptions on agents' risk neutrality, short sales etc, we can align the QP framework with CP based frameworks on asset trading under divergence of beliefs.

[8] In physics such a contextual disturbance on a system is known as "signalling".

In the absence of contextual disturbance the probability distribution for realization of a random variable X_1 (a question, a price observation, etc) would be independent from the context of its measurement, for instance, a previous measurement on a random variable Y_1. It is very difficult to control for signalling in psychological observations, but also in the original quantum physical experiments. A most well-known form of such contextuality in decision making and judgement is alluded to as "order effect". This effect persists in the collected statistics, violating the joint distribution of CP measure, see, e.g., Busemeyer and Bruza (2012), Wang and Busemeyer (2013). Another type of violations, known as violation of Bell inequality (and its various types), is due to correlations between pairwise measured observables being too strong to be described by CP. One specific form of Bell-type inequality is known as CHSH-inequality, derived by Clauser et al. (1969). This inequality was tested for psychological statistics with some dichotomous outcomes, in order to ascertain if some decision making observables obey the CHSH threshold for linear combination of correlations between the outcomes.[9] QP can serve as remedy for a contextual representation of the observed statistics and provide a mechanism to model non-jointly distributed variables through non-commuting observables. These are given by operators that do share the same bases of vectors and hence, yield different squared amplitudes that are translated into contextual probabilities. Contextually of measurements is also characterized by another important feature known as *entanglement*. Loosely speaking, the outcomes of these measurements are exhibiting very strong correlations, and the results of measurements are "non-separable", in the sense that the measurement on some state ψ_1 affects another state ψ_2. In social science, the notion of non-separability of preferences and judgements is also actively explored in politics, economics and behavioural finance. Recent contributions utilizing quantum contextuality to model such preferences borrow from the operational representation of non-separability from quantum physics to model the non-classical interdependence of some random variables (with some examples, such as realization of price states, voting preferences, concept combinations, questions, etc., see Aerts and Sozzo (2014), Busemeyer and Bruza (2012), Haven and Khrennikov (2013), Khrennikova and Haven (2016), and Khrennikova (2016).

---

[9]Cognitive experiments are performed on two pairs of questions related to measurement of four random variables. Some of the questions cannot be answered jointly, showing contextuality of human judgements, see first experiments by Asano et al. (2014), Conte et al. (2008). These experiments were followed by other controlled experiments, including different combinations of random variables and their outcomes, Cervantes and Dzhafarov (2018, 2019). A recent experiment by Basieva et al. (2019) offers evidence that there is contextuality in human judgements that cannot be captured with the aid of CP calculus.

4   Concluding Remarks

We presented a short summary of the advances of QP in modelling contextual beliefs and preferences of economic agents. Given the wide range of revealed behavioural anomalies in investment that are often associated with non-classical information processing by investors and a state dependence in their trading preferences, QP can serve as a complete tool to incorporate agents' belief indeterminacy that feeds into the price dynamics of the traded assets. We should also note that QP based dynamical models can be applied to decision theory and asset pricing as a well-grounded complement to the classical stochastic calculus in capturing the evolution of some financial random variables. We can point to Schrödinger's equation that describes the state evolution, as well as its deterministic counterpart adopted from Bohmian interpretation of quantum physics, Haven (2002) and a review in Haven and Khrennikov (2013).

The main motivation for application of the QP mathematical framework as a mechanism of probability calculus under non-neutral ambiguity attitudes among agents, coupled with a state dependence of their evolution, is its ability to mathematically quantify ambiguity perception and provide an interdependence of beliefs. Finally, we can ascertain that QP calculus provides a complete operational model for contextual (prior) beliefs as well as a non- Bayesian update of these beliefs. Non CP information processing is formalized via state update, where the Born rule allows to compute a posterior probability, given a sequence of state updates that violate the conjunction rule of CP.

We also aim to motivate for a broader application of QP to contextual preferences (rather than using any generalization of CP) given the growing body of studies that explore quantum contextuality in cognitive data.

We hope that the growing body of experimental studies in cognition and decision making will further motivate the application of QP as a theory of subjective belief formation. We are aware that QP is still perceived in economics and finance as a nascent tool to model beliefs and preferences, hence additional controlled and field experiments are essential to "refine" the contexts, in which QP can serve as complete descriptive and also predictive framework of agents' contextual behaviour.

References

Aerts, D., Haven, E., & Sozzo, S. (2018). A proposal to extend expected utility in a quantum probabilistic framework. *Economic Theory, 65*, 1079–1109.

Aerts, D., & Sozzo, S. (2014). Quantum entanglement in concept combinations. *International Journal of Theoretical Physics, 53*, 3587–3603.

Allais, M. (1953). Le comportement de l'homme rationnel devant le risque: critique des postulats et axiomes de l'cole américaine. *Econometrica, 21*, 503–536.

Asano, M., Basieva, I., Khrennikov, A., Ohya, M., & Tanaka, Y. (2017). A quantum-like model of selection behavior. *Journal of Mathematical Psychology, 78*, 2–12.

Asano, M., Hashimoto, T., Khrennikov, A., Ohya, M., & Tanaka, T. (2014). Violation of contextual generalisation of the Legget-Garg inequality for recognition of ambiguous figures. *Physica Scripta, T163*, 014006.

Bagarello, F. (2007). Stock markets and quantum dynamics: a second quantized description. *Physica A, 386*, 283–302.

Basieva, I., Cervantes, V.H., Dzhafarov, E.N., Khrennikov, A. (2019) True contextuality beats direct influences in human decision making. Journal of Experimental Psychology: General. (forthcoming).

Benartzi, S., & Thaler, R. H. (1995). Myopic loss aversion and the equity premium puzzle. *Quarterly Journal of Economics, 110*(1), 73–92.

Bodie, Z., Kane, A., & Marcus, A. J. (2014). *Investments*. UK: McGrawHill Education.

Busemeyer, J., & Bruza, P. (2012). *Quantum models of cognition and decision*. Cambridge University Press.

Busemeyer, J. R., Wang, Z., & Townsend, J. T. (2006). Quantum dynamics of human decision making. *Journal of Mathematical Psychology, 50*, 220–241.

Cervantes, V. H., & Dzhafarov, E. N. (2018). Snow Queen is evil and beautiful: Experimental evidence for probabilistic contextuality in human choices. *Decision, 5*, 193–204.

Cervantes, V. H., & Dzhafarov, E. N. (2019). True contextuality in a psychological experiment. *Journal of Mathematical Psychology, 91*, 119–127.

Clauser, J. F., Horne, M. A., Shimony, A., & Holt, R. A. (1969). Proposed experiment to test local hidden-variable theories. *Physical Review Letters, 23*, 880–884.

Conte, E., Khrennikov, A., Todarello, O., & Federici, A. (2008). A preliminary experimental verification on the possibility of Bell inequality violation in mental states. *Neuroquantology, 6*(3), 214–221.

Ellsberg, D. (1961). Risk, ambiguity and the Savage axioms. *Quarterly Journal of Economics, 75*, 643–669.

Epstein, L. G., & Schneider, M. (2008). Ambiguity, information quality and asset pricing. *Journal of Finance, LXII, 1*, 197–228.

Gilboa, I. (2009) Theory of decision under uncertainty. Econometric Society Monographs.

Gilboa, I., & Schmeidler, D. (1989). Maxmin expected utility with non-unique prior. *Journal of Mathematical Economics, 18*(14118), 141–153.

Gneezy, U., Kapteyn, A., & Potters, J. (2003). Evaluation periods and asset prices in a market experiment. *Journal of Finance, LVIII, 2*, 821–837.

Haven, E., Khrennikova, P. (2018) A quantum probabilistic paradigm: Non-consequential reasoning and state dependence in investment choice. *Journal of Mathematical Economics*. https://doi.org/10.1016/j.jmateco.2018.04.003.

Haven, E. (2002). A discussion on embedding the Black-Scholes option pricing model in a quantum physics setting. *Physica A, 304*, 507–524.

Haven, E., & Khrennikov, A. (2013). *Quantum social science*. Cambridge: Cambridge University Press.

Haven, E., Khrennikov, A., Ma, C., & Sozzo, S. (2018). Introduction to quantum probability theory and its economic applications. *Journal of Mathematical Economics, 78*, 127130.

Haven, E., & Sozzo, S. (2016). A generalized probability framework to model economic agents' decisions under uncertainty. *International Review of Financial Analysis, 47*, 297–303.

Ho., J., Keller, L.R., Keltyka, P.,. (2002). Effects of outcome and probabilistic ambiguity on managerial choices. *Journal of Risk and Uncertainty, 24*, 47–74.

Kahneman, D. (2003). Maps of bounded rationality: psychology for behavioral economics. *American Economic Review, 93*(5), 1449–1475.

Kahneman, D., & Tversky, A. (1972). Subjective probability: A judgement of representativeness. *Cognitive Psychology, 3*(3), 430–454.

Kahneman, D., & Tversky, A. (1979). Prospect theory: An analysis of decision under risk. *Econometrica, 47*, 263–291.

Khrennikov, A. (2010). *Ubiquitous quantum structure*. Springer.

Khrennikova, P. (2016). Application of quantum master equation for long-term prognosis of asset-prices. *Physica A, 450,* 253–263.

Khrennikova, P., & Haven, E. (2016). Instability of political preferences and the role of mass-media: a dynamical representation in a quantum framework. *Philosophical Transactions of the Royal Society A., 374,* 20150106.

Khrennikova, P., & Patra, S. (2018). Asset trading under non-classical ambiguity and heterogeneous beliefs. *Physica A, 521,* 562–577.

Khrennikov, A., Basieva, I., Dzhafarov, E. N., & Busemeyer, J. R. (2014). Quantum models for psychological measurements: An unsolved problem. *PLoS ONE, 9,* e110909.

Kolmogorov, A. N. (1933) Grundbegriffe der Warscheinlichkeitsrechnung. Berlin: Springer [(1956) *Foundations of the probability theory*. New York: Chelsea Publishing Company].

Langer, T., & Weber, M. (2001). Prospect theory, mental accounting, and the differences in aggregated and segregated evaluation of lottery portfolios. *Management Science, 47*(5), 716–733.

Mukerji, S., & Tallan, J. M. (2001). Ambiguity aversion and incompleteness of financial markets. *Review of Economic Studies, 68,* 883–904.

Pothos, M. E., & Busemeyer, J. R. (2009). A quantum probability explanation for violations of rational decision theory. *Proceedings of the Royal Society B, 276*(1665), 2171–2178.

Pothos, E. M., & Busemeyer, J. R. (2013). Can quantum probability provide a new direction for cognitive modeling? *Behavioral and Brain Sciences, 36*(3), 255–274.

Roca, M., Hogarth, R. M., & Maule, A. J. (2006). Ambiguity seeking as a result of the status quo bias. *Journal of Risk and Uncertainty, 32,* 175–194.

Sarin, R. K., & Weber, M. (1993). Effects of ambiguity in market experiments. *Management Science, 39,* 602–615.

Savage, L. J. (1954). *The foundations of statistics*. US: John Wiley & Sons Inc.

Scheinkman, J., & Xiong, W. (2003). Overconfidence and speculative bubbles. *Journal of Political Economy, 111,* 1183–1219.

Schemeidler, D. (1989). Subjective probability and expected utility without additivity. *Econometrica, 57*(3), 571–587.

Shafir, E. (1994). Uncertainty and the difficulty of thinking through disjunctions. *Cognition, 49,* 11–36.

Shiller, R. (2014). Speculative asset prices. *American Economic Review, 104*(6), 1486–1517.

Shubik, M. (1999). Quantum economics, uncertainty and the optimal grid size. *Economics Letters, 64*(3), 277–278.

Thaler, R. H., & Johnson, E. J. (1990). Gambling with the house money and trying to break even: the effects of prior outcomes on risky choice. *Management Science, 36*(6), 643–660.

Trueblood, J. S., & Busemeyer, J. R. (2011). A quantum probability account of order effects in inference. *Cognitive Science, 35,* 1518–1552.

Tversky, D., & Kahneman, D. (1992). Advances in prospect theory: cumulative representation of uncertainty. *Journal of Risk and Uncertainty, 5,* 297–323.

Von Neumann, J., & Morgenstern, O. (1944). *Theory of games and economic behaviour*. Princeton, NJ: Princeton University Press.

Wang, Z., & Busemeyer, J. R. (2013). A quantum question order model supported by empirical tests of an a priori and precise prediction. *Topics in Cognitive Science, 5,* 689–710.

How to Make a Decision Based on the Minimum Bayes Factor (MBF): Explanation of the Jeffreys Scale

Olga Kosheleva, Vladik Kreinovich, Nguyen Duc Trung, and Kittawit Autchariyapanitkul

Abstract In many practical situations, we need to select a model based on the data. It is, at present, practically a consensus that the traditional p-value-based techniques for such selection often do not lead to adequate results. One of the most widely used alternative model selection techniques is the Minimum Bayes Factor (MBF) approach, in which a model is preferred if the corresponding Bayes factor—the ratio of likelihoods corresponding to this model and to the competing model—is sufficiently large for all possible prior distributions. Based on the MBF values, we can decide how strong is the evidence in support of the selected model: weak, strong, very strong, or decisive. The corresponding strength levels are based on a heuristic scale proposed by Harold Jeffreys, one of the pioneers of the Bayes approach to statistics. In this paper, we propose a justification for this scale.

1 Formulation of the Problem

Why Minimum Bayes Factor. In many practical situations, we have several possible models M_i of the corresponding phenomena, and we would like to decide, based on the data D, which of these models is more adequate. To select the most appropriate model, statistics textbooks used to recommend techniques based on p-values. However, at present, it is practically a consensus in the statistics community that

O. Kosheleva · V. Kreinovich (✉)
University of Texas at El Paso, El Paso, TX 79968, USA
e-mail: vladik@utep.edu

O. Kosheleva
e-mail: olgak@utep.edu

N. D. Trung
Banking University HCMC, Ho Chi Minh City, Vietnam
e-mail: trungnd@buh.edu.vn

K. Autchariyapanitkul
Maejo University, Maejo, Thailand
e-mail: kittar3@hotmail.com

N. Ngoc Thach et al. (eds.), *Data Science for Financial Econometrics*,
Studies in Computational Intelligence 898,
https://doi.org/10.1007/978-3-030-48853-6_8

the use of p-values often results in misleading conclusions; see, e.g., Nguyen (2016, 2019), Wasserstein and Lazar (2016).

To make a more adequate selection, it is important to take prior information into account, i.e., to use Bayesian methods. It is reasonable to say that the model M_1 is more probable than the model M_2 if the likelihood $P(D \mid M_1)$ of getting the data D under the model M_1 is larger than the likelihood $P(D \mid M_2)$ of getting the data D under the model M_2, i.e., if the *Bayes factor*

$$K \stackrel{\text{def}}{=} \frac{P(D \mid M_1)}{P(D \mid M_2)}$$

exceeds 1. Of course, if the value is only slightly larger than 1, this difference may be caused by the randomness of the corresponding data sample. So, in reality, each of the two models can be more adequate. To make a definite conclusion, we need to make sure that the Bayes factor is sufficiently large—and the larger the factor K, the more confident we are that the model M_1 is indeed more adequate.

The numerical value of the Bayes factor K depends on the prior distribution π: $K = K(\pi)$. In practice, we often do not have enough information to select a single prior distribution. A more realistic description of the expert's prior knowledge is that we have a *family* F of possible prior distributions π. In such a situation, we can conclude that the model M_1 is more adequate than the model M_2 if/ the corresponding Bayes factor is sufficiently large for all possible prior distributions $\pi \in F$, i.e., equivalently, that the *Minimum Bayes Factor*

$$\text{MBF} \stackrel{\text{def}}{=} \min_{\pi \in F} K(\pi)$$

is sufficiently large; see, e.g., Nguyen (2019), Page and Satake (2017).

Jeffreys scale. In practical applications of Minimum Bayes Factor, the following scale is usually used; this scale was originally proposed in Jeffreys (1989):

- when the value of MBF is between 1 and 3, we say that the evidence for the model M_1 is barely worth mentioning;
- when the value of MBF is between 3 and 10, we say that the evidence for the model M_1 is substantial;
- when the value of MBF is between 10 and 30, we say that the evidence for the model M_1 is strong;
- when the value of MBF is between 30 and 100, we say that the evidence for the model M_1 is very strong;
- finally, when the value of MBF is larger than 100, we say that the evidence for the model M_1 is decisive.

Remaining problem and what we do in this paper. Jeffreys scale has been effectively used, so it seems to be adequate, but why? Why do we select, e.g., 1 to 3 and not 1 to 2 and 1 to 5?

In this paper, we provide a possible explanation for the success of Jeffreys scale. This explanation is based on a general explanation of the half-order-of-magnitude scales provided in Hobbs and Kreinovich (2006).

2 Our Explanation

Towards the precise formulation of the problem. A scale means, crudely speaking, that instead of considering all possible values of the MBF, we consider discretely many values

$$\ldots < x_0 < x_1 < x_2 < \ldots$$

corresponding to different levels of strength. Every actual value x is then approximated by one of these values $x_i \approx x$.

What is the probability distribution of the resulting approximation error $\Delta x \overset{\text{def}}{=} x_i - x$? This error is caused by many different factors. It is known that under certain reasonable conditions, an error caused by many different factors is distributed according to Gaussian (normal) distribution (see, e.g., Sheskin (2011); this result—called the *Central Limit Theorem*—is one of the reasons why Gaussian distributions are ubiquitous). It is therefore reasonable to assume that Δx is normally distributed.

It is known that a normal distribution is uniquely determined by its two parameters: its average μ and its standard deviation σ. For situations in which the approximating value is x_i, let us denote:

- the mean value of the approximation error Δx by Δ_i, and
- the standard deviation of the approximation error by σ_i.

Thus, when the approximate value is x_i, the actual value $x = x_i - \Delta x$ is distributed according to the Gaussian distribution, with the mean $x_i - \Delta_i$ (which we will denote by $\widetilde{x}_i$), and the standard deviation σ_i.

For a Gaussian distribution with mean μ and standard deviation σ, the probability density is everywhere positive, so theoretically, we can have values which are as far away from the mean value μ as possible. In practice, however, the probabilities of large deviations from μ are so small that the possibility of such deviations can be safely ignored. For example, it is known that the probability of having the value outside the "three sigma" interval $[\mu - 3\sigma, \mu + 3\sigma]$ is $\approx 0.1\%$ and therefore, in most applications in science and engineering, it is assumed that values outside this interval are impossible.

There are some applications where we cannot make this assumption. For example, in designing computer chips, when we have millions of elements on the chip, allowing 0.1% of these elements to malfunction would mean that at any given time, thousands of elements malfunction and thus, the chip would malfunction as well. For such critical applications, we want the probability of deviation to be much smaller than 0.1%, e.g., $\leq 10^{-8}$. Such small probabilities (which practically exclude any possibility of an error) can be guaranteed if we use a "six sigma" interval $[\mu - 6\sigma, \mu + 6\sigma]$. For this interval, the probability for a normally distributed variable to be outside it is indeed $\approx 10^{-8}$.

In accordance with the above idea, for each x_i, if the actual value x is within the "three sigma" range $I_i = [\widetilde{x}_i - 3\sigma_i, \widetilde{x}_i + 3\sigma_i]$, then it is reasonable to take x_i as the corresponding approximation.

What should be the standard deviation σ_i of the approximation error? We are talking about a very crude approximation, when, e.g., all the values from 1 to 3 are assigned the same level. Thus, the approximation error has to be reasonably large. The only limitation on the approximation error is that we want to make sure that all values that we are covering are indeed non-negative, i.e., that for every i, even the extended "six sigma" interval $[\widetilde{x}_i - 6\sigma_i, \widetilde{x}_i + 6\sigma_i]$ only contains non-negative values. Other than that, there should not be any other limitations on the approximation error—i.e., the value σ_i should be the largest for which the above property holds.

We want to cover all possible values x, so that each positive real number x be covered by one of the intervals I_i. In other words, we want the union of all these intervals to coincide with the set of all positive real numbers. We also want to make sure that to each value x, we assign exactly one strength level, i.e., that the intervals I_i corresponding to different strength levels do not intersect—except maybe at the borderline point.

Thus, we arrive at the following definitions.

Definition 1

- We say that an interval $I = [\mu - 3\sigma, \mu + 3\sigma]$ is *reliably non-negative* if every real number from the interval $[\mu - 6\sigma, \mu + 6\sigma]$ is non-negative.
- We say that an interval $I = [\mu - 3\sigma, \mu + 3\sigma]$ is *realistic* if for the given μ, the corresponding value σ is the largest for which the corresponding interval is reliably non-negative.
- We say that a set of realistic intervals $\{I_i = [\underline{x}_i, \overline{x}_i]\}$ with

$$\ldots \leq \underline{x}_1 \leq \underline{x}_2 \leq \ldots$$

describes strength levels if these intervals form a partition of the set $\mathbb{R}^+$ of all positive real numbers: $\bigcup_i I_i = \mathbb{R}^+$ and for each $i \neq j$, the intersection $I_i \cap I_j$ is either an empty set or a single point.

Proposition 1 *A set of realistic intervals $I_i = [\underline{x}_i, \overline{x}_i]$ describes strength levels if and only if these intervals have the form $[\underline{x}_i, \overline{x}_i] = [3^i \cdot x_0, 3^{i+1} \cdot x_0]$.*

Discussion. In other words, we have intervals

$$[x_0, 3 \cdot x_0], \quad [3 \cdot x_0, 9 \cdot x_0], \quad [9 \cdot x_0, 27 \cdot x_0], \ldots$$

This is (almost) what the Jeffreys scale recommends, with $x_0 = 1$—the only difference is that in the Jeffreys scale, we have 10 instead of 9. Modulo this minor issue, we indeed have an explanation for the empirical success of the Jeffreys scale.

Proof of the Proposition. Each interval

$$I_i = [\underline{x}_i, \overline{x}_i] = [\mu_i - 3\sigma_i, \mu_i + 3\sigma_i]$$

is realistic. This means that when the value μ_i is fixed, the corresponding value σ_i is the largest for which all the numbers from the interval $[\mu_i - 6\sigma_i, \mu_i + 6\sigma_i]$ are non-negative. One can easily see that this largest value corresponds to the case when $\mu_i - 6\sigma_i = 0$, i.e., when $\sigma_i = \dfrac{1}{6} \cdot \mu_i$. For this value σ_i, we have $\underline{x}_i = \mu_i - 3\sigma_i = \dfrac{1}{2} \cdot \mu_i$ and $\overline{x}_i = \mu_i + 3\sigma_i = \dfrac{3}{2} \cdot \mu_i$. Thus, for each realistic interval $I_i = [\underline{x}_i, \overline{x}_i]$, we have

$$\overline{x}_i = 3 \cdot \underline{x}_i.$$

In particular, this is true for $i = 0$, so we have $\overline{x}_0 = 3x_0$, where we denoted $x_0 \overset{\text{def}}{=} \underline{x}_0$. Let us prove, by induction, that for every i, we have $\underline{x}_i = 3^i \cdot x_0$ and $\overline{x}_i = 3^{i+1} \cdot x_0$. Indeed, we have just proved these equalities for $i = 0$, i.e., we have the induction base.

Let us now prove the induction step. Suppose that

$$I_i = [\underline{x}_i, \overline{x}_i] = [3^i \cdot x_0, 3^{i+1} \cdot x_0].$$

The intervals I_i form a partition, so the next interval I_{i+1} intersects with I_i at exactly one point: $\underline{x}_{i+1} = \overline{x}_i = 3^{i+1} \cdot x_0$. Since the interval I_{i+1} is realistic, we have

$$\overline{x}_{i+1} = 3 \cdot \underline{x}_{i+1} = 3^{(i+1)+1} \cdot x_0.$$

The induction step is thus proven, and so is the proposition.

Acknowledgments This work was supported in part by the National Science Foundation grants 1623190 (A Model of Change for Preparing a New Generation for Professional Practice in Computer Science) and HRD-1242122 (Cyber-ShARE Center of Excellence).

References

Hobbs, J., & Kreinovich, V. (2006). Optimal choice of granularity in commonsense estimation: Why half-orders of magnitude. *International Journal of Intelligent Systems, 21*(8), 843–855.

Jeffreys, H. (1989). *Theory of probability*. Oxford: Claredon Press.

Nguyen, H. T. (2016). Why p-values are banned? *Thailand Statistician, 24*(2), i–iv.

Nguyen, H. T. (2019). How to test without p-values? *Thailand Statistician, 17*(2), i–x.

Page, R., & Satake, E. (2017). Beyond p-values and hypothesis testing: using the Minimum Bayes Factor to teach statistical inference in undergraduate introductory statistics courses. *Journal of Education and Learning*, 6(4), 254–266.

Sheskin, D. J. (2011). *Handbook of parametric and nonparametric statistical procedures*. Boca Raton, Florida: Chapman and Hall/CRC.

Wasserstein, R. L., & Lazar, N. A. (2016). The American Statistical Association's statement on p-values: context, process, and purpose. *American Statistician*, 70(2), 129–133.

An Invitation to Quantum Probability Calculus

Hung T. Nguyen, Nguyen Duc Trung, and Nguyen Ngoc Thach

Abstract This paper is about quantum probability for econometricians. The intent is to lay down necessary material on quantum probability *calculus* to develop associated statistics for economic applications, and not for physics or quantum mechanics.

Keywords Behavioral economics · Functional analysis · Hilbert space · Linear operator · Path integral · Quantum entropy · Quantum mechanics · Quantum probability · Quantum probability space · Self adjoint operator · Spectral measure · Wave function

1 Introduction

In view of current research efforts in this first quarter of this 21st century aiming at promoting the use of quantum probability (QP) as behavioral probability in all aspects of human decision-making, especially in economics, it is about time for applied researchers, such as statisticians and econometricians to get familiar with quantum probability calculus, just like they did with standard probability (SP) calculus which supported all associated statistical procedures they used in applications. Now, while introductory, tutorial writings on QP calculus did exist in the literature, this fundamental information (for applications) did not reach the most important

H. T. Nguyen (✉)
Department of Mathematical Sciences, New Mexico State University,
Las Cruces, NM 88003, USA

Faculty of Economics, Chiang Mai University, Chiang Mai, Thailand
e-mail: hunguyen@nmsu.edu

N. Duc Trung · N. Ngoc Thach
Banking University of Ho Chi Minh City Ho Chi , Minh City, Vietnam
e-mail: trungnd@buh.edu.vn

N. Ngoc Thach
e-mail: thachnn@buh.edu.vn

N. Ngoc Thach et al. (eds.), *Data Science for Financial Econometrics*,
Studies in Computational Intelligence 898,
https://doi.org/10.1007/978-3-030-48853-6_9

community, namely econometricians. The reasons are multifolds. Econometricians might be eager to know QP for applications, but they ran into several obstacles. The mentioned writings are mostly written for physicists (who konw quantum mechanics), or psychologists, in one hand, and on the other hand, the underlying mathematics (functional analysis/ the language of quantum mechanics) is not well presented (as opposed to real analysis in SP). In view of this so important issue (for applying QP in behavioral economics), is it possible to write some introductory material to attract econometricians, without requiring some knowledge of quantum mechanics and functional analysis? and yet, they will be able to use it to pursue further knowledge when needed, as well as starting to see that it is possible to consider applications?

It is trying to accomplish such an important task that this paper is written.

Well, we should begin, for econometricians who are not yet aware of the current, and intensive research in the literature on quantum economics, by motivating the appropriate use of QP in economics, especially in decision-making under uncertainty which is central to investigate economic issues , as economic activities are caused by economic agents' decisions. Specifically, we should begin by pointing out empirical facts which cannot be explained by the approach of the neoclassical economics, namely that eoconmic agents are rational and they base their decisions on their expected utilities. However, if we do so, we should write another kind of paper!. The focus of this present paper is to elaborate, in a friendly language, on the technical part of QP. To accomodate for this, we will give main *references* of the fallacies in decision-making discovered by psychologists which triggered the current flourishing field of *behavioral economics* (with three Nobel prizes in economics lately), and which motivated the use of QP as *behavioral probabilities*, as a step further in improving behavioral economics. a sample of such references is Busemeyer and Bruza (2012), Haven (2013), Herzog (2015), Nguyen (2019), Pasca (2015), Patra (2019). The interested reader is encouraged to read these references before embarking on technical issues.

In the following, we devote the whole paper on functional analysis *at the most elementary level,* just enough to initiate the reader to the language needed to lay down the foundations for quantum probability. Mathematically oriented readers should consult a text such as Parthasarathy (1992). It is our hope that once applied econometricians have gotten a clear idea about quantum probability calculus, they will start applying it to real-world problems.

2   A Mathematical Language

Just like the invention of *calculus* as the mathematical language to study Newton mechanics, we need another language to study quantum mechanics, and hence quantum probability. While standard probability theory, as formulated by Kolmogorov in 1933, which forms the backbone for statistics, and is founded on real analysis in mathematics (measure and function theories), the language that we are going to elaborate is *functional analysis* developed by von Neumann (2018).

Essentially, we are going to generalize familiar Euclidean spaces $\mathbb{R}^n$ to their infinitely dimensional counterparts. As we will see later, there is a need to consider complex numbers (in view of the wave function in the Schrodinger's equation) so that, in fact, we will extend $\mathbb{C}^n$, where $\mathbb{C}$ denotes the complex plane. $\mathbb{C}^n$ is a finitely dimensional (n) complex, and separable Hilbert space. The final goal is to define infinitely dimensional, complex and separable Hilbert spaces and (linear) operators on them.

By $\mathbb{C}^n$, we mean the space consisting of (column) vectors with complex components, say, $z = (z_1, z_2, \ldots, z_n)'$ where $(.)'$ denotes transpose of vectors (i.e., row vectors), and $z_j \in \mathbb{C}$.

It is a ($n-$ dimensional) vector space over $\mathbb{C}$. The canonical (orthonormal) basis of $\mathbb{C}^n$ is $\{e_j, j = 1, 2, \ldots, n\}$ where e_j is the column vector with 1 at the jth component, and zero else.

It is clearly a separable space (since it has a countable orthogonal basis), where separability means that $\mathbb{C}^n$ (as a topological space) has a countable dense subset.

On $\mathbb{C}^n$, we have the inner (or scalar) product $< ., . >: \mathbb{C}^n \times \mathbb{C}^n \to \mathbb{C}$, namely, for $x = (x_1, x_2, \ldots, x_n)'$, $y = (y_1, y_2, \ldots, y_n)'$, $< x, y >= \sum_{j=1}^{n} x_j \bar{y}_j$, where $\bar{y}_j$ denotes the complex conjugate of y_j. The norm of $x \in \mathbb{C}^n$ is $||x|| = \sqrt{< x, x >}$ and with respect to this norm, $\mathbb{C}^n$ is a complete metric (topological) space. An Hilbert space is an abstraction of Euclidean spaces. Specifically, a *Hilbert space* is a vector space H (over $\mathbb{R}$ or $\mathbb{C}$) having an inner product whose corresponding metric makes it a complete topological space. It is a complex and separable Hilbert space when the underlying field is $\mathbb{C}$, and it has a countable orthogonal basis (so that elements of H can be written in terms of a such basis). Note that, by an "inner product" in general, we mean a map $< ., . >: H \times H \to \mathbb{C}$ such that $< ., y >$ is linear for each y, $< x, y >=< y, x >^*$(conjugate symmetry), and $< x, x >\geq 0$, $< x, x >= 0 \iff x = 0$. For example, $L^2(\mathbb{R}^3, \mathscr{B}(\mathbb{R}^3), dx)$ is an infinitely dimensional, separable Hilbert space. It is separable since the Borel $\sigma-$field $\mathscr{B}(\mathbb{R}^3)$ is countably generated, and the Lebesgue measure dx is $\sigma-$ finite.

While we will consider $\mathbb{C}^n$ as a simple (and "concrete") Hilbert space (finitely dimensional) in our introduction to the language of quantum probability, we have in mind also general (infinitely dimensional, complex and separable) Hilbert spaces H. This is so since, unlike classical mechanics, quantum mechanics is intrinsically random so that, in one hand, there is no "phase space" (say $\mathbb{R}^6$) to consider, and on the other hand, the state of a quantum system (e.g., the motion of an electron) is described by a "wave function" $\psi(x, t)$ (in the Schrodinger's equation, counterpart of Newton's equation of motion) which turns out to be an element of a general Hilbert space. Note that the concept of quantum probability that we try to grasp came from quantum mechanics: Without being able to describe the dynamics of particles in a deterministic way, quantum physicists have to be content with computing probabilities of quantum states: Quantum mechanics is a (natural) probabilistic theory. The uncertainty here is intrinsic and probabilities involved are "objective". As Feynman (1951) pointed out, while the meaning of probability is the same in a common sense, its *calculus* (i.e., the way that nature combines, manipulates probabilities, at the subatomic level) is different. To have a feel on this, let us say this. In classical mechanics, Newton revealed to us the (deterministic) law of motion. In quantum mechanics, Schrodinger revealed to us an "equation" whose solution $\psi(x, t) \in L^2(\mathbb{R}^3, \mathscr{B}(\mathbb{R}^3), dx)$ is used to compute probabilities of quantum events, namely its amplitude $|\psi(x, t)|^2$ acts as a probability density function (of position $x \in \mathbb{R}^3$, for each given time t) to predict the particle position in a neighborhood of x at time t.

For concreteness, an infinitely dimensional, separable, complex Hilbert space is $H = L^2(\mathbb{R}^3, \mathscr{B}(\mathbb{R}^3), dx)$ of square integrable (with respect to the Lebesgue measure dx), complex-valued functions defined on $\mathbb{R}^3$, and with scalar product defined by

$$< x, y > = \int_{\mathbb{R}^3} x(t) \bar{y}(t) dt$$

Let's use this $H = L^2(\mathbb{R}^3, \mathscr{B}(\mathbb{R}^3), dx)$ to spell out everything needed on a Hilbert space (no stones lelf unturned!).

The norm is

$$||x|| = \sqrt{< x, x >} = (\int_{\mathbb{R}^3} |x(t)|^2 dt)^{\frac{1}{2}}$$

and the distance (metric): $d(x, y) = ||x - y||$.

A sequence $x_n \in H$ converges strongly (in norm) to $x \in H$ if $\lim_{n \to \infty} ||x_n - x|| = 0$. $x_n \to x$ weakly if, for any $y \in H$, $< x_n, y > \to < x.y >$.

(H, d) is a complete metric space, i.e., any x_n, such that $\lim_{n,m \to \infty} d(x_n, x_m) = 0$ (a Cauchy sequence) $\Leftrightarrow$ there exists x such that $\lim_{n \to \infty} d(x_n, x) = 0$ (a convergent sequence). This is Riesz-Fischer's Theorem.

By definition of scalar product, for each fixed $y \in H$, the map $x \in H \to < x, y > \in \mathbb{C}$ is linear. Moreover, $| < x, y > | \leq ||y|| ||x||$ with $||y|| < \infty$ being the infimum of all $c > 0$ such that $| < x, y > | \leq c ||x||$. We denote by $|| < ., y > ||$ that infimum and call it the norm of the map. A linear map $T : H \to \mathbb{C}$ with $||T|| < \infty$ (bounded) is called simply a *functional*.

It turns out that functionals on a Hilbert space are only of this form, i.e., if T is a functional on H, then there exists a unique $y \in H$ such that $Tx = < x, y >$ for all $x \in H$ (Specifically, $y = ||T||z$, where the unique $z \in H$ has $||z|| = 1$) (*Riesz Representation Theorem*).

As we will see later when using the mathematical framework that we are developing in computing quantum probabilities, we need to represent elements of a Hilbert space H, such as $L^2(\mathbb{R}^3, \mathscr{B}(\mathbb{R}^3), dx)$, with respect to an orthongonal basis of H. So let's elaborate a bit on it.

A sequence $\varphi_n \in H$, $n \in \mathbb{Z}$, is said to be orthonormal when its elements are mutually orthogonal and have unit norm, i.e., $< \varphi_n, \varphi_m > = 0$, for $n \neq m$, and $< \varphi_n, \varphi_n > = ||\varphi_n|| = 1$, for all n. For example, for $H = L^2(0, 1)$, $\varphi_n(t) = e^{2\pi i n t}$, $n \in \mathbb{Z} = \{0, \mp 1, \mp 2, \ldots\}$ is an orthonormal system. In fact, this system is "complete" in the sense that it determines the whole space H entirely, i.e., any element $x(.)$ of H can be approximated by elements of this system, so that it is an orthonormal basis for H.

Let's elaborate on the above statement! Suppose we have an finite orthonormal system, say, $\varphi_1, \varphi_2, \ldots, \varphi_n$. Let's try to approximate $x \in H$ by a linear combination of the $\varphi's$, i.e., by $\sum_{k=1}^{n} a_k \varphi_k$ in the best possible way. Specifically, find the coefficients $a'_k s$ to minimize the distance $||x - \sum_{k=1}^{n} a_k \varphi_k||$

We have

$$||x - \sum_{k=1}^{n} a_k \varphi_k||^2 = < x - \sum_{k=1}^{n} a_k \varphi_k, x - \sum_{k=1}^{n} a_k \varphi_k >=$$

$$< x, x > - \sum_{k=1}^{n} \bar{a}_k < x, \varphi_k > - \sum_{k=1}^{n} a_k < \varphi_k, x > + \sum_{k=1}^{n} a_k \bar{a}_k =$$

$$||x||^2 - \sum_{k=1}^{n} | < x, \varphi_k > |^2 + \sum_{k=1}^{n} | < x, \varphi_k > -a_k||^2$$

so that the minimum is attained when the last term is zero, i.e., when $a_k = < x, \varphi_k >$, $k = 1, 2, \ldots, n$.

We write $x = \sum_{k=1}^{n} < x, \varphi_k > \varphi_k$. Clearly, for $H = L^2$, we need an infinite sequence of orthogonal system for $x = \sum_{k=1}^{\infty} < x, \varphi_k > \varphi_k = \lim_{n \to \infty} \sum_{k=1}^{n} < x, \varphi_k > \varphi_k$ (where the limit is with respect to $||.||$).

Remark. Such a countable orthonormal basis exists since L^2 is separable. L^2 is said of infinitely dimensional. An (orthonormal) basis is a maximal orthonormal system.

Note that a Hilbert space is separable if and only if it has a countable orthonormal basis.

Bounded linear operators on abstract Hilbert spaces are abstractions of square matrices on Euclidean spaces.

Complex matrices $n \times n$ are bounded linear transformations on the finitely dimensional, complex, separable Hilbert space $\mathbb{C}^n$. The extension (which is needed for quantum mechanics) to infinitely dimensional, complex, separable Hilbert spaces (e.g., L^2) from finite to "infinite" matrices leads to the concept of linear operators.

A linear map A from H to itself is called an operator on H. The norm of A is $||A|| = \sup\{||Ax|| : ||x|| = 1\}$. If $||A|| < \infty$, then A is said to be bounded. Since we are dealing only with bounded maps, for simplicity, by an *operator* we mean a bounded linear map from H to itself. A square $n \times n$ complex matrix is an operator on $\mathbb{C}^n$. We denote by $\mathcal{B}(H)$ the space of all operators on H.

Complex symmetric matrices are used to represent quantum variables on finitely dimensional spaces (as generalizations of diagonal matrices). We need to extend them to the infinitely dimensional setting. Thus, we will focus on the notion of *self adjoint operators on Hilbert spaces.*

How to generalize the concept of complex symmetric (Hermitian) matrices?

Recall that the transpose of a $n \times n$ complex matrice $A = [a_{jk}]$ is the matrix defined by $A^* = [\bar{a}_{kj}]$. Since A^* is a $n \times n$ matrix, it is also an operator like A. The operator A is said to be complex symmetric, or Hermitian, or just self adjoint, if $A = A^*$. Thus, first, if A is an operator on an arbitrary Hilbert space H, we need to figure out the "counterpart" of A^* in the finitely dimensional case, i.e., an operator A^* on H having all characteristics of the A^* in the finitely dimensional case.

Now, observe that the (conjugate) transpose matrix A^* is related to A as, for any $x, y \in \mathbb{C}^n$, $< Ax, y > = < x, A^*y >$ (by verification). Thus, the counterpart of the (conjugate) transpose of a matrix should be an operator defined by such a relation: The adjoint of an operator A on H is the unique operator A^* satisfying $< Ax, y > = < x, A^*y >$, for any $x, y \in H$. We proceed to show its existence now.

For each $y \in H$, the map $T_y(x) = < Ax, y >$ is a linear functional on H. As such, according to Riesz Representation Theorem, there is a unique $z \in H$ such that $T_y(x) = < Ax, y > = < x, z >$. It suffices to verify that if we let $A^* : H \to H$ be defined by $A^*(y) = z$, then A^* is linear (use the uniqueness part of Riesz's Theorem).

Thus, the adjoint of an operator A on H is the unique operator A^* such that $< Ax, y > = < x, A^*y >$, for any $x, y \in H$.

When $A = A^*$, the operator A is said to be self adjoint.

Recall that the trace of a $n \times n$ matrix A is $tr(A) = \sum_{j=1}^{n} < Ae_j, e_j >$ where $\{e_1, e_2, \ldots, e_n\}$ is any orthonormal basis of $\mathbb{C}^n$. For a separable Hilbert space H, with a countable orthonormal basis $\{e_j, j \geq 1\}$, the trace of an operator A is $tr(A) = \sum_j < Ae_j, e_j >$.

Let H be a Hilbert space. The set of (bounded, linear) operators on H is denoted as $\mathscr{B}(H)$, and that of self adjoint operators, $\mathscr{S}(H)$.

(i) *Positive operators.* $A \in \mathscr{B}(H)$ is called a positive operator if $< Ax, x > \geq 0$, for all $x \in H$. and is denoted as $A \geq 0$. It is an exercise for students to verify that positive operators are necessarily self adjoint. A partial order relation on $\mathscr{B}(H)$ is $A \geq B$ if $A - B \geq 0$.

(ii) *Projections.* $A \in \mathscr{B}(H)$ is called a projection if A is idempotent (i.e., $A \circ A = A^2 = A$) and self adjoint ($A = A^*$). The set of projections is denoted as $\mathscr{P}(H)$. Note that "projection" means othogonal projection onto a closed subspace. Verify that a projection A is a self adjoint, positive operator such that $0 \leq A \leq 1$.

The so-called spectral theorem of Hermitian matrices is this. Let A is a $n \times n$ complex matrix. If A is Hermitian (complex symmetric) then $A = U^*DU$, where D is a diagonal matrix (with real-valued diagonal terms which are the eigenvalues of A) and U is a unitary matrix whose column vectors are eigenvectors of A. If we write $D = diag\{\lambda_1, \lambda_2, \ldots, \lambda_n\}$, and u_j as the jth column vector of U, then $A = \sum_{j=1}^{n} \lambda_j u_j u_j^*$. Observe that the matrix $u_j u_j^* = P_j$ is the projection onto the eigenspace of λ_j, i.e. $\{x \in \mathbb{C}^n : Ax = \lambda_j x\}$. Thus, $A = \sum_{\lambda \in \sigma(A)} \lambda P_\lambda$ which is referred to as the "spectral decomposition" of the matrix A. The spectral decomposition of a self adjoint operator on a infinitely dimensional, complex and separable Hilbert space is the analogue of the above.

In the finitely dimensional cases, such as for $H = \mathbb{C}^n$, a Hermitian matrix A has the *spectral representation* $A = \sum_{\lambda \in \sigma(A)} \lambda P_\lambda$, where $\sigma(A)$ is the set of eigenvalues of A, called the *spectrum* of A, and P_λ is the projection onto the eigenspace $S_\lambda = \{x \in \mathbb{C}^n : Ax = \lambda x\}$, and the set of distinct eigenvectors forms an orthonormal basis. Since $\sigma(A) \subseteq \mathbb{R}$, the map defined on $\mathscr{B}(\mathbb{R})$ by $\zeta_A(B) = \sum_{\lambda \in B} \lambda P_\lambda$ take values in $\mathscr{P}(\mathbb{C}^n)$ and acts like a "projection-valued" measure. This projection-valued measure, associated with the Hermitiam matrix A, is called the *spectral measure* of A.

The upshot is that this situation can be extended to the infinitely dimensional case. Specifically, any (bounded) self adjoint operator A on a Hilbert space H admits a unique spectral measure ζ_A and $A = \int_{\sigma(A)} \lambda d\zeta_A(\lambda)$ where the integral is defined as a Lebesgue-Stieltjes integral (here of the function $f(\lambda) = \lambda$).

Let's elaborate on this a bit. In genral, for $A \in \mathscr{B}(H)$, an element $x \in Dom(A)$, and $x \neq 0$, is called an eigenvector of A if there is $\lambda \in \mathbb{C}$ such that $Ax = \lambda x$, and in this case, λ is called an eigenvalue of A. Since λ is an eigenvalue of A if and only if $A - \lambda I$ is not injective, the spectrum of A is $\sigma(A) = \{\lambda \in \mathbb{C} : A - \lambda I \quad \text{not invertible}\}$. If A is self adjoint, then $\sigma(A) \subseteq \mathbb{R}$. In particular, if A is a positive operator, then $\sigma(A) \subseteq \mathbb{R}^+$. The spectral theorem for self adjoint operators on a Hilbert space (von Neumann) is an extension to the infinitely dimensional case of the one familiar for matrices mentioned above.

Clearly, now $\sigma(A)$ is not finite or infinite countable, so that $\sum_{\lambda \in \sigma(A)} \lambda P_\lambda$ must be replaced by some sort of integral $\int_{\sigma(A)} \lambda d\zeta(\lambda)$, where the "measure" $d\zeta$ should be "projection-valued"! Thus, first, a spectral measure ζ, on a measurable space $(\Omega, \mathscr{A})$, is a "measure" taking values in $\mathscr{P}(H)$, in the sense that (i) $\zeta(\Omega) = I$, (ii) for any sequence of disjoitn sets B_n in $\mathscr{A}$, $\zeta(\cup_n B_n) = \sum_n \zeta(B_n)$, where the convergence of the series is taken in the strong topology.

With a spectral measure ζ, the expression $\int_{\sigma(A)} \lambda d\zeta(\lambda)$ should be an operator on H (as an extension of $\sum_{\lambda \in \sigma(A)} \lambda P_\lambda$), and when A is a self adjoint operator on H, $\int_{\sigma(A)} \lambda d\zeta(\lambda)$ should be self adjoint, and we wish to write $A = \int_{\sigma(A)} \lambda d\zeta_A(\lambda)$, where ζ_A should be a unique spectral measure associated with A (the spectral measure of A). And this will be the analogue of the spectral representation of Hermitian matrices in the finitely dimensional case, noting that for self adjoint operators, $\sigma(A) \subseteq \mathbb{R}$, so that $(\Omega, \mathscr{A})$ will be taken to be $(\mathbb{R}, \mathscr{B}(\mathbb{R}))$.

So let's carry out this "program"! We note that $\int_{\sigma(A)} \lambda d\zeta(\lambda)$ looks like the integral for the function $f : \mathbb{R} \to \mathbb{R}$, $f(\lambda) = \lambda$, with respect to the "measure" ζ. So let see how a spectral measure on $(\Omega, \mathscr{A})$ can "integrate" measurable functions. We follow measure theory steps. If $f : \Omega \to \mathbb{C}$ is a simple function, i.e., of the form $\sum_{j=1}^n a_j 1_{B_j}$ ($a_j \in \mathbb{C}$, $B_j \in \mathscr{A}$, pairwise disjoint), then define

$$\int_\Omega f(x) d\zeta(x) = \sum_{j=1}^n a_j \zeta(B_j)$$

the rest is a routine from Lebesgue integral construction.

Without going into details, we note that the adjoint of $\int_\Omega f(\lambda) d\zeta(\lambda)$ is $\int_\Omega f^*(\lambda) d\zeta(\lambda)$, so that if f is real-valued, $\int_\Omega f(\lambda) d\zeta(\lambda)$ is self adjoint.

The *von Neumann's spectral theorem* is this. The map $\zeta \rightarrow \int_{\mathbb{R}} \lambda d\zeta(\lambda)$ is a bijection between spectral measures on $(\mathbb{R}, \mathscr{B}(\mathbb{R}))$ and self adjoint operators on H.

Thus, if A is a self adjoint operator on H, there is a unique spectral measure ζ_A on $(\mathbb{R}, \mathscr{B}(\mathbb{R}))$ (a projection-valued measure) such that $A = \int_{\mathbb{R}} \lambda d\zeta(\lambda)$. We call ζ_A the spectral measure of A. Thus, if A is a self adjoint operator on H, there is a unique spectral measure ζ_A on $(\mathbb{R}, \mathscr{B}(\mathbb{R}))$ (a projection-valued measure) such that $A = \int_{\mathbb{R}} \lambda d\zeta(\lambda)$. We call ζ_A the spectral measure of A.

3 Quantum Probability Spaces

A quantum probability space is a noncommutative generalization of a standard (Kolmogorov) probability space. The noncommutativity (of observables) will become clear as we proceed.

The (general) generalization procedure can be motivated by considering the simplest case of a finite standard probability space $(\Omega, \mathscr{A}, P)$ with the sample space Ω being finite, say, $\Omega = \{1, 2, \ldots, n\}$, $\mathscr{A} = 2^{\Omega}$, and $P : \mathscr{A} \rightarrow [0, 1]$ being a probability measure (or, equivalently, its probability density function $\rho(.) : \Omega \rightarrow [0, 1]$, $\rho(j) = P(\{j\})$.

In this setting, all functions of interest can be identified as elements of the euclidean space $\mathbb{R}^n$. Indeed, an event $A \in \mathscr{A}$ is equivalent to its indicator function $1_A(.) : \Omega \rightarrow \{0, 1\}$ by $j \in A \Longleftrightarrow 1_A(j) = 1$ which, in turn, is equivalent to the (column) vector $(1_A(1), 1_A(2), \ldots, 1_A(n))' \in \mathbb{R}^n$; the density $\rho(.)$ is determined by $(\rho(1), \rho(2), \ldots, \rho(n))' \in \mathbb{R}^n$; and a random variable $X(.) : \Omega \rightarrow \mathbb{R}$ is represented by $(X(1), X(2), \ldots, X(n))' \in \mathbb{R}^n$. Now, each element $x = (x_1, x_2, \ldots, x_n)' \in \mathbb{R}^n$ is identified (represented) by a $n \times n$ diagonal matrix $[x]$ whose diagonal terms are $x_1, x_2, \ldots, x_n$. The space of all $n \times n$ diagonal matrices is a commutative subalgebra of the noncommutative algebra of symmetric matrices, denoted as $\mathscr{S}(\mathbb{R}^n)$. Since an $n \times n$ matrix M is a linear map from $\mathbb{R}^n$ to $\mathbb{R}^n$, it is an operator on $\mathbb{R}^n$.

Now, from the mathematical language (functional analysis) outlined in the previous section, the above identification of mappings $M(.) : \Omega \rightarrow \mathbb{R}$ by diagonal $n \times n$ matrices $[M]$, we recognize that events $A \in \mathscr{A}$, represented by $[1_A]$, are projection operators on $\mathbb{R}^n$; probability density functions $\rho(.)$, represented by $[\rho]$, are positive operators with unit trace; and random variables $X(.) : \Omega \rightarrow \mathbb{R}$, represented by $[X]$, are symmetric operators. All of them are symmetric matrices. Thus, we obtain an equivalence of standard probability space $(\Omega, \mathscr{A}, P)$, namely $(\mathbb{R}^n, \mathscr{P}(\mathbb{R}^n), \rho)$. Moreover, we recognize that the range of a random variable X is precisely the set of the diagonal terms of $[X]$, i.e., the spectrum $\sigma([X])$ which is the set of the eigenvalues on $[X]$; also

$$E_\rho(X) = \sum_{j=1}^{n} X(j)\rho(j) = tr(\rho[X])$$

which will allow us to use the trace operator (in general Hilbert spaces) at the place of integral.

To generalize $(\Omega, \mathscr{A}, P)$ to a noncommutative (finite) probability space (which we will refere to as a quantum probability space), it suffices to extend diagonal matrices to arbitrary symmetric matrics (or self adjoint operators on $\mathbb{C}^n$, noting that eigenvalues of self adjoint matrices are real-valued), so that $(\mathbb{R}^n, \mathscr{P}(\mathbb{R}^n), \rho)$ is a triple in which $\mathbb{R}^n$ is a (finitely dimensional) separable Hilbert space, $\mathscr{P}(\mathbb{R}^n)$ is the set of (all arbitrary) projections on $\mathbb{R}^n$, and ρ is an arbitrary positive operator with unit trace (called a "density matrix", playing the role of "probability distribution of a random variable").

In summary, the noncommutative setting consists of (arbitrary) self adjoint operators on a Hilbert space. It is noncommutative since the algebra of self adjoint operators is noncommutative: by extending the notion of random variables to "observables", which are represented by self adjoint operators, observables (as operators) are not commutative in general, but still have real values as their outcomes.

For the general case, let H be an infinitely dimensional, complex and separable Hilbert space (replacing $\mathbb{C}^n$ in the previous simple setting), such as $L^2(\mathbb{R}^3, \mathscr{B}(\mathbb{R}^3), dx)$. Then a general quantum probability space is a triple $(H, \mathscr{P}(H), \rho)$ where $\mathscr{P}(H)$ is the non distributive lattice of projections on H, and ρ is a positive operator with unit trace. Observables are self adjoint operators $\mathscr{S}(H)$ on H. Note that a such $(H, \mathscr{P}(H), \rho)$ is a "general" quantum probability space (as opposed to $(\Omega, \mathscr{A}, P)$ where, in general, P is a probability *measure*, and not a probability *density function*) is due to Gleason's theorem (see Parthsa.......) which says that, for $\dim(H) \geq 3$, there is a bijection between density matrices ρ and probability measures μ on $\mathscr{P}(H)$ via $\mu(p) = tr(\rho p)$, i.e., $\mu(.) = tr(\rho(.))$. We will elaborate on "quantum probability μ" in the next section.

Let's elaborate a bit on the *non-boolean* algebraic structure of quantum events $\mathscr{P}(H)$. In view of the bijection between projections and closed subspaces of H, we have, for $p, q \in \mathscr{P}(H)$, $p \wedge q$ is taken to be the projection corresponding to the closed subspace $\mathscr{R}(p) \cap \mathscr{R}(q)$, where $\mathscr{R}(p)$ denotes the range of p; and $p \vee q$ is taken to be the projection corresponding to the smallest closed subspace containing $\mathscr{R}(p) \cup \mathscr{R}(q)$. We can check that $p \wedge (q \vee r) \neq (p \wedge p) \vee (\rho \wedge r)$, unless they commute. Note also that, for $p, q \in \mathscr{P}(H)$, we have $pq \in \mathscr{P}(H)$ if and only if p, q commute, i.e., $pq = qp$. If we set $\mu(p) = tr(\rho p)$, then by additivity of the trace operator, we have

$$p \vee q - p - q + p \wedge q = 0 \Longrightarrow tr(\rho(p \vee q - p - q + p \wedge q)) =$$

$$\mu(p \vee q) - \mu(p) - \mu(q) + \mu(p \wedge q) = 0$$

which is the analogue of additivity for $\mu(.)$, for commuting p, q. In general, i.e., when p, q do not commute, i.e., their "commutator" $[p, q] = pq - qp \neq 0$, $\mu(.)$ is not additive. Indeed, since $[p, q] = (p - q)(p \vee q - p - q + p \wedge q)$, we have $[p, q] \neq 0 \Longleftrightarrow p \vee q - p - q + p \wedge q \neq 0$.

4   Quantum Probability Calculus

Let $(H, \mathscr{P}(H), \rho)$ be a quantum probability space. The map $\mu_\rho(.) : \mathscr{P}(H) \to [0, 1]$ defined by $\mu_\rho(p) = tr(\rho p)$ has $\mu_\rho(I) = 1$ where I is the identity operator on H. It is, in fact, a quantum probability measure, i.e., "σ−additivity" in a sense to be specified.

If X is an observable, i.e., a self adjoint operator on H, then it has its spectral measure $\zeta_X(.) : \mathscr{B}(\mathbb{R}) \to \mathscr{P}(H)$. Thus, for $B \in \mathscr{B}(\mathbb{R})$, $\zeta_X(B) \in \mathscr{P}(H)$ so that the probability that X takes a value in B should be $\mu_\rho(\zeta_X(B))$, where $\mu_\rho \circ \zeta_X(.) : \mathscr{B}(\mathbb{R}) \to [0, 1]$. For this set function to be $\sigma-$ additive, i.e., for $B_j \in \mathscr{B}(\mathbb{R})$, pairwise disjoint, $\mu_\rho(\zeta_X(\cup_j B_j)) = \sum_j \mu_\rho(\zeta_X(B_j))$, $\mu_\rho(.)$ must satisfy the condition: for $p_j \in \mathscr{P}(H)$, pairwise orthogonal (i.e., $p_j p_k = 0$ for $j \neq k$), $\mu_\rho(\sum_j p_j) = \sum_j \mu_\rho(p_j)$ (convergence in norm). This condition is indeed satisfied by Gleason's theorem which says that "For H with $\dim(H) \geq 3$, the map $\mu(.) : \mathscr{P}(H) \to [0, 1]$ is $\sigma-$ additive if and only if there exists a density matrix ρ such that $\mu(.) = tr(\rho(.))$". Specifically, any quantum probability measure $\mu(.)$ on $\mathscr{P}(H)$ is written as $\mu(p) = \sum_j \alpha_j < pu_j, u_j >$ for $\{u_j\}$ being an orthonormal basis of H (separable), and $\alpha_j \geq 0$ with $\sum_j \alpha_j = 1$. Thus, as in standard probability, replacing $p \in \mathscr{P}(H)$ by observable X, we have

$$E_\rho(X) = \int_\mathbb{R} \lambda d\mu_{\rho, X}(d\lambda) = \int_\mathbb{R} \lambda tr(\rho \zeta_X(d\lambda)) = tr(\rho X)$$

Indeed, $\rho = \sum_j a_j |u_j><u_j|$ for any orthonomal basis $\{u_j\}$ of H (separable), and $a_j > 0$ with $\sum_j a_j = 1$. For $B \in \mathscr{B}(\mathbb{R})$,

$$\mu_\rho(B) = tr(\rho \zeta_X(B)) = \sum_j < \rho \zeta_X(B) u_j, u_j >= \sum_j a_j < \zeta_X(B) u_j, u_j >$$

Thus, since $X = \int x \zeta_X(dx)$ (spectral representation of X), we have

$$E_\rho(X) = \int_\mathbb{R} x \mu_\rho(dx) = \sum_j a_j \int_\mathbb{R} x < \zeta_X(dx) u_j, u_j >=$$

$$\sum_j a_j < X u_j, u_j >= \sum_j < \rho X u_j, u_j >= tr(\rho X)$$

When ρ is induced by the normalized wave function ψ on $\mathbb{R}^3$ as $\rho = \sum_j a_j |u_j><u_j|$ with $a_j = |c_j|^2$, $c_j =< \psi, u_j >$, we have

$$E_\psi(X) = \int_{\mathbb{R}^3} \psi^*(x) X \psi(x) dx$$

Remark. In quantum mechanics, the density matrix ρ (playing the role of P in $(\Omega, \mathscr{A}, P)$) is obtained from a normalized wave function of the Schrodinger's equation, or by *path integral* (see Feynman and Hibbs (1965)). The meaning of Dirac's notation $< . |$ and $| . >$ used in the above analysis is this. The inner (scalar) product on the Hilbert space H is the "bracket" $< ., . >: H \times H \to \mathbb{C}$, $< x, y >= \sum_j x_j \bar{y}_j$. There is another "outer product" on the finitely dimensional Hilbert space $\mathbb{C}^n$, namely the multiplication of a column vector x (a $n \times 1$ matrix) with a row vector y (a $1 \times n$ matrix) resulting in a $n \times n$ matrix, which is an operator on $\mathbb{C}^n$. The notation for this outer product, on general Hilbert space H, is given by Dirac as follows. An element $x \in H$ is written as a "ket" $|x >$, whereas the "bra" $< x|$ is an element of the dual of H, i.e., a map from H to $\mathbb{C}$, so that the "bracket" is obtained as $< x|y >$, and the outer product of $x, y \in H$ is $|x >< y|$ which is an operator on H: for $z \in H$, $|x >< y|(z) =< y, z > x \in H$.

A bit of quantum mechanics Unlike statistical mechanics, quantum mechanics reveals the randomness believed to be caused by nature itself. As we are going to examine whether economic fluctuations can be modeled by quantum uncertainty, we need to take a quick look at quantum mechanics.

The big picture of quantum mechanics is this. A particle with mass m, and potential energy $V(x_o)$ at a position $x_o \in \mathbb{R}^3$, at time $t = 0$, will move to a position x at a later time $t > 0$. But unlike Newtonian mechanics (where moving objects obey a law of motion and their time evolutions are deterministic trajectories, with a state being a point in $\mathbb{R}^6$/position and velocity), the motion of a particle is not deterministic, so that at most we can only look for the probability that it could be in a small neighborhood of x, at time t. Thus, the problem is: How to obtain such a probability? According to quantum mechnanics, the relevant probability density $f_t(x)$ is of the form $|\psi(x, t)|^2$ where the (complex) "probability amplitude" $\psi(x, t)$ satisfies the *Schrodinger equation* (playing the role of Newton's law of motion in macrophysics)

$$ih\frac{\partial \psi(x, t)}{\partial t} = -\frac{h^2}{2m}\Delta_x \psi(x, t) + V(x)\psi(x, t)$$

where h is the Planck's constant, $i = \sqrt{-1}$, and Δ_x is the Laplacian $\Delta_x \psi = \frac{\partial^2 \psi}{\partial x_1^2} + \frac{\partial^2 \psi}{\partial x_2^2} + \frac{\partial^2 \psi}{\partial x_3^2}$, $x = (x_1, x_2, x_3) \in \mathbb{R}^3$.

Solutions of the Schrodinger equation are "wave-like", and hence are called wave functions of the particle (the equation itself is called the wave equation). Of course, solving this PDE equation, in each specific situation, is crucial. Richard Feynman (1948) introduced the concept of path integral to solve it.

For a solution of the form $\psi(x, t) = \varphi(x)e^{it\theta}$, $|\psi(x, t)|^2 = |\varphi(x)|^2$ with $\varphi \in L^2(\mathbb{R}^3, \mathscr{B}(\mathbb{R}^3), dx)$, in fact $||\varphi|| = 1$. Now, since the particle can take any path from $(x_o, 0)$ to (x, t), its "state" has to be described probabilistically. Roughly speaking, each φ (viewed as a "vector" in the complex, infinitely dimensional Hilbert space $L^2(\mathbb{R}^3, \mathscr{B}(\mathbb{R}^3), dx)$) represents a state of the moving particle. Now $L^2(\mathbb{R}^3, \mathscr{B}(\mathbb{R}^3), dx)$ is separable so that it has a countable orthornormal basis, φ_n, say, and hence

$$\varphi = \sum_n <\varphi, \varphi_n> \varphi_n = \sum_n c_n \varphi_n = <\varphi_n|\varphi|\varphi_n>$$

where $< ., . >$ denotes the inner product in $L^2(\mathbb{R}^3, \mathscr{B}(\mathbb{R}^3), dx)$, and the last notation on the right is written in popular Dirac's notation, noting that $||\varphi||^2 = 1 = \sum_n |c_n|^2$, and

$\sum_n |\varphi_n><\varphi_n| = I$ (identity operator on $L^2(\mathbb{R}^3, \mathscr{B}(\mathbb{R}^3), dx)$)

where $|\varphi><\psi|$ is the operator: $f \in L^2(\mathbb{R}^3, \mathscr{B}(\mathbb{R}^3), dx) \to <\varphi, f><\psi| \in L^2(\mathbb{R}^3, \mathscr{B}(\mathbb{R}^3), dx)$.

From the solution $\varphi(x)$ of Schrodinger equation, the operator

$$\rho = \sum_n |c_n|^2 |\varphi_n><\varphi_n|$$

is positive with unit trace $(tr(\rho) = \sum_n <\varphi_n|\rho|\varphi_n> = 1)$, which is the (canonical) decomposition of the "state" ρ in terms of the wave function φ. In other words, the wave function induces a state.

Indeed, $\rho : L^2 \to L^2 : \rho(f) = \sum_n [|c_n|^2 <\varphi_n, f>]\varphi_n$, so that $<\rho f, f> = \sum_n |c_n|^2 |<\varphi_n, f>|^2 \geq 0$, for any $f \in L^2$; $tr(\rho) = \sum_n <\rho\varphi_n, \varphi_n> = \sum_n |c_n|^2 = 1$.

Not that ρ is called a "state" since it is derived from the wave function (as above) which describes the state of a particle, at a given time. The state of a quantum system is a function, an element of an infinitely dimensional space. It contains all information needed to derive associated physical quantities like position, velocity, energy, etc. but in a probability context. In quantum mechanics, ρ is known, and hence probability distributions of associated physical quantities will be known as well. There is no need to consider models (as opposed to statistical modeling) ! We got the probability laws of quantum random variables. The situation is like we get a game of chance, but designed by nature (not man-made!). Put it differently, randomness in quantum systems is created by nature, but we found out correctly its law! However, nature's games of chance are different than man-made ones: they are essentially non commutative!

Thus it plays the role of the classical probability density function. By separability of $L^2(\mathbb{R}^3, \mathscr{B}(\mathbb{R}^3), dx)$, we are simply in a natural extension of finitely dimensional euclidean space setting, and as such, the operator ρ is called a *density matrix* which represents the "state" of a quantum system.

This "concrete setting" brings out a general setting (which generalizes Kolmogorow probability theory), namely, a complex, infinitely dimensional, separable, Hilbert space $H = L^2(\mathbb{R}^3, \mathscr{B}(\mathbb{R}^3), dx)$, and a density matrix ρ which is a (linear) positive definite operator on H (i.e., $< f, \rho f > \geq 0$ for any $f \in H$, implying that it is self adjoint), and of unit trace.

Remark. At a given time t, it is the entire function $x \to \psi(x, t)$ which describes the state of the quantum system, and not just one point! The wave function $\psi(x, t)$ has a probabilistic interpretation: its amplitude gives the probability distribution for the position, a physical quantity of the system, namely, $|\psi(x, t)|^2$.

Now, observe that for $\varphi(p)$ arbitrary, where $p = mv$ is the particle momentum, a solution of Schrodinger's equation is

$$\psi(x, t) = \int_{\mathbb{R}^3} \varphi(p) e^{-\frac{i}{\hbar}(Et - <p, x>)} dp / (2\pi h)^{\frac{3}{2}}$$

(where $E = \frac{\|p\|^2}{2m}$), i.e., ψ is the Fourier transform of the function $\varphi(p) e^{-\frac{i}{\hbar}(Et)}$, and hence, by Parseval-Plancherel,

$$\int_{\mathbb{R}^3} |\psi(x, t)|^2 dx = \int_{\mathbb{R}^3} |\varphi(p)|^2 dp$$

Thus, it suffices to choose $\varphi(.)$ such that $\int_{\mathbb{R}^3} |\varphi(p)|^2 dp = 1$ (to have all wave functions in $L^2(\mathbb{R}^3, \mathscr{B}(\mathbb{R}^3), dx)$, as well as $\int_{\mathbb{R}^3} |\psi(x, t)|^2 dx = 1$. In particular, for stationary solutions of Schrodinger'equation $\psi(x)e^{-iEt/h}$, describing the same stationary state. Here, note that $\|\psi\| = 1$.

Measuring physical quantities
Physical quantities are numerical values associated to a quantum system, such as position, momentum, velocity, and functions of these, such as energy.

In classical mechanics, the result on the measurement of a physical quantity is just a number at each instant of time. In quantum mechanics, at a given time, repeated measurements under the same state of the system give different values of a physical quantity A: There should exist a probability distribution on its possible values, and we could use its expected (mean) value.

For some simple quantities, it is not hard to figure out their probability distributions, such as position x and momentum p (use Fourier transform to find the probability distribution of p) from which we can carry out computations for expected values of functions of then, such as potential energy $V(x)$, kinetic energy (function of p alone). But how about, say, the mechanical energy $V(x) + \frac{p^2}{2m}$, which is a function of both position x and momentum p? Well, its expected value is not a problem, as you can take $E(V(x) + \frac{p^2}{2m}) = EV(x) + E(\frac{p^2}{2m})$, but how to get its distribution when we need it? Also, if the quantity of interest is not of the form of a sum where the knowledge of $E(x)$, $E(p)$ is not sufficient to compute its expectation?

If you think about classical probability, then you would say this. We know the marginal distributions of the random variables x, p. To find the distribution of $V(x) + \frac{p^2}{2m}$, we need the joint distribution of (x, p). How? *Copulas* could help? But are we in the context of classical probability!?

We need a general way to come up with necessary probability distributions for all physical quantities, soly from the knowledge of the wave function $\psi(x, t)$ in the Schrodinger's equation. It is right here that we need mathematics for physics!

For a spacial quantity like position X (of the particle), or $V(X)$ (potential energy), we know its probability distribution $x \to |\psi(x, t)|^2$, so that its expected valued is given by

$$EV(X) = \int_{\mathbb{R}^3} V(x)|\psi(x,t)|^2 dx = \int_{\mathbb{R}^3} \psi^*(x,t)V(x)\psi(x,t)dx$$

If we group the term $V(x)\psi(x,t)$, it looks like we apply the "operator" V to the function $\psi(.,t) \in L^2(\mathbb{R}^3)$, to produce another function of $L^2(\mathbb{R}^3)$. That operator is precisely the multiplication $A_V(.) : L^2(\mathbb{R}^3) \to L^2(\mathbb{R}^3) : \psi \to V\psi$. It is a bounded, linear map from a (complex) Hilbert space H to itself, which we call, for simplicity, an operator on H.

We observe also that $EV(X)$ is a real value (!) since

$$EV(X) = \int_{\mathbb{R}^3} V(x)|\psi(x,t)|^2 dx$$

with $V(.)$ being real-valued. Now,

$$\int_{\mathbb{R}^3} \psi^*(x,t)V(x)\psi(x,t)dx = <\psi, A_V\psi>$$

is the inner product on $H = L^2(\mathbb{R}^3)$. We see that, for any $\psi, \varphi \in H$, $<\psi, A_V\varphi> = <A_V\psi, \varphi>$, since V is real-valued, meaning that the operator $A_V(.) : \psi \to V\psi$ is *self adjoint*.

For the position $X = (X_1, X_2, X_3)$, we compute the vector mean $EX = (EX_1, EX_2, EX_3)$, where we can derive, for example, EX_1 directly by the observables of $Q = X_1$ as $A_{X_1} : \psi \to x_1\psi$ (multiplication by x_1), $A_{x_1}(\psi)(x,t) = x_1\psi(x,t)$.

Remark. The inner product in the (complex) Hilbert space $H = L^2(\mathbb{R}^3)$ (complex-valued functions on $\mathbb{R}^3$, squared integrable wrt to Lebesgue measure dx on $\mathscr{B}(\mathbb{R}^3)$) is defined as

$$<\psi, \varphi> = \int_{\mathbb{R}^3} \psi^*(x,t)\varphi(x,t)dx$$

where $\psi^*(x,t)$ is the complex conjugate of $\psi(x,t)$. The *adjoint operator* of the (bounded) operator A_V is the unique operator, denoted as A_V^*, such that $<A_V^*(f), g> = <f, A_V(g)>$, for all $f, g \in H$ (its existence is guaranteed by Riesz theorem in functional analysis). It can be check that $A_V^* = A_{V^*}$, so that if $V = V^*$ (i.e., V is real-valued), then $A_V^* = A_V$, meaning that A_V is self adjoint. Self adjoint operators are also called *Hermitian* (complex symmetry) operators, just like for complex matrices. The property of self adjoint for operators is important since eigenvalues of such operators are real values, and as we will see later, which correspond to possible values of the physical quantities under investigation, which are real valued.

As another example, let's proceed directly to find the probability distribution of the momentum $p = mv$ of a particle, at time t, in the state $\psi(x,t)$, $x \in \mathbb{R}^3$, and from it,.compute, for example, expected values of functions of momentum, such as $Q = \frac{||p||^2}{2m}$.

The Fourier transform of $\psi(x, t)$ is

$$\varphi(p, t) = (2\pi h)^{-\frac{3}{2}} \int_{\mathbb{R}^3} \psi(x, t) e^{-\frac{i}{h}<p,x>} dx$$

so that, by Parseval-Plancherel, $|\varphi(p, t)|^2$ is the probability density for p, so that

$$E(\frac{||p||^2}{2m}) = \int_{\mathbb{R}^3} \frac{||p||^2}{2m} |\varphi(p, t)|^2 dp$$

But we can obtain this expectation via an appropriate operator A_p as follows. Since

$$\psi(x, t) = (2\pi h)^{-\frac{3}{2}} \int_{\mathbb{R}^3} \varphi(p, t) e^{\frac{i}{h}<p,x>} dp$$

with $x = (x_1, x_2, x_3)$, we have

$$\frac{h}{i} \frac{\partial}{\partial x_1} \psi(x, t) = (2\pi h)^{-\frac{3}{2}} \int_{\mathbb{R}^3} p_1 \varphi(p, t) e^{\frac{i}{h}<p,x>} dp$$

i.e., $\frac{h}{i} \frac{\partial}{\partial x_1} \psi(x, t)$ is the Fourier transform of $p_1 \varphi(p, t)$, and since ψ is the Fourier transform of φ, Parveval-Plancherel implies

$$E(p_1) = \int_{\mathbb{R}^3} \varphi^*(p, t) p_1 \varphi(p, t) dp = \int_{\mathbb{R}^3} \psi^*(p, t) [\frac{h}{i} \frac{\partial}{\partial x_1}] (\psi(x, t) dx$$

we see that the operator $A_p = \frac{h}{i} \frac{\partial}{\partial x_1}(.)$ on H extracts information from the wave function ψ to provide a direct way to compute the expected value of the component p_1 of the momentum vector $p = (p_1, p_2, p_3)$ (note $p = mv$, with $v = (v_1, v_2, v_3)$) on one axis of $\mathbb{R}^3$. For the vector p (three components), the operator $A_p = \frac{h}{i} \nabla$, where $\nabla = \left(\frac{\partial}{\partial p_1}, \frac{\partial}{\partial p_2}, \frac{\partial}{\partial p_3} \right)$.

As for $Q = \frac{||p||^2}{2m}$, we have

$$EQ = \int_{\mathbb{R}^3} \psi^*(x, t) [(\frac{-h^2}{2m}) \Delta] (\psi(x, t) dx$$

where Δ is the Laplacian. The corresponding operator is $A_Q = (\frac{-h^2}{2m}) \Delta$.

Examples, as the above, suggest that, for each physical quantity of interest Q (associated to the state ψ of a particle) we could look for a self adjoint operator A_Q on H so that

$$EQ = < \psi, A_Q \psi >$$

A such operator extracts information from the state (wave function) ψ for computations on Q. This operator A_Q is refered to as the *observable* for Q.

Remark. If we just want to compute the expectation of the random variable Q, without knowledge of its probability distribution, we look for the operator A_Q. On the surface, it looks like we only need a weaker information than the complete information provided by the probability distribution of Q. This is somewhat similar to a situation in statistics, where getting the probability distribution of a *random set S*, say on $\mathbb{R}^3$ is difficult, but a weaker and easier information about S can be obtained, namely it coverage function $\pi_S(x) = P(S \ni x)$, $x \in \mathbb{R}^3$, from which the expected value of the measure $\mu(S)$ can be computed, as $E\mu(S) = \int_{\mathbb{R}^3} \pi_S(x) d\mu(x)$, where μ is the Lebesgue measure on $\mathscr{B}(\mathbb{R}^3)$. See e.g., Nguyen (2006).

But how to find A_Q for Q in general? Well, a "principle" used in quantum measurement is this. Just like in classical mechanics, all physical quantities associated to a dynamical systems are functions of the system state, i.e., position and momentum (x, p), i.e., $Q(x, p)$, such as $Q(x, p) = \frac{\|p\|^2}{2m} + V(x)$. Thus, the observable corresponding to $Q(x, p)$ should be $Q(A_x, A_p)$, where A_x, A_p are observables corresponding to x and p which we already know in the above analysis. For example, if the observable of Q is A_Q, then the observable of Q^2 is A_Q^2.

An interesting example. What is the observable A_E corresponding to the energy $E = \frac{\|p\|^2}{2m} + V$?

We have $A_V = V$ $(Q_V(f) = Vf$, i.e., multiplication by the function V: $(A_V(f)(x) = V(x)f(x))$.

$$A_p = \frac{h}{i}\nabla = \frac{h}{i}\begin{pmatrix} \frac{\partial}{\partial x_1} \\ \frac{\partial}{\partial x_2} \\ \frac{\partial}{\partial x_3} \end{pmatrix}$$

so that

$$A_p^2(f) = (A_p \circ A_p)(f) = A_p(A_p(f)) = A_p \begin{bmatrix} \frac{h}{i} \frac{\partial f}{\partial x_1} \\ \frac{\partial f}{\partial x_2} \\ \frac{\partial f}{\partial x_3} \end{bmatrix}$$

$$= (\frac{h}{i})^2 \begin{bmatrix} \frac{\partial f^2}{\partial x_1^2} \\ \frac{\partial f^2}{\partial x_2^2} \\ \frac{\partial f^2}{\partial x_3^2} \end{bmatrix} = -h^2 \begin{bmatrix} \frac{\partial f^2}{\partial x_1^2} \\ \frac{\partial f^2}{\partial x_2^2} \\ \frac{\partial f^2}{\partial x_3^2} \end{bmatrix}$$

Thus, the observable of $\frac{\|p\|^2}{2m}$ is $\frac{-h^2}{2m}\Delta$, and that of $E = \frac{\|p\|^2}{2m} + V$ is $A_E = \frac{-h^2}{2m}\Delta + V$, which is an operator on $H = L^2(\mathbb{R}^3)$.

By historic reason, this observable of the energy (of the quantum system) is called the *Hamiltonian* of the system (in honor of Hamilton, 1805-1865) and denoted as

$$\mathscr{H} = \frac{-h^2}{2m}\Delta + V$$

Remark. Since

$$E(V) = \int_{\mathbb{R}^3} \psi^*(x,t)V(x)\psi(x,t)dx = \int_{\mathbb{R}^3} \psi^*(x,t)(M_V\psi)(x,t)dx$$

it follows that $A_V = V$.

The Laplacian operator is

$$\Delta f(x) = \frac{\partial f^2}{\partial x_1^2} + \frac{\partial f^2}{\partial x_2^2} + \frac{\partial f^2}{\partial x_3^2}$$

where $x = (x_1, x_2, x_3) \in \mathbb{R}^3$.

Now, if we look back at Schrodinger's equation

$$ih\frac{\partial}{\partial t}\psi(x,t) = -\frac{h^2}{2m}\Delta\psi(x,t) + V(x)\psi(x,t)$$

with (stationary) solutions of the form $\psi(x,t) = \varphi(x)e^{-i\omega t}$, then it becomes

$$(-i^2)h\omega\varphi(x)e^{-i\omega t} = -\frac{h^2}{2m}\Delta\varphi(x)e^{-i\omega t} + V(x)\varphi(x)e^{-i\omega t}$$

or

$$h\omega\varphi(x) = -\frac{h^2}{2m}\Delta\varphi(x) + V(x)\varphi(x)$$

With $E = h\omega$, this is

$$-\frac{h^2}{2m}\Delta\varphi(x) + V(x)\psi(x) = E\varphi(x)$$

or simplt, in terms of the Hamiltonian,

$$\mathcal{H}\varphi = E\varphi$$

Putting back the term $e^{-i\omega t}$, the Schrodinger's equation is written as

$$\mathcal{H}\psi = E\psi$$

i.e., the state ψ (solution of Schrodinger's equation) is precisely the *eigenfunction* of the Hamiltonian $\mathcal{H}$ of the system, with corresponding *eigenvalue* E. In other words, the wave function of a quantum system (as described by Schrodinger's equation) is an eigenfunction of the observable of the system energy.

In fact, the Schrodinger equation is

$$ih\frac{\partial}{\partial t}\psi(x,t) = \mathcal{H}\psi(x,t)$$

with $\mathcal{H}$ as an operator on a complex Hilbert space H in a general formalism, where the wave function is an element of H: The Schrodinger's equation is an "equation" in this "Operators on Complex Hilbert spaces" formalism. This equation tells us clearly: It is precisely the observable of the energy that determines the time evolution of states of a quantum system. On the other hand, being an element in a separable Hilbert space, a wave function ψ can be decomposed as a linear surperposition of stationary states, corresponding to the fact that energy is quantified (i.e., having discrete levels of energy, corresponding to stationary states) . Specifically, the states (wave functions in the Schrodinger's equation) of the form $\psi(x,t) = \varphi(x)e^{-i\omega t}$ are *stationary states* since $|\psi(x,t)| = |\varphi(x)|$, independent of t, so that the probability density $|\varphi(x)|^2$ (of finding the particle in a neighbirhood of x) does not depend on time, resulting in letting anything in the system unchanged (not evoluting in time). That is the meaning of stationarity of a dynamical system (the system does not move). To have motion, the wave function has to be a linear superposition of stationary states in interference (as waves). And this can be formulated "nicely" in Hilbert space theory! Indeed, let φ_n be eigenfunctions of the Hamiltonian, then (elements of a separable Hilbert space have representations with respect to some orthonormal basis) $\psi(x,t) = \sum_n c_n \varphi_n(x)e^{-iE_n t/h}$, where $E_n = h\omega_n$ (energy level). Note that, as seen above, for stationary states $\varphi_n(x)e^{-iE_n t/h}$, we have $\mathcal{H}\varphi_n = E_n\varphi$, i.e., φ_n is an eigenfunction of $\mathcal{H}$. Finally, note that, from the knowledge of quantum physics where energy is quantified, the search for (discrete) energy levels $E_n = h\omega_n$ corresponds well to this formalism.

We can say that *Hilbert spaces and linear operators* on them form the *language of quantum mechanics*.

Thus, let's put down an abstract definition: *An observable is a bounded, linear, and self adjoint operator on a Hilbert space.*

We have seen that multiplication operator $M_f : g \in H = L^2(\mathbb{R}^3) \to M_f(g) = fg$ is self adjoint when f is real-valued. In particular, for $f = 1_B$, $B \in \mathcal{B}(\mathbb{R}^3)$, M_{1_B} is a (orthogonal) *projection* on H, i.e., satisfying $M_{1_B} = (M_{1_B})^2$ (idempotent) $= (M_{1_B})^*$, which is a special self adjoint operator. This will motive the space $\mathcal{P}(H)$ of all projections on H as the set of "events".

Each observable A is supposed to represent an underlying physical quantity. So, given a self adjoint operator A on H, what is the value that we are interested in, in a given state ψ? Well, it is $< \psi, A\psi >$ (e.g., $\int_{\mathbb{R}^3} \psi^*(x,t)A(\psi)(x,t)dx$), with, by abuse of language, is denoted as $< A >_\psi$. Note that $< \psi, A\psi > \in \mathbb{R}$, for any $\psi \in H$, since A is self adjoint, which is "consistent" with the fact that physical quantities are real-valued.

Remark. If we view the observable A as a random variable, and the state ψ as a probability measure on its "sampling space" H, in the classical setting of probability theory, then $< A >_\psi$ plays the role of expectation of A wrt the probability measure ψ. But here is the fundamental difference with classical probability theory: as operators, the "quantum random variables" do not necessarily commute, so that we are facing a *noncommutative probability theory*. This is compatible with the "matrix" viewpoint of quantum mechanics, suggested by Heisenberg, namely that numerical measurements in quantum mechanics should be matrices which form a noncommutative algebra.

A final remark. Now as the "density matrix" ρ on a general quantum probability space plays the role of an ordinary probability density function f (whose ordinary entropy is $-\int f(x)\log f(x)dx$), its *quantum entropy* (as defined by von Neumann von Neumann (2018)) is $-tr(\rho\log\rho)$. As maximum (Kolmogorov) entropy principle provides equilibrium models in statistical mechanics or other stochastic systems, it also enters financial econometrics as the most diversified portfolio selection, see e.g. Bera (2008), Golan et al. (1996).

The von Neunman's quantum entropy in a simple case, e.g., when $H = \mathbb{C}^n$, is this.

For a density matrix ρ (extension of a probability density function) on $(\mathbb{C}^n, \mathscr{P}(\mathbb{C}^n))$, $\rho\log\rho$ is a $n \times n$ self adjoint matrix (operator) which is defined as follows (by using spectral theorem). The spectral therem says this. Since ρ is a self adjoint operator on $\mathbb{C}^n$, there exists an orthonomal basis of $\mathbb{C}^n$, $\{u_1, u_2, \ldots, u_n\}$ consisting of eigenvectors of ρ, with associated eigenvalues $\{\lambda_1, \lambda_2, \ldots, \lambda_n\}$ (the spectrum of ρ). If we let P_j be the projector onto the closed subspace spanned by u_j, then $\rho = \sum_{j=1}^n \lambda_j P_j$.

For $g(.): \mathbb{R} \to \mathbb{R}$, $g(x) = x\log x$, the (self adjoint) operator $g(\rho) = \rho\log\rho$ is defined by $\sum_{j=1}^n g(\lambda_j)P_j$ whose trace is

$$tr(\rho\log\rho) = \sum_{j=1}^n <u_j, (\rho\log\rho)u_j> = \sum_{j=1}^n \lambda_j\log\lambda_j$$

so that the quantum entropy of ρ is $-tr(\rho\log\rho) = -\sum_{j=1}^n \lambda_j\log\lambda_j$ which depends only on the eigenvalues of ρ.

References

Bera, A., & Park, S. (2008). Optimal portfolio diversification using maximum entropy principle. *Econometric Review, 27*, 484–512.

Busemeyer, J. R., & Bruza, P. D. (2012). *Quantum models of cognition and decision*. Cambridge: Cambridge University Press.

Feynman, R. (1951). The concept of probability in quantum mechanics. In *Berkeley Symposium on Mathematical Statistics* (pp. 533–541). University of California.

Feynman, R., & Hibbs, A. (1965). *Quantum mechanics and path integrals*. New York: Dover.

Golan, A., Judge, G., & Miller, D. (1996). *Maximum entropy econometrics*. New York: Wiley.

Haven. E., & Khrennikov. A. (2013). *Quantum social science*. Cambridge: Cambridge University Press.

Herzog, B. (2015). Quantum models of decision-making in economics. *Journal of Quantum Information Science*, *5*, 1–5.

Nguyen, H. T. (2019). Toward improving models for decision making in economics. *Asian Journal of Economics and Banking*, *3*(01), 1–19.

Nguyen, H. T. (2006). *An introduction to random sets*. Boca Raton: Chapman and Hall/ CRC Press.

Parthasarathy, K. R. (1992). *An introduction to quantum stochastic calculus*. Basel: Springer.

Pasca, L. (2015). A critical review of the main approaches on financial market dynamics modeling. *Journal of Heterodox Economics*, *2*(2), 151–167.

Patra, S. (2019). A quantum framework for economic science: New directions (No. 2019-20). Economics.

von Neumann, J. (2018). *Mathematical foundations of quantum mechanics* (New ed.). Princeton: Princeton University Press.

Vukotic, V. (2011). Quantum economics. *Panoeconomics*, *2*, 267–276.

Extending the A Priori Procedure (APP) to Address Correlation Coefficients

Cong Wang, Tonghui Wang, David Trafimow, Hui Li, Liqun Hu, and Abigail Rodriguez

Abstract The present work constitutes an expansion of the a priori procedure (APP), whereby the researcher makes specifications for precision and confidence and APP equations provide the necessary sample size to meet specifications. Thus far, APP articles have mostly focused on the results of true experiments, where it is possible to randomly assign participants to conditions designed to create variance. But researchers in fields such as education and psychology often are faced with problems that are not amenable to the random assignment of participants to conditions, either because of ethical or feasibility issues. In such cases, researchers usually obtain sample correlation coefficient estimators to help them estimate corresponding population correlation coefficients. Unfortunately, there are no published APP articles to help these researchers determine the necessary sample size. The present work introduces new APP equations that remedy the lack.

Keywords A priori procedure · Correlation coefficient · Precision · Confidence

C. Wang · T. Wang (✉) · H. Li · L. Hu
Department of Mathematical Sciences, New Mexico State University, Las Cruces, USA
e-mail: twang@nmsu.edu

C. Wang
e-mail: cong960@nmsu.edu

H. Li
e-mail: huili@nmsu.edu

L. Hu
e-mail: hlq1994@nmsu.edu

D. Trafimow · A. Rodriguez
Department of Psychology, New Mexico State University, Las Cruces, USA
e-mail: dtrafimo@nmsu.edu

A. Rodriguez
e-mail: abigail9@nmsu.edu

© The Editor(s) (if applicable) and The Author(s), under exclusive license
to Springer Nature Switzerland AG 2021
N. Ngoc Thach et al. (eds.), *Data Science for Financial Econometrics*,
Studies in Computational Intelligence 898,
https://doi.org/10.1007/978-3-030-48853-6_10

1 Introduction

The a priori procedure (APP) developed as an alternative to the traditional null hypothesis significance testing procedure. The null hypothesis significance testing procedure results in a decision about whether to reject the null hypothesis, though there has been much debate about the validity of the procedure (see Hubbard 2016; Ziliak et al. 2016 for well-cited reviews). For present purposes, it is not necessary to engage this debate. It is merely necessary to acknowledge that it is possible to ask different questions than about whether to reject the null hypothesis. The APP is based on the idea that a desirable characteristic of sample statistics is that they are good estimates of corresponding population parameters. To see that this is crucial, imagine a scenario where sample means have absolutely nothing to do with corresponding population means. In this ridiculous scenario, the usefulness of sample means would be highly questionable. The simple fact of the matter is that scientists do expect sample means to be relevant to corresponding population means, differences in sample means to be relevant to corresponding differences in population means, and so on. But what do we mean when we say that sample statistics ought to be relevant to corresponding population parameters? There are at least two meanings. First, sample statistics hopefully are close to corresponding population parameters. Second, given some arbitrarily designated degree of closeness, there is hopefully a considerable probability that the sample statistics will be close to corresponding population parameters. In short, there are two preliminary questions always worth asking when planning an experiment, bullet-listed below. Precision: How close do I want the sample statistics to be to their corresponding population parameters? Confidence: With what probability do I want the sample statistics to be within the precision specification? Given that the researcher has specified the desired degree of precision and the desired degree of confidence, APP equation can be used to determine the necessary sample size to meet precision and confidence specifications. To clarify, consider a simple example. Suppose a researcher plans to collect a single group of participants and is interested in using the sample mean to estimate the population mean. The researcher wishes to have a 95% probability of obtaining a sample mean within 0.15 of a standard deviation of the population mean. Assuming normality and random and independent sampling, how many participants does the researcher need to collect to meet specifications? Trafimow (2017) proved that Eq. 2.1 can be used to answer the question, where n is the minimum sample size necessary to meet specifications for precision f and confidence, with z_c denoting the z-score that corresponds with the specified level of confidence c:

$$n = \left(\frac{z_c}{f}\right)^2.$$

Of course, researchers rarely are content with a single mean. Trafimow and MacDonald (2017) showed how to expand Eq. 2.1 to address any number of means. Trafimow et al. (in press) showed how to address differences in means, in contrast to the means themselves. Wang et al. (2019a, b) showed that it is possible to perform similar calculations under the larger family of skew normal distributions, and for locations rather than means. Trafimow et al. (under submission) have recently extended the APP to address binary data, where the statistic of interest is not a mean but rather a proportion. Or if there are two groups, the researcher might be interested in the difference in proportions across the two groups. Nevertheless, despite these APP advances, there remains an important limitation. Often in educational or psychological research, researchers are unable to randomly assign participants to conditions. Consequently, there is no way to assess the effects on means or differences between means. Instead, researchers may have to settle for naturally occurring relationships between variables. In short, the typical statistic of interest for researchers who are unable to perform experiments is the correlation coefficient. Unfortunately, there currently is no way for researchers to calculate the number of participants needed to be confident that the correlation coefficient to be obtained is close to the population correlation coefficient. The present goal is to remedy this deficiency. Specifically, our goal is to propose APP equations to enable researchers to determine the minimum sample size they need to obtain to meet specifications for precision and confidence for using the sample-based correlation coefficient estimator $\hat{\rho}$ to estimate the population correlation coefficient ρ.

2 Some Properties of Normal Distribution

Lemma 2.1 *Let X_1, X_2, ...,X_n be independent and identically distributed (i.i.d.) random variables from $N(\mu, \sigma^2)$, normal population with mean μ and variance σ^2. Let $S^2 = \frac{1}{n-1}\sum_{i=1}^{n}(X_i - \bar{X})^2$ be the sample variance. Then $\frac{(n-1)S^2}{\sigma^2} \sim \chi_{n-1}^2$, the chi-squared distribution with $n - 1$ degrees of freedom.*

Proposition 2.1 *Suppose that $(X_i, Y_i)'$ for $i = 1, ..., n$ be i.i.d random vectors from $N_2(\mu, \Sigma)$ where*

$$\mu = \begin{pmatrix} \mu_1 \\ \mu_2 \end{pmatrix}, \quad \Sigma = \begin{pmatrix} 1 & \rho \\ \rho & 1 \end{pmatrix}.$$

Consider $D_i = X_i - Y_i$. Then D_i's are i.i.d. random variables distributed $N(\mu_D, \sigma_D^2)$, where $\mu_D = \mu_1 - \mu_2$ and $\sigma_D^2 = 2(1 - \rho)$.

Let $\bar{D} = \frac{1}{n}\sum_{i=1}^{n} D_i$ and $S_D^2 = \frac{1}{n-1}\sum_{i=1}^{n}(D_i - \bar{D})^2$. By Lemma 2.1, we obtain

$$\frac{(n-1)S_D^2}{2(1-\rho)} \sim \chi_{n-1}^2. \tag{2.1}$$

Proposition 2.2 *Under the assumption in Proposition 2.1, if we denote $A_i = \frac{X_i+Y_i}{2}$, then A_i's are i.i.d. random variables distributed $N(\mu_A, \sigma_A)$, where $\mu_A = \frac{\mu_1+\mu_2}{2}$ and $\sigma_A^2 = \frac{1+\rho}{2}$*

Let $\bar{A} = \frac{1}{n}\sum_{i=1}^{n} A_i$ and $S_A^2 = \frac{1}{n-1}\sum_{i=1}^{n}(A_i - \bar{A})^2$. By Lemma 2.1, we obtain

$$\frac{2(n-1)S_A^2}{1+\rho} \sim \chi_{n-1}^2. \tag{2.2}$$

Now we consider the multivariate normal case. Let $\mathbf{X_1}$, $\mathbf{X_2}$, ..., $\mathbf{X_n}$ be i.i.d. random vectors from $N_p(\boldsymbol{\mu}, \Sigma)$, the p-dimensional normal distribution with mean vector μ and covariance matrix Σ, where

$$\boldsymbol{\mu} = \begin{pmatrix} \mu_1 \\ \mu_2 \\ \vdots \\ \mu_n \end{pmatrix}, \qquad \Sigma = \begin{pmatrix} 1 & \rho_{12} & \cdots & \rho_{1p} \\ \rho_{21} & 1 & \cdots & \rho_{2p} \\ \vdots & \vdots & \ddots & \vdots \\ \rho_{p1} & \rho_{p2} & \cdots & 1 \end{pmatrix},$$

where the random vector $\mathbf{X_k} = (X_{1k}, X_{2k}, , X_{pk})'$ for $k = 1, 2, \ldots, n$, and ρ_{ij}'s are unknown parameters to estimate for $i, j = 1, \ldots, p(i \neq j)$. For estimating ρ_{ij}, we consider the bivariate normal random vector $(X_k, Y_k)' \equiv (X_{ik}, Y_{jk})'$, which has

$$\begin{pmatrix} X_k \\ Y_k \end{pmatrix} \sim N_2\left[\begin{pmatrix} \mu_i \\ \mu_j \end{pmatrix}, \begin{pmatrix} 1 & \rho_{ij} \\ \rho_{ji} & 1 \end{pmatrix}\right],$$

for all $k = 1, 2, \ldots, n$ and $i, j = 1, \ldots, p(i \neq j)$. If we consider $D_k = X_k - Y_k$ and $A_k = \frac{X_k + Y_k}{2}$, the by Proposition 2.1 and 2.2, Equation (2.1) and (2.2) can be rewritten, respectively, as

$$\frac{(n-1)S_D^2}{2(1-\rho_{ij})} \sim \chi_{n-1}^2 \tag{2.3}$$

and

$$\frac{2(n-1)S_A^2}{1+\rho_{ij}} \sim \chi_{n-1}^2, \tag{2.4}$$

where $\bar{D} = \frac{1}{n}\sum_{k=1}^{n} D_k$, $S_D^2 = \frac{1}{n-1}\sum_{k=1}^{n}(D_k - \bar{D})^2$, $\bar{A} = \frac{1}{n}\sum_{k=1}^{n} A_k$ and $S_A^2 = \frac{1}{n-1}\sum_{k=1}^{n}(A_k - \bar{A})^2$.

Remark 2.1 Without loss of generality, we only consider the bivariate normal case for estimating ρ (or ρ_{ij}), given in next section below.

3   The Minimum Sample Size Needed for a Given Sampling Precision for Estimating the Correlation ρ

To estimate the correlation in a bivariate normal population, we can first determine the desired sample size if the difference of S_D^2 (or S_A^2) and its mean is given with respect to some confidence.

Theorem 3.1   *Under the assumption given in Proposition 2.1, let c be the confidence level and f be the precision which are specified such that the error associated with estimator S_D^2 is $E = f\sigma_D^2$. More specifically, if we set*

$$P\left[f_1\sigma_D^2 \le S_D^2 - E(S_D^2) \le f_1\sigma_D^2\right] = c, \tag{3.1}$$

where f_1 and f_2 are restricted by $\max\{|f_1|, f_2\} \le f$, and $E(S_D^2)$ is the mean of S_D^2, then the minimum sample size n required can be obtained by

$$\int_L^U f(z)dz = c \tag{3.2}$$

such that $U - L$ is minimized, where $f(z)$ is the probability density function of the chi-squared distribution with $n - 1$ degrees of freedom, and

$$L = (n-1)(1+f_1), \qquad U = (n-1)(1+f_2).$$

Proof. It is clear that $E(S_D^2) = 2(1-\rho)$. Then by simplifying (3.1), we obtain

$$P\left[(n-1)(1+f_1) \le Z \le (n-1)(1+f_2)\right] = c,$$

where $Z = \frac{(n-1)S_D^2}{2(1-\rho)}$ is of the chi-squared distribution with $n - 1$ degrees of freedom. If we denote

$$L = (n-1)(1+f_1), \qquad U = (n-1)(1+f_2),$$

then the required n can be solved through the integral equation (2.2). Solving for ρ from (2.1), we obtain

$$P\left[1 - \frac{(n-1)S_D^2}{2L} \le \rho \le 1 - \frac{(n-1)S_D^2}{2U}\right] = c. \tag{3.3}$$

Remark 3.1   Similarly, if conditions in Theorem 3.1 are satisfied for S_A^2, then we obtain the same sample size n so that

$$P\left[\frac{2(n-1)S_A^2}{U} - 1 \le \rho \le \frac{2(n-1)S_A^2}{L} - 1\right] = c. \tag{3.4}$$

Remark 3.2 The expected lengths for confidence intervals given in (3.3) and (3.4) are, respectively,

$$(n-1)(1-\rho)(\frac{1}{L}-\frac{1}{U}) \quad \text{and} \quad (n-1)(1+\rho)(\frac{1}{L}-\frac{1}{U}).$$

Terefore, if $\rho \le 0$, Equation (2.4) should be used, otherwise, (2.3) is preferred.

4 The Simulation Study

We perform computer simulations to support the derivation in Sect. 2. Without loss of generality, we assume, in this paper, that the sample is from the bivariate normal population in Proposition 2.1. For given confidence level $c = 0.95, 0.9$, Table 1 below shows the desired sample size and the lengths of the intervals U-L for shortest (SL) and equal-tail (EL) cases for different precisions $f = 0.2, 0.4, 0.6, 0.8, 1$.

Using the Monte Carlo simulations, we account relative frequency and the corresponding average point estimates for different valve of ρ. Tables 2 and 3 show the result for the relative frequency of 95% confidence intervals constructed by S_D^2 and S_A^2, respectively, for $f = 0.2, 0.4, 0.6, 0.8, 1$, and $\rho = 0.9, 0.5, 0$, and all results are illustrated with a number of simulation runs M = 10000.

Based on Tables 2, 3, 4 and 5 list the average length of the 95% confidence intervals of ρ after 10000 runs. It is instructive to compare across the two tables. When the user-defined correlation coefficients are positive, the method featuring S_D^2 gives shorter sample-based confidence intervals and is therefore to be preferred. However, when the user-defined correlation coefficients are negative, the method featuring S_A^2 gives shorter sample-based confidence intervals and is therefore to be

Table 1 The value of sample size n under different precision f for the given confidence $c = 0.95, 0.9$ where SL and EL are the corresponding shortest and equal-tail lengths, respectively

f	c	n	SL	EL
0.2	0.95	197	0.58	0.58
	0.9	134	0.59	0.6
0.4	0.95	51	0.77	0.78
	0.9	35	0.78	0.81
0.6	0.95	23	0.94	0.97
	0.9	16	0.98	1.02
0.8	0.95	14	1.11	1.16
	0.9	9	1.18	1.25
1	0.95	9	1.11	1.16
	0.9	5	1.18	1.25

Table 2 The relative frequency (r.f.) and the corresponding average point estimates of different values of ρ (p.e.) related to the confidence interval constructed by S_D^2 for different precision with $c = 0.95$

f	n	$\rho = -0.9$		$\rho = -0.5$		$\rho = 0$		$\rho = 0.5$		$\rho = 0.9$	
		r.f.	p.e.	r.f.	p.e.	r.f.	p.e.	r.f.	p.e.	r.f.	p.e.
0.2	197	0.9482	−0.8990	0.9523	−0.4991	0.9466	−0.0001	0.9479	0.5000	0.9495	0.8997
0.4	51	0.9528	−0.9042	0.9493	−0.5003	0.9496	0.0035	0.9500	0.4990	0.9502	0.9000
0.6	23	0.9491	−0.8992	0.9475	−0.5018	0.9549	0.0017	0.9457	0.4975	0.9509	0.9001
0.8	14	0.9494	−0.9101	0.9458	−0.5037	0.9516	−0.0019	0.9502	0.5008	0.9498	0.8996
1	9	0.9493	−0.8963	0.9493	−0.5055	0.9529	0.0035	0.9521	0.4974	0.9521	0.9001

Table 3 The relative frequency (r.f.) and the corresponding average point estimates of different values of ρ (p.e.) related to the confidence interval constructed by S_A^2 for different precision with $c = 0.95$

f	n	$\rho = -0.9$		$\rho = -0.5$		$\rho = 0$		$\rho = 0.5$		$\rho = 0.9$	
		r.f.	p.e.	r.f.	p.e.	r.f.	p.e.	r.f.	p.e.	r.f.	p.e.
0.2	197	0.9482	−0.8998	0.9479	−0.5006	0.9510	−0.0014	0.9499	0.4977	0.9490	0.9022
0.4	51	0.9513	−0.8999	0.9486	−0.5002	0.9505	−0.004	0.9486	0.5014	0.9511	0.8980
0.6	23	0.9546	−0.8997	0.9487	−0.5012	0.9514	−0.0032	0.9449	0.5032	0.9497	0.9072
0.8	14	0.9485	−0.9006	0.9476	−0.5003	0.9516	−0.0008	0.9478	0.5065	0.9509	0.9008
1	9	0.9492	−0.8992	0.9494	−0.4987	0.9501	−0.0008	0.9503	0.4943	0.9509	0.8973

Table 4 The average length of the 95% confidence intervals constructed using the S_D^2 method

f	$\rho = -0.9$	$\rho = -0.5$	$\rho = 0$	$\rho = 0.5$	$\rho = 0.9$
0.2	0.7753	0.6121	0.4083	0.2041	0.0408
0.4	1.6895	1.3311	0.8842	0.4445	0.0887
0.6	3.0080	2.3786	1.5811	0.7959	0.1582
0.8	4.9829	3.9228	2.6137	1.3023	0.2619
1	9.7293	7.7242	5.1127	2.5787	0.5126

preferred. When the user-defined correlation coefficient is zero, the sample-based confidence intervals are similar for the two methods.

We present Figs. 1 and 2 to reinforce that the method featuring S_D^2 is preferred for positive population correlation coefficients, the method featuring S_A^2 is preferred for negative population correlation coefficients, and there is methodological indifference when the population correlation coefficient is zero. To demonstrate this last, the next graphs (Figs. 1 and 2) show the density curves of $\hat{\rho}$ constructed by S_D^2 and S_A^2, respectively, for precision $f = 0.6$, $c = 0.95$ when $\rho = 0$, 0.5 .

Table 5 The average length of the 95% confidence intervals constructed using the S_A^2 method

f	$\rho = -0.9$	$\rho = -0.5$	$\rho = 0$	$\rho = 0.5$	$\rho = 0.9$
0.2	0.0409	0.2039	0.4077	0.6115	0.7767
0.4	0.0888	0.4435	0.8872	1.3321	1.6840
0.6	0.1589	0.7900	1.5788	2.3808	3.0207
0.8	0.2593	1.3036	2.6066	3.9301	4.9587
1	0.5172	2.5720	5.1266	7.6667	9.7344

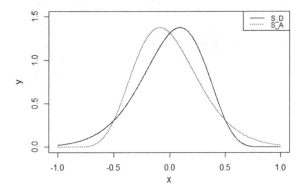

Fig. 1 The density curves of $\hat{\rho}$ constructed by S_D^2 (S_D) and S_A^2 (S_A) for $f = 0.6$ and $c = 0.95$ when $\rho = 0$

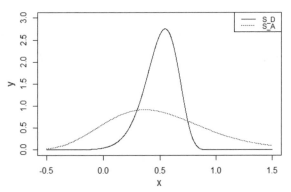

Fig. 2 The density curves of $\hat{\rho}$ constructed by S_D^2 (S_D) and S_A^2 (S_A) for $f = 0.6$ and $c = 0.95$ when $\rho = 0.5$

References

Hubbard, R. (2016). *Corrupt research: The case for reconceptualizing empirical management and social science*. Los Angeles, California: Sage Publications.

Trafimow, D. (2017). Using the coefficient of confidence to make the philosophical switch from a posteriori to a priori inferential statistics. *Educational and Psychological Measurement*, 77(5), 831–854. https://doi.org/10.1177/0013164416667977.

Trafimow, D., & MacDonald, J. A. (2017). Performing inferential statistics prior to data collection. *Educational and Psychological Measurement*, 77(2), 204–219. https://doi.org/10.1177/0013164416659745.

Trafimow, D., Wang, C., & Wang, T. (in press). Making the a priori procedure (APP) work for differences between means. *Educational and Psychological Measurement.*

Wang, C., Wang, T., Trafimow, D., & Myuz, H. A. (2019). Necessary sample size for specfied closeness and confidence of matched data under the skew normal setting. *Communications in Statistics-Simulation and Computation.* https://doi.org/10.1080/03610918.2019.1661473.

Wang, C., Wang, T., & Trafimow, D. (2019). Extending a priori procedure to two independent samples under skew normal setting. *Asian Journal of Economics and Banking, 03*(02), 29–40. ISSN 2588-1396.

Ziliak, S. T., & McCloskey, D. N. (2016). *The cult of statistical significance: How the standard error costs us jobs, justice, and lives.* Ann Arbor, Michigan: The University of Michigan Press.

Variable Selection and Estimation in Kink Regression Model

Woraphon Yamaka

Abstract Recently, regression kink model has gained an increasing popularity as it provides a richer information than the ordinary linear model in the light of an economic structural change. However, as the number of parameters in the kink regression model is larger than that of the linear version, the traditional least squares estimates are not valid and may provide infinite solutions, especially when the number of observations is small and there are many coefficients. To deal with this problem, the LASSO variable selection method is suggested to estimate the unknown parameters in the model. It not only provides the estimated coefficients, but also shrinks the magnitude of all the coefficients and removes some whose values have been shrunk to zero. This process helps decrease variance without increasing the bias of the parameter estimates. Thus, LASSO could play an important role in the kink regression model building process, as it improves the result accuracy by choosing an appropriate subset of regression predictors.

Keywords TLASSO · Kink regression · Variable selection method

1 Introduction

After it was introduced by Card et al. (2015) and extended by Hansen (2017) as a modification of the discontinuous threshold model with unknown threshold, the continuous kink regression model has gained an increasing popularity due to its two main advantages. First, it is a special case of the threshold regression with a constraint that the regression function must be continuous everywhere, whereas the threshold regression model is discontinuous at the threshold. Hansen (2017) showed successfully that the continuous kink regression model is appealing for empirical applications where there is no reason to expect a discontinuous regression response at

W. Yamaka (✉)
Faculty of Economics, Center of Excellence in Econometrics,
Chiang Mai University, Chiang Mai 50200, Thailand
e-mail: woraphon.econ@gmail.com

© The Editor(s) (if applicable) and The Author(s), under exclusive license
to Springer Nature Switzerland AG 2021
N. Ngoc Thach et al. (eds.), *Data Science for Financial Econometrics*,
Studies in Computational Intelligence 898,
https://doi.org/10.1007/978-3-030-48853-6_11

the threshold (because the data requires that the broken slopes be connected). Second, it provides a relatively richer information regarding the effects of the predictors on the response variable in different regimes compared to the regular linear regression. These two appealing features of kink regression result in its application in many studies such as Maneejuk et al. (2016), Sriboochitta et al. (2017), Tibprasorn et al. (2017), Lien et al. (2017), to name just a few. Consider the simple two-regime kink regression model.

$$
\begin{aligned}
Y_i = \beta_0 &+ \beta_1{}^-(x_{1,i} - \gamma_1)_- + \beta_1{}^+(x_{1,i} - \gamma_1)_+ +, \ldots, \\
&+ \beta_p{}^-(x_{p,i} - \gamma_p)_- + \beta_p{}^+(x_{p,i} - \gamma_p)_+ + \varepsilon_i,
\end{aligned}
\tag{1}
$$

where Y_i is $(n \times 1)$ response variable of sample i, $x_{j,i}$ is $(n \times 1)$ regime dependent predictor variable j of sample i, $j = 1, \ldots, p$. The relationship between Y_i and $x_{j,i}$ is nonlinear. Therefore, the relationship of $x_{j,t}$ with Y_i changes at the unknown location or threshold or kink point γ_j. We use $(x_{j,i})_- = \min[x_{j,i}, 0]$ and $(x_{j,i})_+ = \max[x_{j,i}, 0]$ to denote the "negative part" and "positive part" of an $x_{j,i}$. Therefore, β_j^- is the negative part coefficient of $x_{j,t}$ for value of $x_{j,t} \leq \gamma_j$, while, β^+ is the positive part coefficient of $x_{j,i}$ for value of $x_{j,i} \leq \gamma_j$. Thus, there is $P = (1 + (p \times 2))$ vector of the estimated coefficients in this model. Hansen (2017), the response variables are subject to regime-change at unknown threshold or kink point. Therefore, the model can separate the data into two or more regimes. ε_i denotes the noise (or error) term for sample i which are independent and identically distributed random variables with mean zero and finite variance σ^2.

This study is conducted with an interest to estimate the kink regression model when P is large or even larger than n and the regime dependent coefficients are sparse in the sense that many of them are zero. In the practical situation, there is likely a nonlinear causal effect of predictors on response variable; and under this scenario, the traditional linear regression may not be appropriate. Although the kink regression model may be proposed to apply for the practical reason and it could fit well when the structural change exists, it may still face two main problems. First, there is a large number of parameter estimates which can reduce the accuracy and efficiency of the estimation. For example, given a j predictor $x_{j,i}$ where $j = 1, \ldots, 10$, there will be $P = (1 + (10 \times 2)) = 21$ parameters to be estimated. Second, kink regression can take many forms depending on what happens at the threshold Fong et al. (2017) and it could not answer the question which would be the most likely one. For example, for each predictor, there may exist the zero slope either before or after the kink point. Thus, if there are 10 predictors; some may have the nonlinear relationship with the response variable, while the rest have a linear relationship. As the model contains many predictors, it becomes necessary to test whether or not each predictor $x_{j,i}$ has a nonlinear relationship with Y_i. Followings Hansen (2017), testing for a threshold effect can be done by comparing the linear and the kink model. He proposed using the F-test to test the hypothesis $H_0 : \beta_j^- = \beta_j^+$. This testing is useful, and can be applied to examine the threshold effect in the model. This null hypothesis however involves only one parameter; thus, each predictor has to be tested individually. This means a lot of effort is needed for testing hypothesis for each pair of predictor and response. Also, in practice, researchers may misspecify their models or may include in their models redundant variables not truly related to the response.

With these two problems, estimating a kink regression model with large P is challenging and the traditional least squares method may not be applicable. Hence, predictor variable selection and threshold effect testing become the essential parts of kink regression model. Over the years, various procedures have been developed for variable selection in regression model such as forward and backward stepwise selection Froymson (1960), Ridge regression Hoerl and Kennard (1970) and Least Absolute Shrinkage and Selection Operator (LASSO) Tibshirani (1996). These methods have been introduced to solve the traditional least squares model and find the best subset of predictors for the final model. In this study, only Lasso approach is considered for estimating kink regression. It is believed that the problems of predictor variable selection and threshold effect testing could be simultaneously solved within this approach. Compared to the classical variable selection methods and Ridge, the LASSO has several advantages. 1) LASSO involves a penalty factor that determines how many coefficients are retained; using cross-validation to choose the penalty factor helps assure that the model will generalize well to future data samples. In contrast, the stepwise selection chooses coefficients with the lowest p-values. This technique has many drawbacks. Hastie et al. (2005) noted that these methods cannot be used for high dimensional data. Specifically, when the number of predictor variables are large or if the predictor variables are highly correlated. Backward selection is limited to be used when $P > n$ and forward selection is not computationally possible if the amount of data is large Tibshirani (1996), Fokianos (2008). In addition, the validity of p-value has been questioned in many works and recently banned by the American Statistical Association in 2016 (see, Wasserstein and Lazar 2016. 2) Ridge regression does not attempt to select coefficients at all. It instead uses a penalty applied to the sum of the residual squares in order to force the least squares method applicable when $P > n$. As a result, ridge regression is restricted to include all predictor variables in the model. Thus, when p is very large, the model is difficult to interpret.

In this study, the problem of estimating kink regression model is reframed as a model selection problem, and the LASSO method is introduced to give a computationally feasible solution. The estimation can be performed in high dimensional data as the ℓ_1−penalization can set some coefficients to exactly zero, thereby excluding those predictors from the model. Thus, the kink regression can be solved even the number of predictors exceeds the sample size. This study will also show that the number and the location of the kink points or thresholds can be consistently estimated.

The remainder of this article is organized as follows. The estimation procedure is described in Sect. 2. Simulation studies are presented in Sect. 3. The proposed methods are applied for a real data analysis in Sect. 4. The conclusions are provided in Sect. 5.

2 Estimation

This section introduces a one-step procedure for consistent and computationally efficient kink estimation using LASSO. The estimation is made on a set of potential kink parameters for all predictor variables as well as their corresponding coefficients.

2.1 LASSO Estimation for Kink Regression

According to the kink regression model Eq. (1), the vector of kink or threshold parameters $\gamma = \gamma_1, \ldots, \gamma_p$ are non-zero and there is a sparse solution to the high dimension $\beta = (\beta_0, \beta_1^-, \ldots, \beta_p^-, \beta_1^+, \ldots, \beta_p^+)$ in Eq. (1). Thus, γ and β can be estimated by the following group LASSO equation:

$$S_n(\beta, \gamma) = \frac{1}{n} \sum_{i=1}^n \left(Y_i - \beta' x_i(\gamma) \right)^2 + \lambda \sum_{j=1}^P |\beta_j|, \tag{2}$$

where

$$x(\gamma) = \begin{pmatrix} x_{j,i} - \gamma_j \\ \vdots \\ x_{p,i} - \gamma_p \end{pmatrix}. \tag{3}$$

and $\lambda \geq 0$ is a nonnegative regularization parameter. The second term in Eq. (2) is the LASSO penalty which is crucial for the success of the LASSO method. The LASSO shrinks kink regression coefficients towards zero as λ increases. Note that when $\lambda = 0$ the penalized loss function equals the loss function without a penalty (LASSO reverses to OLS estimation) while $\lambda = \infty$ yields an empty model, where all coefficients become zero. In this sense, λ can be viewed as the tuning parameter for controlling the overall penalty level.

From the LASSO estimation point of view, $(\widehat{\beta}, \widehat{\gamma})$ is the joint minimizer of $S_n(\beta, \gamma)$:

$$(\widehat{\beta}, \widehat{\gamma}) = \underset{\beta \in R^P, \gamma \in \Gamma}{arg \min} \ S_n(\beta, \gamma) \tag{4}$$

In the view of Eq. (4), we penalize only β and opted to penalize γ since it is worth nothing when γ is shrunk to zero. As a remark, if the group of predictor variables is highly correlated to each other, the LASSO tends to randomly select some predictor variables of that group and neglect the remaining predictors. Under this minimization problem Eq. (4), the parameter estimates and the set of selected variables depend on the regularization parameter λ which cannot be obtained directly from the minimization in Eq. (4). Thus, various methods such as Information criteria and Cross-validation are often adopted to find the optimal $\widehat{\lambda}$. The selection of the method mostly depends on the objectives and the setting in particular, the aim of the analysis (prediction or model identification), computational constraints, and if and how the *i.i.d.* assumption is violated. In this study, the aim is just to introduce the LASSO to estimate and find the best specification structure of the kink regression. Therefore, only the Information criteria are considered to search for the optimal as it provides a better choice than cross-validation. Stone (1977) suggested that information criteria, such as Akaike information criterion (AIC) and Bayesian information criterion (BIC) are superior in practice as they involve a relatively lower computational cost compared to cross-validation. In addition, information criteria and cross-validation have asymptotically equivalent assumptions e.g. homoskedasticity are satisfied.

As there is no closed-form solution of the lasso optimization. Thus, a simple modification of the LARS algorithm Efron et al. (2004) is considered to solve the problem of Eq. (4).

2.2 Tuning Parameter Selection Using Information Criteria

This section describes how the information criterion is used to find the regularization parameter λ and the fitted model. Information criterion is a validation method that is used to examine the performance of a model and compare the model with the alternatives. It is a widely used method for model selection. It is easy to compute once the coefficient estimates are obtained. Thus, it seems natural to utilize the strengths of information criterion as model selection procedure to select the penalization level. There are various types of information criteria and the most popular ones are AIC and BIC. However, both are not appropriate in the case of $P > n$ Ahrens et al. (2019). Therefore, the extended BIC of Chen and Chen (2008) is considered to find the optimal $\widehat{\lambda}$. The Extended BIC is defined as

$$EBIC(\lambda) = n \log \left(\widehat{\sigma}^2(\lambda) \right) + v(\lambda) \log(n) + 2\xi v(\lambda) \log(p), \qquad (5)$$

where $v(\lambda)$ is the degrees of freedom, which is a measure of model complexity and computed by the number of observations minus the number of non-zero coefficients. $\xi \in [0, 1]$ is the parameter controlling the size of the additional penalty. $\widehat{\sigma}^2(\lambda)$ is the variance of the kink regression model which is computed by

$$\widehat{\sigma}^2 = n^{-1} \sum_{i=1}^{n} \left(\widehat{\varepsilon}_i(\lambda) \right)^2 \qquad (6)$$

where $\widehat{\varepsilon}_i(\lambda)$ is the residuals conditional on candidate λ. The optimal $\widehat{\lambda}$ corresponds to the lowest $EBIC(\lambda)$.

3   Monte Carlo Experiments

In this section, simulation studies are conducted to evaluate the finite sample performance of the LASSO for fitting kink regression model. Simulations for LS and Ridge are also run to make the comparison with the LASSO. Hence this section involves three different experiment studies with different scenarios.

3.1   First Experiment Study: Accuracy of the LASSO Estimation

In the first study, the investigation is made on the accuracy of the LASSO estimation for kink regression and also the effect of sample size. The simulated data are drawn from the following model

$$Y_i = \beta_0 + \beta_1^- \left(x_{1,i} - \gamma_1\right)_- + \beta_1^+ \left(x_{1,i} - \gamma_1\right)_+ + \varepsilon_i \qquad (7)$$

First $x'_{1,t} \sim Unif[0, 5]$ is simulated and given the value of the kink or threshold parameter $\gamma_1 = 3$. The true values for coefficient parameters are $\beta_0 = 1$, $\beta_1^- = 2$, and $\beta_1^+ = -1$. The error ε_i is simulated from $N(0, 1^2)$. Sample sizes $n = 100$, $n = 300$ and $n = 1000$ are considered. Then, the performance of the estimator is evaluated in terms of the absolute Bias and Mean Squared Error of each parameter based on 100 replications.

Table 1 contains the results of the first experiment for LASSO, Ridge and LS over the 100 simulated data sets with three different dimensions for the sample size. The most important finding in Table 1 is that the LASSO estimation provides the reliable parameter estimates as the Biases and MSEs are close to zero. In addition, the Biases and MSEs seem to be lower as the sample size increases. The performance of the LASSO is also comparable with the robust methods. Apparently, these three estimations perform a little differently. The Bias of LASSO seems to be higher than Ridge and LS for some parameters. However, LASSO shows the best results regarding MSE. This is somewhat not surprising since the LS and Ridge methods are usually expected to be unbiased but to have relatively high variance, and Lasso can make up the flaw from its higher bias with its smaller variance.

Table 1 Simulation results with different sample sizes

	LASSO		Ridge		LS	
$n = 100$	Bias	MSE	Bias	MSE	Bias	MSE
β_0	0.0066	0.0296	0.0031	0.0361	0.003	0.0366
β_1^-	0.0323	0.0428	0.0263	0.043	0.0261	0.0431
β_1^+	0.0473	0.0719	0.053	0.1003	0.0532	0.1018
γ_1	0.0289	0.0514	0.0241	0.0521	0.024	0.0525
$n = 300$						
β_0	0.007	0.0234	0.0033	0.0326	0.0032	0.0327
β_1^-	0.0274	0.0393	0.0236	0.0434	0.0236	0.0435
β_1^+	0.0504	0.0833	0.0541	0.098	0.0542	0.0983
γ_1	0.0273	0.0395	0.0239	0.0435	0.0239	0.0436
$n = 1000$						
β_0	0.0082	0.0207	0.0055	0.0256	0.0055	0.0256
β_1^-	0.0269	0.0323	0.0244	0.0422	0.0244	0.0423
β_1^+	0.0525	0.0622	0.0544	0.0818	0.0544	0.1019
γ_1	0.0271	0.0373	0.0213	0.0423	0.0243	0.0423

3.2 Second Experiment Study: Accuracy of the LASSO Estimation Which Corresponds to the Sparse Case

This second part deals with the performance of LASSO when the irrelevant variables are included in the true model. To this end, consider a Kink regression model as follows:

$$
\begin{aligned}
Y_i = {} & \beta_0 + \beta_1^-(x_{1,i} - \gamma_1)_- + \beta_1^+(x_{1,i} - \gamma_1)_+ + \\
& \beta_2^-(x_{2,i} - \gamma_2)_- + \beta_2^+(x_{2,i} - \gamma_2)_+ + \\
& \beta_3^-(x_{3,i} - \gamma_3)_- + \beta_3^+(x_{3,i} - \gamma_3)_+ + \\
& \beta_3^-(x_{3,i} - \gamma_3)_- + \beta_3^+(x_{3,i} - \gamma_3)_+ + \\
& \beta_5^-(x_{5,i} - \gamma_5)_- + \beta_5^+(x_{5,i} - \gamma_5)_+ + \varepsilon_i.
\end{aligned}
\tag{8}
$$

The data are generated from the model Eq. (8) where $\beta = (1, 2, 1, 2, 1, 2, 1, 0, 0, 0, 0)$. The threshold parameter is set to $\gamma = (1, 1.5, 0.8, 1.2, 0.5)$. The error ε_i is simulated from $N(0, 1^2)$. In this simulation, what is not examined is the effect of the sample size on the performance of the LASSO. Here only the sample size $n=500$ is considered and run 100 trials.

Figure 1 summarizes the results of the second simulation stud by plotting the boxplots of each $\hat{\beta}$ for each of the methods. According to the result, the LASSO method generally outperforms OLS (Green boxplot) and Ridge (Yellow boxplot) especially when the true coefficients are zero. The average of 100 parameter estimates are close to the true value and the standard deviations are small, especially when the

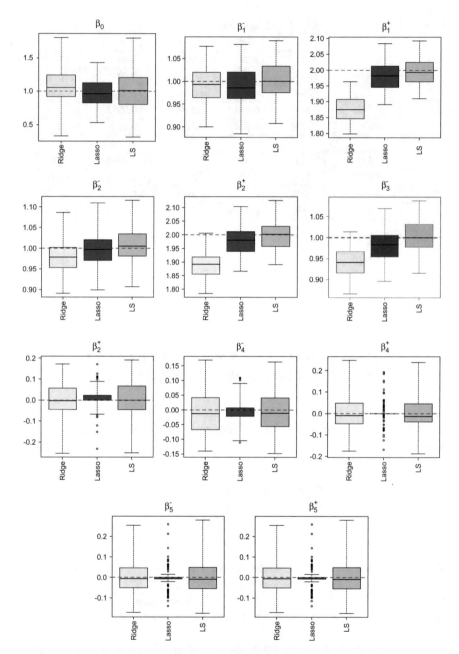

Fig. 1 Boxplots of the kink regression parameters: comparison among the ridge, LASSO and LS estimates of the kink regression parameters ($n = 500$ with 100 repetitions). The averages of the parameter estimates are shown with a black horizontal line in each boxplot. The red dashed horizontal line represents the true value of the kink regression parameter. (the threshold parameter is omitted in this figure)

true coefficients are zero. Thus. It can be concluded that the LASSO estimates the kink regression parameters quite well.

3.3 Third Experiment Study: Accuracy of the LASSO Estimation Under Different Number of Predictors

Finally, an experiment is conducted to prove the accuracy of variable selection. The number of true regression coefficients is varied to check if the variable selection method will perform differently while the number of variables changes. The simulation set-up is similar to those of the first two simulation studies. The data are generated from the model Eq. (1). To simplify the experiment, only one simulated dataset is used for performance comparison. Consider the following three different settings for β:

- **Simulation I**: The case of $P < n(P = 50, \ n = 100)$. The two-regime coefficients are set to pairwise be 0 and 1 for β^- and be 0 and 2 for β^+, Thus

$$\beta^- = (1, 1, 1, 1, 1, 0, 0, 0, 0, 0, 1, 1, 1, 1, 1, 0, 0, 0, 0, 0, ...)$$

$$\beta^+ = (2, 2, 2, 2, 2, 0, 0, 0, 0, 0, 2, 2, 2, 2, 2, 0, 0, 0, 0, 0, ...).$$

This means that the components 1–5 of β^- and β^+ are 1 and 2, respectively, the components 6–10 of β^- and β^+ are 0, the components 11–15 of β^- and β^+ are 1 and 2, respectively, again, and so on. In this simulation,
- **Simulation II**: The case of $P = n(P = 100, \ n = 100)$ given $\beta^- = 1$ and the remaining coefficients of regime 2 to be zero, $\beta^+ = 0$. Thus,

$$\beta^- = (1, \ldots, 1)$$

$$\beta^+ = (0, \ldots, 0).$$

- **Simulation III:** The case of $P > n(P = 120, \ n = 100)$. Similar to Simulation I, the two-regime coefficients are set to pairwise be 0 and 1 for β^- and be 0 and 2 for β^+, but the coefficients are split into groups: In regime 1, the first 30 β^- coefficients are assigned to be 1 and the remaining 30 β^- coefficients to be 0. For the coefficients of regime 2, the first 30 β^+ coefficients are assigned to be 2 and the remaining 30 β^+ coefficients to be 0. Thus,

$$\beta^- = (1, \ldots, 1, 0, \ldots, 0)$$

$$\beta^+ = (2, \ldots, 2, 0, \ldots, 0).$$

Plotted in Fig. 2 are the coefficient index and the value of the estimated parameter on the X-axis and Y-axis, respectively, for simulation data presented in this second experiment. For the first simulation case, $P < n$, the estimated coefficient results of LASSO, Ridge and LS are not much different. However, we could see that LASSO

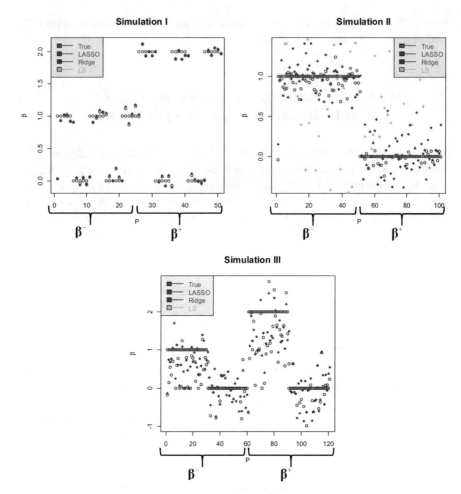

Fig. 2 The coefficients plot for true values (red circle), LASSO (blue circle), Ridge (purple solid circle) and LS (orange dot) under different number of predictors

sets three out of twenty coefficients to exactly zero, thereby excluding these predictors from the model (see, Table 2). According the Simulation I result, when the number of predictors is less than the number of observations, the similar performance of LASSO, Ridge and LS are obtained. For simulation II, we can see that LASSO generally performs better than the other two methods as Ridge and LS estimates are far away from the true values. In this case, LASSO sets twenty out of fifty coefficients to exactly zero. Finally, when $P > n$, the LS is as expected no longer a unique least squares coefficient estimate and its variance is found to be infinite so the LS estimation cannot be used and consequently its results are not reported. From simulation III results, the performance comparison is made only between Ridge and LASSO. Clearly, LASSO performs better than Ridge, especially when the true

Table 2 Simulation study, comparison of Ridge, LASSO, and LS

$P < n$	BIC	Number $\beta = 0$	True Number of $\beta = 0$
Ridge	−12.3261	0	20
LASSO	−13.2044	3	20
LS	−12.6301	0	20
$P = n$			
Ridge	138.496	0	50
LASSO	113.716	20	50
LS	250.015	0	50
$P > n$			
Ridge	433.527	0	60
LASSO	326.272	30	60
LS	NA	0	60

coefficient is zero as 30 out of 60 coefficients are set to be zero. One can also observe that the coefficient estimates from Ridge show high variability (Purple solid circle). Furthermore, the summary performance of the three estimations are shown in Table 2.

4 Data Example

This section compares the performance of the three methods LASSO, Ridge, and LS using the R built-in dataset "state.x77" in "EBglmnet" package. as an example, which includes a matrix with 50 rows and 8 columns giving the following measurements in the respective columns: Population (POP), Income (INC), Illiteracy (IL), Life Expectancy ($LIFE$), Murder Rate (M), High School Graduate Rate (HS), Days Below Freezing Temperature (DAY), and Land Area (A) Huang et al. (2016). The regression model under the present consideration has the following form

$$
\begin{aligned}
LIFE_i = \beta_0 &+ \beta_1^-(POP_i - \gamma_1)_- + \beta_1^+(POP_i - \gamma_1)_+ + \\
&\beta_2^-(INC_i - \gamma_2)_- + \beta_2^+(INC_i - \gamma_2)_+ + \\
&\beta_3^-(IL_i - \gamma_3)_- + \beta_3^+(IL_i - \gamma_3)_+ + \\
&\beta_4^-(M_i - \gamma_4)_- + \beta_4^+(M_i - \gamma_4)_+ + \\
&\beta_5^-(HS_i - \gamma_5)_- + \beta_5^+(HS_i - \gamma_5)_+ + \\
&\beta_6^-(DAY_i - \gamma_6)_- + \beta_6^+(DAY_i - \gamma_6)_+ + \\
&\beta_7^-(A_i - \gamma_7)_- + \beta_7^+(A_i - \gamma_7)_+ + \varepsilon_i.
\end{aligned}
\tag{9}
$$

There are 22 parameters to be estimated by the three estimation methods (7 parameters in β^-, 7 parameters in β^+, 7 parameters in γ, and the intercept term β_0). Table 3 summarizes the model selection and reports the estimated parameters obtained from the three estimations.

Table 3 Model selection and estimation results for kink regression specification

Parameter	Ridge	LASSO	LS
β_0	4.2655	4.2751	4.2668
$(POP - \gamma_1)_-$	0.008	0.0012	0.0131
$(POP - \gamma_1)_+$	0.0542	0.0173	0.0338
$(INC - \gamma_2)_-$	−0.0011	**0.0000**	0.0017
$(INC - \gamma_2)_+$	−0.0156	−0.0072	−0.0173
$(IL - \gamma_3)_-$	0.0249	0.0174	0.0244
$(IL - \gamma_3)_+$	−0.0061	−0.0011	−0.0111
$(M - \gamma_4)_-$	−0.0023	**0.0000**	−0.0052
$(M - \gamma_4)_+$	0.001	**0.0000**	0.0012
$(HS - \gamma_5)_-$	−0.0166	**0.0000**	−0.0365
$(HS - \gamma_5)_+$	−0.0066	**0.0000**	−0.0131
$(DAY - \gamma_6)_-$	−0.0247	−0.0253	−0.0247
$(DAY - \gamma_6)_+$	0.0386	**0.0000**	0.0001
$(A - \gamma_7)_-$	0.0009	**0.0000**	0.0064
$(A - \gamma_7)_+$	0.0038	**0.0000**	0.0094
γ_1	7.6537 (2.3389)	7.1658 (0.0454)	7.6539 (2.2352)
γ_2	8.2705 (0.8127)	8.3849 (0.0315)	8.2706 (1.0236)
γ_3	0.0344 (25.9482)	0.0321 (0.0505)	0.0375 (43.7741)
γ_4	1.8882 (426.0850)	1.7281 (0.0323)	1.8781 (59.2516)
γ_5	4.0628 (1.2579)	4.3563 (0.0536)	4.2625 (4.1158)
γ_6	4.1715 (73.74620)	4.3989 (0.0236)	3.9519 (40.2843)
γ_7	10.3127 (27.9538)	10.662 (0.0289)	10.2681 (40.5378)
$EBIC$	−419.378	−424.049	−413.7434

Note ( ) denotes standard error and the bolded number is the shrinkage coefficient

The main empirical findings are as follows. First, LASSO outperforms Ridge and LS in this real dataset as it shows the lowest EBIC. Therefore, the performance of the kink regression is enhanced with the removal of the irrelevant variables. Second, the variable selection result obtained from LASSO shows that 8 out of 15 coefficients (β_2^-, β_4^-, β_4^+, β_5^-, β_5^+, β_6^+, β_7^-, and β_7^+), are shrunk to zero. This means that LASSO has an ability to remove the irrelevant regime dependent variables from the model. For the remaining coefficients, the results from all the three methods are similar in terms of coefficient value and the same for coefficient sign. Last, the lasso with the kink regression model specification is superior to that with the linear specification considering its lower EBIC (see, Table 4). This indicates that a model employing Lasso can result in the poor estimation if the structural change from real data set is not included in it.

Table 4 Model selection and estimation result for linear regression specification

Parameter	Ridge	LASSO	LS
β_0	4.1571	4.7141	4.1701
POP	0.0038	**0.0000**	0.0045
INC	−0.0036	**0.0000**	−0.0088
IL	−0.0041	**0.0000**	−0.0023
M	−0.0208	−0.0149	−0.0228
HS	0.042	0.0288	0.0491
DAY	−0.0048	**0.0000**	−0.0049
A	−0.0002	**0.0000**	−0.0001
EBIC	−154.6928	**−131.6541**	−155.4809

5 Conclusion

As the kink regression model contains a lot of coefficients thus it may lead to the over-parameterization problem which could bring a high variance to the model. Furthermore, the conventional estimation tools like ordinary least squares may not be applicable well to estimate the large number of the unknown parameters, especially when the number of parameters is larger than the number of the observations. Thus, this study suggests using the LASSO method for variable selection in the kink regression model. We expect that LASSO can reduce the number of coefficients to be estimated as well as the variance from the kink model.

Both simulation and real application studies were conducted in this study to examine the performance of the LASSO estimator for fitting the kink regression model. Numerical simulation studies show that the LASSO method is superior to the traditional Ridge and LS, in terms of the Bias, MSE and point estimates of the regression coefficients, especially when $P > n$. In addition, LASSO behaves well in terms of variable selection as it can remove non-important regime dependent coefficients.

Consequential to the present study, suggestions are some areas of future research that might shed light on the important role of machine learning approach in kink regression estimation. First, it would be interesting to extend other penalized estimators (e.g. the adaptive lasso and elastic net) to see whether they could improve the performance of the kink regression estimation method. Second, an extension to more than two-regime kink regression is also an important research topic.

Acknowledgements The author is grateful to Prof.Hung T.Nguyen for a number of helpful suggestions and discussions. Thank also goes to Dr. Laxmi Worachai for her helpful comments.

References

Ahrens, A., Hansen, C. B., & Schaffer, M. E. (2019). Lassopack: Model selection and prediction with regularized regression in Stata. arXiv preprint arXiv:1901.05397.

Card, D., Lee, D. S., Pei, Z., & Weber, A. (2015). Inference on causal effects in a generalized regression kink design. *Econometrica, 83*(6), 2453–2483.

Chen, J., & Chen, Z. (2008). Extended Bayesian information criteria for model selection with large model spaces. *Biometrika, 95*(3), 759–771.

Efron, B., Hastie, T., Johnstone, I., & Tibshirani, R. (2004). Least angle regression. *The Annals of Statistics, 32*(2), 407–499.

Fokianos, K. (2008). Comparing two samples by penalized logistic regression. *Electronic Journal of Statistics, 2*, 564–580.

Fong, Y., Huang, Y., Gilbert, P. B., & Permar, S. R. (2017). Chngpt: Threshold regression model estimation and inference. *BMC bioinformatics, 18*(1), 454.

Froymson, M. A. (1960). Multiple regression analysis. In A. Ralston & H. S. Wilf (Eds.), *Mathematical methods for digital computers*. New York: Wiley.

Hansen, B. E. (2017). Regression kink with an unknown threshold. *Journal of Business & Economic Statistics, 35*(2), 228–240.

Hastie, T., Tibshirani, R., Friedman, J., & Franklin, J. (2005). The elements of statistical learning: data mining, inference and prediction. *The Mathematical Intelligencer, 27*(2), 83–85.

Hoerl, A. E., & Kennard, R. W. (1970). Ridge regression: Biased estimation for nonorthogonal problems. *Technometrics, 12*(1), 55–67.

Huang, A., Liu, D., & Huang, M. A. (2016). Package 'EBglmnet'.

Lien, D., Hu, Y., & Liu, L. (2017). Subjective wellbeing and income: A reexamination of satiation using the regression Kink Model with an unknown threshold. *Journal of Applied Econometrics, 32*(2), 463–469.

Maneejuk, P., Pastpipatkul, P., & Sriboonchitta, S. (2016). Economic growth and income inequality: evidence from Thailand. In *International Symposium on Integrated Uncertainty in Knowledge Modelling and Decision Making* (pp. 649–663). Cham: Springer.

Sriboochitta, S., Yamaka, W., Maneejuk, P., & Pastpipatkul, P. (2017). A generalized information theoretical approach to non-linear time series model. In *Robustness in Econometrics* (pp. 333–348). Cham: Springer.

Stone, M. (1977). An asymptotic equivalence of choice of model by cross-validation and Akaike's criterion. *Journal of the Royal Statistical Society: Series B (Statistical Methodology), 39*(1), 44–47.

Tibprasorn, P., Maneejuk, P., & Sriboochitta, S. (2017). Generalized information theoretical approach to panel regression kink model. *Thai Journal of Mathematics*, 133–145.

Tibshirani, R. (1996). Regression shrinkage and selection via the lasso. *Journal of the Royal Statistical Society: Series B (Methodological), 58*(1), 267–288.

Wasserstein, R. L., & Lazar, N. A. (2016). The ASA's statement on p-values: Context, process, and purpose. *The American Statistician, 70*(2), 129–133.

Practical Applications

Performance of Microfinance Institutions in Vietnam

Nguyen Ngoc Tan and Le Hoang Anh

Abstract The present study was aimed to analyze the technical and scale efficiency of the Microfinance Institutions (MFIs) in Vietnam. Input oriented Data Envelopment Analysis (DEA) was applied to estimate the technical, pure technical, and scale efficiency of these MFIs. The data of a sample of 26 MFIs during the period of 2013–2017 provided by Mix Market was utilized. The results of the present study demonstrated that the profit in the MFI holdings could be potentially increased further by 46% by following the best practices of efficient MFIs. The study also revealed that approximately 69% of the MFIs were not operating at an optimal scale or even close to optimal scale. The results also indicated that the number of employees input was used excessively in the sample MFIs. The findings of the present study would be useful for policymakers in improving the current levels of technical and scale efficiencies of MFIs.

Keywords Data envelopment analysis · Constant Returns to Scale (CRS) · Variable Returns to Scale (VRS) · Microfinance institutions

1 Introduction

Poverty remains a reality in most of the developing countries. The causes of poverty include a lack of economic diversity, asset inequality, income distribution, and poor management (Abdulai and Tewari 2017). The access to opportunities for financial expansion and stability of the financial system have promoted savings and investments, which was important for a thriving market economy (Abdulai and Tewari 2017). The access to financial services is important, in particular, for the poor

N. N. Tan · L. H. Anh (✉)
Banking University of Ho Chi Minh city, Ho Chi Minh City, Vietnam
e-mail: lehoanganhct@yahoo.com

N. N. Tan
e-mail: tanvpubtp@gmail.com

N. Ngoc Thach et al. (eds.), *Data Science for Financial Econometrics*, Studies in Computational Intelligence 898, https://doi.org/10.1007/978-3-030-48853-6_12

167

people because it assists them in using financial services further conveniently and improves their living standards. In other words, financial services (even involving small amounts and in various forms) may bring positive changes in the economic conditions of poor people. However, financing for the poor has been a global concern, as a consequence of failures in the formal credit market (Hulme and Mosley 1996). The risk of high repayment and lack of collateral have prevented the poor from accessing financial services (Hermes et al. 2011). In this context, microfinance has played an important role in the socio-economic development, especially poverty reduction and social development, in the developing countries (Ledgerwood 1998; Morduch and Haley 2002; Nguyen Kim Anh et al. 2011).

In the initial stages of development of the microfinance market, huge funds were drawn from the donors; however, as the numbers of MFIs entering into the market is increasing, the shares of donor funds are becoming smaller. Therefore, The efficient use of internal resources by the MFIs is becoming increasingly important. The present study attempted to analyze the efficiency of microfinance institutions in Vietnam. Using the Data Envelopment Analysis (DEA), efficiency analysis was performed for these MFIs in order to identify the MFIs following the best practices. The results of the present study may assist the policymakers in identifying appropriate policies and strategies for improving the efficiency of MFIs.

2 Literature Review

A significant number of studies measuring the efficiency of MFIs are available in the literature. However, the methods available for measuring efficiency are not well-defined. Baumann (2005) used the measures of borrower per staff and saver per staff for measuring efficiency. Higher levels of these measures represent that high productivity of the staff in the MFIs assists in accomplishing the two operational goals of financial sustainability and borrower outreach. Therefore, higher levels of these measures may result in high levels of efficiency in MFIs. Other studies conducted on MFIs have employed variables used typically in the studies conducted on banking efficiencies. For instance, Farrington (2000) used administrative expense ratio, number of loans, and loans to total staff members for examining the MFI efficiencies. Moreover, the author considered loan size, lending methodology, sources of funds, and salary structure as the drivers for efficiencies.

Neither of the aforementioned two studies used any parametric or non-parametric approach to evaluate the efficiencies of MFIs. In addition to conventional financial ratios, the assessment may be performed using the efficiency analysis of MFIs (Nawaz 2010). There exists a significant amount of literature reporting the assessment of the efficiency of traditional financial institutions by employing non-parametric techniques, for example Data Envelopment Analysis (DEA), which has been employed widely in recent times. Although DEA has been normally associated with the efficiency analysis of the traditional banking sector, certain researchers have successfully replicated it for the efficiency analysis of MFIs.

With the application of non-parametric method of DEA, Sufian (2006) attempted to analyze the efficiency of NFBIs in Malaysia during the period between the years 2000 and 2004, and reported that only 28.75% of the 80 observations were efficient. It was also reported that the size and the part of the market exerted a negative effect on efficiency. Finally, the author concluded that the NFBIs which exhibit higher efficiency tend to be more profitable.

Another important study on MFIs was the one conducted by Hassan and Tufte (2001), who used a parametric approach referred to as stochastic frontier analysis (SFA) and observed that Grameen Bank's branches with a staff of female employees operated more efficiently than their counterparts with a male employee staff. Charnes et al. (1978) also used parametric approaches to study the efficiency of cooperative rural banks in the Philippines, and observed that cooperative rural banks with good governance exhibited higher efficiency in comparison to their counterparts that encountered bad governance. Leon (2001) reported the productivity of resources, governance, and business environment to be the contributing factors to the cost-efficiency of Peruvian municipal banks.

Masood and Ahmad (2010) analyzed the Indian MFIs by employing Stochastic Frontier Analysis. Hartarska and Mersland (2009) analyzed a sample of MFIs across the world using SFA and a semi-parametric smooth coefficient cost function. Caudill et al. (2009) estimated a parametric cost function of MFIs in eastern Europe and Central Asia and accounted for the unobserved heterogeneity using a mixture model.

3 Research Methods

3.1 Data Envelopment Analysis

Financial institutions offer a wide range of financial products and services, although their effectiveness is currently under assessment. Nowadays, besides the approach to financial indicators, the method of marginal efficiency analysis is used commonly in the assessment of the business performance of financial institutions. The marginal efficiency analysis method calculates the relative efficiency index on the basis of comparison of the distance of financial institutions with the firm performing the best on the margin (which is calculated from the dataset). This tool allows the calculation of the overall performance index of each financial institution on the basis of their performance, in addition to allowing to rank the business performance of these financial institutions. Moreover, this approach allows the managers to identify the best current performance in order to evaluate the financial institution's system and to enhance the best possible performance. Overall, it would result in an improvement in the business performance of the financial institutions studied.

The marginal effect analysis method may be used by following two approaches: parametric approach and non-parametric approach. The parametric approach involves determining a specific type of function for efficient boundaries, ineffective distribution, or random errors. However, if the function format is incorrect, the results of the calculation would negatively affect the efficiency index. In contrast, the non-parametric approach does not require the constraints for efficient boundaries and ineffective distribution in the data as required in the parametric approach, with the exception that the value range of the efficiency index must be between 0 and 1, and assumes no random errors or measurement errors in the data. Data Envelopment Analysis (DEA) is a commonly used method when using the non-parametric approach. This method was originally developed by Farrell (1957), and further developed by Charnes et al. (1978), Banker et al. (1984), and several other scientists in order to evaluate the economic efficiency of a unit (Decision Making Unit—DMU).

DEA allows the determination of the relative efficiency of firms in a complex system. According to DEA, the best performing firm would have an efficiency index value of 1, while the index for ineffective firms is calculated on the basis of ineffective units and the effective boundary. In the case of each ineffective firm, DEA provides a set of benchmarks for the other entities so that the values obtained for the assessed firm are comparable.

Linear planning techniques derived from the study conducted by Charnes et al. (1978) have been applied to determine technical efficiency. Fare et al. (1994) in their study decomposed technical efficiency into efficiency by the scale and other components. In order to achieve separate estimations of scale effectiveness, technical efficiency is measured on inputs to meet the following two different types of scale behavior: Constant Returns to Scale (CRS) and Variable Returns to Scale (VRS). The calculation of efficiency under the assumption of constant returns to scale (CRS) provides the overall technical efficiency score, while the assumption of variable returns to scale (VRS) allows the calculation of one component of this total efficiency score which is referred to as pure technical efficiency.

With the assumption that there are N microfinance institutions, m outputs, and n inputs, the efficiency index of each microfinance institution is calculated as follows:

$$e_s = \frac{\sum_{i=1}^{m} u_i y_{is}}{\sum_{j=1}^{n} v_j x_{js}}$$

where, y_{is} is the amount of output i of the microfinance institutions; x_{js} is the amount of input j used by the microfinance institutions; u_i is the weight of the output; and v_j is the weight of the input. The e_s ratio is then maximized to obtain the optimal weights with constraints, as follows:

$$\frac{\sum_{i=1}^{m} u_i y_{ir}}{\sum_{j=1}^{n} v_j x_{jr}} \leq 1, \ r = \overline{1, N}$$

$$u_i \geq 0, \ v_i \geq 0$$

The first constraint ensures that the maximum effective measure is 1, while the second constraint ensures that the weights of the inputs and outputs are non-negative.

However, the issue with the above-stated problem is that countless solutions exist for this problem.

In order to overcome this situation, Charnes et al. (1978) introduced certain additional constraints, as follows:

$$\sum_{j=1}^{n} v_j x_{js} = 1$$

Therefore, the above equation could be converted into a linear programming equation, as follows:

$$\text{Max}_{uv} e_s = \sum_{i=1}^{m} u_i y_{is}$$

with the following constraints:

$$\sum_{j=1}^{n} v_j x_{js} = 1$$

$$\sum_{i=1}^{m} u_i y_{ir} - \sum_{j=1}^{n} v_j x_{jr} \leq 0, \quad r = \overline{1, N}$$

$$u_i \geq 0, \; v_j \geq 0; \; \forall i, j$$

When $\sum_{j=1}^{n} v_j x_{js} \neq 1$, it is Variable Returns to Scale (VRS).

The scale efficiency of individual MFIs was estimated by calculating the ratio between the technical efficiency scores of the CRS and VRS models using the following equation:

$$e_s = \frac{e_{s(CRS)}}{e_{s(VRS)}}$$

where, $e_{(s(CRS))}$ is technical efficiency under CRS and $e_{(s(VRS))}$ is technical efficiency under VRS.

3.2 Data

The data of a sample of 26 MFIs during the period of 2013–2017 provided by Mix Market was utilized. In order to evaluate the business performance of the microfinance institutions in Vietnam, the present study applied the DEA method to calculate the technical efficiency (TE). An important step during the application of the DEA method is to establish a model with input and output variables to suit the business characteristics. The following inputs and outputs were selected, on the basis of the reviewed studies, for the present study.

Table 1 Description of the input and output variables used in the DEA analysis

Variables	Definition	Units (per MFI)
Input Variable		
Operating expense	Including interest and equivalent payments, staff expenses, non-interest expenses excluding employees' expenses showing equipment, technical facilities	VND
Number of employees	Includes all employees working at microfinance institutions	No.
Output Variable		
Gross Loan Portfolio	All outstanding principals due to all outstanding client loans	VND
Number of active borrowers	The numbers of individuals or entities who currently have an outstanding loan balance with the MFI	No.

Input variables: according to the studies conducted by Gutierrez-Nieto et al. (2007) and Bolli et al. (2012), the input consisted of two variables representing the input resources of a microfinance institution, such as:

Operating expenses: which included interest and equivalent payments, staff expenses, non-interest expenses excluding the employees' expenses showing equipment, and technical facilities.

Number of employees: which included all the employees working at the microfinance institutions being studied.

Output variables: according to the studies conducted by Berger and Humphrey (1997) and Bolli et al. (2012), the output consisted of two variables reflecting the business results of a microfinance institution, such as:

Gross Loan Portfolio: which included all the outstanding principals due to all outstanding client loans, including the current, delinquent, and renegotiated loans, and excluding the loans that have been written off.

Number of active borrowers: which included the numbers of individuals or entities who currently have an outstanding loan balance with the MFI being studied (Table 1).

4   Research Results

The results of the DEA of the MFIs performed in the present study have been presented in Table 2. In the period between the years 2013 and 2017, the overall technical efficiency of the Vietnam Bank for Social Policies was observed to be the highest

Table 2 The DEA results of MFIs

ID	CRSTE			VRSTE			SE		
	Mean	Min	Max	Mean	Min	Max	Mean	Min	Max
1	0.28	0.19	0.41	1	0.99	1	0.28	0.19	0.41
2	0.26	0.18	0.35	0.34	0.25	0.38	0.76	0.68	0.92
3	0.32	0.19	0.47	0.35	0.21	0.48	0.9	0.7	0.97
4	0.71	0.66	0.77	1	1	1	0.71	0.66	0.77
5	0.93	0.82	1	1	1	1	0.93	0.82	1
6	0.92	0.62	1	0.94	0.7	1	0.97	0.88	1
7	0.4	0.21	0.93	0.89	0.7	1	0.44	0.25	0.93
8	0.81	0.59	1	0.83	0.59	1	0.97	0.93	1
9	0.91	0.64	1	1	1	1	0.91	0.64	1
10	0.58	0.23	1	0.64	0.29	1	0.91	0.78	1
11	0.39	0.22	0.56	0.79	0.56	1	0.55	0.3	0.88
12	0.33	0.21	0.46	0.5	0.43	0.55	0.66	0.41	0.83
13	0.54	0.23	1	0.83	0.56	1	0.62	0.31	1
14	0.53	0.22	0.94	0.61	0.33	1	0.81	0.68	0.99
15	0.45	0.36	0.58	0.81	0.63	0.97	0.57	0.44	0.92
16	0.41	0.21	1	0.44	0.25	1	0.9	0.83	1
17	0.69	0.44	1	0.83	0.67	1	0.82	0.65	1
18	0.37	0.23	0.44	0.44	0.26	0.59	0.87	0.74	1
19	0.53	0.37	0.77	0.7	0.53	0.8	0.74	0.55	0.95
20	0.38	0.3	0.49	0.64	0.48	0.78	0.6	0.51	0.63
21	0.47	0.38	0.59	0.52	0.44	0.6	0.91	0.72	1
22	0.35	0.2	0.47	0.42	0.3	0.5	0.82	0.68	0.95
23	0.38	0.22	0.65	0.46	0.29	0.68	0.82	0.61	0.99
24	0.83	0.66	1	0.97	0.83	1	0.86	0.66	1
25	0.37	0.33	0.41	0.57	0.36	0.66	0.69	0.5	0.99
26	1	1	1	1	1	1	1	1	1

Source Analysis results from DEAP software

among all the microfinance institutions in Vietnam. In contrast, the overall technical efficiency of the ANH CHI EM program was observed to be the lowest.

It may be inferred from Table 3 that approximately 15.38% of the sample MFIs were included under the efficiency group (efficiency greater than 90%) when using the assumption of constant return to scale (CRS), while the sample firms included under the least efficiency group (efficiency lower than 50%) constituted 53.85% of all the sample MFIs. This finding indicated that most of the firms in the study area were not technically efficient in terms of input usage.

Moreover, the overall technical efficiency of the sample MFIs ranged from 0.26 to 1.00, with a mean efficiency score of 0.54. Similarly, the pure technical efficiency score ranged from 0.34 to 1.00, with a mean efficiency score of 0.71, and the scale

Table 3 Efficiency level and summary statistics of overall technical efficiency, pure technical efficiency, and scale efficiency of MFIs

Efficiency level	CRSTE	VRSTE	SE
Below 0.50	14	6	2
	−53.85%	−23.08%	−7.69%
0.50–0.60	4	3	2
	−15.38%	−11.54%	−7.69%
0.60–0.70	1	4	4
	−3.85%	−15.38%	−15.38%
0.70–0.80	1	1	3
	−3.85%	−3.85%	−11.54%
0.80–0.90	2	5	7
	−7.69%	−19.23%	−26.92%
Above 0.90	4	7	8
	−15.38%	−26.92%	−30.77%
Total No. of MFIs	**26**	**26**	**26**
Mean	**0.54**	**0.71**	**0.77**
Standard Deviation	**0.22**	**0.22**	**0.17**
Minimum	**0.26**	**0.34**	**0.28**
Maximum	**1**	**1**	**1**

Figures in parentheses are percentage to total MFIs; CRSTE- Technical Efficiency under Constant Return to Scale; VRSTE- Technical Efficiency under Variable Return to Scale; SE—Scale Efficiency
Source Authors' calculation results

efficiency score ranged from 0.28 to 1.00, with a mean efficiency score of 0.77. Therefore, the mean level of overall technical inefficiency was estimated to be 46%. This result revealed the fact that the MFIs were not utilizing their input resources efficiently, indicating that maximal output was not being obtained from the given levels of inputs available with these MFIs. In other words, the technical efficiency of the sample MFIs could be increased by 46% through the adoption of the best practices of efficient MFIs. In regard to scale efficiency, approximately 31% of MFIs were performing either at the optimum scale or at a level close to the optimum scale (MFIs exhibiting scale efficiency values equal to or greater than 0.90).

The percentages of mean input slacks and excess input use have been listed in Table 4. Since a slack indicates an excess of an input, an MFI could reduce its expenditure on this input by the amount of slack, without having to reduce its output. As visible in Table 4, the number of employees input was used excessively, and the greatest slack in this input was observed in 2016. This finding demonstrated that the labor force in the microfinance institutions in Vietnam was not being used effectively.

Table 4 Mean input slack and number of MFIs using inputs excessively

Year	Input	Mean input slack	Number of MFIs using input excessively
2013	Operating expense	0	
	Number of employees	8.953	3
			−11.54%
2014	Operating expense	0	
	Number of employees	2.357	1
			−3.85%
2015	Operating expense	0	
	Number of employees	14.579	2
			−7.69%
2016	Operating expense	0	
	Number of employees	25.179	7
			−26.92%
2017	Operating expense	0	
	Number of employees	6.399	7
			−26.92%

Figures in parentheses are percentage to total MFIs
Source Authors' calculations

5 Conclusion

According to the results of the present study, the mean technical efficiency of the MFIs indicated that there exists a potential of a 46% increase in the profit of these MFIs if the production gap between the average and the best-practicing MFIs is reduced. The present study revealed that approximately 31% of the studied MFIs exhibited optimal scale efficiency, which indicated that the majority of MFIs were not operating at the optimal scale and were far away from the efficiency frontier. In addition, the overall technical inefficiency of the MFIs was attributed more to the scale inefficiency rather than pure technical inefficiency. It is noteworthy that the number of employees input was used excessively by the sample MFIs of the present study.

References

Abdulai, A., & Tewari, D. D. (2017). Trade-off between outreach and sustainability of microfinance institutions: evidence from sub-Saharan Africa. *Enterprise Development and Microfinance, 28*(3), 162–181.

Banker, R. D., Charnes, A., & Cooper, W. W. (1984). Some models for estimating technical and scale inefficiencies in data envelopment analysis. *Management Science, 30*(9), 1078–1092.

Baumann, T. (2005). Pro poor microcredit in South Africa: Cost efficiency and productivity of South African pro-poor microfinance institutions. *Journal of Microfinance*, 7(1), 95–118.

Berger, A. N., & Humphrey, D. B. (1997). Efficiency of financial institutions: International survey and directions for future research. *European Journal of Operational Research*, 98(2), 175–212.

Bolli, T., & Thi, A. V. (2012). On the estimation stability of efficiency and economies of scale in microfinance institutions. *KOF Swiss Economic Institute Working Paper* (296).

Caudill, S. B., Gropper, D. M., & Hartarska, V. (2009). Which microfinance institutions are becoming more cost effective with time? Evidence from a mixture model. *Journal of Money, Credit and Banking*, 41(4), 651–672.

Charnes, A., Cooper, W.W., & Rhodes, E. (1978). Measuring the efficiency of decision making units. *European Journal of Operational Research*, 2(6), 429–444. https://doi.org/10.1016/0377-2217(78)90138-8.

Desrochers, M., & Lamberte, M. (2003). Efficiency and expense preference behavior in Philippines, cooperative rural banks. *Centre interuniversitairesur les risque, les politiques economiques et al'emploi (CIRPÉE.)* Cahier de recherche/Working paper 03-21.

Fare, R., Grosskopf, S., & Lowell, C. A. K. (1994). *Production frontiers*. Cambridge: CUP.

Farrell, M. J. (1957). The measurement of productive efficiency. *Journal of the Royal Statistical Society*, 120(3), 253–290.

Farrington, T. (2000). Efficiency in microfinance institutes. *Microbanking Bulletin*, 20–23.

Gutierrez-Nieto, B., Serrano-Cinca, C., & Molinero, C. M. (2007). Social efficiency in microfinance institutions. *Journal of the Operational Research Society*, 60(1), 104–119.

Hartarska, V., & Mersland, R. (2009). Which governance mechanisms promote efficiency in reaching poor clients? *Evidence from rated microfinance institutions, European Financial Management*, 18(2), 218–239.

Hermes, N., Lensink, R., & Meesters, A. (2011). Outreach and efficiency of microfinance institutions. *World Development*, 39(6), 938–948.

Hulme, D., & Mosley, P. (1996). *Finance against poverty, 1 and 2*. London: Routledge.

Kabir Hassan, M., & Tufte, D. R. (2001). The x-efficiency of a group-based lending institution: The case of the Grameen Bank. *World Development*, 29(6), 1071–1082.

Ledgerwood, J. (1998). *Microfinance handbook: An institutional and financial perspective*. World Bank Publications.

Leon, J. V. (2001). *Cost frontier analysis of efficiency: an application to the Peruvian Municipal Banks*. Ohio State University.

Masood, T., & Ahmad, M. (2010). Technical Efficiency of Microfinance Institutions in India-A Stochastic Frontier Approach. *Technical Efficiency of Microfinance Institutions in India-A Stochastic Frontier Approach*.

Morduch, J., & Haley, B. (2002). *Analysis of the effects of microfinance on poverty reduction* (Vol. 1014, p. 170). NYU Wagner working paper.

Nawaz, A. (2010). *Efficiency and productivity of microfinance: incorporating the role of subsidies*. ULB–Universite Libre de Bruxelleso.

Nguyen Kim Anh., Ngo Van Thu., Le Thanh Tam & Nguyen Thi Tuyet Mai (2011). Microfinance with Poverty Reduction in Vietnam - Testing and Comparison. Statistical publisher.

Sufian, F. (2006). The efficiency of non-bank financial institutions: empirical evidence from Malaysia. *International Journal of Finance and Economics*, 6.

Factors Influencing on University Reputation: Model Selection by AIC

Bui Huy Khoi

Abstract The purpose of this study was to identify the factors that influenced the University Reputation by AIC. This is a new point having the difference with previous researches in other countries. Survey data was collected from 1538 respondents living in HCM City, Vietnam. The research model was proposed from the studies of University Reputation. The reliability and validity of the scale were evaluated by Cronbach's Alpha, Average Variance Extracted (Pvc) and Composite Reliability (Pc). The model selection of AIC showed that University Reputation was impacted by the six components of the University Reputation included: social contributions (SCN), environments (EN), leadership (LE), funding (FU), research and development (RD), and students guidance (SG).

Keywords Vietnam · University reputation · Pc · Pvc · AIC

1 Introduction

Reputation referred to total opinion, which people construct in their thought about something or someone (Chen and Esangbedo 2018). The reputation involved the beliefs, emotions, customs, views, suitable behaviors, and the impression that someone had of a thing, a person, or a business. Moreover, university reputation (UR) is a construct characterized by the attitude of students or staff of the institution as well as the public, which makes the distinguishing and similar evaluation of features (Delgado-Márquez et al. 2013). UR is the shared knowledge that people had on the university and how it should work. Also, UR could be regarded as recognized outer fame, corporate reputation, and identification (Pérez and Torres 2017). The current improvements in university had made its reputation ever more critical to colleges because the resources for changing the course of the current education system are limited and the associated risk involved. Top universities, reputable universities, have

B. H. Khoi (✉)
Industrial University of Ho Chi Minh City, Ho Chi Minh City, Vietnam
e-mail: buihuykhoi@iuh.edu.vn

© The Editor(s) (if applicable) and The Author(s), under exclusive license
to Springer Nature Switzerland AG 2021
N. Ngoc Thach et al. (eds.), *Data Science for Financial Econometrics*,
Studies in Computational Intelligence 898,
https://doi.org/10.1007/978-3-030-48853-6_13

a greater chance of been selected to first participate in these changing times. The different studies on the impression of the mark on higher education showed that the UR fits the principal symbol that defines the exceptionality of the institution (Rachmadhani et al. (2018; Keh and Xie 2009). Although UR is becoming a necessary part of higher education in globalization, it attracts students, staff, and research investments (Chen and Esangbedo 2018), the extent of UR is a matter of contention, due largely to a lack of consensus regarding the relationship of reputation. The purpose of this study was to identify the factors that influenced the University Reputation in Vietnam by AIC. This is a new point having the difference with previous researches in other countries.

2   Literature Review

A reputable university can be described as one having a good stand within the community, parents and guardians have confidence in the university, the leadership of the university is well-respected, and students believe that the university image has a positive influence on the value of their degree. Based on the factors affecting reputation, some researchers analyzed the performances of different universities and presented models of evaluation. Rachmadhani et al. (2018) showed the growing competition between state and private universities in the context of the commercialization of higher education in Indonesia. Consequently, this competition makes it increasingly challenging for a state university to recruit a new student. An apparent knowledge of why and how student prefer universities were investigated to improve the brand recognition enrichment policies of the state university. The combination of higher education performance dimensions and brand awareness signs were applied to improve the variables. Plewa et al. (2016) evaluated the reputation of high education students using both local and international students as the output. Chen and Esangbedo (2018) presented a two-level hierarchical model for evaluating UR with a case study of Chinese university.

2.1   Social Contributions (SCN)

Social Contributions (SCN) is the university giving back to society as a response to the need of the community with the activities of the university have an impact in the community (Chen and Esangbedo 2018). SCN is also the university's ethical contribution, and responsibility to the society as the university's human resource contribution to industries since graduates are the testament of the education received as students (Chen and Esangbedo 2018; Calitz et al. 2016). The feedback of a university to the community, such as services (Esangbedo and Bai 2019; Liu et al. 2019), consists of an institutional moral contribution and responsibility (Chen and Esangbedo 2018; Verčič et al. 2016). Alumni associations are results of the solidity of the relationship strengthen from their studentship experience by reporting on the effect of the UR on their degree (Chen and Esangbedo 2018; Plewa et al. 2016). The SDG can, by no means, least at this time, be isolated from SCN.

2.2 Environments (EN)

Environments (EN) are the educational conditions that influence the improvement of students and lecturers. Also, the university environment is where the major learning and administrative process occurs, which is under the care of the university as its duty. A safe, clean, and pleasant environment for students to study was a primary expectation (Chen and Esangbedo 2018; Badri and Mohaidat 2014). A reputable university should have the ability to protect students from danger (Chen and Esangbedo 2018; Badri and Mohaidat 2014). A respectable university should exceed its physical environment to mobilize students from far and near, as well as assume the social responsibility to support student image while maintaining a good quality education using up-to-date learning materials (Chen and Esangbedo 2018; Sarwari and Wahab 2016). Interestingly, the internet has a big row in today's learning experience by giving of electronic studying platform over the network that highlights the university was globally renowned (Chen and Esangbedo 2018; Verčič et al. 2016).

2.3 Leadership (LE)

A clear vision for development is mandatory for a reputable university because it shows competence and good organization (Chen and Esangbedo 2018; Esangbedo and Bai 2019; Verčič et al. 2016, 2007). This could be observed in the quality of teaching resources (Chen and Esangbedo 2018; Ahmed et al. 2010), duties in the form of various assessments in the year of study, and the academic staff that presents these materials. The student's prospect had to be considered as part of what counts as the university activities from the admission procedures to the graduating process can be depicted the university services to attract prospective students (Chen and Esangbedo 2018).

2.4 Funding (FU)

UR had a financial role in several cases. Reputable schools should be capable of obtaining government grants and some funding contributions from the salaries of parents and sponsors as well as reduced tuition fees from funding bodies as a scholarship for the student (Chen and Esangbedo 2018; McPherson and Schapiro 1999). Since most parents are the primary sponsors of children's education, their income/level has is a factor that plays in the kind of education their children receive (Chen and Esangbedo 2018; Saleem et al. 2017; West et al. 2017). Also, the primary cost of education are indicated as tuition fees (Chen and Esangbedo 2018; Verčič et al. 2016; Burgess et al. 2018), and the ability for the university to attract top student through scholarship provisioning (Chen and Esangbedo 2018; Flavell et al. 2018). To reiterate, funding has some relationship with parents/sponsor's income, tuition, and scholarships.

2.5 Research and Development (RD)

Universities should be able to maximize funded research to encourage knowledge transfer, reform, invention, innovation, and national development. Industrial development and university research output are interrelated in today's knowledge-based economy that helps in enhancing scientific and technological breakthrough (Chen and Esangbedo 2018; D'Este and Patel 2007; Hamdan et al. 2011). Also, key research plans from the governments were extended to the universities to provide solutions, where universities were concerned in the key project (Chen and Esangbedo 2018; Nelson and Rosenberg 1993). The research performance of universities could be measured by many publications, cited-publications, and international and industry-university publications (Chen and Esangbedo 2018; Frenken et al. 2017). Furthermore, the aims, results, and successes of these researches should be available to the academic community (Chen and Esangbedo 2018; Kheiry et al. 2012). Therefore, RD can be expressed as an industrial linkage to the university in the form of key projects, as evident in academic publications.

2.6 Student Guidance (SG)

Few students have complete knowledge of university while leaving secondary schools. SG can be considered advice received from the guidance counselor as well as individual evaluating the university using their knowledge and available information (Chen and Esangbedo 2018). University's academic quality could predict the university's proposals, which is a form of the UR (Chen and Esangbedo 2018; Pedro et al. 2016). The trust developed by students in the university has a chance to influence their friends on their choice for university enrolment, and this can be in the form of observed service quality of the university (Chen and Esangbedo 2018; Shamma 2012; Twaissi and Al-Kilani 2015). Student guardians understand their needs by making interaction with them easier since the guardian is available for consultation and provides the support to help the student succeed academically. These guardians care about their wards experience as a student and can communicate issues that concern them.

In summary, there were many factors in university reputation. In this paper, we designed an empirical study in the context of education in Vietnam to examine factors on university reputation as function 1:

$$UR = \beta_0 + \beta_1 SCN + \beta_2 LE + \beta_3 EN + \beta_4 FU + \beta_5 RD + \beta_6 SG + e$$

Code: UR: University Reputation, social contributions (SCN), environments (EN), leadership (LE), funding (FU), research and development (RD), and student's guidance (SG). **Function 1**. The theoretical model.

3 Methodology

3.1 Sample and Data

In this study, questionnaires were used to obtain data for analysis. The survey was conducted in English and then translated into Vietnamese since all of the respondents are Vietnamese. The research method was performed in two stages: a qualitative and quantitative analysis. The qualitative analysis was carried out with a sample of 61 peoples. We divided into four groups to discuss observations and variables, and then modified them to be suitable in the context of education in Vietnam. The first group is from Vietnam national academy of education management. They are five experts (four professors and one Ph.D. lecturer) in the educational field. They have many years of experience in teaching and educational research. This group used bilateral discussion within 30 min/person. The second group is from the Industrial University of HCM City. There are two economic, educational, and management specialists (one professor, three Ph.D. lecturers) within 60 min/person. The third group is 22 lecturers at the FBA in Industrial University of HCM City, which includes three professors. The fourth group is 30 graduates in Vietnam that have intentions to enroll in a master's program in the future. There are 11 people working at some banks in HCM city. Others are unemployed.

Subsequently, quantitative research had two phases. First, a pilot test of the questionnaire on a small sample to receive quick feedback that was used for improving the questionnaire. The second phase was an official study conducted soon after the questions were revised from the test results. Respondents were chosen by convenient methodology with a sample dimension of 1538 students from Vietnam universities. The questionnaire replied by respondents was the main instrument to collate the data—the questionnaire comprised questions about their graduated university and year. The survey was carried out in 2017, 2018, and 2019. The questionnaire answered by respondents was the main tool to obtain the data. The questionnaire included questions about the position of the determinants that influence UR in universities in Vietnam about their choice of a master's program in Vietnam and their personal information. A Likert-scale type questionnaire was utilized to discover those factors measured from (1) *"Strongly disagree"* to (7) *"Strongly agree."* Respondents were chosen by convenient methods with a sample size of 1538 students. There were 736 males (47.9%) and 802 females (52.1%) in this survey. Every independent variable (SCN, EN, LE, FU, RD, and SG) includes a sum of five answers, and a dependent variable (UR) measures a sum of three answers with the value from 1 to 7. Their questionnaire related variables Table 1.

Table 1 Questionnaire for related variables

Variables	Measurement items
SCN	Q_1: This university has a positive influence on society Q_2: This university is committed and involved in community services Q_3: Graduates from this university are well equipped for the workplace Q_4: This university name positively influences the value of my degree Q_5: During my time at the university, I have learned how to be more adaptable
EN	Q_6: This university is internationally renowned Q_7: This university is a safe place to study Q_8: The university's physical facilities are visually appealing Q_9: The physical environment of the university is pleasant Q_{10}: This university provides up-to-date University equipment
LE	Q_{11}: The lecturers stimulated my interest in my course Q_{12}: This university employs prestigious professors Q_{13}: This university has a clear vision for development Q_{14}: Courses are designed in this university to make use of the latest technology Q_{15}: This university provides the support I need to help me succeed academically
FU	Q_{16}: The cost of living is this university is reasonable Q_{17}: I sometimes feel pressurized by financial worries Q_{18}: This university receives funds from the government to gives scholarships to the student Q_{19}: This university provides grants for researches done by students Q_{20}: Tuition fees are competitive with other similar universities
RD	Q_{21}: This university follows technological trends in conveying knowledge Q_{22}: This university takes part in key national projects Q_{23}: This university is innovative in its publications Q_{24}: Laboratory equipment is in good working condition and properly maintained Q_{25}: The library is provided with up-to-date books and sources
SG	Q_{26}: The university is well-liked or admired by friends and family Q_{27}: Our guardians understand my needs Q_{28}: Our guardians provide the support to help the student succeed academically Q_{29}: My friends, relatives, or siblings attended this university Q_{30}: Our guardians care about their wards experience as a student
UR	Q_{31}: This university has good prestige within the community Q_{32}: This university is a well-respected one Q_{33}: This university's reputation positively influences the value of my degree

SCN: Social Contributions, **LE**: Leadership, **EN**: Environments, **FU**: Funding, **RD**: Research and development, **SG**: Students guidance, **UR**: University reputation

3.2 Blinding

All study personnel and participants were blinded to treatment for the duration of the study. No people had any contact with study participants.

3.3 Datasets

We validate our model on three standard datasets for the factors affecting University reputation in Vietnam: SPSS.sav, R, and Smartpls.splsm. Dataset has seven variables: six independent variables and one variable. There are 1538 observations and 33 items in a dataset. SPSS.sav was used for descriptive statistics and Smartpls.splsm, R for advanced analysis.

3.4 Data Analysis

Data processing and statistical analysis software are used by Smartpls 3.0 Software developed by SmartPLS GmbH Company in Germany. The reliability and validity of the scale were tested by Cronbach's Alpha, Average Variance Extracted (Pvc), and Composite Reliability (Pc). Cronbach's alpha coefficient greater than 0.6 would ensure scale reliability (Nunnally and Bernstein 1994; Ngan and Khoi 2019; Khoi and Ngan 2019). Composite Reliability (Pc) is better than 0.6 and Average Variance Extracted and rho_A must be greater than 0.5 (Ngan and Khoi 2019; Khoi and Ngan 2019; Wong 2013; Latan and Noonan 2017; Khoi and Tuan 2018; Hair et al. 2006, 2016; Khoi et al. 2019). AIC (Akaike's Information Criteria) was used for model selection in the theoretical framework. AIC method can handle many independent variables, even when multicollinearity exists. AIC can be implemented as a regression model, predicting one or more dependent variables from a set of one or more independent.

4 Empirical Results

4.1 Reliability and Validity

The measurement model analyzed data reliability and validity. The Cronbach criteria, composite reliability, and average variance criteria had been used to validate the internal data reliability. On the other hand, Heterotrait-Monotrait ratio of correlations (HTMT) was employed to validate the data validity. However, according to Hair et al. (2016, 2017), Cronbach alpha and composite reliability values should be more

Table 2 Cronbach's alpha, composite reliability (Pc), rho_A, and AVE values (Pvc)

Factor	Cronbach's alpha	rho_A	Pc	Pvc	Decision
BI	0.810	0.818	0.887	0.723	Accepted
EN	0.747	0.793	0.824	0.490	Accepted
FU	0.691	0.759	0.786	0.440	Accepted
LE	0.803	0.815	0.864	0.561	Accepted
RD	0.778	0.839	0.841	0.520	Accepted
SC	0.855	0.925	0.892	0.628	Accepted
SCN	0.800	0.828	0.858	0.550	Accepted
SG	0.740	0.765	0.820	0.480	Accepted

Table 3 Constructs validity results (HTMT: Fornell-Larcker Criterion)

Construct	EN	FU	LE	RD	SCN	SG	UR
EN	0.700						
FU	0.415	0.664					
LE	0.448	0.498	0.749				
RD	0.191	0.327	0.344	0.721			
SCN	0.543	0.463	0.423	0.322	0.742		
SG	0.380	0.398	0.511	0.364	0.393	0.693	
UR	0.418	0.472	0.530	0.351	0.496	0.454	0.741

than 0.60, and AVE values should be more than 0.50 for the validation of construct reliability. On the other hand, in terms of construct validity, according to Hair et al. (2016, 2017), HTMT values should be less than 1.0 in Table 3. The present found that all construct values were less than threshold values. The result of the construct's reliability and validity can be seen in Tables 2 and 3. Furthermore, the values of AVE can be seen in Table 2.

Table 2 showed that Pc varied from 0.786 to 0.892, Cronbach's alpha from 0.691 to 0.855, Pvc from 0.440 to 0.723, and rho_A from 0.759 to 0.925 which were above the preferred value of 0.5. This proved that the model was internally consistent. To check whether the indicators for variables display convergent validity, Cronbach's alpha was used. From Table 2, it could be observed that all the factors were reliable (>0.60) and Pvc, rho_A > 0.5 (Wong 2013). Factor EN, FU, and SG had Pvc lower than 0.5, but other standards were greater than 0.5, so they were accepted, and we would continue them in AIC.

Table 4 Akaike's Information Criteria

Variable	Sum of Sq	RSS	AIC
EN	57.58	7783.7	2505.9
RD	86.42	7812.6	2511.6
FU	134.98	7861.1	2521.2
SG	185.38	7911.5	2531.0
LE	478.89	8205.0	2587.0
SCN	491.23	8217.4	2589.3

4.2 Akaike's Information Criteria (AIC)

Akaike's Information Criteria (AIC) was used on the Lasso regression (Table 4). AIC method could handle many independent variables, even when multicollinearity exists. AIC could be implemented as a regression model, predicting one or more dependent variables from a set of one or more independent variables or it could be implemented as a path model.

AIC results in Table 4 showed that the model was the best. The University reputation was affected by five factors. In the AIC analysis in Table 4, the variables associated with University reputation. The most important factor for University reputation was Leadership aspects with the Beta equals 0.14515 with the function as follows.

$$UR = 2.61132 + 0.05316RD + 0.08399FU + 0.11514SCN$$
$$+ 0.04199EN + 0.07067SG + 0.14515LE$$

Code: SCN: Social Contributions, **LE**: Leadership, **EN**: Environments, **FU**: Funding, **RD**: Research and development, **SG**: Students guidance, **UR**: University reputation. **Function 2**. The practical model.

Especially, the consequences had established the six components of the UR combined Social Contributions (SCN), Leadership (LE), Environments (EN), Funding (FU), Research and development (RD), Student guidance (SG). There were six components strongly significant to the UR in order of significance: (1) Leadership, (2) Contribution, (3) Funding, (4) Student guidance, (5) Research and development, and (6) Environments. All six components contributed importantly to the UR in order of significance: (1) Leadership (LE), (2) Social Contributions (SCN), (3) Students guidance (SG), (4) Funding (FU), (5) Environments (EN) and (6) Research and development (RD).

Higher education institutions are frequently recognized as an essential driver for the improvement of sustainable communities and to incorporate sustainable development into the university. Most universities around the world are incorporating sustainable development into their programs and processes (Zutshi et al. 2019). Therefore, it is necessary to develop sustainable UR. In contemporaneous higher education, reputation has enhanced a principal relation for universities. Thus, the issue of how to create organizational reputation warrants in higher education guarantee inclusive and equal quality education and encourage lifelong learning chances for all.

Developing globalization of colleges is often attributed to the opinion of somebody with many advantages that the educational area has to face. The varieties in an opinion by university administrators and several analysts recognize that struggle between universities has increased over the last few years (Chen and Esangbedo 2018; Plewa et al. 2016). By the approval of the idea of trademark, reputation enhances more significant to promote the identification of the university and its reputation in the sustainable competing global environment.

Corporate reputation could lead sustainable to competitive advantages (Kanto et al. 2016) since organizations may participate in the "rankings game" (Aula 2015; Corley and Gioia 2000) and initiate brand-building activities (Aula 2015) to establish their reputation. UR is purposefully and actively constructed, promoted, and defended in sustainability. The choice of the students to maintain their study in the aspired area of the field of study plays an essential part in their future progress. The intended struggle made each university aware of the necessity to entirely employ its assets to maximize achievement and to improve competing advantage. One form to achieve this was to grow a university name so that it has a reliable reputation in the people's hearts and increase trust (Heffernan et al. 2018). Therefore, a good reputation could lead sustainable to competitive advantages.

5 Conclusion

Reputation had become essential for higher educating institutes, and universities have been working harder to improve their reputation. As the higher education market had enhanced more accessible, the state schools that previously performed within tight national systems compete for resources now. Creating and maintaining a reputation in higher education is significant but does guarantee inclusive and equal quality education to lifelong studying chance for all. This paper investigated the factors that influence UR in Vietnam. UR as a bi-dimensional construct was explored. This paper confirmed the reliability, validity of measurement scales, and verification of the relationship among constructs in the proposed research model by AIC. This is a new point having the difference with previous researches in other countries. Expert research helped adjust the measurement scales to match the research context. The author had also provided scientific evidence and practical application in the study. Thus, this paper provides more incites to researchers in Vietnam of both academic and applied reputation for higher education. Future research could explore factors associated with educational quality in this model.

Acknowledgements This research is funded by Industrial University of Ho Chi Minh City, Vietnam.

References

Ahmed, I., Nawaz, M. M., Ahmad, Z., Ahmad, Z., Shaukat, M. Z., Usman, A., & Ahmed, N. (2010). Does service quality affect student's performance? Evidence from Institutes of Higher Learning. *African Journal of Business Management, 4*(12), 2527–2533.

Aula, H.-M. (2015). *Constructing reputation in a university merger* (Vol. 184). Department of Management Studies: School of Business, Aalto University publication series, Finland.

Badri, M. A., & Mohaidat, J. (2014). Antecedents of parent-based school reputation and loyalty: An international application. *International Journal of Educational Management, 28*(6), 635–654.

Burgess, A., Senior, C., & Moores, E. (2018). A 10-year case study on the changing determinants of university student satisfaction in the UK. *PloS One, 13*(2), e0192976.

Calitz, A. P., Greyling, J., & Glaum, A. (2016). *Cs and is alumni post-graduate course and supervision perceptions*. Springer: Annual Conference of the Southern African Computer Lecturers' Association, 2016, (pp. 115–122).

Chen, C., & Esangbedo, M. O. (2018). Evaluating university reputation based on integral linear programming with grey possibility. *Mathematical Problems in Engineering, 2018*, 17.

Corley, K., & Gioia, D. (2000). The rankings game: Managing business school reputation. *Corporate Reputation Review, 3*(4), 319–333.

D'Este, P., & Patel, P. (2007). University-industry linkages in the UK: What are the factors underlying the variety of interactions with industry? *Research Policy, 36*(9), 1295–1313.

Delgado-Márquez, B. L., Escudero-Torres, M. Á, & Hurtado-Torres, N. E. (2013). Being highly internationalised strengthens your reputation: An empirical investigation of top higher education institutions. *Higher Education, 66*(5), 619–633.

Esangbedo, M. O., & Bai, S. (2019). Grey regulatory focus theory weighting method for the multi-criteria decision-making problem in evaluating university reputation. *Symmetry, 11*(2), 230.

Flavell, H., Roberts, L., Fyfe, G., & Broughton, M. (2018). Shifting goal posts: The impact of academic workforce reshaping and the introduction of teaching academic roles on the scholarship of teaching and learning. *The Australian Educational Researcher, 45*(2), 179–194.

Frenken, K., Heimeriks, G. J., & Hoekman, J. (2017). What drives university research performance? An analysis using the Cwts Leiden ranking data. *Journal of Informetrics, 11*(3), 859–872.

Hair, J. F., Black, W. C., Babin, B. J., Anderson, R. E., & Tatham, R. L. (2006). *Multivariate data analysis (vol. 6)*. Upper Saddle River, NJ: Pearson Prentice Hall.

Hair, Jr, J. F., Hult, G. T. M., Ringle, C., & Sarstedt, M. (2016). *A primer on partial least squares structural equation modeling (pls-sem)*. Sage Publications.

Hair, J. F., Hult, G. T. M., Ringle, C. M., Sarstedt, M., & Thiele, K. O. (2017). Mirror, mirror on the wall: A comparative evaluation of composite-based structural equation modeling methods. *Journal of the Academy of Marketing Science*, 1–17.

Hamdan, H., Yusof, F., Omar, D., Abdullah, F., Nasrudin, N., & Abullah, I. C. (2011). University industrial linkages: Relationship towards economic growth and development in Malaysia. *World Academy of Science, Engineering and Technology, 5*(10), 27–34.

Heffernan, T., Wilkins, S., & Butt, M. M. (2018). Transnational higher education: The importance of institutional reputation, trust, and student-university. *International Journal of Educational Management, 32*(2), 227–240.

Kanto, D. S., de Run, E. C., & Bin Md Isa, A. H. (2016). The reputation quotient as a corporate reputation measurement in the Malaysian banking industry: A confirmatory factor analysis. *Procedia-Social and Behavioral Sciences, 219*, 409–415.

Keh, H. T., & Xie, Y. (2009). Corporate reputation, and customer behavioral intentions: The roles of trust, identification and commitment. *Industrial Marketing Management, 38*(7), 732–742.

Kheiry, B., Rad, B. M., & Asgari, O. (2012). University intellectual image impact on satisfaction and loyalty of students (Tehran selected universities). *African Journal of Business Management, 6*(37), 10205–10211.

Khoi, B. H., & Ngan, N. T. (2019). Factors impacting to smart city in Vietnam with smartpls 3.0 software application. *IIOAB, 10*(2), 1–8.

Khoi, B. H., & Van Tuan, N. (2018). Using smartpls 3.0 to analyze internet service quality in Vietnam. *Studies in Computational Intelligence, 760*, 430–439 (Springer).

Khoi, B. H., Dai, D. N., Lam, N. H., & Van Chuong, N. (2019). The relationship among education service quality, university reputation and behavioral intention in Vietnam. *Studies in Computational Intelligence, 809*, 273–281 (Springer).

Latan, H., & Noonan, R. (2017). *Partial least squares path modeling: Basic concepts, methodological issues, and applications.* Springer.

Liu, Y., Esangbedo, M. O., & Bai, S. (2019). Adaptability of inter-organizational information systems based on organizational identity: Some factors of partnership for the goals. *Sustainability, 11*(5), 1–20.

McPherson, M. S., & Schapiro, M. O. (1999). *The student aid game: Meeting need and rewarding talent in American higher education,* vol. 31. Princeton University Press.

Nelson, R. R., & Rosenberg, N. (1993). Technical innovation and national systems. *National innovation systems: A comparative analysis, 322.*

Ngan, N. T., & Khoi, B. H. (2019). Empirical study on intention to use bike-sharing in Vietnam. *IIOAB, 10*(Suppl 1), 1–6.

Nunnally, J. C., & Bernstein, I. (1994). The assessment of reliability. *Psychometric Theory, 3*(1), 248–292.

Pedro, E., Leitão, J., & Alves, H. (2016). Does the quality of academic life matter for students' performance, loyalty and university recommendation? *Applied Research in Quality of Life, 11*(1), 293–316.

Pérez, J. P., & Torres, E. M. (2017). Evaluation of the organizational image of a university in a higher education institution. *Contaduría Y Administración, 62*(1), 123–140.

Plewa, C., Ho, J., Conduit, J., & Karpen, I. O. (2016). Reputation in higher education: A fuzzy set analysis of resource configurations. *Journal of Business Research, 69*(8), 3087–3095.

Rachmadhani, A. P., Handayani, N. U., Wibowo, M. A., Purwaningsih, R., & Suliantoro, H. (2018). Factor identification of higher education choice to enhance brand awareness of state university. *MATEC Web of Conferences, EDP Sciences, 2018,* 01051.

Saleem, S. S., Moosa, K., Imam, A., & Khan, R. A. (2017). Service quality and student satisfaction: The moderating role of university culture, reputation, and price in education sector of Pakistan. *Iranian Journal of Management Studies, 10*(1), 237–258.

Sarwari, A. Q., & Wahab, N. (2016). The role of postgraduate international students in the process of internationalization of higher education. *IIUM Journal of Educational Studies, 4*(1), 28–45.

Shamma, H. M. (2012). Toward a comprehensive understanding of corporate reputation: Concept, measurement and implications. *International Journal of Business and Management, 7*(16), 151.

Twaissi, N. M., & Al-Kilani, M. H. (2015). The impact of perceived service quality on students' intentions in higher education in a Jordanian governmental university. *International Business Research, 8*(5), 81.

Verčič, A. T., Verčič, D., & Žnidar, K. (2016). Exploring academic reputation–is it a multidimensional construct? *Corporate Communications: An International Journal, 21*(2), 160–176.

Vidaver-Cohen, D. (2007). Reputation beyond the rankings: A conceptual framework for business school research. *Corporate Reputation Review, 10*(4), 278–304.

West, A., Lewis, J., Roberts, J., & Noden, P. (2017). Young adult graduates living in the parental home: Expectations, negotiations, and parental financial support. *Journal of Family Issues, 38*(17), 2449–2473.

Wong, K.K.-K. (2013). Partial least squares structural equation modeling (pls-sem) techniques using smartpls. *Marketing Bulletin, 24*(1), 1–32.

Zutshi, A., Creed, A., & Connelly, B. (2019). Education for sustainable development: Emerging themes from adopters of a declaration. *Sustainability, 11*(1), 156.

Impacts of Internal and External Macroeconomic Factors on Firm Stock Price in an Expansion Econometric model—A Case in Vietnam Real Estate Industry

Dinh Tran Ngoc Huy, Vo Kim Nhan, Nguyen Thi Ngoc Bich, Nguyen Thi Phuong Hong, Nham Thanh Chung, and Pham Quang Huy

Abstract After the global economic crisis 2007–2011 and the recent post-low inflation 2014–2015, Viet Nam economies, its financial and stock market as well as real estate market experienced indirect and direct impacts on their operation, system and stock price. Although some economists have done researches on the relationship among macro economic factors such as: consumer price index (CPI), inflation, GDP…, this paper aims to consider the interaction between macro economic factors such as Viet Nam inflation and GDP growth rate, US inflation, exchange rate, risk free rate and other macro factors, and esp. their impacts on stock price of Vingroup (VIC), a big Vietnam real estate firm, in the context Viet Nam and the US economies receive impacts from global economic crisis. This is one main objective of this research paper. This research paper finds out VIC stock price has a negative correlation with risk free rate in VN and deposit rate of VN commercial banks, but has a positive correlation with lending rate in Vietnam. And the statistical analysis will generate results which help us to suggest macro policies in favor of the local stock and financial market. Real estate industrial risk over years has been affected much by macro economic risk, credit risk, and legal risk; therefore, government bodies need to issue proper macro economic legal, financial and credit policies in order

D. T. N. Huy (✉)
Banking University Ho Chi Minh City, Ho Chi Minh City, Vietnam
e-mail: dtnhuy2010@gmail.com

GSIM, International University of Japan, Niigata, Japan

V. K. Nhan · N. T. N. Bich · N. T. P. Hong · N. T. Chung
University of Economics Ho Chi Minh City, Ho Chi Minh City, Vietnam
e-mail: vokimnhan@gmail.com

N. T. P. Hong
e-mail: hongntp@ueh.edu.vn

P. Q. Huy
School of Accounting, University of Economics Ho Chi Minh City, Ho Chi Minh City, Vietnam
e-mail: Pqh.huy@gmail.com

© The Editor(s) (if applicable) and The Author(s), under exclusive license
to Springer Nature Switzerland AG 2021
N. Ngoc Thach et al. (eds.), *Data Science for Financial Econometrics*,
Studies in Computational Intelligence 898,
https://doi.org/10.1007/978-3-030-48853-6_14

to stimulate, develop stock market and reduce workload pressure for Vietnam bank system.

Keywords Inflation · Stock market · Risk free rate · Lending rate · Exchange rate · Stock price · Real estate industry

JEL G12 · G17 · L85 · M21

1 Introduction

Viet Nam economy has become active and growing recently with GDP growth rate as one of the fastest economic growth levels in South East Asia (GDP growth in 2019 expected at around 6.6–6.8%) and it is affected by both internal and external factors such as global economic crisis 2007–2009 and post-low (L) inflation environment 2014–2015. Esp., Vietnam real estate market is growing so fast with many big names in real estate industry such as: Vingroup (VIC), Nam Long, FLC, REE... Hence, after the global crisis, the US economy has certain impacts on Vietnam economies and financial market. Therefore, real estate firms share price such as VIC stock price are also affected by external factors such as the inflation from US economy, S&P500 index, USD/VND exchange rate, as well as internal macro factors such as Vietnam inflation, GDP growth, VN Index, lending rate, deposit rate and risk free rate (government bond). After the crisis time in 2008–2010 in real estate industry, in recent years Vietnam real estate market and firms have been growing very fast with lots of big projects in big cities in the country and even island, including but not limited to: HCM city, Ha Noi, Binh Duong, Nha Trang, Da Lat, Binh Dinh, Phu Quoc, Phan Thiet, Phan Rang, Hue, Quang Ngai, Ba Ria Vung Tau, Bac Ninh, Thanh Hoa, Da Nang, Long An, Tien Giang, Ben Tre, Daklak, Gia Lai, Binh Phuoc, Tay Ninh, Quang Binh, Ha Tinh, Phu Yen, Can Tho, Dong Nai, Vinh Phuc, Dong Thap, Soc Trang, Ca Mau etc.

The below chart shows us that VIC stock price moves in the same direction as VN Index and S&P 500 (with little fluctuation) during the period 2014–2019, and how much is the correlation between them is forecasted in the below sections. We do not focus on analysis of impacts of VnIndex on VIC stock price in this paper (as we stated in the below discussion for further researches section); however, the below chart helps us to recognize that the stock price of VIC follows the trend of SP500 and VnIndex (Chart 1).

In this research, we will consider Vingroup (VIC) stock price in Viet Nam stock market, are affected by both external and internal factors - six (6) major variables (inflation in USA and Viet Nam, S&P500, USD/VND exchange rate, lending rate, and risk free rate):

Y (VIC stock price) = f $(x_1, x_2, x_3, x_4, x_5, x_6, x_7)$ = $ax_1 + bx_2 + cx_3 + dx_4 + ex_5 + fx_6 + gx_7 + k$.

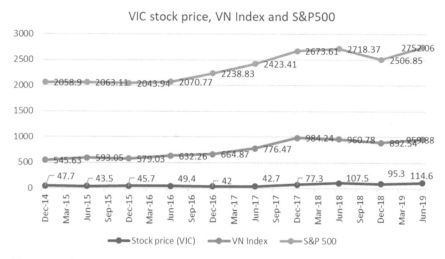

Chart 1 VIC stock price versus VN Index and S&P500

Note x_1: inflation in Viet Nam, x_2: lending rate in Viet Nam, x_3: risk free rate, x_4: inflation in the US, x_5: SP500, x_6: USD/VND exchange rate, x_7: VNIndex.

In following sections, this paper will present research issues, research methods, data explanation, research results, discussion and policy suggestion.

2 Research Issues

Because U.S. economy factors and internal Vietnam economic factors, both of them have certain impacts on stock market and Vingroup stock price during the post-global financial crisis period 2014–2019, this paper will find out:

Research issue 1: estimate the relationship between internal macro economic factors/variables in Viet Nam and Vingroup stock price.

Research issue 2: estimate the relationship between external macro economic factors/variables in the U.S., internal macro economic factors in Viet Nam and Vingroup stock price.

3 Literature Review

Nicholas (2003) revealed that among various factors, the variation of real housing prices has been affected highest by housing loan rate, followed by inflation and employment.

Real estate market is one of the most active financial markets and economic sectors that contribute considerably for economic and social development. It is affected by a number of macro-economic variables including but not limited to: legal, financial, credit, inflation, unemployment, salary, households, etc.

Then, Emilia et al. (2011) figured out that population density, employment, and so on have a significant impact on the market, i.e. the number of employees in the region, as people without economic activity cannot afford to buy housing. Moreover, birth or the creation of new families also affect real estate market.

Real estate industry and companies have been developing lots of new products, houses, departments, residential block, houses for the low income, houses combining of office, houses in real estate municipal or urban projects etc. Miroslaw (2014) showed that social and economic changes, during periods of instability, is driven by the real estate market.

Next, Bijan et al. (2016) mentioned that a distinct pattern that applies to France, Greece, Norway and Poland, where the price of real estate observed statistically significantly associated with unemployment. Olatunji et al. (2016) suggested that because macroeconomic policy could adversely affect property returns, policy-makers should understand the future implications of them.

So far, there is no researches that have been done on both internal and external macro economic variables on real estate firm and their stock price, esp. in Viet Nam market. This is research gap.

4   Overview on Vietnam Macro Economy and Stock Market

The below graphs describe inflation in Viet Nam, credit growth rate, VN Index and Vingroup stock price over past years. The Chart 4 shows, VIC stock price reached a peak on 2nd April, 2018.

The Chart 2 shows us inflation in Vietnam stayed in a steady rate of about 6% from 2012–2013, then reduce in 2014–2015 (low inflation). Chart 4 tells us VN index lied in low levels from 2014–2016, then after 2016 until now the stock market has recovered. One of main reasons is shown in the Chart 3 where it stated credit/loan growth rate has increased during 2014–2018 period.

We also see that in the above Chart 5 from 2017–2019, investors invest in VIC stock price will receive big profits. In general, VIC stock price movement follows the same direction with VN Index.

5   Conceptual Theories

Firm stock price not only is affected by company performance, operation and dividend, but also by macro economic variables. Inflation rate may harm investment values and therefore affect negatively on stock price. During high inflation time,

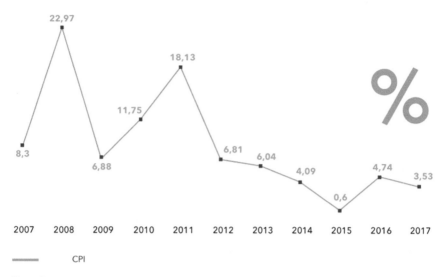

Chart 2 Inflation, CPI over past 10 years (2007–2017) in Vietnam

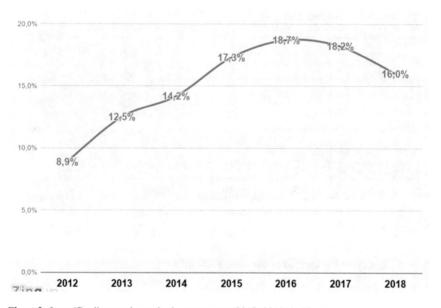

Chart 3 Loan/Credit growth rate in the past years (2012–2018) in Vietnam

purchasing power and consuming decline, as well as investment decreases. Or during low inflation periods, companies revenues and profits are inflated and values decrease. That's why the US, Fed, tried in many years to keep inflation and GDP growth at low level just because real return equals to (=) nominal return minus (−) inflation. The empirical date from Vietnam stock exchange during 2007–2009 crisis showed us

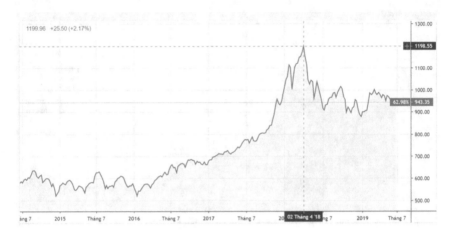

Chart 4   VNIDEX over past 5 years (2014–2019)

Chart 5   History stock price chart of Vingroup 2015–2019

that inflation may harm stock market and negatively affect stock price. Other macro variables such as exchange rate might have positive and long term impact on firm stock price and later on, we will soon figure out its impact result in the below regression section. From investors viewpoint in stock exchange, knowing the relationship between stock price and various macro-economic factors is a benefit because they can predict their returns on investment and take advantages of investment opportunities in financial and stock market.

6   Research Method and Data

In this research, analytical method is used with data from the economy such as inflation in Vietnam and USA (data from Bureau statistics), lending rate (data from Vietnam commercial banks), exchange rate (from VN commercial banks) and risk free rate (from government bond maturity 5 years). S&P 500 index and VN Index and VIC stock price data are included from 2014 to 2019 with semi-annual data (10 observations in total). Beside, econometric method is used with the software Eview. It will give us results to suggest policies for businesses and authorities.

Therefore, our econometric model is established as in the introduction part. Vingroup stock price in Viet Nam is a function with 7 variables: x_1: inflation in Viet Nam, x_2: lending rate in Viet Nam, x_3: risk free rate, x_4: inflation in the US, x_5: SP500, x_6: USD/VND exchange rate, x_7: VnIndex (see below regression equation in Sect. 8).

7   General Data Analysis

The Chart 6 shows us that inflation in VN has a positive corelation with inflation in the US:

Then, we see the relationship between a set of internal macro factors including inflation, GDP growth rate, VnIndex and VIC stock price from 2014–2019. The Chart 7 shows us that VIC stock price, and then second, VN Index, have bigger volatility from June 2015 to June 2019.

Chart 6 Inflation in Viet Nam and in the US *Source* Vneconomy, mof.gov.vn

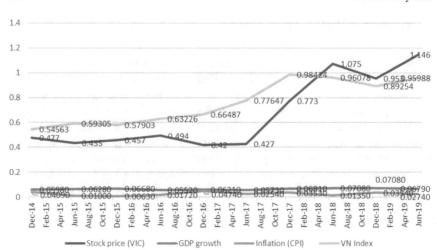

Chart 7 Inflation, VNIndex and GDP growth in Viet Nam versus VIC stock price *Source* Vneconomy, tradingeconomic

Another set of internal economic factors (lending rate, deposit rate and risk free rate in Vietnam) are shown in the below chart, together with VIC stock price from 2014 to 2019. And the Chart 8 shows us that from June 2017 to June 2019, although there is not much fluctuation in other macro factors, VIC stock price has bigger fluctuation (increase, then decrease and increase again in June 2019).

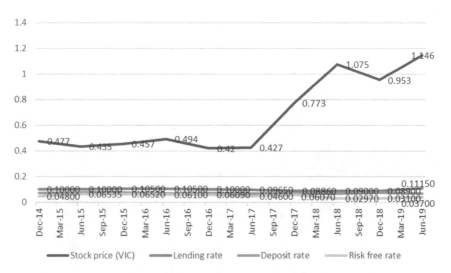

Chart 8 Viet Nam Risk free rate, lending rate and deposit rate vs. VIC stock price in Viet Nam, inflation in VN and in the US (VIC stock price unit: 100.000 VND, risk free rate: for bonds with maturity 5 years)

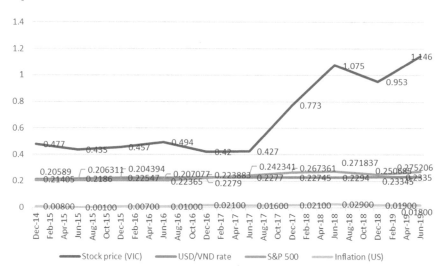

Chart 9 Exchange rate, S&P 500, Inflation in the US versus VIC stock price (macro factors calculated at the equivalent ratio for easy comparison)

This research sample uses data (GDP growth, inflation, risk free rate...) during 5 years from 2014 to 2019. The global crisis starting from 2007 and low-inflation period from 2014–2015 have impacts on Viet Nam economy. Therefore, we could assume Vingroup stock price in Viet Nam as a function depending on these macro-economic factors in the US and in VN.

Next we see the visual relationship between external factors (macro factors in the US) and VIC stock price in the Chart 9. It shows us that, in around June 2015, VIC stock price has a little decrease.

Last but not least, we see a mix of internal and external factors and their relation-ship to VIC stock price in the Chart 10. It shows us that, VIC stock price followed the same trend as other macro factors in VN and in USA and esp. In June 2016, and June 2018, VIC stock price has increases little much.

On the other hand, we could see statistical results with Eview in the below table with 3 variables.

Table 1 shows us standard deviation of VIC stock price in Vietnam is the highest (0.29), standard deviation of S&P 500 in the US is the second highest (0.03) and standard deviation of USD/VND exchange rate is the lowest (0.006).

If we want to see correlation matrix of six (6) internal macro variabes, Eview generate the below result in Table 2.

Table 2 shows us that correlation among six macro variables (total internal macro economic factors). An increase in GDP growth and increase in inflation might lead to a decrease in VN Index. Whereas an increase in deposit interest rates might lead to decrease in investment and reduce GDP growth rate. Furthermore, when deposit rate increases, risk free rate tends to decline.

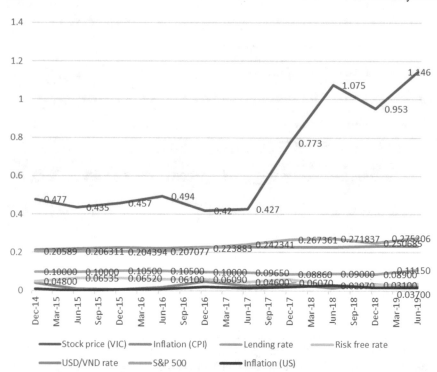

Chart 10 In the US versus VIC stock price (macro factors calculated at the equivalent ratio for easy comparison)

Table 1 Statistical results for internal and external macro-economic variables

	Stock price (VIC)	inflation (CPI)	Lending rate	Risk free rate	USD/VND rate	S&P 500	Inflation (US)
Mean	0.6657	0.02588	0.09856	0.050485	0.226117	0.235499	0.015
Median	0.4855	0.0264	0.1	0.05435	0.227575	0.233112	0.017
Maximum	1.146	0.0474	0.1115	0.06535	0.2335	0.275206	0.029
Minimum	0.42	0.0063	0.0886	0.0297	0.21405	0.204394	0.001
Stand. dev	0.29278	0.013884	0.007636	0.014066	0.006102	0.029493	0.008353

Table 3 shows us that correlation among six (6) macro economic variables (both internal and external macro factors). Inflation in VN has a negative correlation with inflation in the U.S., lending rate in VN and risk free rate (government bond) and has a positive correlation with exchange rate and S&P 500 index. Whereas lending rate in Viet Nam has a negative correlation with inflation in VN, risk free rate and exchange rate and it has a positive correlation with inflation in the U.S. and S&P 500. Last but not least, risk free rate has a negative correlation with inflation in VN

Table 2 Covariance matrix for six (6) internal macro-economic variables (GDP growth, inflation in VN, VN index, risk free rate, lending and deposit rates)

Coefficient Covariance MAtrix

	GDPGrowth	Inflation	VNIndex	Rf_rate	Lending rate	Deposit rate	C
GDP Growth	1605766	102734.8	-47.0	42884.2	36204.4	-1641944	41710.6
Inflation		80715.5	-2.79	10557	29443.8	-89321	-3603.5
VNIndex			0.002	4.78	6.1	70.05	-4.6
Rf_rate				116345	517.3	-11672.9	-11743
Lending rate					311173.8	128855.3	-47631
Deposit rate						3490767	-207295
C							21057

Table 3 Correlation matrix for 6 macro economic variables (inflation in VN and the U.S., lending rate, risk free rate, exchange rate, S&P500)

Coefficient Covariance MAtrix

	GDPGrowth	Inflation	VNIndex	Rf_rate	Lending rate	Deposit rate	C
GDP Growth	1605766	102734.8	-47.0	42884.2	36204.4	-1641944	41710.6
Inflation		80715.5	-2.79	10557	29443.8	-89321	-3603.5
VNIndex			0.002	4.78	6.1	70.05	-4.6
Rf_rate				116345	517.3	-11672.9	-11743
Lending rate					311173.8	128855.3	-47631
Deposit rate						3490767	-207295
C							21057

and exchange rate and it has a positive correlation with inflation in the US and S&P 500.

Hence, a reduction in risk free rate may lead to an increase in lending rate. A reduction in risk free rate might stimulate VN inflation. And an increase in inflation may lead to a reduction in lending rate to stimulate demand and investment.

8 Regression Analysis and Main Results

In this section, we will find out the relationship between macro economic factors such as inflation in Viet Nam or Myanmar, inflation in USA and unemployment rates in Viet Nam or Myanmar.

Scenario 1 Regression model with 1-6 variables: Inflation (CPI) in Viet Nam and VIC stock price.

Note: inflation in Viet Nam (INFLATION_CPI), C: constant.

Using Eview give us the below results:

	Six scenarios					
	1 variable	3 variables	3 variables	6 variables	6 variables (mix)	7 variables
Inflation_coefficient	36.9	-126.7		-45.7	-245	(see below)
GDPgrowth_coefficient		1763.6		2559		
VNIndex_coefficient		0.1		0.05		
Risk free rate			-1540.7	-767	-815	
Lending rate			131.4	969	437	
Deposit rate			-1517.3	-1527		
USD_VND					0.015	
SP500					0.07	
Inflation_US					-372	
R-squared	0.0003	0.8	0.6	0.95	0.84	
SER	31,04	15.7	21.7	10.9	19.7	

We can analyze as below:

A. In case 3 variables: Vingroup stock price has a negative correlation with inflation in Vietnam, but has a positive correlation with GDP growth rate and VN Index in Viet Nam. Esp., it is highly positively affected by GDP growth rate and slightly positively affected by VNindex.

In case other 3 variables:

B. We find out Vingroup stock price has a negative correlation with risk free rate in VN and deposit rate of VN commercial banks, but has a positive correlation with lending rate in Vietnam (within a range of 10 observations from 2014–2019). When Rf and borrowing rate increase, VIC stock price tend to decrease due to investment capital decline into stock market. High lending rate, on the contrary, will not encourage firms to borrow money from banks, so they might invest in stock market.

C. In case 6 variables: Therefore we see that Vingroup stock price has a negative correlation with risk free rate, inflation in VN and deposit rate of VN commercial banks, but has a positive correlation with GDP growth rate, VN Index and lending rate in Vietnam (within a range of 10 observations from 2014–2019). GDP growth, lending rate, deposit rate and Rf have highly impact on VIC stock price.

D. Incase mix of 6 internal and external variables: The above coefficients enables us to figure out that Vingroup stock price has a negative correlation with inflation, risk free rate in Vietnam, and inflation in the US, while it has a positive correlation with Lending rate in Vietnam, SP500 in the U.S. and exchange rate (USD/VND) (within a range of 10 observations from 2014–2019). In interactive global financial markets, both inflation in VN and US has negatively affect on VIC stock price. It means that high inflation might harm the investment values and may not encourage the increase in stock price of Vingroup, and therefore the whole stock market. Rf, Inflation in these 2 countries, and Lending rate are among factors that affect much more on Vingroup stock price. Looking at the standard errors of these coefficients (values in parentheses), we recognize that the variation or dispersion of sample means of CPI in Vietnam and Rf, measured by standard deviation, are smaller than those of inflation in the US and lending rate in Vietnam.

Scenario 2 A mix of seven (7) internal and external factors model: regression model with 4 internal variables: VNIndex, risk free rate in Vietnam (government bonds maturity 5 years), lending rate and inflation of Vietnam, together with 3 external macro variables: SP 500, inflation in the US and USD/VND rate:

Using Eview gives us the result:

```
┌──────────────────────────────────────────────────────────────────┐
│ ☐ File  Edit  Object  View  Proc  Quick  Options  Window  Help    │
│ View│Proc│Object│ Print│Name│Freeze│ Estimate│Forecast│Stats│Resids│ │
└──────────────────────────────────────────────────────────────────┘
```

Dependent Variable: VICSTOCKPRICE
Method: Least Squares
Date: 10/29/19 Time: 23:08
Sample: 1 10
Included observations: 10

Variable	Coefficient	Std. Error	t-Statistic	Prob.
VNINDEX	0.336552	0.305683	1.100982	0.3857
USD_VND_RATE	-0.012294	0.021630	-0.568378	0.6271
SP500	-0.122518	0.181779	-0.673997	0.5698
RF_RATE	-1069.702	661.5752	-1.616902	0.2473
LENDINGRATE	1508.947	1407.896	1.071775	0.3960
INFLATION_US	431.2986	1808.198	0.238524	0.8337
INFLATION_CPI	-146.9585	517.3195	-0.284077	0.8031
C	280.2972	508.0517	0.551710	0.6366

R-squared	0.905989	Mean dependent var	66.57000
Adjusted R-squared	0.576950	S.D. dependent var	29.27802
S.E. of regression	19.04310	Akaike info criterion	8.721849
Sum squared resid	725.2793	Schwarz criterion	8.963917
Log likelihood	-35.60924	F-statistic	2.753438
Durbin-Watson stat	2.366044	Prob(F-statistic)	0.292168

Therefore, Stock price_VIC $= 0.336*$VNIndex $- 146.95 *$ Inflation_CPI $+ 1508.9 *$ Lendingrate $- 1069.7*$Rf_rate $-0.01*$ USD_VND_rate $- 0.12*$SP500 $+ 431.3*$Inflation_US $+ 280.29$ (7.7), $R^2 = 0.9$, SER $= 19.04$.

(0.3) (517.3) (1407.8) (661.5)

(0.02) (0.18) (1808.2)

Finally, the above equation shows that Vingroup stock price has a negative correlation with inflation, risk free rate in Vietnam, and S&P500 in the U.S., while it has a positive correlation with Lending rate in Vietnam, Inflation in the US, VNIndex and exchange rate (USD/VND) (within a range of 10 observations from 2014–2019). Here, inflation in VN and risk free rate has negatively affect on VIC stock price. It means that high inflation might harm the investment values and may not encourage the increase in stock price of Vingroup, and therefore the whole stock market. Similar to Eq. 7.6, here, Rf, Inflation in these 2 countries, and Lending rate are among factors that affect much more on Vingroup stock price. Also, looking at the standard errors of these coefficients (values in parentheses), we recognize that the variation or dispersion of sample means of CPI in Vietnam and Rf, measured by standard deviation, are smaller than those of inflation in the U.S. and lending rate in Vietnam.

9   Limitation of the Model

Eview has advantages such as: analyzing data quickly, mange it efficiently, and good for econometric and statistical analysis. On the other hand, Eview can not give the absolutely correct correlation between variables in the model. Therefore, in this model, Eview can only provide us with results for reference.

10   Discussion for Further Research

We can add one more factor into our regression model, for example, unemployment rate. Or we can add foreign investment growth rate over years into our regression model in order to see the effects of these macro factors.

11   Conclusion and Policy Suggestion

The government and authorities of Vietnam might consider controlling inflation and issuing proper policies in which risk free rate is controlled and not increasing too much in order to support stock price, in specific, and stock market in general. To do this, they need to pay attention to not only treasury bonds but also corporate bonds to created proper mechanisms and policies to attract more capital.

Macro economic and financial policies need to consider impacts of macro factors such as inflation in their countries and outside factors such as inflation in the US. Further more, lending interest rates and exchange rate need to be increased at a rational level, because stock price, and therefore stock market, has a positive correlation with lending rate and exchange rate. Hence, in credit/loan policy, bank system and relevant government companies need to control and balance the capital source to limit the using of short-term capital to finance long-term projects.

Because inflation in the US has a positive correlation with stock price (see Eq. 7.7) and therefore stock market in Viet Nam, the government of Viet Nam need to implement suitable macro policies if inflation in the US increases or decreases. One lesson from Vingroup company development is balancing the importance of both internal and foreign investment capital. In order to develop real estate market, it needs not only bank capital, investors' pre-payment, but also stock market channel, corporate bond market, and foreign investment.

Viet Nam stock market has been becoming a competitive capital channel compared to commercial bank system and attracted more than 86,000 b VND for companies in 2018, together with 2,75 b USD from net foreign investment flow. Hence, the government and authorities in Viet Nam can issue proper policies which can protect their market economy, encourage and develop stock market to reduce pressure on bank system, increase its size to 100% GDP in period from 2020–2022 and reduce negative impacts from the global recession.

Last but not least, this study provides a regression model with its equation which helps investors to seek financial gains in the real estate industry and stock market based on observing fluctuations in macro policies and macro economic (internal and external) variables or data as we saw in the above equations. This paper also opens new direction for further researches, for instance, we can add 3 more variables into a 10 variable model: VN GDP growth, unemployment rate, deposit rate.

Exhibit 1 GDP growth rate past 10 years (2007–2018) in Vietnam.

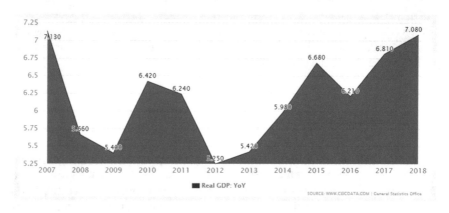

Acknowledgements I would like to take this opportunity to express my warm thanks to Board of Editors, Professors and Colleagues at Universities and companies. Lastly, thank you very much for my family, co-workers, and brother in assisting convenient conditions for my research paper.

References

Apergis, N. (2003). Housing prices and macroeconomic factors: Prospects within the European monetary union. *International Real Estate Review, 6*(1), 63–74.

Belej, M., & Cellmer, R. (2014). The effect of macroeconomic factors on changes in real estate prices—response and interaction. *ACTA Oeconomia, 13*(2), 5–16.

Chatterjea, A., Cherian, J. A., & Jarrow, R. A. (2001). *Market manipulation and corporate finance: A new perspectives, 1994 annual meeting review.* San Antonio, TX, USA: SouthWestern Finance Association.

Fang, H., Chang, T. Y. F., Lee, Y. H., & Chen, W. J. (2016). The impact of macroeconomic factors on the real estate investment trust index return on Japan, Singapore and China. *Investment Management and Financial Innovation, 13*(4), 242–253.

Grum, B., & Govekar, D. K. (2016). Influence of macroeconomic factors on prices of real estate in various cultural environments: Case of Slovenia Greece, France, Poland and Norway. *Procedia Economics and Finance, 39,* 597–604.

Gunarathna, V. (2016). How does financial leverage affect financial risk? An empirical study in Sri Lanka, Amity. *Journal of Finance, 1*(1), 57–66.

Mahyar, H. (2017). The effect of inflation on financial development indicators in Iran. *Studies in Business and Economics, 12*(2), 53–62.

Olatunji, I. A., Wahab, B. M., Ajayi, M. T. A., & Liman, H. S. (2016). Influence of macroeconomic factors on residential property returns in Abuja Nigeria. *ATBU Journal of Environmental Technology, 10*(1), 67–83.

Puarattanaarunkorn, O., Kiatmanaroch, T., & Sriboonchitta, S. (2016). Dependence between volatility of stock price index returns and volatility of exchange rate returns under QE programs: Case studies of Thailand and Singapore. *Studies in Computational Intelligence, 622,* 415–435.

ADB and Vietnam Fact Sheet. (2010).

https://www.ifc.org/ifcext/mekongpsdf.nsf/Content/PSDP22.

https://www.construction-int.com/article/vietnam-construction-market.html.

https://fia.mpi.gov.vn/Default.aspx?ctl=Article&MenuID=170&aID=185&PageSize=10& Page=0.

https://kientruc.vn/tin_trong_nuoc/nganh-bat-dong-san-rui-ro-va-co-hoi/4881.html.

https://www.bbc.co.uk/vietnamese/vietnam/story/2008/12/081226_vietnam_gdp_down.shtml.

https://www.mofa.gov.vn/vi/.

https://www.ceicdata.com/en/indicator/vietnam/real-gdp-growth.

How Values Influence Economic Progress? Evidence from South and Southeast Asian Countries

Nguyen Ngoc Thach

Abstract This study analyzes the effects of Hofstede's six cultural indices on real GDP per capita in the South and Southeast Asian countries. Applying the Bayesian normal linear regression against the least-square regression based on the frequentist approach, the results of this study demonstrate that cultural values have strong effects on economic development in the studied countries. Power distance and uncertainty avoidance indices are negatively correlated with real GDP per capita, whereas masculinity, long-term orientation, and indulgence versus restraint indices are positively correlated with the latter. Interestingly, this study found that individualism index shows a negative impact on the real GDP per capita in contrast to most of the early studies. A possible explanation of this outcome is that collectivism is considered as one of the most important determinants of economic progress in Confucian societies.

Keywords Hofstede's cultural dimensions · Bayesian analysis

1 Introduction

The determination of that has made some countries to become rich, while many other ones have been in a stagnant state for a long time has attracted the attention of researchers since Adam Smith's era. Up to now, much has been clarified on this issue, although not absolutely. In particular, the role of culture in socio-economic development continues to be of great interest of many scholars.

With regard to the relationship between culture and economic performance, there are four perspectives:

N. N. Thach (✉)
Institute for Research Science and Banking Technology, Banking University HCMC, 39 Ham Nghi Street, District 1, Ho Chi Minh City, Vietnam
e-mail: thachnn@buh.edu.vn

N. Ngoc Thach et al. (eds.), *Data Science for Financial Econometrics*, Studies in Computational Intelligence 898, https://doi.org/10.1007/978-3-030-48853-6_15

(i) *Cultural determinism*: Cultural values affect economic performance (Barro 2004; Franke et al. 1991; Harrison and Huntington 2001; Landes 1999; McClelland 1961; Sowell 1994; Weber 1930; Hofstede 1984).

(ii) *Economic determinism*: Economic changes bring cultural changes (Marx 1976; Bell 1973).

(iii) *A middle ground between economic determinism and cultural determinism*: A causal relationship exists between culture and the economy (Inglehart (1977, 1997).

(iv) *Disingenuous observations*: Some theorists do not support any of the above perspectives and argue that all reported associations are weak or suffer from methodological and conceptual flaws (Smith and Bond 1998; Yeh and Lawrence 1995).

With regard to this discussion, the author believes that there is an intercorrelation between cultural values and economic progress. It was almost impossible to combine two manuscripts related to some degree but with different conceptual frameworks and model characteristics into a single long paper. Therefore, the whole concept is presented in two distinct articles. In the first article, the impact of culture on economic progress in South and Southeast Asian countries is investigated. This concept is based on the fact that South and Southeast Asian countries are influenced by the Confucian ideology as a cultural feature of this region at different levels. In this ideology, the long-term objective is one of the core attributes of Confucian ideology which shows a positive impact on economic development in terms of savings and investment growth. However, this relationship will be discussed in a separate paper, also published in the upcoming proceedings (Thach et al. 2020), because of the nonlinear, complex impact of economic performance on cultural changes.

In addition, previous quantitative studies on the relationship between culture and economic performance were conducted by applying frequentist probabilistic methods. This study uses the Bayesian framework instead of the least-square regression.

2   Theoretical Background

2.1   Definition of Culture

In order to determine how culture impacts economic activity, first and foremost, it is necessary to understand what is culture. Culture can be defined in many different ways. According to (Hofstede 1984, p. 21), culture is "the collective programming of the mind that distinguishes the members of one human group from another." According to Schein (1985), culture is "a pattern of basic conditions that the group learned as it solved its problems of external adaptation and internal integration, that has worked perfectly to be considered valid and that, therefore, is learned to new members as the correct approach to perceive, think, and feel with regard to those problems." According to the GLOBE research program (House et al. 2004, p. 15), culture is "shared ideas, values, beliefs, identities, and interpretations or explanations of significant events that result from common experiences of members of collectives that are transformed across generations." According to Harrison (1985, 1997), culture defines attitude and values that regulate human action in a society, which is a simple definition of culture. According to North (1990), culture is the knowledge acquired or imparted from a generation to another through teaching and learning by imitation, which is derived after analyzing culture in the realm of institutions and institutional changes.

In the academic literature, several frameworks for measuring national cultural dimensions can be found (Hofstede 1984; Schwartz 1994; Trompenaars and Hampden-Turner 1997; House et al. 2004). In order to assess the dimensions of national culture, Hofstede's study represented the most common framework over the past three decades (McSweeney 2002).

Based on the Hofstede survey of 80,000 IBM employees in 66 countries, the following four dimensions of national culture are established: power distance versus closeness, uncertainty avoidance, individualism versus collectivism, and masculinity versus femininity.

However, McSweeney (2002) reports that measurement reliability and validity are the principal drawbacks of the dimensions of Hofstede. Moreover, Hofstede has not provided the values of Cronbach's alpha for the items that create the indices. In addition, the average number of questionnaires used per country by Hofstede was small (McSweeney 2002). According to Schwartz (1994), the two main drawbacks of the framework of Hofstede are his sample plan and historical change. First, data are obtained from employees of only a single multinational corporation (IBM) by Hofstede. Second, the analyses of Hofstede were based on data collected from 1967 to 1973.

Based on the above comments, Hofstede added two new dimensions in his framework. First, long-term (versus short-term) orientation was included in his model taking into consideration the Chinese Value Survey. Second, the sixth and last dimension indulgence versus restraint was included in 2010 based on the World Values Survey.

Finally, the dimensions of Hofstede are considered useful for the analysis of the relationship between cultural values and economic performance despite some limitations thanks to their clarity, parsimony, and resonance with managers (Kirkman et al. 2006, p. 286). Moreover, the cultural indices in his framework contain fairly extensive information about the cultural attributes of various nations. Therefore, six cultural dimensions of Hofstede are used in this study.

2.2 Hofstede's Cultural Dimensions

In order to determine cultural differences across nations, Hofstede (1984) applied the following six indices.

Power distance (versus closeness) index: It is defined as the extent to which less powerful members of an organization or an institution (or family) accept unfairly allocated power. In this case, less powerful people unconditionally perceive inequity and power concentration. Therefore, a clearly defined and enforced distribution of power in society is indicated by a high power distance society without any doubt. On the other hand, people hardly accept unfairness in power allocation in a low power distance society.

Individualism (versus collectivism) index: This index represents "the degree of individual integration with a collective and a community." A high individualist society often has a relatively weak level of binding. They focus on the "me" subject rather than "us." On the contrary, collectivism represents a society with close relationships between families and other institutions and groups. Team members receive absolute loyalty and support of other members in each dispute with other groups and associations.

Masculinity (versus femininity) index: In this case, "male rights" is defined as "social respect for achievement, material rewards, and success which individuals gain." On the contrary, femininity refers to the importance of collaboration, humility, concern for difficult individuals as well as life quality. Women in society are respected and express different values. However, women, despite being talented and competitive, are often less respected than men in a masculine society.

Uncertainty avoidance index: It is defined as "social acceptance of ambiguity", when people accept or prevent something unexpected, unclear, and different compared to the usual status. A strong uncertainty avoidance index demonstrates the level of engagement of community members with the standards of behavior, rules, guiding documents, and they often believe in the absolute truth or "righteousness" in all aspects. On the contrary, a weak uncertainty avoidance index indicates openness and acceptance of a variety of opinions. Moreover, low uncertainty avoidance societies are often less regulated and they tend to let everything grow freely and take risks.

Long-term (versus short-term) orientation index: This aspect describes the relationship between the past, present, and future actions. A low index represents the short-term orientation of a society when traditions are treasured and consistency is appreciated. Meanwhile, a society with long-term orientation often focuses on long-term processes, cares about adaptation, and shows pragmatism when solving problems. In the short-term orientation, economic development will be difficult for a poor country. On the contrary, countries with long-term orientation are often more interested in development.

Indulgence (versus restraint) index: This concept measures happiness, whether or not there is self-satisfaction of simple joys. "The permission of the society to freely satisfy human basic and natural needs, such as enjoying life", is called indulgence. Similarly, the concept of "restraint" reflects "social control by stereotypes, strict norms for personal enjoyment." A society that allows enjoyment often makes individuals believe that they control their lives and emotions, whereas people in a restraint society believe that there are other factors, apart from themselves, that control their own lives and emotions.

3 Research Method and Data

3.1 Method

In this study, the effects of six cultural indices of Hofstede on the real GDP per capita of South and Southeast Asian countries are analyzed by applying both OLS and Bayesian regression.

Based on the framework of the OLS regression, this relationship is expressed as follows:

$$\text{Log\_capita\_GDP} = \beta_0 + \beta_1 X1 + \beta_2 X2 + \beta_3 X3 + \beta_4 X4 + \beta_5 X5 + \beta_6 X6 + \varepsilon,$$

where the independent variables X1 (power distance index), X2 (individualism index), X3 (masculinity index), X4 (uncertainty avoidance index), X5 (long-term orientation index), and X6 (indulgence versus restraint index) represent six cultural dimensions of Hofstede. The dependent variable is the log of a country's average GDP per capita (Log_capita_GDP) between 2005 and 2017 in terms of 2011 US dollars. The logarithmic form of GDP per capita is used to normalize the highly skewed economic variable so that it is more amenable to least-square statistics because the cultural scores of Hofstede have been standardized between 0 and 120 (Hofstede et al. 2010). Here, ε is a random error with zero mean and variance σ^2.

The fact that Bayesian analysis has many advantages over the frequentist approach (the OLS regression in our case) is widely known (Anh et al. 2008; Kreinovich et al. 2019). A Bayesian model exhibits higher robustness to sparse data, i.e., analysis is not limited by the sample size. Nguyen et al. (2019) report that instead of testing hypotheses using p-values in social investigations, nonfrequentist probabilistic methods such as Bayesian analysis should be evaluated. Thach et al. (2019) applied the Bayesian approach in their research on economic dynamics.

Given the observed data, the Bayesian analysis depends on the posterior distribution of model parameters. According to the Bayes rule, the posterior distribution results from combining prior knowledge with evidence from the observed data. Incorporating prior knowledge into the model makes the inferential result more robust. By using Monte Carlo Markov Chain (MCMC) methods, such as Metropolis–Hasting (MH) methods and Gibbs sampler, many posterior distributions are simulated. Samples that are simulated by applying MCMC methods are correlated. The smaller is the correlation, the more efficient is the sampling process. Most of the MH methods typically provide highly correlated draws, whereas the Gibbs sampler leads to less-correlated draws. As a special case of the MH algorithm, despite a high computational cost, the Gibbs method has the main advantage of high efficiency.

The Bayesian general model is expressed as follows:

$$\text{Posterior} \propto \text{Likelihood} \times \text{Prior} \tag{1}$$

Assume that the data vector y is a sample from a model with unknown parameter vector β. Based on the available data, some characteristics of β are inferred. In Bayes statistics, y, β are considered as random vectors. Therefore, Eq. (1) is expressed formally as follows:

$$p\left(\beta|y\right) \propto L\left(y; \beta\right) \pi(\beta) \tag{2}$$

where $p\left(\beta|y\right)$ represents the posterior distribution of β given y; $L\left(y; \beta\right)$ denotes the probability density function of y given β; and $\pi\left(\beta\right)$ represents a prior distribution of β.

The posterior distribution is rarely available, so it requires to be estimated via simulation except in the case of some special models. For this purpose, MCMC methods are considered useful.

In order to perform sensitivity analysis, three different Bayesian linear normal regression models are compared with informative and uninformative priors.

Fitting a Bayesian parametric model, the likelihood function and prior distributions for all model parameters are specified. Our Bayesian linear model includes the following eight parameters: seven regression coefficients and the variance of the data. Before specifying Model 3 with uninformative priors, informative priors are incorporated in Models 1 and 2. Conjugate and multivariate are considered as two different informative prior distributions specific to Bayesian normal linear regression.

Model 1: Let us consider a conjugate prior in the first model, where all regression coefficients are independently and identically distributed as normal with different means μ_0 but the same variance σ^2. The variance parameter shows an inverse-gamma distribution with the shape parameter of ϑ_0 and the scale parameter of $\vartheta_0\sigma_0^2$.

Model 1 is expressed as follows:

Likelihood model:

$$\text{Log\_capita\_GDP} \sim N\left(\mu, \sigma^2\right)$$

Prior distributions:

$$\beta|\sigma^2 \sim N\left(\mu_0, \sigma^2\right)$$

$$\sigma^2 \sim \text{InvGamma}\left(\vartheta_0, \vartheta_0\sigma_0^2\right)$$

where β represents the vector of coefficients, σ^2 represents the variance for the error term, μ represents the mean of the normal distribution of Log_capita_GDP, μ_0 represents the vector of prior means, ϑ_0 represents the prior degree of freedom, and σ_0^2 represents the prior variance for the inverse-gamma distribution.

Hence, our model is the following:

$$\beta|\sigma^2 \sim N\left(\mu_0, \sigma^2\right)$$

$$\sigma^2 \sim \text{InvGamma}\left(6, 6\right),$$

where $\mu_{x1} = -0.03$, $\mu_{x2} = -0.01$, $\mu_{x3} = 0.01$, $\mu_{x4} = -0.02$, $\mu_{x5} = 0.02$, $\mu_{x6} = 0.04$, $\vartheta_0 = 12$, $\sigma_0^2 = 1$ are obtained from Table 1.

Model 2: Let us consider multivariate g-prior of Zellner, which is considered as one of the widely used informative priors for coefficients in Bayesian normal linear regression, whereas the prior for the variance parameter is also distributed as an inverse-gamma distribution. Model 2 is expressed as follows:

$$\beta|\sigma^2 \sim \text{zellnersg} \left(p, g, \mu_0, \sigma^2\right)$$

$$\sigma^2 \sim \text{InvGamma} \left(\vartheta_0, \vartheta_0\sigma_0^2\right),$$

where β represents the vector of coefficients, p represents the dimension of the distribution (the number of regression coefficients), g represents the prior degree of freedom, μ_0 represents a vector of prior means, and σ^2 represents the prior variance for the inverse-gamma distribution.

Therefore, we obtain the following expression:

$$\beta|\sigma^2 \sim \text{zellnersg} \left(7, 12, -0.03, -0.01, 0.01, -0.02, 0.02, 0.04, 9.5, \sigma^2\right)$$

$$\sigma^2 \sim \text{InvGamma} \left(6, 6\right)$$

Model 3: Let us consider a normal distribution for our outcome, Log_capita_GDP, the flat prior, a prior with the density of 1, for all regression coefficients, and the noninformative Jeffreys prior for the variance. Generally, the uninformative priors are used in a normal model with unknown mean and the Jeffrey priors are considered as a variance.

Model 3 is expressed as follows:

$$\beta \sim 1 \text{ (flat)}$$

$$\sigma^2 \sim \text{Jeffreys}$$

In comparison with the specified models, we primarily apply the Bayes factor, which is expressed as follows:

$$p\left(mo_1|y\right)/p\left(mo_2|y\right) = \{p\left(mo_1\right)|p\left(mo_2\right)\} * \{p\left(y|mo_1\right)/p(y|mo_2),$$

where mo_1 and mo_2 are Model 1 and Model 2, respectively. A large value of this ratio indicates a preference for mo_1 over mo_2.

3.2   Research Data

During the period 2005–2017, data on real GDP per capita of South and Southeast Asian countries are obtained from the IMF Database (IMF 2019), which is derived by dividing the constant price purchasing power parity (PPP) GDP by the total population. Moreover, GDP is measured in terms of constant international dollars per person. Primarily for the pre-2005 period, data on cultural indices of Hofstede are taken from Hofstede et al. (2010) based on the IBM Database plus extensions. The scores of six cultural dimensions of Hofstede range from 0 to 120. For instance, Bangladesh, China, Hong Kong, India, Indonesia, Japan, Malaysia, Philippines, South Korea, Singapore, Taiwan, Thailand, and Vietnam are considered as the selected countries and regions of South and Southeast Asian countries. These are the countries and regions on which we can gain access according to data of Hofstede.

4   Results of Bayesian Regression

4.1   Bayesian Simulations

With regard to the results achieved from the OLS regression, all the coefficients for variables X1 to X6 are not found statistically significant (Table 1). This is assumed because of a small sample size. As discussed above, one of the main advantages of the Bayesian approach over the frequentist methods is that the results of the Bayesian analysis are robust to sparse data. Hence, Bayesian regression is required to perform instead.

Let us analyze the simulation results of our Bayesian-specified models.

First, the acceptance rate and the efficiency of MCMC are examined. These are different concepts, although they indicate the efficiency of an MCMC sample. An acceptance rate is defined as the number of accepted proposals among the total number of proposals, whereas the efficiency means the mixed properties of the MCMC sample.

In the Bayesian analysis, the verification of the MCMC convergence is considered an essential step. Convergence diagnostics ensures that the MCMC chain has converged to its stationary distribution. The stationary distribution is considered the true posterior distribution. Therefore, the Bayesian inference based on an MCMC sample is observed to be valid. The acceptance rate of the chains helps in demonstrating the signs of nonconvergence. The efficient MH samplers should have an optimal acceptance rate ranging from 0.15 to 0.5 (Roberts and Rosenthal 2001). Besides Roberts and Rosenthal (2001), Gelman et al. (1997) claim that the acceptance rate of 0.234 is considered optimal in the case of multivariate distribution, and the optimal value is considered as 0.45 in the case of univariate. The acceptance rate of Models 1, 2, and 3 does not indicate problems with convergence according to our results (Table 2). An MCMC sample is considered effective when the simulation efficiency exceeds

Table 1 Estimation results by the OLS regression

Source	SS	df	MS	Number of obs = 13		
Model	7.1454884	6	1.19091473	F(6, 6) = 1.22		
Residual	5.84586192	6	0.974310321	Prob > F = 0.4068		
Total	12.9913503	12	1.08261253	R-squared = 0.5500		
				Adj R-squared = 0.1000		
				Root MSE = 0.98707		
Log_capita_GDP	Coef.	Std. err.	t	P > |t|	[95% Conf. interval]	
XI	−0.030432	0.0311182	−0.98	0.366	−0.1065754	0.0457114
X2	−0.0050037	0.0403999	−0.12	0.905	−0.1038586	0.0938513
X3	0.0097084	0.0283317	0.34	0.744	−0.596167	0.0790336
X4	−0.015862	0.0148429	−1.07	0.326	−0.0521814	0.204574
X5	0.0184572	0.0175768	1.05	0.334	−0.0245516	0.0614661
X6	0.0428237	0.0245644	1.74	0.132	0.0172832	0.1029305
_cons	9.532424	3.267073	2.92	0.027	1.538184	17.52666

Source The authors' calculation

Table 2 Simulation efficiency of specified models

	Model 1	Model 2	Model 3
Acceptance rate	0.345	0.3097	0.3931
Efficiency: min	0.008834	0.009858	0.006592
Avg	0.02223	0.02845	0.01858
max	0.04443	0.04107	0.05815

Source The author's calculation

0.01. The average efficiency of 0.02, 0.03, and 0.02 is obtained for Models 1, 2, and 3, respectively. All these values are found to be reasonable for the MH algorithm.

Multiple chains are often applied to inspect the MCMC convergence besides examining the MCMC sample efficiency. This analysis is performed for Model 2, which is selected as the most appropriate model based on the results of the Bayes factor test and the model test carried out below.

Based on multiple chains, a numerical convergence summary is provided by the Gelman–Rubin convergence diagnostic. The diagnostic Rc values greater than 1.2 for any of the model parameters should point out nonconvergence according to Gelman and Rubin (1992) and Brooks and Gelman (1998). However, researchers often apply a more stringent rule of $Rc < 1.1$ for indicating convergence. All the parameters of Model 2 have the maximum Rc value below 1.1 as shown in Table 3. Therefore, the MCMC chain of this model indicated that it is converged.

Furthermore, the estimation results are discussed for the three models. As mentioned above, two different informative priors are used in Models 1 and 2. MCSE is used to measure the precision of the posterior mean estimates. Incorporating an informative conjugate prior distribution in Model 1, the estimation table reports that the MCSE of the estimated posterior means accurate to at least a decimal place

Table 3 Gelman-Grubin diagnostics for Model 2

	Rc
Log_capita_GDP	
_cons	1.07705
XI	1.042624
X5	1.023346
var	1.01291
Log_capita_GDP	
X3	1.004659
X2	1.004634
X4	1.004017
X6	1.003667
Convergence rule: Rc < 1.1	

Source The author's calculation

Table 4 Simulation results for Model 1

	Mean	Std. dev.	MCSE	Median	Equal-tailed [95% cred. interval]	
Log_capita_GDP						
XI	−0.0280788	0.0158468	0.001686	−0.0284438	−0.0584901	0.0026883
X2	−0.0025256	0.0325248	0.001977	−0.0033507	−0.0666507	.0608859
X3	0.0072776	0.0239937	0.001724	0.0075348	−0.0415108	0.0514524
X4	−0.0155835	0.0118547	0.00069	−0.0156043	−0.0392642	0.0083366
X5	0.0191708	0.0114872	0.000719	0.018915	−0.0033368	0.0413312
X6	0.0423059	0.0210876	0.001993	0.0419336	0.000136	0.0826592
_cons	9.362929	0.8434279	0.077327	9.377796	7.660844	10.97042
var	0.7762553	0.2404736	0.011409	0.7289863	0.436871	1.374283

Source The author's calculation

(Table 4). What are the credible intervals? These intervals reflect a straightforward probabilistic interpretation unlike the confidence intervals in frequentist statistics. For instance, the probability that the coefficient for X1 is −0.03 is between −0.1 and 0.02 is about 95%.

That is also good enough as most of the posterior mean estimates are accurate to two decimal positions in Models 2 and 3 (Tables 5 and 6).

The posterior mean estimates are found to be very close to the OLS estimates because of completely noninformative priors incorporated in Model 3 (Tables 1 and 6). However, they are not found exactly the same because the posterior means of the model parameters are estimated taking into consideration the MCMC sample simulated from its posterior distribution instead of a known formula used.

Table 5 Simulation results for Model 2

	Mean	Std. dev.	MCSE	Median	Equal-tailed [95% cred. interval]	
Log_capita_GDP						
XI	−0.028228	0.0271666	0.001661	−0.0284774	−0.0812555	0.027536
X2	−0.0082664	0.0344626	0.001714	−0.0100474	−0.0730729	0.0628087
X3	0.0114741	0.023355	0.001937	0.0115405	−0.0364623	0.056359
X4	−0.0155084	0.0130209	0.00077	−0.0152562	−0.0437057	0.0108185
X5	0.0188724	0.0149979	0.000791	0.0185392	−0.0103889	0.0491954
X6	0.0394508	0.0212327	0.002139	0.039503	−0.0049523	0.0824417
_cons	9.424615	2.692317	0.154472	9.45576	4.285304	14.95797
var	0.7798531	0.2464239	0.012159	0.7338822	0.4344051	1.391522

Source The author's calculation

Table 6 Simulation results for Model 3

	Mean	Std. dev.	MCSE	Median	Equal-tailed [95% Cred. Interval]	
Log_capita_GDP						
XI	−0.0364588	0.0439984	.004569	−0.0338305	−0.1286974	0.0400708
X2	−0.0095761	0.0510602	0.003418	−0.0078885	−0.1213598	0.082165
X3	0.0115415	0.0364309	0.003312	0.009484	−0.0528229	0.0879534
X4	−0.0169551	0.0181943	0.001285	−0.0172971	−0.0519089	0.0195967
X5	0.0172563	0.0250456	0.00242	0.0191631	−0.0380527	0.0606834
X6	0.0429551	0.032232	0.001337	0.0424647	−0.01991	0.1045762
_cons	10.11639	4.527025	0.464849	9.806533	2.182876	20.11608
var	1.555054	1.568798	0.193226	1.127898	0.4074877	5.593312

Source The author's calculation

Table 7 Bayesian information criteria

	DIC	log(ML)	log(BF)
Mol	40.00501	−39.70595	
mo2	40.99943	−23.10297	16.60297
mo3	44.80716	−28.29432	11.41162

Source The author's calculation

4.2 Model Selection

In the Bayesian inference, model selection is considered an important step. In order to select the best model in this study, three candidate models are compared with various prior specifications. For this purpose, the Bayesian information criteria are used (Table 7).

The log (ML) and log (BF) values of Model 2 are the highest out of the three candidate models. DIC contains no prior information like BIC and AIC, whereas the

Table 8 Bayesian model test

	log(ML)	P(M)	P(M|y)
mol	−39.7059	0.3333	0.0000
mo2	−23.1030	0.3333	0.9945
mo3	−28.2943	0.3333	0.0055

Source The author's calculation

Bayes factor incorporates all the information about the specified Bayesian model. Based on these criteria, Model 2 performs best among all the considered models.

In addition, one more criterion could be applied. The actual probability associated with each of the candidate models was computed. As a result, Model 1 obtains the highest probability of 0.99 (Table 8).

Based on the above results, Model 2 is selected, which best fits the given data.

The above analysis shows that the three Bayesian models are fitted with different informative and noninformative priors. Moreover, the selection of our model is sensitive to prior specification, but economic meanings are completely robust to prior selection.

4.3 Discussion

The results of the Bayesian linear regressions indicate that the variables X1 (power distance index), X2 (individualism index), and X4 (uncertainty avoidance index) are negatively correlated with the real GDP per capita, whereas the variables X3 (masculinity index), X5 (long-term orientation index), and X6 (indulgence versus restraint index) show a positive impact on the latter in the selected South and Southeast Asian countries. The results of the five dimensions conform to several studies except for the individualism index (for instance, Greness 2015; Papamarcus and Waston 2006; Husted et al. 1999; Leiknes 2009).

Based on the empirical results, some details are presented below.

First, high power distance is considered a great cultural trait of Asian societies. The power distance index is negatively correlated with the real GDP per capita. In the high power distance countries, firms exist with a high hierarchy of authority and income inequality between the employees. In addition, societies with high scores on the power distance dimension are found to be more corrupted. All these factors weaken labor motivation, decrease labor supply, and as a consequence, slow long-term economic growth.

Second, the fact that the individualism index shows a negative impact on the real GDP per capita may be explained as follows. The majority of the South and Southeast Asian countries are collectivist societies, where people are mutually supportive in life and business, and economic success is usually achieved thanks to collective strength rather than individual talents. Thus, the less the individualism index (in contrast, the more the collectivism score), the higher economic performance.

Third, the masculinity index is positively correlated with the real GDP per capita. This result is consistent with other studies based on the role of culture in economic development in the West. Moreover, achievements, material success, and competitiveness are highly estimated and respected in high masculinity societies. These cultural characteristics significantly encourage economic progress.

Fourth, the uncertainty avoidance index shows a negative impact on the real GDP per capita in the South and Southeast Asian countries. It is easy to understand that high uncertainty avoidance is related to the desire to maintain the status quo, so it negatively affects economic development.

Fifth, a positive relationship is observed between the long-term orientation index and economic progress in the Confucian countries, where saving and investment are strongly stimulated. This result is found to be consistent with most of the studies on the role of cultural factors in economic development.

Sixth, the indulgence versus restraint index shows a positive impact on the real GDP per capita in the selected Asian countries. Moreover, individuals and firms are motivated for innovative activities in the societies that allow free pleasures related to enjoying life and having fun, including R&D that leads to technological progress and thus accelerates economic growth.

5 Conclusion

In the countries affected by Confucian ideology, such as South and Southeast Asia, culture plays a very important role in economic development. This work is conducted toward cultural determinism. As a result, the study examines the impact of cultural values on the real GDP per capita in 13 South and Southeast Asian countries. The six cultural dimensions of Hofstede data were primarily collected for the pre-2005 period and used to represent national cultural indicators. The average real GDP data for the period from 2005 was selected because the objective of the study is to assess the impact of culture on economic performance. The OLS regression considering the framework of the frequentist approach was estimated first, but the results related to the regression coefficients are not found statistically significant. This is because of a small sample size. Hence, the Bayesian approach is used in this study. One of the advantages of the Bayesian statistics as compared to the frequentist approach is that it allows the use of a small sample but the inferential result is still found to be robust. Three Bayesian linear regression models with various prior specifications are considered in this study. The results of formal tests demonstrated that the model with informative g-prior of Zellner fits best. In addition, economic significance remains unchanged despite different priors are selected.

Empirical results from the Bayesian linear regression demonstrate that power distance, individualism, and uncertainty avoidance indices show a negative impact on the average real GDP, whereas masculinity (versus femininity), long-term (versus short-term) orientation, and indulgence versus restraint indices show a positive correlation with the average real GDP. Generally, these results conform to the results of many studies except for the individualism index. The individualism index is found to be negatively correlated with the average real GDP, indicating that South and Southeast Asian countries are collectivist societies in which economic success is generally achieved by collective strength rather than individual efforts.

Finally, it is concluded that South and Southeast Asian countries require better policies in order to maintain positive cultural values such as collectivism, long-term thinking, and cultural reforms for improving attributes, such as a large power distance, fear of risk, and fear of adventure, that show a negative impact on development.

References

Anh, L.H., Le, S.D., Kreinovich, V., & Thach N.N., eds. (2018). Econometrics for Financial Applications. Cham: Springer, https://doi.org/10.1007/978-3-319-73150-6

Barro, R. J. (2004). Spirit of capitalism. *Harvard International Review*, *25*(4), 64–68.

Bell, D. (1973). *The coming post-industrial society*. New York: Basic Books.

Brooks, S. P., & Gelman, A. (1998). General methods for monitoring convergence of iterative simulations. *Journal of Computational and Graphical Statistics*, *7*, 434–455.

Franke, R. H., Hofstede, G., & Bond, M. H. (1991). Cultural roots of economic performance: A research note. *Strategic Management Journal*, *12*, 165–173.

Gelman, A., & Rubin, D. B. (1992). Inference from iterative simulation using multiple sequences. *Statistical Science*, *7*, 457–472.

Gelman, A., Gilks, W. R., & Roberts, G. O. (1997). Weak convergence and optimal scaling of random walk Metropolis algorithms. *Annals of Applied Probability*, *7*, 110–120.

Greness, T. (2015). National culture and economic performance: A cross-cultural study of culture's impact on economic performance across the 27 member countries of the European Union. *Journal of International Doctoral Research*, *4*(1), 69–97.

Harrison, L. E. (1985). *Underdevelopment is a state of mind* (2nd ed.). New York: University Press of America.

Harrison, L. E. (1997). *The Pan American Dream-Do Latin America's cultural values discourage true partnership with the United States and Canada* (1st ed.). New York: Basic Books.

Harrison, L. E., & Huntington, S. P. (2001). *Culture matters: How values shape human progress*. New York: Basic Books.

Hofstede, G. H. (1984). *Culture consequences: International differences in work- related values*. Beverly Hills: Sage.

Hofstede, G. H., Hofstede, G. J., & Minkov, M. (2010). *Cultures and organizations: Software of the mind*. Revised and expanded third Edition. New York, USA: McGraw-Hill.

House, R. J., Hanges, P. J., Javidan, M., Dorfman, P. W., & Gupta, V. (2004). *Culture, leadership, and organizations: The GLOBE study of 62 societies*. Sage Publication.

Husted, B. W. (1999). Wealth, culture, and corruption. *Journal of International Business Studies*, *30*(2), 339–359.

IMF: World Economic Outlook database, April 2019. https://www.imf.org/external/pubs/ft/weo/2019/01/weodata/download.aspx.

Inglehart, R. (1977). *The silent revolution: Changing values and political styles*. Princeton, NJ: Princeton University Press.

Inglehart, R. (1997). *Modernization and postmodernization: Cultural, economic, and political change in 43 societies*. Princeton, NJ: Princeton University Press.

Kirkman, B., Lowe, K., & Gibson, C. (2006). A quarter century of Culture's consequences: A review of empirical research incorporating Hofstede's cultural values framework. *Journal of International Business Studies, 37*(3), 285–320.

Kreinovich, V., Thach, N.N., Trung, N.D., & Thanh, D.V., eds. (2019). Beyond Traditional Probabilistic Methods in Economics. Cham: Springer, https://doi.org/10.1007/978-3-030-04200-4

Landes, D. S. (1999). *The wealth and poverty of nations: Why some are so rich and some so poor.* New York: Norton.

Leiknes, T. (2009). *Explaining economic growth: The role of cultural variables.* Master Thesis. The Norwegian School of Economics and Business Administration

Marx, K. (1976). *Capital* (Vol. 1). Harmondsworth: Penguin. (Original work published 1867).

McClelland, D. C. (1961). *The achieving society.* New York: Free Press.

McSweeney, B. (2002). Hofstede's Model of national cultural differences and their consequences: A triumph of faith - a failure of analysis. *Human Relations, 55*(1), 89–118.

Nguyen, H. T., Trung, N. D., & Thach, N. N. (2019). Beyond traditional probabilistic methods in econometrics. In V. Kreinovich, N. Thach, N. Trung & D. Van Thanh (Eds.), *Beyond traditional probabilistic methods in economics, ECONVN 2019.* Studies in Computational Intelligence (Vol. 809). Cham: Springer.

North, D. (1990). *Institutions, institutional change, and economic performance.* Cambridge: Cambridge University Press. http://dx.doi.org/10.1017/CBO9780511808678.

Papamarcus, S. D., & Waston, G. W. (2006). Culture's consequences for economic development. *Journal of Global Business and Technology, 2*(1), 48–57.

Roberts, G. O., & Rosenthal, J. S. (2001). Optimal scaling for various Metropolis-Hastings algorithms. *Statistical Science, 16*, 351–367.

Schein, E. H. (1985). *Organizational culture and leadership* (1st ed.). San Francisco: Jossey–Bass Publishers.

Schwartz, S. H. (1994). Beyond individualism/collectivism: New cultural dimensions of values. In U. Kim, H. C. Triandis, C. Kagitcibasi, S. C. Choi, & G. Yoon (Eds.), *Individualism and collectivism: Theory method and applications* (pp. 85–119). Thousand Oaks, California: Sage Publication.

Smith, P., & Bond, M. H. (1998). *Social psychology across cultures.* London: Prentice Hall.

Sowell, T. (1994). *Race and culture: World view.* New York: Basic Books.

Thach, N. N., Anh, L. H., & An, P. T. H. (2019). The effects of public expenditure on economic growth in Asia countries: A Bayesian model averaging approach. *Asian Journal of Economics and Banking, 3*(1).

Thach, N. N, Nam, O. V., & Linh, N. T. X. (2020). Economic changes matter for cultural changes: A study of selected Asian countries. In N. Trung, N. Thach & V. Kreinovich (Eds.), *Data science for financial econometrics, ECONVN 2020.* Studies in Computational Intelligence. Cham: Springer (to appear).

Trompenaars, A., & Hampden-Turner, C. (1997). *Riding the waves of culture: Understanding cultural diversity in global business* (2nd ed.). New York: McGraw-Hill.

Weber, M. (1930). *The Protestant ethic and the spirit of capitalism.* London: Allen and Unwin. (Original work published 1905).

Yeh, R., & Lawrence, J. J. (1995). Individualism and Confucian dynamism: A note on Hofstede's cultural root to economic growth. *Journal of International Business Studies, 26*, 655–669.

Assessing the Determinants of Interest Rate Transmission in Vietnam

Nhan T. Nguyen, Yen H. Vu, and Oanh T. K. Vu

Abstract This paper estimates the magnitude and speed of the interest rate pass-through and assesses the determinants of interest rate transmission in Vietnam for period 2006–2017. The finding shows that the interest rate pass-through from policy rate to interbank rate and from interbank rate to market rates are not perfect. Particularly, the effect of interbank rate on deposit and lending rates is weak, not as high as SBV's expectation. The speed of interest rate adjustment is relatively slow. Normally, it takes around one month for interest rate to adjust to the long-term equilibrium. The primary causes of limitations come from different factors, such as the degree of financial dollarization, lack of exchange rate flexibility, low liquidity and asset quality ratios of the banking system, the concentration and underdevelopment of the financial system, poor legal system quality, government budget deficit, the poor independence of the State Bank of Vietnam-SBV (Central Bank of Vietnam) and the overwhelming dominance of the fiscal policy.

Keywords Monetary policy transmission · Interest rate · Central bank

1 Introduction

Monetary policy is one of the important macroeconomic policies, including the overall measures that the central bank uses to regulate the monetary conditions of the economy to implement the objectives such as currency stability, inflation control, economic growth support and macroeconomic stability. The monetary transmission

N. T. Nguyen (✉)
Vietcombank Human Resources Development and Training College, Hanoi, Vietnam
e-mail: nhannt.ho@vietcombank.com.vn

Y. H. Vu · O. T. K. Vu
Banking Faculty, Banking Academy of Vietnam, Hanoi, Vietnam
e-mail: yenvh@hvnh.edu.vn

O. T. K. Vu
e-mail: oanhvtk@hvnh.edu.vn

© The Editor(s) (if applicable) and The Author(s), under exclusive license
to Springer Nature Switzerland AG 2021
N. Ngoc Thach et al. (eds.), *Data Science for Financial Econometrics*,
Studies in Computational Intelligence 898,
https://doi.org/10.1007/978-3-030-48853-6_16

mechanism is built upon the approach of factors affecting monetary demand through a system of transmission channels including interest rate channel, exchange rate channel, asset price channel and credit channel (Mishkin 2013). With the effectiveness and appropriateness both in theory and practice, interest rate targeting has been chosen by countries around the worlds and the mechanism of monetary policy transmission via interest rate channel has attracted many researchers about the importance of this channel in implementing monetary policy, especially in countries with developed markets.

Interest rate channel (or interest rate pass-through) is considered as a direct monetary transmission channel which can affect the cost of fund and opportunity cost of consumption. (Mishkin 2013). When Central bank adjusts monetary policy instruments in terms of rediscount rate and refinancing rate, this leads to the change in monetary base and the overnight interbank rate, *ceteris paribus*. On the basis of term structure of interest rate, the overnight interbank rate in turn will transfer the effects on market rates (deposit rates and lending rates) in the same direction. This will change the investment decision as well as the consumption decision, thereby, affecting the aggregate demand (Fig. 1).

In fact, the efficiency of monetary policy and interest rate channel in particular are mainly affected by external factors beyond the Central Bank as the international market; quality of the banking system balance sheet; financial market development; budget status and dominance of fiscal policy (Cottarelli and Kourelis 1994; Cecchetti 1999; Ehrmann et al. 2001; De Bondt 2002; Sorensen and Werner 2006; Leiderman et al. 2006; Betancourt et al. 2008; Frisancho-Marischal and Howells 2009; Mishra and Peter (2013); Saborowski and Weber 2013; Leroy and Yannick 2015). These external factors can make monetary policy hardly to achieve its objectives, and thus reduce the policy efficiency.

The paper aims to examine how much policy rates transmit to lending and deposit rates in Vietnam, i.e. the short-term interest rate pass-through and to assess the effectiveness of the interest-rate transmission mechanism in Vietnam.

This study shows that the interest-rate transmission is rather weak and slow in Vietnam according to different factors, such as the degree of financial dollarization,

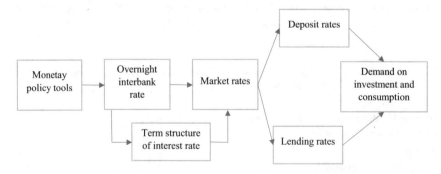

Fig. 1 Interest rate transmission mechanism

lack of exchange rate flexibility, low liquidity and asset quality ratios of the banking system, the concentration and underdevelopment of the financial system, poor legal system quality, government budget deficit; the poor independence of the State Bank of Vietnam-SBV (Central Bank of Vietnam) and the overwhelming dominance of the fiscal policy.

The paper is organized as follows. Section 2 reviews the economic literature on determinants of interest-rate transmission. Section 3 presents estimates of the interest-rate transmission mechanism in Vietnam, investigates the speed of transmission, and examines the factors that determine the interest-rate pass-through. A summary of the findings and key policy recommendations for Vietnam are provided in Sect. 4.

2 Literature Review

The degree and speed of interest rate pass-through have significant impact on the effectiveness of monetary policy transmission mechanisms (Kamin et al. 1998). The stronger and faster response of lending rates and deposit rates to changes of monetary market interest rates makes the transmission from the central bank's policy rate to the real economy variables more effective (Loayza and Schmidt-Hebbel 2002). There are some factors affecting the effectiveness of monetary policy through the mechanism of interest rates.

A high degree of dollarization may affect interest rate pass through by reducing the impact of central bank interest rate adjustment to the local currency rate of the banking system. When the balance sheet is overwhelmed by the foreign currency, the fluctuation in the exchange rate, in particular the major depreciation of the domestic currency, may narrow the balance sheet, leading to the fear of floating or to a larger margin of exchange rate fluctuation (Calvo and Carmen 2002; Leiderman et al. 2006). When capital is freely circulated, monetary policy can not be independent in case of fixed exchange rate regime (impossible trinity theory), therefore the effectiveness of interest rate transmission is greatly influenced by the degree of flexibility of the exchange rate and the flow of capital. Under the floating exchange rate mechanism, the central bank's operating rates are regarded as the main instruments of monetary policy to deliver clear policy signals to market participants, and to enhance monetary policy independence. The flexible exchange rate also shows the two-way exchange-rate risks, encouraging the development of risk insurance contracts, thereby reducing foreign currency deviations and dollarization (Freedman and Otker-Robe 2010). The study of Medina Cas et al. (2011) in Central America, Saborowski and Weber (2013) in the 2000–2011 in 120 countries suggest that interest rate transmission has a negative relationship with the dollarization, and a positive relationship with exchange rate flexibility. The interest rate transmission mechanism will increase from 25 to 50 basis points when the exchange rate regime is switched from pegged to floating exchange rate regime.

Banks with low asset quality can react to expansionary monetary policy by consolidating liquidity rather than expanding credit, thus changes in central bank's rate may have a limited impact on market rates. Empirical studies have confirmed the negative relationship between holding high liquid assets (due to lack of investment opportunities) and the effectiveness of interest rate channel. When the banking system has strong liquidity, the change in the central bank interest rate has little impact on changing the supply or demand of capital though it does impact on the lending interest rate of the banking system. In this case, liquidity is regarded as a buffer against market volatility and currency shocks (De Graeve et al. 2007; Kashyap and Stein 2000). The study of Saborowski and Weber's (2013) shows the negative impact of liquidity and bad debt ratio on interest rate transmission in developing countries. The authors confirm that an increase in banking system liquidity from 20 to 80 percent is associated with a reduction of about 20 basis points in the interest rate transmission mechanism. The study of Stephanie et al. (2011) in Central America countries and Avci and Yucel (2016) in Turkey also show similar conclusion with the study of Saborowski and Weber (2013). The studies of Wrobel and Pawlowska (2002), Chmielewski (2003) in Poland, the study of Bredin et al. (2001) in Ireland in the period of 1985–2001 have resulted in positive influence of bank profits, negative effects of risks on the effectiveness of interest rate transmission in these countries. In addition, big credit institutions seem to adjust lending rates faster than the small ones. Small banks with lower liquidity ratios will adjust interest rates according to the Central Bank's monetary policy faster than large banks with higher excess reserve ratios. The study of Gigineishvili (2011) also concludes the inverse relationship between the efficiency of the transmission channel and the degree of market volatility and the bank's excess reserve ratio.

The efficiency of the interest rate pass-through might decrease when the low level of competition and the separation among sectors of the financial system reduces the banks' response to the central bank's interest rate adjustment. The study of Bondt (2002), Kok Sørensen and Werner (2006), Gropp et al. (2007), Michiel et al. (2008) in European countries show that the strong competition in the financial markets of these countries leads to the faster transmission of interest rates. Moreover, a highly competitive banking market associated with the availability of alternative financial instruments in the stock market provided by financial intermediaries will speed up the interest rate channel. Some studies also prove that in the context of low competition, the response of deposit and lending rate to the adjustment of central bank rate is also low as the study of Heffernan (1997) about the interest rate channel in the UK, the study of Weth (2002) about the transmission of interest rates in Germany, De Graeve et al. (2004) about the determinants of interest rate transmission in Belgium.

Concerning the influence of competition on how banks adjust their interest rates, Berger and Hannan (1991) shows that deposit rates are quite rigid in the centralized market, especially in the period when the central bank increases policy rate sharply, banks tend not to raise deposit rates, which could be a sign of collusion among banks. In cross-country studies, Cottarelli and Kourelis (1994), Borio and Fritz (1995) find the important influence of competition on monetary policy transmission and lending rate tends to be more rigid when banks operate in a less competitive environment.

This is also confirmed in the study of Cottarelli et al. (1995). Mojon (2000) experiments the impact of banking competition on the interest rate transmission in the Euro area and find that higher competition tends to put pressure on banks to adjust lending rates faster when interest rates in the money market fall. Moreover, higher competition tends to reduce the ability of banks to increase lending rates when the money market interest rates increase, and vice versa for deposit rates. Similar results of this asymmetry are confirmed in Scholnick (1996), Heinemann and Schuler (2002), Sander and Kleimeier (2002 and 2004a, b) and Gropp et al. (2007).

The development of the financial system enhances the effectiveness of the transmission mechanism of monetary policy. In the developed money market and interbank market, the transmission from the central bank's rates to interbank rates increases - the first step in the mechanism of interest rate transmission, therefore transmission from central bank's rates to deposit and lending rates of the banking system improves (Yang et al. 2011). In a thin and underdeveloped financial market, when the central bank increases money supply or reduce interest rate, commercial banks increase surplus reserves or increase provisioning, which reduces flexibility and and dynamics of the interbank market, and reduce the efficiency of interest rate channel (IMF 2010). Singh et al. (2008) studied 10 industrialized countries and developing Asian countries and found that the development of the financial market generally leads to a stronger transmission of interest rates both in the medium and long term. Mishra and Peter (2013) argued that transmission from central bank rates to lending rates of the banking system in low-income countries is very weak and unreliable due to the limited level of financial development in these countries.

Innovations in the financial sector may shorten the link between short-term interest rates and lending rates. Axel et al. (2009), Gropp et al. (2007) conclude that progress in risk management technology is one of the factors that can accelerate the speed of interest rate transmission. Kamin et al. (1998) mention the access of households and companies to alternative sources of capital, such as the stock market that also determine lending rates and deposit rates in accordance with the domestic banking system. The higher the level of diversification of capital sources, the faster the rate of interest rate transmission. The study of Esman and Lydia (2013) on financial and monetary policy reform in Kenya in the period of 1998–2012 emphasizes the positive impact of financial innovation on the interest rate transmission channels of monetary policy. The study of Roseline et al. (2011) contradicts these studies when concluding that financial innovation poses challenges and complicates the monetary policy operation, thus financial innovation weakens interest rate channel in the mechanism of monetary policy transmission.

Volatility in the money market is a factor determining the speed and extent of the interest rate transmission. According to Gigineishvili (2011), Cottarelli and Kourelis (1994), Mojon (2000), and Sander and Kleimeier (2004a, b), the volatile currency market tends to weaken the transmission of interest rates. Gigineishvili (2011) states that money market interest rates bear reliable information about the need to adjust bank's deposit rates and lending rates. The fluctuations in the money market lead to uncertainty in market signals, banks become more cautious in offering interest to their customers and often wait until these fluctuations be reduced. Lopes (1998) also argues that in the case of high inflation, fluctuations in inflation should be considered when determining real interest rates and Lopes also notes that when stabilization policies can reduce fluctuation of inflation, the transmission of interest rate channel will be stronger.

The quality of the institutional system can affect the efficiency of monetary policy through the activities of participants in the macro economy. With the same policy move, banks and investors will react in different ways depending on the constraints of the institutional environment, thus affect the efficiency of policy conduct. The poor quality of the regulatory system impacts the effective transmission of interest rates because it cause asymmetric information problems and increase the cost of contract performance, thereby increasing costs of financial intermediaries (Mishra and Peter 2013). Cecchetti (1999), Cottarelli and Kourelis (1994), Ehrmann et al. (2001) and Kamin et al. (1998) point out that the financial structure of the economy and the quality of governance are key to the efficiency of the interest rate channel. Gigineishvili (2011) studies 70 countries of different development level in the 2006–2009 period and concludes that macroeconomic, institutional and legal characteristics as well as financial structure are factors affecting efficiency of interest rate channel.

The dominance of fiscal policy, budget deficits and independence of the central bank also affect the efficiency of the interest rate channel and the effectiveness of monetary policy. This is found in the studies of Sargent and Wallace (1981), Michele and Franco (1998), Bernard and Enrique (2005), Andrew et al. (2012). These studies show the negative impact of fiscal dominance, budget deficit and the lack of independence of the Central Bank on the effectiveness of interest rate transmission in the implementation of monetary policy. Eliminating the dominance of fiscal policy will increase the effectiveness of the interest rate channel as central bank could independently fulfill their objectives regardless of fiscal policy objectives (IMF 2010). Direct lending to the government undermines the balance sheets of central banks and limits the ability of central banks to provide clear signal of changes in monetary policy (such as raising interest rates and implementing open market operations) (Laurens 2005). Monetary policy will become effective if the public understands the objectives as well as tools to operate policies and believes in the central bank's commitments to these goals.

3 Interest Rate Transmission in Vietnam

3.1 Correlation of the Policy Rate with Market Rates

i. **Model**

This paper applies the pass-through model based on cost of capital approach in De Bondt (2002), which uses money market interest rate to examine the change in lending and deposit rates. This model is commonly used in measuring and evaluating the effectiveness of interest rate transmission mechanism. The pass-through process consits of two stages:

- Stage 1: Examine the effect of policy rates (rediscount rate and refinancing rate) on money market interest rate (overnight interbank rate)
- Stage 2: Examine the impact of money market interest rate (overnight interbank rate) on retail rates (deposit and lending rates of banking system).

Pass-through model in long term

The long term relation between policy rate and money market interest rate as well as the the relation between money market rate and retail rates of banking system are expressed by OLS regression model as follow:

$$y_t = \alpha + \beta x_t \tag{1}$$

y_t is money market rate and retail rate respectively for the case that x_t is policy rate (stage 1) and money market rate (stage 2); α is constant and β is coefficient which measures the degree of pass-through in long term. This coefficient shows the response of interbank rate, deposit rate and lending rate in long term against the movement of policy rate and interbank rate respectively. The greater level of the coefficient β indicates the higher degree of pass-through between two pair of interest rates. If the coefficient β is lower than 1, this implies the incompleted transmission in interest rate. If the coefficient β equals to 1, this is perfect case in that 1% change in the policy rate and interbank rate will lead to 1% change in interbank rate and market rates respectively. The case in that coefficient β is higher than 1 is excessive transmission and rarely occurs in practice. (Coricelli et al. 2006).

Pass-Through Model in Short Term and Moving Average Lags

The degree of interest rate transmission in short term is measured by error correction model (ECM) where employing the lags of interest rates in line with VAR model.

$$\Delta y_t = \mu + \rho\left(y_{t-1} - \alpha - \beta x_{t-1}\right) + \gamma \Delta x_t + \varepsilon \tag{2}$$

ρ measures the speed of adjustment of correction error and ρ also indicates the speed of self-balancing of y_t. γ measures the level of direct pass-through in short run and γ implies the response of interest rate of y against the change in interest rate of x in the same period. When γ equals to 1, the transmission is perfect and when γ is lower than 1, the process is imperfect.

According to Hendry (1995), Adjustment Mean Lags (AML) of perfect pass-through in ECM model is computed as follow:

$$AML = (\gamma - 1)/\rho \tag{3}$$

Therefore, AML, which is weighted average of all lags, measures the response of interbank rate and market rates against the change in policy rate and interbank rate respectively. AML also implies the time required for interbank rate and market rates to adjust to the long-term equilibrium. The higher AML is, the slower response of interest rates is. On the other hand, smaller AML implies the higher flexibility of interbank rate, market rates according to the changes of policy rates and interbank rate (stage 1 and stage 2 respectively). Many studies demonstrate that the short term adjustment is asymmetric, in other words, the speed of interest rate adjustment is different when interest rates are above and below the equilibrium level. (Chong 2005 and Scholnick 1996).

ii. Data

In our model, data is collected monthly for the period from January 2006 to December 2018. For the purpose of measuring the interest rate pass-through, five variables are employed into model, including:

- Discount rate and refinancing rate which are used as proxies for policy rate in monetary policy in Vietnam, are collected from the data source of SBV.
- Due to the limitation of Vietnam's financial market, an average interbank rate is employed to represent for money market interest rate. The average overnight interbank rates are collected and calculated from the data source of SBV.
- Retail rates including deposit rates and lending rates are collected from data source of International Monetary Fund (IMF).

Table 1 Interest rates data

Type of interest rates	Notation	Source
Discount rate (monthly)	DIR	State Bank of Vietnam
Refinancing rate	RER	State Bank of Vietnam
Average overnight interbank rate	IB1	State Bank of Vietnam
Average deposit rate	DR	IMF database
Average lending rate	LR	IMF database

iii. Defection detection

Augmented Dickey-Fuller (ADF test) and Phillips-Perron (PP test) are conducted and they prove that all input variables are stationary at the first difference I(1) rather than their own levels I(0). However, unit root test demonstrated that two pairs of interest rates: discount rate and overnight interbank rate, refinancing rate and overnight interbank rate, overnight interbank rate and deposit rate, overnight interbank rate and lending rate are cointegrated at 1% level of significance. Therefore, following Engle and Ganger (1987), OLS and ECM models can be employed to measure the pass-through in long term and short term respectively.

iv. Lag optimal selection

Sequential Modified LR, Final Prediction Error, Akaike Information Criterion, Shcwarz Information Criterion and Hannan-Quinn Information Criterion are used to determine lag optimal for model. The lag of two-period is chosen, as recommended by Akaike Information Criterion (AIC).

v. Results

For the measurement of interest rate transmission effect from policy rate to interbank rate during 2006–2017, the appropriate models are conducted and have results which is shown in Tables 1 and 2 (see Appendix 1–4).

Interest rate transmission from discount rate to interbank rate

In long term, the pass-through effect of discount rate on interbank rate is relatively high (0.85) and this impact is just lower (0.71) in short term. The adjustment mean lag shows that it takes about 0.53 month (equivalent to 15.9 days) for the overnight interbank rate being back to the long term equilibrium.

Interest rate transmission from refinancing rate to interbank rate

Under perspective of refinancing rate, the transmission on interbank rate in long term is lower than the pass-through effect of discount rate (0.79). The direct transmission in short term achieves the level of 0.7. Adjustment mean lag that it spends 0.54 month (approximately above 16 days) for overnight interbank rate moving to the long term equilibrium.

It is found that the overnight interbank rate in Vietnam is relatively sensitive to the change of discount rate and refinancing rate in both short term and long term. This transmission, however, is imperfect and it seems that the overnight interbank rate is more sensitive to discount rate than to refinancing rate.

The empirical result of the transmission model which measures impacts of inter-bank rate on deposit and lending rate of banking system during 2006–2017 is shown in Table 3 (see Appendix 5–8).

Table 2 Pass-through effect of policy rate on interbank rate

	DIR					RER					
	Short term			Long term		Short term			Long term		
	Direct pass-through coefficient (γ)	Correction error coefficient (ρ)	AML	Pass-through coefficient (β)		Direct pass-through coefficient (γ)	Correction error coefficient (ρ)	AML	Pass-through coefficient (β)		
IB1	0.706941	−0.551416	0.53	0.853769		0.700419	−0.550310	0.54	0.791762		

Table 3 Pass-through effect of the interbank rate on deposit and lending rates

	IB1			
	Short term			Long term
	Direct pass-through coefficient (γ)	Correction error coefficient (ρ)	AML	Pass-through effect coefficient (β)
DR	0.135835	−0.856049	1.01	0.788684
LR	0.145884	−0.863684	0.98	0.805528

Interest rate transmission from the interbank rate to deposit rate

In long term, the pass-through effect of interbank rate on deposit rate is relatively low (0.788) and this impact is even lower (0.136) in short term. The adjustment mean lag shows that it takes approximately one month for deposit rate being back to the long term equilibrium.

Interest rate transmission from the interbank rate to lending rate

The transmission of interbank rate on lending rate in long term is relatively low (0.806) but is still higher than on deposit rate. In short term, the level of direct transmission is around 0.146. Adjustment mean lag indicates that it spends 29 days for lending rate adjusting to long term equilibrium.

Therefore, it can be seen that the transmission effect of interbank rate on market rates is relatively loosing, in which, lending rate transmission is closer than deposit rate.

3.2   Interest Rate Transmission: Empirical Evidence

i. Model

In Sect. 3.1, we employ the interest rate transmission efficiency model based on cost of capital approach in De Bondt (2002) to evaluate independently the impact of policy rates (x_t) on lending rate of banking system (y_t). But in this Sect. 3.2, we apply the model following the previous studies including Mishra and Peter (2013), Stephanie M. C., Alejandro C. M. and Florencia F. (2011) to measure the influence of driven factors on the effectiveness of interest rate transmission in Vietnam for the period from 2006 to 2017. The model is expressed as follow:

$$y_t = \alpha_0 + \beta_1 y_{t-1} + \beta_2 x_t + \mu z'_t + \theta x_t z'_t + \varepsilon_t$$

In the model, there is an assumption that the effect of policy rates implemented by Central Bank (x_t) on banking system's lending rate (y_t) is driven by factors (z'_t) including: global market, the quality of banking system balance sheet, infrustruc-true of financial market, the degree of policy market's implementation in financial market and other macroeconomic variables such as State budget deficit, Central Bank independence and fiscal policy dominance in relation with interest rate policy $(x_t z'_t)$.

Due to multicollinearity by using OLS model, we decide to use Generalized Method of Momments model (GMM) generated by Hansen (1982) and developed by Arellano and Bond (1991) to overcome this error.

ii. Variables and tests

The time-series data of all variables is collected quarterly from International Monetary Fund (IMF), statistics reports and information disclosure issued by Vietnam's General Statistics Office (GSO) and Ministry of Finance during 2006–2017.

For the measurement of interest rate pass-through, this model uses lending rate (LR) of banking system as dependent variable. Moreover, our study employs some independent variables including:

- Discount rate (quarterly) (DRATE) is representative of policy rate of Central Bank.
- The flexibility of exchange rate (EXF) and the degree of dollarization (DOLLARIZATION) are used as measurement of global market's influence. Following the suggestion of IMF, dollarization is measured by foreign currency/M2. Similarly, to Virginie Coudert et al. (2010), the flexibility of exchange rate is calculated by taking logarithm of the change of exchange rate in the market. The high degree of dollarization and the less flexibility of exchange rate are expected to lower the efficiency of interest rate transmission channel.
- The quality of bank's asset is mainly measured by non-performing loan ratio (NPL). For Vietnam's banking system, loan portfolio always accounts for largest proportion in total assets since this is primarily profitable assets which generate a large income for banks. Therefore, the quality of lending portfolio relatively reflects the quality of banking assets. Additionally, many empirical studies normally use non-performing loan ratio as measurement of the quality of loan portfolio. An increase in bad debt ratio leads to the decrease in asset quality of banking system and thus is expected to lower the effectiveness of interest rate pass-through.
- The development of financial system (FDEV) which is measured by total deposit/GDP is used as a proxy for business environment where financial market operates. This is caused by there is no quantitative indicator which can measure the seperation of market segments as well as the administrative intervention of regulators. Competition indicators (H index, Boone) are not used since they can not exactly reflect the degree of competition in the market. The ratio of total deposit over GDP is used in accordance with the suggestion of World Bank (WB), and is expected to be positively correlated to effectiveness of interest rate transmission.
- The level of budget deficit (GBUDGET) and the fiscal dominance (FDOM) are added into the model as proxies for other factors. Due to the continuous budget deficit as well as fiscal dominance, the effectiveness of monetary policy is restricted in Vietnam during research period. The former is measured by

the ratio of budget deficit over GDP and the latter is calculated by credit for Government/M2 as in IMF's study.

Augmented Dickey-Fuller (ADF test) and Phillips-Perron (PP test) indicates that most of input variables are stationary at their first difference I (1) rather than at the own levels I (0).

iii. Result

The regression result is given in Table 4. The relevance of the GMM model is confirmed by the coefficient of the lag of dependent variables (DLR).

According to Table 4, some conclusion is made as follow.

First of all, the interaction of independent variables with policy rates of SBV is proper as our expectation. However, the budget deficit variable as well as its interaction are not statistically significant. This can be explained by the limitations of quarterly data that does not reflect the nature of the budget deficit.

Secondly, the interaction of dollarization and interest rate indicates that the high level of dollarization erodes the effectiveness of interest rate pass-through.

Thirdly, the flexibility of exchange rate positively improves the interest rate transmission.

Fourthly, the NPL ratio remarkably and negatively affects the transmission of interest rate in the market. This proves that the NPL ratio is the strongest impact factor of interest rate transmission mechanism.

Fifthly, the financial development which is represented by the ratio of total deposit over GDP is positively related to effectiveness of interest rate channel. Its impact on interest rate pass-through is lower in comparison to bad debt ratio's influence.

Lastly, it shows a strong evidence that fiscal dominance has negative impact on interest rate transmission.

4   Conclusion and Recommendation

The paper has analyzed the effectiveness of interest rate transmission mechanism in implementing monetary policy in Vietnam from 2006 to 2017 and found that the interest rate pass-through from policy rate to interbank rate and from interbank rate to market rates are not perfect. Particularly, the pass-through effect of discount rate on interbank rate is 0.85 and 0.71 in long term and short term respectively. The effect of interbank rate on deposit and lending rates, however, is much weaker with 0.146 in short term and 0.806 in long term, not as high as SBV's expectation. The speed of interest rate adjustment is relatively slow. Normally, it takes around 16 days for one month for the overnight interbank rate and 30 days for market rates adjusting to the long-term equilibrium. The primary causes of limitations come from many factors in which the low level of asset quality of banking system has the most significant impact on interest rate pass-through. The other factors including the level of dollarization and the lack of flexibility in exchange rate, the concentrated and

Table 4 Regression Result and relation between minimum Bayes factors and the effect of such evidence on the probability of the null hypothesis
Dependent variable: DLR

	Minimum Bayes factor	Decrease in probability of the null hypothesis, %		Strength of evidence	Coefficient
		From	To no less than		
DLR(-1)	0.103 (1/10)	90 50 25	48.1 9.3 3.3	Moderate	0.159141
DDRATE	0.005 (1/190)	90 50 25	4.5 0.5 0.2	Strong to very strong	0.538822
DDOLLARIZATION	0.078 (1/13)	90 50 25	41.2 7.2 2.5	Moderate	−0.180242
EXF	0.012 (1/82)	90 50 25	9.9 1.2 2.5	Moderate to strong	0.001071
DNPL	0.219 (1/5)	90 50 25	66.3 17.9 6.7	Weak	−0.288442
DFDEV	0.102 (1/10)	90 50 25	48 9.3 3.3	Moderate	0.018632
DFDOM	0.387 (1/3)	90 50 25	77.7 27.8 11.3		−0.250388
DGBUBGET	0.842 (1/0.8)	90 50 25	88.3 45.7 21.7		−0.000322
DDRATE*DDOLLARIZATION	0.028 (1/35)	90 50 25	20.3 2.8 0.9	Moderate to strong	−2.185450
DDRATE*EXF	0.005 (1/200)	90 50 25	3.9 0.4 0.1	Strong to very strong	0.494485
DDRATE*DNPL	0.004 (1/225)	90 50 25	3.8 0.4 0.2	Strong to very strong	−8.417918
DDRATE*DFDEV	0.005 (1/200)	90 50 25	3.9 0.4 0.1	Strong to very strong	2.040165

(continued)

Table 4 (continued)

	Minimum Bayes factor	Decrease in probability of the null hypothesis, %		Strength of evidence	Coefficient
		From	To no less than		
DDRATE*DFDOM	0.003 (1/360)	90 50 25	2.4 0.3 0.1	Strong to very strong	−3.410694
DDRATE*DGBUBGET	0.009 (1/109)	90 50 25	7.7 0.9 0.3	Strong to very strong	−0.095607
C	0.032 (1/31)	90 50 25	22.5 3.1 1.1	Moderate to strong	−0.003745
AR(1)	0.005 (1/190)	90 50 25	4.5 0.5 0.2	Strong to very strong	−0.423787

Source The authors

underdeveloped financial market and poor legal system; chronic budget deficit, the poor independence of SBV and the dominance of fiscal policy also have influences on interest rate transmission.

Based on the current effectiveness of interest rate pass-through in Vietnam in the period from 2006 up to present and in order to enhance the efficiency of interest rate transmission mechanism, the SBV should conduct some implications including: (i) Completing the interest rate targeting framework, (ii) Improving the quality of banking system's balance sheets, (iii) Enhancing the competitiveness of banking system and the effectiveness of financial market, (iv) Reducing the dollarization and implementing the exchange rate policy in a way of more flexibility, (v) Limiting the fiscal dominance, (vi) Transforming the economic growth model, (vii) Increasing the effectiveness of public investment, decreasing the Government regular spending and building a sustainable State budge, (viii) Strengthening the regulations on information system, reports, disclosure and accountability of policy makers and (ix), Improving the quality of legal system.

Appendix 1: Result of Pass-Through Model from DIR to IB1 in Long Run

Dependent variable: IB1

	Coefficient	Std. Error	Minimum Bayes factor	Decrease in Probability of the null hypothesis, %		Strength of evidence
				From	To no less than	
DIR	0.853769	0.075672	0.0045 (1/222)	90 50 25	3.89 0.45 0.15	Strong to very strong
C	0.511502	0.519480	0.1538 (1/6.5)	90 50 25	58.1 13.3 22.9	Weak

Appendix 2: Result of Pass-Through Model from RER to IB1 in Long Run

Dependent variable: IB1

	Coefficient	Std. Error	Minimum Bayes factor	Decrease in probability of the null hypothesis, %		Strength of evidence
				From	To no less than	
RER	0.791762	0.081900	0.0045 (1/224)	90 50 25	3.89 0.45 0.15	Strong to very strong
C	−1.492649	0.701152	0.1937 (1/5.2)	90 50 25	63.54 16.23 6.00	Weak

Appendix 3: Result of ECM Model from DIR to IB1 in Short Run

Dependent variable: D(IB1)

	Coefficient	Std. Error	Minimum Bayes factor	Decrease in probability of the null hypothesis, %		Strength of evidence
				From	To no less than	
D(IB1(-1))	−0.083535	0.196395	0.010733 (1/93.2)	90 50 25	8.81 1.06 0.35	Moderate to strong
D(DIR(-1))	0.706941	0.233277	0.022243 (1/45)	90 50 25	16.68 2.18 0.73	Moderate to strong
ECM1(-1)	−0.551416	0.218417	0.082233 (1/12.2)	90 50 25	42.53 7.6 2.64	Moderate
C	−0.032966	0.163508	0.218196 (1/5)	90 50 25	66.26 17.91 6.71	Weak

Appendix 4: Result of ECM Model from RER to IB1 in Short Run

Dependent variable: D(IB1)

	Coefficient	Std. Error	Minimum Bayes factor	Decrease in probability of the null hypothesis, %		Strength of evidence
				From	To no less than	
D(IB1(-1))	−0.083191	0.202375	0.01371 (1/72.9)	90 50 25	10.98 1.35 0.45	Moderate to strong
D(RER(-1))	0.700419	0.252242	0.044502 (1/22.47)	90 50 25	28.59 4.26 1.45	Moderate to strong
ECM2(-1)	−0.550310	0.225062	0.098719 (1/10.12)	90 50 25	47.04 8.98 3.15	Moderate

(continued)

(continued)

Dependent variable: D(IB1)

	Coefficient	Std. Error	Minimum Bayes factor	Decrease in probability of the null hypothesis, %		Strength of evidence
				From	To no less than	
C	−0.033046	0.164352	0.219943 (1/5)	90 50 25	66.43 18.02 6.76	Weak

Appendix 5: Result of Pass-Through Model from IB1 to LR in Long Run

Dependent variable: LR

	Coefficient	Std. Error	Minimum Bayes factor	Decrease in probability of the null hypothesis, %		Strength of evidence
				From	To no less than	
IB1	0.805528	0.052246	0.0045 (1/222)	90 50 25	3.89 0.45 0.15	Strong to very strong
C	6.582150	0.335674	0.0045 (1/222)	90 50 25	3.89 0.45 0.15	Strong to very strong

Appendix 6: Result of Pass-Through Model from IB1 to DR in Long Run

Dependent variable: DR

	Coefficient	Std. Error	Minimum Bayes factor	Decrease in probability of the null hypothesis, %		Strength of evidence
				From	To no less than	
IB1	0.788684	0.048537	0.0045 (1/222)	90 50 25	3.89 0.45 0.15	Strong to very strong

(continued)

(continued)

Dependent variable: DR						
	Coefficient	Std. Error	Minimum Bayes factor	Decrease in probability of the null hypothesis, %		Strength of evidence
				From	To no less than	
C	3.927042	0.311841	0.0045 (1/222)	90 50 25	3.89 0.45 0.15	Strong to very strong

Appendix 7: Result of ECM Model from IB1 to DR in Short Run

Dependent variable: D(DR)						
	Coefficient	Std. Error	Minimum Bayes factor	Decrease in probability of the null hypothesis, %		Strength of evidence
				From	To no less than	
D(DR(-1))	1.049713	0.251028	0.000992 (1/1007)	90 50 25	0.885 0.099 0.033	Strong to very strong
D(IB1(-1))	0.135835	0.032471	0.000992 (1/1007)	90 50 25	0.885 0.099 0.033	Strong to very strong
ECM3(-1)	−0.856049	0.277389	0.019454 (1/51.4)	90 50 25	14.90 1.91 0.64	Moderate to strong
C	0.004242	0.064458	0.248557 (1/4.02)	90 50 25	69.1 19.9 7.58	

Appendix 8: Result of ECM Model from IB1 to LR in Short Run

Dependent variable: D(LR)

	Coefficient	Std. Error	Minimum Bayes factor	Decrease in probability of the null hypothesis, %		Strength of evidence
				From	To no less than	
D(LR(-1))	1.033892	0.234043	0.0045 (1/222)	90 50 25	3.89 0.45 0.15	Strong to very strong
D(IB1(-1))	0.145884	0.032024	0.0045 (1/222)	90 50 25	3.89 0.45 0.15	Strong to very strong
ECM4(-1)	−0.863684	0.258858	0.009211 (1/109)	90 50 25	7.65 0.91 0.30	Strong to very strong
C	0.005599	0.064717	0.176187 (1/6)	90 50 25	61.3 14.97 5.49	Weak

References

Andrew, F., Madhusudan, M., & Ramon, M. (2012). *Central bank and government debt management: Issues for monetary policy.* BIS Papers No 67.

Arellano, M., & Bond, S. (1991). Some tests of specification for panel data: monte carlo evidence and an application to employment equations. *Review of Economic Studies, 58*(2), 277–297.

Avci, S. B., & Yucel, E. (2016). *Effectiveness of monetary policy: Evidence from Turkey.* MPRA Paper 70848.

Axel, A. W., Rafael, G., & Andreas, W. (2009). *Has the monetary transmission process in the euro area changed? Evidence based on VAR estimates.* BIS Working Papers No.276.

Berger, A., & Hannan, T. (1991). The price-concentration relationship in banking. *The Review of Economics and Statistics, 71*(2).

Bernanke, B. S., Reinhart, V. R., & Sack, B. P. (2004). *Monetary policy alternatives at the zero bound: An empirical assessment*, Brookings Papers on Economic Activity, Economic Studies Program, The Brookings Institution, vol. 35(2).

Bernard, L., & Enrique, G. D. P. (2005). *Coordination of monetary and fiscal policies.* IMF Working Paper 05/25.

Bondt, G. D. (2002). Retail Bank interest rate pass-through: New evidence at the Euro area level. *European Central Bank Working Paper Series, 136.*

Borio, C., & Fritz, W. (1995). The response of short-term bank lending rates to policy rates: a cross-country perspective. *BIS Working Paper, 27.*

Bredin, D., Fitzpatrick, T., & Reilly, G. (2001). *Retail interest rate pass-through: THE Irish experience.* Central Bank of Ireland, Technical paper 06/RT/01.

Calvo, G. A., & Carmen, M. R. (2002). Fear of floating. *Quarterly Journal of Economics, 107.*

Cecchetti, S. (1999). Legal structure, financial structure, and the monetary policy transmission mechanism, Federal Reserve Bank of New York. *Economic Policy Review, 5, 2.*

Chmielewski, T. (2003). *Interest rate pass-through in the polish banking sector and bank-specific financial disturbances.* MPRA Paper 5133, University Library of Munich, Germany.

Chong, B. S., Liu, M. H., & Shrestha, K. (2006). Monetary transmission via the administered interest rate channel. *Journal of Banking and Finance, 5.*

Cottarelli, C., & Kourelis, A. (1994). Financial structure, banking lending rates, and the transmission mechanism of monetary policy. *IMF Staff Papers, 41*(4), 587–623.

Courvioisier, S., & Gropp, R. (2002). Bank concentration and retail interest rates. *Journal of Banking and Finance, 26.*

De Graeve, F., De Jonghe, O., & Vennet, R. V. (2007). Competition, transmission and bank pricing policies: Evidence from Belgian loan and deposit markets. *Journal of Banking and Finance, 31*(1).

Ehrmann, M., Gambacorta, L., Martinez-Pages, J., Sevestre, P., & Worms, A. (2001). *Financial systems and the role of banks in monetary policy transmission in the Euro area.* ECB Working Paper.

Engle, R., & Granger, C. (1987). Co-integration and error correction: Representation, estimation, and testing. *Econometrica, 55*(2), 251–276.

Esman, N., & Lydia, N. N. (2013). *Financial innovations and monetary policy in Kenya.* MPRA Paper No.52387

Freedman, C., & Ötker-Robe, I. (2010). *Important elements for inflation targeting for emerging economies.* IMF Working Paper, WP/10/113

Gigineishvili, N. (2011). *Determinants of interest rate pass-through: Do macroeconomic conditions and financial market structure matter?* IMF Working Paper No. 11/176

Gropp, J., Sorensen, R., & Lichtenberger, C. K. (2007). The dynamics of bank spread and financial structure. *ECB Working Paper Series No. 714.*

Hansen, L. P. (1982). Large sample properties of generalized method of moments estimators. *Econometrica, 50*(4), 1029–2054.

Heffernan, S. A. (1997). *Modelling British interest rate adjustment: An error correction approach* (p. 64). Issue: Economica.

Heinemann, F., & Schuler, M. (2002). *Integration benefits on EU retail credit markets—evidence from interest rate pass-through.* Discussion Paper No. 02–26, Centre for European Economic Research (ZEW), Mannheim.

Hendry, D. F. (1995). Dynamic Econometrics. Oxford University Press, Oxford.

IMF. (2010). *Monetary policy effectiveness in sub-saharan Africa, regional economic outlook.* October 2010.

Kashyap, A. K., & Stein, C. J. (2000). *What do a million observations on banks say about the transmission of monetary policy?* NBER Working Paper.

Kok Sørensen, C., & Werner, T. (2006). *Bank interest rate pass-through in the euro area: a crosscountry comparison.* ECB Working Paper Series No. 580.

Laurens, B. (2005). *Monetary policy implementation at different stages of market development.* IMF Occasional Paper No. 244

Leroy, A., & Yannick, L. (2015). *Structural and cyclical determinants of bank interest rate pass-through in Eurozone.* NBP Working Paper No: 198, Narodowy Bank Polski.

Loayza, N., & Schmidt-Hebbel, K. (2002). *Monetary policy functions and transmission mechanisms: An overview.* Santiago, Chile: Central Bank of Chile.

Lopes. (1998). *The transmission of monetary policy in emerging market economies.* BIS Policy Papers No.3.

Medina, C. S., Carrión-Menéndez, A., & Frantischek, F. (2011). *Improving the monetary policy frameworks in central America.* IMF Working Paper, No. 11/245, Washington.

Michele, F., & Franco, S. (1998). Fiscal dominance and money growth in Italy: the long record. New York: AEA Meetings.

Michiel, V. L., Christoffer, K. S., Jacob, A. B., & Adrian, V. R. (2008). Impact of bank competition on the interest rate pass-through in the Euro area. *ECB Working Paper Series No. 885.*

Mishkin, F. S. (2013). *The economics of money, banking, and financial markets* (10th ed.). New York: Pearson Education.

Mojon, B. (2000). Financial structure and the interest rate channel of ECB monetary policy. *European Central Bank, Working paper No.4.*

Mishra, P., & Peter J. M. (2013). How effective is monetary transmission in developing countries? A survey of the empirical evidence. *Economic System, 37*(2), 187–216.

Roseline, N. M., Esman, M. N., & Anne, W. K. (2011). Interest rate pass-through in Kenya. *International Journal of Development Issues, 10*(2).

Saborowski, C., & Weber, S. (2013). Assessing the determinants of interest rate transmission through conditional impulse response functions. *IMF Working Paper.*

Sander, H., & Kleimeier, S. (2004a), Convergence in Eurozone retail banking? what interest rate pass-through tells us about monetary policy transmission competition and integration. *Journal of Inernational Money and Finance, 23.*

Sander, H., & Kleimeier, S. (2004b). *Interest rate pass-through in an Enlarged Europe: The role of banking market structure for monetary policy transmission in transition economies.* METEOR Research Memoranda No. 045, University of Maastricht, The Netherlands.

Sargent, T., & Wallace, N. (1981). Some unpleasant monetarist arithmetic. *Quarterly Review, 5*(3).

Scholnick, B. (1996). Asymmetric adjustment of commercial bank interest rates: Evidence from Malaysia and Singapore. *Journal of International Money and Finance,15.*

Singh, S., Razi, A., Endut, N., Ramlee, H. (2008). Impact of financial market developments on the monetary transmission mechanism. *BIS Working Paper Series, 39.*

Stephanie, M. C., Alejandro, C. M., & Florencia, F. (2011). The policy interest-rate pass-through in central America. IMF Working Paper WP11/240.

Weth, M. (2002). *The pass-through from market interest rates to bank lending rates in Germany* (p. 11). Discussion Paper No: Economic Research Centre of The Deustche Bundesbank.

Wróbel, E., & Pawłowska, M. (2002). *Monetary transmission in Poland: Some evidence on interest rate and credit channels* (p. 24). Materiały i Studia, Paper No: National Bank of Poland.

Applying Lasso Linear Regression Model in Forecasting Ho Chi Minh City's Public Investment

Nguyen Ngoc Thach, Le Hoang Anh, and Hoang Nguyen Khai

Abstract Forecasting public investment is always an issue that has attracted much attention from researchers. More specifically, forecasting public investment helps budget planning process take a more proactive approach. Like many other economic variables such as oil prices, stock prices, interest rates, etc., public investment can also be forecasted by different quantitative methods. In this paper, we apply the Ordinary Least Square (OLS), Ridge, and Lasso regression models to forecasting Ho Chi Minh City's future public investment. The most effective forecasting method is chosen based on two performance metrics – root mean square error (RMSE) and mean absolute percentage error (MAPE). The empirical results show that the Lasso algorithm has superiority over the two other methods, OLS and Ridge, in forecasting Ho Chi Minh City's Public Investment.

Keywords Ridge regression · Lasso regression · Public investment

1 Introduction

Investment is an indispensable factor of economic growth that contributes greatly to a country's GDP (Cooray 2009). In particular, public investment plays an important role in the direction of transforming macroeconomic structure, balancing the commodity market, helping a national economy to develop harmoniously. Public

N. N. Thach (✉) · L. H. Anh
Banking University HCMC, 39 Ham Nghi street, District 1, Ho Chi Minh City, Vietnam
e-mail: thachnn@buh.edu.vn

L. H. Anh
e-mail: lehoanganhct@yahoo.com

H. N. Khai
Faculty of Business Administration, Ho Chi Minh City University of Technology,
475A Dien Bien Phu Street, Binh Thanh District, Ho Chi Minh City, Vietnam
e-mail: hn.khai@hutech.edu.vn

© The Editor(s) (if applicable) and The Author(s), under exclusive license
to Springer Nature Switzerland AG 2021
N. Ngoc Thach et al. (eds.), *Data Science for Financial Econometrics*,
Studies in Computational Intelligence 898,
https://doi.org/10.1007/978-3-030-48853-6_17

investment is defined as 'investment activity of the State in programs, projects and other public investment subjects under the provisions of this Law (Public Investment Law)' (Thach et al. 2019). Through capital formation in several different sectors of an economy, public investment has contributed significantly to GDP of countries. Specifically, the use of the state budget to invest in infrastructure projects has created a favourable environment for businesses in all areas (Nguyen et al. 2019). Infrastructure projects often require large mobilized capital, but the capital recovery period is long. So, the state budget can afford to pay them. Therefore, this financial source is invested in the construction of basic infrastructures such as national electrical system, roads and communications etc., thereby forms good conditions for enterprises development. Because of the above-mentioned reasons, public investment needs forecasting so that social resources can be effectively distributed. Similar to numerous financial variables, public investment can also be predicted by using various quantitative methods, such as multivariate regression models, time series regression models.... In recent years, the forecasting methods resting on multiple regression models have been increasingly improved (Hung et al. 2018). The most effective methods are Ridge and Lasso algorithms in the context of traditional frequentist statistics being outdated (Nguyen et al. 2019). This paper will use the above mentioned methods for forecasting Ho Chi Minh City's public investment. The selection of predictive methods relies on two metrics: RMSE and MAPE.

2 Research Model and Data

2.1 Research Model

Let's consider N observations of the outcome y_i and p explanatory variables $x_i = (x_{i1}, \ldots, x_{ip})$, where x_i and y_i belong to R^p and R, respectively.

Our research purpose is to predict the outcome from the explanatory variables. The outcome will be predicted using the following linear regression model:

$$f(x) = \beta_0 + \sum_{j=1}^{p} x_{ij} \beta_j$$

Thus, the linear regression model is parameterized with weight vector $\beta = (\beta_1, \ldots, \beta_p)^T \in R^p$ and intercept $\beta_0 \in R$. OLS estimates for pair (β_0, β) is based on minimizing the square error as follows:

$$\underset{\beta_0, \beta}{\text{minimize}} \left\{ \frac{1}{N} \sum_{i=1}^{N} \left(y_i - \beta_0 - \sum_{j=1}^{p} x_{ij} \beta_j \right)^2 \right\} = \underset{\beta_0, \beta}{\text{minimize}} \left(\frac{1}{N} \|y - \beta_0 1 - X\beta\|^2 \right) \quad (1)$$

where $y = (y_1, \ldots, y_N)^T$, X is matrix Nxp and $1 = (1, \ldots, 1)^T$. The solution to (1) is as follows:

$$\beta = (X^T X)^{-1} X^T y$$

However, Hastie et al. (2015) claim that there are a number of reasons for the necessity to modify the OLS regression model. The first reason is forecasting accuracy. Estimation by the OLS regression, as usual, has low bias, but a large variance and forecasting accuracy can sometimes be improved by shrinking the number of regression coefficients or setting some coefficients to 0. By this way, we accept some biases in estimating regression coefficients but reduce the variance of predicted values and thus improve prediction accuracy. The second reason is that in the case of sparse explanatory variables or focus on some explanatory variables having the strongest impact on the output variable, this approach will give a better predictive effect.

The modification of the OLS model to the Ridge regression model can be carried out as follows:

$$\text{minimize}_{\beta_0, \beta} \left\{ \frac{1}{N} \sum_{i=1}^{N} \left(y_i - \beta_0 - \sum_{j=1}^{p} x_{ij} \beta_j \right)^2 + \lambda \sum_{j=1}^{p} \beta_j^2 \right\}$$
$$= \text{minimize}_{\beta_0, \beta} \left(\frac{1}{N} \| y - \beta_0 1 - X\beta \|^2 + \lambda \| \beta \|^2 \right) \tag{2}$$

where λ is the regularization parameter.

The OLS model can be modified to the Lasso model as follows:

$$\text{minimize}_{\beta_0, \beta} \left\{ \frac{1}{N} \sum_{i=1}^{N} \left(y_i - \beta_0 - \sum_{j=1}^{p} x_{ij} \beta_j \right)^2 + \lambda \sum_{j=1}^{p} |\beta_j| \right\}$$
$$= \text{minimize}_{\beta_0, \beta} \left(\frac{1}{N} \| y - \beta_0 1 - X\beta \|^2 + \lambda \| \beta \| \right) \tag{3}$$

The important issue in the lasso algorithm is selecting the optimal λ value. The selection of the optimal λ value is done by such methods as Cross-validation, Theory-driven, and Information criteria. Cross-validation selects the optimal λ value by dividing the data into K groups, referred to as folds. Theory-driven relies on an iterative algorithm for estimating the optimal λ value in the presence of non-Gaussian and heteroskedastic errors. The optimal λ value can also be selected using information criteria: Akaike Information Criterion (AIC), Bayesian Information Criterion (BIC), Extended Bayesian information criterion (EBIC) and the corrected AIC (AICc).

In this research, the optimal λ value is selected by cross-validation. The purpose of cross-validation is to assess the out-of-sample prediction performance of the estimator.

The cross-validation method splits the data into K groups, referred to as folds, of approximately equal size. Let n_k denote the number of observations in the kth data partition with $k = 1, ..., K$.

One fold is treated as the validation dataset and the remaining $K - 1$ parts constitute the training dataset. The model is fit to the training data for a given value of λ. The resulting estimate is denoted as $\beta(1, \lambda)$. The mean-squared prediction error for group 1 is computed as:

$$MSPE\,(1, \lambda) = \frac{1}{n_1} \sum_{i=1}^{n_1} \left(y_i - \beta_0 - \sum_{j=1}^{p} x_{ij}\,\beta_j\,(1, \lambda) \right)^2$$

The procedure is repeated for $k = 2, ..., K$. Thus,

$$MSPE(2, \lambda), ..., MSPE(K, \lambda)$$

are calculated.

Then, the K-fold cross-validation estimate of the MSPE, which serves as a measure of prediction performance, is

$$CV\,(MSPE\,(\lambda)) = \frac{1}{K} \sum_{k=1}^{K} MSPE\,(k, \lambda)$$

The above procedures are repeated for a range of λ values. In this research, the data is divided into 10 folders with 100 λ values. The maximum value of λ is $\lambda_{max} = \max_j \left| \frac{1}{N} \langle x_j, y \rangle \right|$. The ratio of minimum to maximum value of λ is $\frac{1}{10000}$. So, the optimal λ value is in the range $[24271.873; 2.427x10^8]$. Then, the authors determine 100 values equidistant from the range $[\ln(24271.873); \ln(2.427 \times 10^8)]$. The λ values are determined by taking the base e of these 100 values.

The selected optimal λ value will have the smallest $CV\,(MSPE\,(\lambda))$.

In this research, the authors forecast public investment in Ho Chi Minh City in the future months under both Ridge and Lasso regression models. The best forecasting method will be selected based on the two metrics proposed by Roy et al. (2015) as RMSE and MAPE:

$$RMSE = \sqrt{\frac{\sum_{i=1}^{N} (y_i - \hat{y}_i)^2}{N}}$$

$$MAPE = \frac{\sum_{i=1}^{N} \frac{|y_i - \hat{y}_i|}{y_i}}{N}$$

where y_i is the actual values of the outcome, $\hat{y}_i$ is the predicted values of the outcome estimated by Ridge and Lasso algorithm, x_i is the vector of the variables including

S1, S2 are dummy variables representing the seasonal factor of investment; Trend is the trend of Ho Chi Minh City's public investment; Lcap is one lag of Ho Chi Minh City's public investment.

The model with the smallest RMSE and MAPE values will be chosen. The beta coefficients estimated from the best model in the period from Jan-2004 to Dec-2018 will be used for forecasting Ho Chi Minh City's public investment through 12 months of 2019.

2.2   Research Data

Our research utilizes monthly data on Ho Chi Minh City's public investment obtained from the Ho Chi Minh City Statistical Office website for the period from Jan-2004 to Dec-2018. The model acquiring the smallest RMSE and MAPE values will be selected to forecast public investment of this city.

3   Empirical Results

Figure 1 shows that the public investment in Ho Chi Minh City through the period from Jan-2004 to Dec-2018 increased from 342,398 million VND to 5,840,655 million VND. On average for this period, Ho Chi Minh City's public investment was 1,578,275 million VND. Figure 1 also shows that the investment trend expresses marked seasonality. Specifically, investment scale dropped was at the lowest level in the months of January and February and was reached at the highest level in December. Therefore, we need to take into accounts the seasonal factor in our forecasting model.

The authors estimate the OLS, Ridge, and Lasso regression models. In order to estimate the Ridge and Lasso models well, it is necessary to specify the optimal value of the regularization parameter λ. This optimal value is achieved by minimizing $CV\,(MSPE\,(\lambda))$. The result of specifying the optimal value of the regularization parameter λ is shown in Table 1.

Table 1 demonstrates that the optimal regularization parameter λ for the Ridge and Lasso models is 24271.873 and 629860.99, respectively. Afterwards, the authors used the dataset from Jan-2004 to Dec-2018 to conduct the estimation of the OLS, Ridge and, Lasso models with the above-mentioned optimal regularization parameter.

The authors estimate 3 models including the lasso model, the Ridge model, and the OLS model. Each model consists of 4 variables: S1, S2, Trend, Lcap. Where, S1 and S2 are dummy variables representing the seasonal factor of investment, they are at the lowest level in January and February and at the highest level in December; Trend is the trend of Ho Chi Minh City's public investment; Lcap is one lag of Ho Chi Minh City's public investment. The estimated results of 3 models are presented in Table 2.

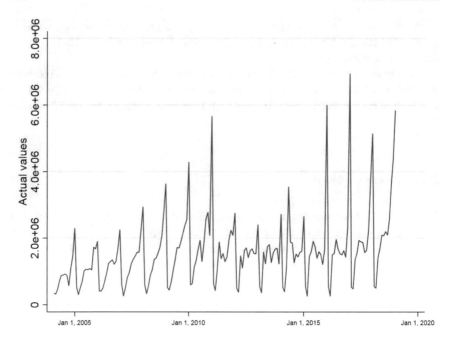

Fig. 1 Ho Chi Minh City's public investment from Jan-2004 to Dec-2018
Source: Ho Chi Minh City Statistical Office

Table 1 Result of specifying the optimum of the regularization parameter λ

	Lasso			Ridge		
	λ	CV(MSPE)	St. dev.	λ	CV(MSPE)	St. dev.
1	2.427×10^8	1.13×10^{12}	2.63×10^{11}	2.427×10^8	1.15×10^{12}	2.44×10^{11}
2	2.212×10^8	1.10×10^{12}	2.64×10^{11}	2.212×10^8	1.15×10^{12}	2.44×10^{11}
...	...	...	...	...	...	...
65	629860.99*	3.85×10^{11}	7.43×10^{10}	629860.99	1.15×10^{12}	2.44×10^{11}
...	...	...	...	...	...	...
100	24271.873	3.85×10^{11}	7.39×10^{10}	24271.873*	1.12×10^{12}	2.40×10^{11}

Source Calculated results from STATA software 15.0

Table 2 Estimation results of OLS, Ridge, and Lasso models

Selected	Lasso	Ridge	OLS
S1	-1.2436296×10^6	−19476.63545	−1248821
S2	2039940.696	35460.20849	2044161
Trend	5167.809608	102.9241925	5190.372
Lcap	0.2140933	0.0045295	0.2156216
Cons	812645.7399	1568956.118	808704.3

Source Calculated results from STATA software 15.0

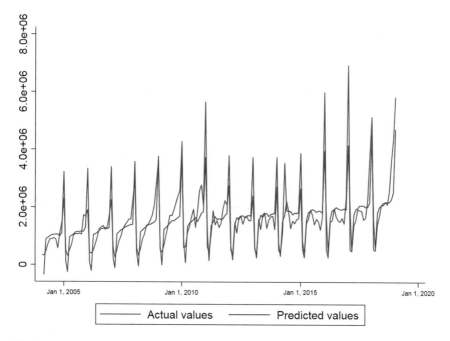

Fig. 2 Ho Chi Minh City's predicted public investment (Lasso)

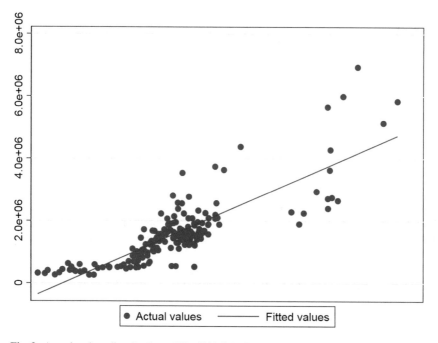

Fig. 3 Actual and predicted values of Ho Chi Minh City's public investment (Lasso)

Table 3 Results of specifying RMSE and MAPE in OLS, Ridge, and Lasso models

Method	RMSE	MAPE
Ridge	1049362.658	0.727428
Lasso	591483.5266	0.338168
OLS	591483.8509	0.338201

Source Calculated results from STATA software 15.0

Table 4 Result of forecasting Ho Chi Minh City's public investment for 12 months of 2019 (Lasso)
Unit: billion VND

Month	1	2	3	4	5	6	7	8	9	10	11	12
Public investment	175	883	1.947	2.18	2.236	2.252	2.261	2.268	2.275	2.282	2.288	4.335

Source Calculated results from STATA software 15.0

The authors forecast Ho Chi Minh City's public investment estimating all the three models. RMSE and MAPE are calculated for the period from Jan-2004 to Dec-2018. Calculation results are presented in Table 3.

Table 3 shows that according to both RMSE and MAPE, the Lasso model provides more accurate predictive result than the Ridge one. However, the predictive results are not much different between the Lasso model and the OLS model.

Hence, the Lasso algorithm should be used to forecast future public investment in Ho Chi Minh City. In this paper, the authors use the Lasso algorithm to forecast Ho Chi Minh City's public investment through 12 months of 2019. The results are presented in Table 4.

4   Conclusion

The empirical results indicate that the Lasso model provides more accurate predictive result than the Ridge one. However, the predictive results are not much different between the Lasso model and the OLS model. This result shows that the lasso algorithm is also a good predictor. Moreover, as demonstrated in Figs. 2 and 3, the predicted values of Ho Chi Minh City's public investment fit the actual ones during January 2004–December 2018 under the Lasso algorithm. This result shows the methodological advantage of the Lasso model over the OLS model. Although the estimation of regression coefficients with the Lasso algorithm can be biased, it reduces the variance of the predicted values and thus improves forecasting accuracy.

References

Cooray, A. (2009). Government expenditure, governance, and economic growth. *Comparative Economic Studies.*, *51*(3), 401–418.

Hastie, T., Tibshirani, R., & Wainwright, M. (2015). *Statistical learning with sparsity: The lasso and generalizations* (1st ed.). New York: Chapman and Hall/CRC.

Hung N.T., Thach N.N., & Anh L.H.: (2018). GARCH Models in Forecasting the Volatility of the World's Oil Prices. In: Anh L., Dong L., Kreinovich V., Thach N. (eds) Econometrics for Financial Applications. ECONVN 2018. *Studies in Computational Intelligence*, vol 760. Springer, Cham, https://doi.org/10.1007/978-3-319-73150-6_53

Nguyen, H. T., Trung, N. D., & Thach, N. N. (2019). Beyond traditional probabilistic methods in econometrics. In V. Kreinovich, N. Thach, N. Trung, & D. Van Thanh (Eds.), *Beyond Traditional Probabilistic Methods in Economics. ECONVN 2019. Studies in Computational Intelligence*. Cham: Springer.

Thach, N. N., Anh, L. H., & An, P. T. H. (2019). The effects of public expenditure on economic growth in Asia Countries: A bayesian model averaging approach. *Asian Journal of Economics and Banking*, *3*(1), 126–149.

Roy, S. S., Mittal, D., Basu, A., & Abraham, A. (2015). Stock market forecasting using LASSO linear regression model. In S. V. Abraham, & A. P. Krömer (Eds.). Cham: Springer. https://doi.org/10.1007/978-3-319-13572-4_31.

Vietnam Parliament. (2019). Public Investment Law. Retrieved from https://thuvienphapluat.vn/van-ban/Dau-tu/Luat-Dau-tu-cong-2019-362113.aspx. (in Vietnamese).

Markov Switching Quantile Regression with Unknown Quantile τ Using a Generalized Class of Skewed Distributions: Evidence from the U.S. Technology Stock Market

Woraphon Yamaka and Pichayakone Rakpho

Abstract The traditional Markov Switching quantile regression with unknown quantile (MS–QRU) relies on the Asymmetric Laplace Distribution (ALD). However, the old fashion ALD displays medium tails and it is not suitable for data characterized by strong deviations from the Gaussian hypothesis. This study compares ALD with two alternative skewed likelihood types for the estimation of MS–QRU, including the skew-normal distribution (SKN) and skew-student-t distribution (SKT). For all the three skewed distribution-based models, we estimated parameter by the Bayesian approach. Finally, we apply our models to investigate the beta risk of individual FAANG technology stock under the CAPM framework. The model selection results show that our alternative skewed distribution performs better than the ALD. Only for two out of five stocks suggest that ALD is more appropriate for MS–QRU model.

Keywords Markov regime-switching quantile regression models · CAPM model · FAANG stocks · Generalized class of skewed distributions

1 Introduction

FAANG refers to the group of top-five most popular and best-performing technology stocks in the United States (U.S.) stock market. It consists of Facebook (FB), Amazon (AMZN), Apple (AAPL), Netflix (NFLX), and Alphabet (GOOG). FAANG stocks have played an important role as the prime contributors to the growth of the U.S. stock market in the recent years. Therefore, many investors have paid special attention to these stocks in their portfolio (Kenton 2019). However, the prices of FAANG stocks are highly volatile and they changed over time. Although the high fluctuation

W. Yamaka · P. Rakpho (✉)
Faculty of Economics, Center of Excellence in Econometrics, Chiang Mai University, Chiang Mai, Thailand
e-mail: pichayakone@gmail.com

W. Yamaka
e-mail: woraphon.econ@gmail.com

N. Ngoc Thach et al. (eds.), *Data Science for Financial Econometrics*, Studies in Computational Intelligence 898, https://doi.org/10.1007/978-3-030-48853-6_18

255

contributes to greater uncertainty prospects, the stocks investment become riskier, and thereby forcing investors to correct their investment strategy with the new level of risk.

In financial practice, the risk can be measured by various methods such as standard deviation, beta risk, value at risk and sharp ratio. In this study, we focus only on the beta risk as the risk measure of the stock returns. This risk measure can be evaluated from the relationship between stock and market returns. According to the Capital Asset Price Model (CAPM), this relationship can be viewed as the risk of the stock (Sharpe 1964). In other words, CAPM attempts to quantify how the risk of a particular stock is related to the risk of the overall stock market.

The estimation of the beta risk or the relationship between excess stock return and excess market return is usually obtained using the linear mean regression. Thus, the beta risk is estimated at the average of the excess stock return and excess market return. However, many studies on capital investment have pointed out that the beta coefficient can behave differently under different market conditions. For examples, Chiang and Li (2012), Atkins and Ng (2014), and Yamaka et al. (2019) suggested that the difference of the beta risks can occur when the variance of the error term varies across different values of the stock returns. Therefore, the Quantile regression which is initially developed by Koenker and Bassett (1978) has been applied to describe the different regression relationships across quantile levels and has recently attracted an increasing research attention in finance. The Quantile approach is a robust regression technique when the typical assumption of normality of the error term might not be strictly satisfied (Koenker and Bassett 1978). From the econometric point of view, the quantile regression seems more practical and realistic compared to the linear mean regression as it can provide a solution to the problems related to the outliers and the fat-tailed error distribution. In addition, its estimation embraces the whole quantile spectrum of the stock returns.

The quantile regression, although logically simple and useful for a CAP modeling, is a static, single-period linear model which can hardly answer the real-world need about risk measurement. Many empirical studies using financial data found the non-linear relationship to exist among model variables due to structural change, such as Huang (2000), Abdymomunov and Morley (2011), Thamprasert et al. (2018). Thus, the linear quantile regression models cannot accommodate many stylized facts such as structural breaks and nonlinearities in financial time series. We thus address this problem by putting quantile CAPM under the Markov regime switching (MRS) framework. This method is called Markov–Switching quantile regression (MS–QR). This model was initially introduced by Liu (2016) and Tungtrakul et al. (2016). Rakpho et al. (2018) and Yamaka et al. (2019) extended this model by considering the quantile level as a parameter to be estimated and showed that the model becomes more flexible to accommodate various relationship structures. They noted that the conventional MS–QR considers the model at some specific quantiles, but, in the practice, we may be interested in obtaining only one risk in each regime or state of the economy; and under this scenario, the traditional MS–QR may not be appropriate.

The aim of this paper is to develop a Markov Switching quantile regression with unknown quantile (MS–QRU) to quantify the risk of the stock return under the CAPM framework. The Bayesian estimation is conducted to obtain the estimated parameters including the quantile parameter. As suggested by Rakpho et al. (2018) and Yamaka et al. (2019), the Bayesian analysis for MS–QRU begins with specifying a likelihood, which can often be specified as the Asymmetric Laplace Distribution (ALD). However, we are concerned about some limitations of this likelihood distribution. Galarza et al. (2017) cautioned that the ALD has the zero-quantile property and a useful stochastic representation; thus it is not differentiable at zero which could lead to problems of numerical instability. Hence, the Laplace density is a rather strong assumption in order to set a quantile regression model through the Bayesian framework. To overcome this deficiency, we used a generalized class of skewed densities (SKD) for the analysis of MS–QRU that provides competing solutions to the ALD-based formulation.

Therefore, this paper suggests a new approach to the analysis of MS–QRU by replacing the ALD likelihood with two skew likelihoods namely skew-normal distribution (SKN) and skew-student-t distribution (SKT). We then apply the model to investigate the relationship between expected return and the systematic risk of an asset of FAANG stocks.

The rest of this paper is structured as follows. Section 2 briefly explains the methodology employed in this study. Section 2 describes the descriptive statistics. Section 3 deals with the empirical estimation results and Sect. 5 presents the conclusions.

2 Methodology

In this section, we firstly describe the three skewed likelihood functions considered in this study. Then, the MS–QRU model is defined. Finally, we explain the Bayesian estimation for MS–QRU model.

2.1 Skew Likelihood Distributions

2.1.1 Asymmetric Laplace Distribution (ALD)

The Asymmetric Laplace Distribution (ALD) is proposed by Koenker and Machado (1999). The density function is

$$\pi\left(y \,|m, \sigma, \tau\right) = \frac{\tau\left(1 - \tau\right)}{\sigma} \exp\left\{-\rho_\tau\left(\frac{y - m}{\sigma}\right)\right\}, \tag{1}$$

where $\rho_\tau\left(\left(y - m\right)/\sigma\right) = \left(\left(y - m\right)/\sigma\right)\left(\tau - I\left(\left(y - m\right)/\sigma < 0\right)\right)$ is the so-called check function. The quantile level τ is the skewness parameter in the distribution, m is the location parameter, and σ is the scale parameter.

2.1.2 Skew-Normal Distribution (SKN)

Wichitaksorn et al. (2014) present a random variable y having skew-normal distribution (SKN) with location parameter m, scale parameter $\sigma > 0$ and skewness parameter $\tau \in (0, 1)$ is given by

$$\pi(y \,|m, \sigma, \tau) = \frac{4\tau(1 - \tau)}{\sqrt{2\pi\sigma^2}} \exp\left\{-2\rho_\tau^2\left(\frac{y - m}{\sigma}\right)\right\}. \tag{2}$$

2.1.3 Skew-Student-t Distribution (SKT)

Galarza et al. (2017) present a random variable y having skew-student-t distribution (SKT) with location parameter m, scale parameter $\sigma > 0$ and skewness parameter. Student t-distribution has the probability density function given by

$$\pi(y \,|m, \sigma, \tau, \nu) = \frac{4\tau(1 - \tau)\,\Gamma\left(\frac{\nu+1}{2}\right)}{\Gamma\left(\frac{\nu}{2}\right)\sqrt{2\pi\sigma^2}} \left\{\frac{4}{\nu}\rho_\tau^2\left(\frac{y - m}{\sigma}\right) + 1\right\}^{-\frac{\nu-1}{2}}, \tag{3}$$

where ν is the number of degrees of freedom and Γ is the gamma function.

2.2 Markov Switching-Quantile Regression with Skewed Distribution

The Markov Switching–quantile regression with skewed distribution is firstly introduced in Rakpho et al. (2018). In the CAPM, given the stock excess return r_{st} and the market excess return r_{mt} for $t = 1, \ldots, T$. The model can be written as

$$r_{st} = \beta_{s_t}^0(\tau) + \beta_{s_t}^1(\tau) r_{mt} + \varepsilon_{s_t,t}, \tag{4}$$

where $\beta_{s_t}^0(\tau)$ and $\beta_{s_t}^1(\tau)$ are regime dependent intercept and the beta risk coefficient at given quantile, respectively. $\beta_{s_t}^1(\tau)$ define a relationship between r_{mt} and the conditional quantile function of r_{st} in the regimes or states $s_t = \{0, 1\}$. s_t is an unobserved discrete-valued indicator variable which can be obtained from the Hamilton's filter process (Hamilton 1989). $\varepsilon_{s_t,t}$ is regime dependent random errors which can be assumed to have $Z \sim ALD\left(0, \sigma_{s_t}^2(\tau)\right)$, $Z \sim SKN(0, 1, \tau)$ or $Z \sim SKT(0, 1, \tau, \nu)$. Note that the quantile $0 < \tau \leq 1$ is the estimated parameter and it is assumed to be regime independent.

In this study, we consider only two regimes, namely bear and bull markets, hence the state variable s_t is an ergodic homogeneous Markov chain on a finite set, with a transition matrix

$$P = \begin{bmatrix} p_{11} & p_{12} \\ p_{21} & p_{22} \end{bmatrix}, \tag{5}$$

where p_{ij} is the probability of switching from state i to state j and $\sum p_{ij} = 1$.

In this model, the parameter set is $\Psi = (\beta_{s_t}(\tau), \sigma_{s_t}(\tau), \tau)$ which can be estimated by the Bayesian estimation. The Bayesian analysis of MS–QRU begins with specifying a likelihood function, therefore, the full likelihood function of the model can be defined as follows:

(1) Asymmetric Laplace likelihood function

$$L\left(\Psi \mid r_{st}, r_{mt}\right) = \sum_{s_t=1}^{2} \left\{ \frac{\tau(1-\tau)}{\sigma_{s_t}} \frac{\left\{ \rho_\tau \left(\frac{r_{st} - \beta_{s_t,0}(\tau) - \beta_{s_t}(\tau)r_{mt}}{\sigma_{s_t}} \right) \right\}}{((s_t \mid \Theta_{t-1}; \Psi))} \right\}, \tag{6}$$

(2) Skew-normal likelihood function

$$L(\Psi \mid r_{st}, r_{mt}) = \sum_{s_t=1}^{2} \left\{ \frac{4\tau(1-\tau)}{\sqrt{2\pi\sigma_{s_t}^2}} \frac{\left\{ -2\rho_\tau^2 \left(\frac{r_{st} - \beta_{s_t,0}(\tau) - \beta_{s_t}(\tau)r_{mt}}{\sigma_{s_t}} \right) \right\}}{((s_t \mid \Theta_{t-1}; \Psi))} \right\}, \tag{7}$$

(3) Skew-student-t likelihood function

$$L(\Psi \mid r_{st}, r_{mt}) = \sum_{s_t=1}^{2} \left\{ \frac{4\tau(1-\tau)\Gamma\left(\frac{\nu+1}{2}\right)}{\Gamma\left(\frac{\nu}{2}\right)\sqrt{2\pi\sigma_{s_t}^2}} \frac{\left\{ \frac{4}{\nu}\rho_\tau^2 \left(\left(\frac{r_{st} - \beta_{s_t,0}(\tau) - \beta_{s_t}(\tau)r_{mt}}{\sigma_{s_t}} \right) + 1 \right) \right\}}{((s_t \mid \Theta_{t-1}; \Psi))} \right\}, \tag{8}$$

where Θ_{t-1} is all available information set at time $t-1$ in model, and $(\Pr(s_t \mid \Theta_{t-1}; \Psi)$ is weighted probabilities computed recursively from the Hamilton's filter algorithm (Hamilton 1989). Thereby, filtered probabilities of each state computed recursively can be shown as follows:

$$\Pr(s_t \mid \Theta_{t-1}; \Psi) = \left\{ p_{11}\Pr(s_t = i \mid \Theta_{t-1}; \Psi) + p_{22}\Pr(s_t = j \mid \Theta_{t-1}; \Psi) \right\}, \tag{9}$$

where

$$Pr(s_t = i \mid \Theta_{t-1}; \Psi) = \frac{L(y_t \mid (s_t = i \mid \Theta_{t-1}; \Psi)(s_t = i \mid \Theta_{t-1}; \Psi)}{\sum_{h=1}^{2} (L(y_t \mid (s_t = h \mid \Theta_{t-2}; \Psi)(s_t = h \mid \Theta_{t-2}; \Psi))}, \tag{10}$$

2.3 Bayesian Inference

In the Bayesian inference for MS–QRU models, it requires a prior distribution to construct the posterior distribution function. According to the Bayes theorem, it generates new variables under the posterior distribution, like fitting the old data with posterior distribution. The new data can be obtained from

$$\Pr(\Psi, P, s_t \,|r_{st}, r_{mt}) \propto \Pr(\Psi, P, s_t) \, L\,(r_{st}, r_{mt} \,|\Psi, P, s_t), \qquad (11)$$

where $L\,(r_{st}, r_{mt} \,|\Psi, P, s_t)$ is the likelihood functions of Eqs. (6), (7) and (8). The rest of the function is the prior distribution $\Pr(\Psi, P, s_t)$, which can be formed as

$$\Pr(\Psi, P, s_t) = \pi\,(\Psi)\,\pi\,(P)\,\pi\,(s_t \,|\Psi, P). \qquad (12)$$

We will employ a Bayesian approach via Markov chain Monte Carlo method (MCMC) for parameter estimation. As the posterior distributions are not of a standard form, the Metropolis-Hasting (MH) sampler is employed to draw the joint posterior distribution of the model and the parameters, given the sample data in Eq. 11. The resulting simulated samples from the parameter space can be used to make inferences about the distribution of the process parameters and regimes. For more detail of MH, our study refers to Chib and Greenberg (1995).

All parameters are drawn by an iterative MH sampling scheme over a partition of parameter groups. We use the following groups: (i) the unknown parameters $\beta_{s_t}\,(\tau)$; (ii) the variance of the model $\sigma_{s_t}\,(\tau)$; (iii) the transition matrix (P) and (iv) the quantile parameter τ. The prior distributions for the parameters in these four groups are assumed to be uninformative priors, thus, the prior distributions for $\beta_{s_t}\,(\tau)$, $\sigma_{s_t}\,(\tau)$, P and τ are assumed to have normal distribution, Inverse gamma, Dirichlet distribution, and uniform distribution, respectively. Therefore, we have

$$\begin{aligned} \beta_{s_t}\,(\tau) &\sim N\,(0, \Sigma), \\ \sigma_{s_t}\,(\tau) &\sim IG\,(0.01, 0.01), \\ P &\sim Dirichlet\,(q), \\ \tau &\sim uniform, \end{aligned} \qquad (13)$$

where Σ is the diagonal variance matrix parameter $\beta_{s_t}\,(\tau)$. We select these three priors since the sign of the $\beta_{s_t}\,(\tau)$ can be either positive or negative, the sign of $\sigma_{s_t}\,(\tau)$ must be positive and P should be persistently staying in their own regime. The MH iterations for all parameters can be described as follows:

1. Starting at an initial parameter value, $\Omega^0 = \Psi^0, P^0$
2. Choosing a new parameter value close to the old value based on proposal function. The proposal function employed in the MH algorithm is a normal distribution with mean at Ω^0 and covariance (C_t), that is Proposal $= (.\,|\Omega^0, \ldots, \Omega^{j-1}, C_t) = N\,(\Omega^{(j-1)}, C_t)$. In MH algorithm, covariance of the proposal distribution, C_t is set as $C_t = \sigma_d \, \text{cov}\,(\Omega^0, \ldots, \Omega^{j-1}) + \sigma_d \varepsilon I_d$ after initial period, where σ_d is a parameter that depends on dimension d and ε is a constant term which is very tiny when compared with the size of the likelihood function.

3. Computing the acceptance probability which is calculated by

$$
\theta_j = \frac{L\left(\Omega^* | r_{st}, r_{mt}\right)\left(\Omega^{j-1}, C_t\right)}{L\left(\Omega^{j-1} | r_{st}, r_{mt}\right)\left(\Omega^* | \Omega^{j-1}, C_t\right)}. \tag{14}
$$

If $\theta_j \geq 1$ then draw trace $\Omega^j = \Omega^{j-1}$. If $\theta_j < 1$ then draw trace Ω^j from a proposal distribution.

4 . Repeat steps 2–3 for $j = 1, \ldots, N$ in order to obtain samples $\Omega^1, \ldots, \Omega^N$.

2.4 Outline of the Estimation Procedure

In the MH algorithm, it is difficult to sample all the parameters together as the candidate parameters obtained from the proposal function may not ensure convergence to the desired target density. To deal with this problem, we separate the parameters into four groups and the algorithm will revolve the repeated generation of variates from their full conditional densities as follows:

$$
\begin{aligned}
\beta_{s_t}(\tau)^{(j+1)} &\leftarrow \beta_{s_t}(\tau)^{(j)}, \sigma_{s_t}(\tau)^{(j)}, P^{(j)}, \tau^{(j)} \\
\sigma_{s_t}(\tau)^{(j+1)} &\leftarrow \beta_{s_t}(\tau)^{(j+1)}, \sigma_{s_t}(\tau)^{(j)}, P^{(j)}, \tau^{(j)} \\
P(\tau)^{(j+1)} &\leftarrow \beta_{s_t}(\tau)^{(j+1)}, \sigma_{s_t}(\tau)^{(j+1)}, P^{(j)}, \tau^{(j)}. \\
\tau^{(j+1)} &\leftarrow \beta_{s_t}(\tau)^{(j+1)}, \sigma_{s_t}(\tau)^{(j+1)}, P^{(j+1)}, \tau^{(j)}
\end{aligned} \tag{15}
$$

3 Data Specification

We analyze five technology stocks from NASDAQ stock market to illustrate our model. We select the data of FAANG stocks consisting of Facebook Inc. (FB), Apple Inc. (AAPL), Amazon.com Inc. (AMZN), Netflix Inc. (NFLX) and Alphabet Inc. (GOOG) over the period from May 2012 to April 2019. All the data are collected from www.Investing.com website. First of all, it is necessary to transform the daily three-month Treasury-bill rate into daily risk free rate $r f_t$, then the excess returns on individual stock and the market are obtained by $r_{st} = \ln(\text{price}_{st}/\text{price}_{st}) - r_{ft}$ and $r_{mt} = \ln(\text{price}_{mt}/\text{price}_{mt}) - r_{ft}$, where price_{st} and price_{mt} are individual stock price and the value of the NASDAQ stock on day t. Table 1 shows the summary statistics of market and stock excess returns. It should be observed that the means of all returns are close to zero. The excess returns on the market and stocks range from -0.2102 to 0.3522. In terms of excess kurtosis, all are greater than 3 and range from 9.133 to 24.0755. This means that the distributions of the six returns have larger, thinner tails than the normal distribution. The Jarque–Bera Normality Test is employed to confirm the characteristics of these returns and the results show that their distributions are not normal. Minimum Bayes factor (MBF) values indicate that the returns are decisive not normally distributed. Unit root test is also conducted to investigate the stationarity

Table 1 Descriptive statistics

	NDAQ	FB	AAGL	AMZA	NFLX	GOOG
Mean	0.0008	0.0008	0.0005	0.0012	0.002	0.0008
Median	0.0008	0.0009	0.0006	0.001	0.0005	0.0004
Maximum	0.0671	0.2593	0.0787	0.1321	0.3522	0.1488
Minimum	−0.137	−0.2102	−0.1318	−0.1165	−0.2878	−0.0834
Std. Dev.	0.0125	0.023	0.016	0.0187	0.0303	0.0143
Skewness	−0.6323	0.5816	−0.5434	0.2088	0.852	0.8757
Kurtosis	12.2731	21.0388	9.133	10.9698	24.0755	15.3519
Jarque-Bera	6357.613	23716.89	2815.933	4622.98	32450.76	11296.84
MBF-JB	0.000	0.000	0.000	0.000	0.000	0.000
MBF-Unit root	0.000	0.000	0.000	0.000	0.000	0.000

Note MBF denotes Minimum Bayes factor which computed by e^{plogp}, where p is $p - value$

of the data and the result shows that all returns are stationary with decisive evidence. Consequently, these variables can be used for statistical inferences.

4 Estimated Results

4.1 Model Selection

In a Bayesian framework, we can compare the performance of the models using Deviance Information Criterion (DIC), which is a generalization of the AIC and is intended for use with MCMC output. The minimum DIC value corresponds to the best model specification with different distributions. The results as shown in Table 2 suggest none of the skew distribution-based models in this paper is absolutely superior to the other ones. For Facebook (FB), its model is best described by skew-normal density. Meanwhile, ALD is best for Apple Inc. (AAPL) and Alphabet Inc. (GOOG) and SKT is so for Amazon.com Inc. (AMZA) and Netflix Inc. (NFLX). Therefore, these evidences confirm the variability of the beta risk across the different skewed distributions.

4.2 Estimated Parameter Results

This part presents parameter estimates obtained from the Bayesian procedure. The best fit models for FAANG stocks are presented in Table 3. All parameters obviously change with regime change. The beta risks for two regimes, namely bull market and

Table 2 Model selection

Variable	DIC		
	ALD	SKN	SKT
FB	−10570.1	−10673.98	−10495.83
AAGL	−10659.94	−10565.33	−10508.72
AMZA	−10693.33	−10675.14	−10723.63
NFLX	−10539.93	−10514.55	−10695.2
GOOG	−10737.61	−10673.98	−10526.33

Table 3 Estimation result of Markov switching quantile regression with unknown quantile

Parameter	FB (SKN)	AAGL (ALD)	AMZA (SKT)	NFLX (SKT)	GOOG (ALD)
$\beta^0_{s_t=0}(\tau)$	0.450 (0.000)	0.0006 (0.000)	0.124 (0.000)	0.348 (0.001)	0.0032 (0.000)
$\beta_{1,s_t=0}(\tau)$	−14.115 (0.000)	0.103 (0.000)	−12.384 (0.000)	−0.024 (0.000)	0.120 (0.000)
$\beta^0_{s_t=1}(\tau)$	0.001 (0.000)	0.0008 (0.000)	0.001 (0.000)	0.0007 (0.000)	−0.0005 (0.000)
$\beta_{1,s_t=1}(\tau)$	0.127 (0.000)	0.262 (0.000)	0.218 (0.000)	0.093 (0.000)	0.441 (0.000)
$\sigma_{s_t=0}(\tau)$	0.184 (0.000)	0.003 (0.001)	0.197 (0.000)	−0.104 (0.000)	0.003 (0.000)
$\sigma_{s_t=1}(\tau)$	0.011 (0.000)	0.005 (0.000)	0.009 (0.000)	0.009 (0.000)	0.004 (0.000)
p_{11}	0.998 (0.000)	0.988 (0.000)	0.999 (0.000)	0.999 (0.000)	0.876 (0.000)
p_{22}	0.947 (0.000)	0.968 (0.000)	0.805 (0.000)	0.935 (0.000)	0.947 (0.000)
τ	0.515 (0.000)	0.499 (0.000)	0.529 (0.000)	0.507 (0.000)	0.503 (0.000)

Note () is MBF denotes Minimum Bayes factor which computed by $e^{p \log p}$, where p is p − value

bear market are found to be different. The beta risks in regime 1 are lower than regime 2. Thus, we can interpret the first regime as a bull market or high-risk market and the second regime as bear market or low-risk market. We observe that the beta risks all are less than one in both regimes, indicating that the FAANG stock returns move less volatilely than the stock market return (NASDAQ index).

Technically, it is found that beta risks of FB, AMZA and NFLX stocks are all negative in bull market but positive in bear market, meaning that there are heterogenous risks occurring in these stocks' returns. Hence, FB, AMZA and NFLX stocks will decrease/increase in value by 14.115%, 12.384%, and 0.024%, respectively for each increase/decrease of 1% in the NASDAQ market, and vice versa. AAPL and GOOG stocks have positive beta risks for both bull market and bear market, meaning that these two stocks have a positive correlation with the NASDAQ market in both market conditions. Thus, AAPL and GOOG stocks will decrease/increase in value by 0.103% and 0.120%, respectively for each decrease/increase of 1% in the NASDAQ market. Considering the transition probabilities of staying in bull market (p_{11}) and bear market (p_{22}) which appear extremely high, one can say that the probability of staying in the same regime is substantially high regardless of the current period being in either bull or bear market.

Focusing on the quantile parameter for all models, we can observe a similar result of the optimal quantile estimates. The values of quantile parameters seem to slightly deviate from the center of the distribution $\tau = 0.5$. This indicates that our model is close to the Markov Switching mean regression as the most information of the data is located in the median of the distribution. However, the result could confirm the usefulness of this model when the quantile level is generally unknown. The possible interpretation for these quantile estimates is that the beta risks of FAANG technology stocks in the U.S. market seem not to deviate from the mean estimation and there exists weak heterogenous effect of the FAANG returns.

5 Conclusion

The challenge of the Markov Switching quantile regression approach is that what is the best quantile or which quantile is the most informative to explain the behavior of the data. Many researchers have encountered a problem in interpreting the result of the parameter estimates at given quantiles. If the model contains many quantile levels, which is the best quantile level for explaining the real behavior of the dependent and independent variables. Our study concerns about this issue and thus the Markov Switching quantile regression with unknown quantile (MS–QRU) is considered. Furthermore, if we do not allow the model to have different skewed distributions, we may miss the true risk of the stocks. As a consequence, in this paper, we have proposed to construct various skew-likelihood distributions for MS–QRU. Three skewed distributions, namely ALD, SKD, and SKT are considered to be the likelihood function of the model. We then illustrate our model for the FAANG technology stocks in NAS-DAQ market. FAANG stocks consist of Facebook (FB), Amazon (AMZN), Apple (AAPL), Netflix (NFLX), and Alphabet (GOOG) in NASDAQ market.

In summary, the results suggest that none of the skewed distribution-class models is found to be absolutely superior to the other ones. This result reminds us that a model which performs very well in a particular market may not be reliable in other markets. In addition, considering the estimated quantile parameter results, we can observe that all quantile levels are close to the center of the distribution. This indicate that the beta risk of individual FAANG technology stock can be explained by the quantile around 0.5.

Acknowledgements The authors would like to thank Dr. Laxmi Worachai for her helpful comments on an earlier version of the paper. The authors are also grateful for the financial support offered by Center of Excellence in Econometrics, Faculty of Economics, Chiang Mai University, Thailand.

References

Abdymomunov, A., & Morley, J. (2011). Time variation of CAPM betas across market volatility regimes. *Applied Financial Economics, 21*(19), 1463–1478.

Atkins, A., & Ng, P. (2014). Refining our understanding of beta through quantile regressions. *Journal of Risk and Financial Management, 7*(2), 67–79.

Chiang, T., & Li, J. (2012). Stock returns and risk: Evidence from quantile. *Journal of Risk and Financial Management, 5*(1), 20–58.

Chib, S., & Greenberg, E. (1995). Understanding the Metropolis-Hastings algorithm. *The American Statistician, 49*(4), 327–335.

Galarza Morales, C., Lachos Davila, V., Barbosa Cabral, C., & Castro Cepero, L. (2017). Robust quantile regression using a generalized class of skewed distributions. *Stat, 6*(1), 113–130.

Hamilton, J. D. (1989). A new approach to the economic analysis of nonstationary time series and the business cycle. *Econometrica: Journal of the Econometric Society*, 357–384.

Huang, H. C. (2000). Tests of regimes-switching CAPM. *Applied Financial Economics, 10*(5), 573–578.

Kenton, W. (2019). *FAANG stocks*. https://www.investopedia.com/terms/f/faang-stocks.asp.

Koenker, R., & Bassett, G. (1978). Regression quantiles. *Econometrica, 46*, 33–50. Mathematical Reviews (MathSciNet): MR474644. https://doi.org/10.2307/1913643.

Koenker, R., & Machado, J. A. (1999). Goodness of fit and related inference processes for quantile regression. *Journal of the American Statistical Association, 94*(448), 1296–1310.

Liu, X. (2016). Markov switching quantile autoregression. *Statistica Neerlandica, 70*(4), 356–395.

Rakpho, P., Yamaka, W., & Sriboonchitta, S. (2018). Which quantile is the most informative? Markov switching quantile model with unknown quantile level. *Journal of Physics: Conference Series, 1053*(1), 012121 (IOP Publishing).

Sharpe, W. F. (1964). Capital asset prices: A theory of market equilibrium under conditions of risk. *The Journal of Finance, 19*(3), 425–442.

Thamprasert, K., Pastpipatkul, P., & Yamaka, W. (2018). Interval-valued estimation for the five largest market capitalization stocks in the stock exchange of Thailand by Markov-Switching CAPM. In *International Econometric Conference of Vietnam* (pp. 916–925). Cham: Springer.

Tungtrakul, T., Maneejuk, P., & Sriboonchitta, S. (2016). Macroeconomic factors affecting exchange rate fluctuation: Markov switching Bayesian quantile approach. *Thai Journal of Mathematics*, 117–132.

Wichitaksorn, N., Choy, S. B., & Gerlach, R. (2014). A generalized class of skew distributions and associated robust quantile regression models. *Canadian Journal of Statistics, 42*(4), 579–596.

Yamaka, W., Rakpho, P., & Sriboonchittac, S. (2019). Bayesian Markov switching quantile regression with unknown quantile: Application to Stock Exchange of Thailand (SET). *Thai Journal of Mathematics*, 1–13.

Investment Behavior, Financial Constraints and Monetary Policy – Empirical Study on Vietnam Stock Exchange

Dinh Thi Thu Ha, Hoang Thi Phuong Anh, and Dinh Thi Thu Hien

Abstract Using the data-set of 200 non-financial companies listed on HOSE and HNX for the period 2006 - 2015, this paper examines the effect of bank financing on corporate investment behavior and the extent to which bank financing affects corporate investment behavior in different monetary policy periods. By applying Bayesian method, the results find that company with more bank financing will reduce the proportion of investment, and in comparison with the period of loosening monetary policy, company tend to reduce its investments by bank loans in the period of tightening monetary policy.

Keywords Investment behavior · Bank financing · Monetary policy · GMM model

1 Introduction

Monetary policy affects the economy through various channels such as credit channels and interest rate channels (Chatelain et al., 2003). Understanding this transmission mechanism is essential for policy makers. In case of traditional interest rate channel, when the interest rate changes, cost of capital changes accordingly, thus affecting investment. Regarding to the credit channel, changes in market interest rates affect the balance sheet and net cash flow of firms in an inefficient market, this net cash flow in return will affect investment.

D. T. T. Ha (✉)
Thai Binh Duong University, Khanh Hoa, Vietnam
e-mail: dttha@tbd.edu.vn

H. T. P. Anh
University of Economics Ho Chi Minh City, Ho Chi Minh City, Vietnam
e-mail: anhtcdn@ueh.edu.vn

D. T. T. Hien
Ho Chi Minh City Open University, Ho Chi Minh City, Vietnam
e-mail: hien.dtt@ou.edu.vn

© The Editor(s) (if applicable) and The Author(s), under exclusive license
to Springer Nature Switzerland AG 2021
N. Ngoc Thach et al. (eds.), *Data Science for Financial Econometrics*,
Studies in Computational Intelligence 898,
https://doi.org/10.1007/978-3-030-48853-6_19

267

The corporate sector has an important influence on the real economy and financial stability through linkages with the banking sector and financial markets. The sensitivity of the corporate sector to the shocks not only affects the business cycle but also affects the capital structure. Enterprises with low debts and collaterals are more sensitive to interest rate shocks and to financial crisis. Moreover, short-term debts can amplify interest rate risks by reinvestment risks.

Understanding the effects of monetary policy on corporate investment decisions contribute to both theoretical and empirical analysis. Based on asymmetric information and agency problems, some theories such as pecking order theory show that the corporate financial model begins with retained earnings, followed by debts, and equity offerings (Myers 1977; Myers and Majluf 1984). Moreover, there are costs and distribution differences when a company use external finance from the bank. Contracts that do not have enough information can limit the ability of enterprises to access external financial funds, so it is difficult to have sufficient funding for profitable investment opportunities (Stigliz and Weiss, 1981, Besanki and Thakor, 1987).

Researches on corporate funding, investment behavior and monetary policy have gained significant attentions from researchers; however most of them focus on developed countries such as Japanese studies (Fuchi et al., 2005), United States (Vijverberg, 2004), UK (Mizen and Vermeulen, 2005; Guariglia, 2008), Europe region (Bond with partners, 2003, Chatelain et al., 2003; De Haan and Sterken, 2011), Canada (Alivazian et al., 2005) and Spain (Gonzales and Lopez, 2007). Research on the linkages between corporate finance, investment and the monetary policy transmission mechanism in developing markets is more limited. However, the widespread of financial crisis in the developing markets have attract more attentions among researchers, policy makers and economists on corporate finance, investment and monetary policy. Among the studies are Borensztein and Lee (2002) for Korea, Rungsomboon (2003) for Thailand, Perotti and Vesnaver (2004) for Hungary, and Firth et al. (2008) for China. Their results support the existence of credit channels in developing markets but shown variations across country and firm specific characteristics.

Numerous researches have been devoted to Vietnamese corporate investment decisions such as study on manager behaviors and investment activities in Vietnam (Nguyen Ngoc Dinh, 2015), effects of cash flow on investment decisions (Le Ha Diem Chi, 2016), the impact of excessive cash accumulation on financial decisions (Nguyen Thi Uyen Uyen, 2015), determinants of investment decisions of foreign enterprises in Kien Giang (Pham Le Thong, 2008). However, the number of studies examining the role of monetary policy in investment decisions is very limited; therefore, this research aims to fill the gaps in literature by examining the relationship between monetary policy and investment behavior of Vietnamese enterprises to provide evidence on corporate funding choices and explain investment behavior when monetary policy change.

2 Literature Review

The monetary policy transmission mechanism has attracted great attention from researchers and policy makers. First, monetary policy affects the economy through traditional interest rate channel: an expansionary monetary policy reduces interest rate; subsequently, cost of using external capital decreases will stimulate investments. Currency is a type of assets, in an open economy, foreign exchange rate is one of the traditional channels through which monetary policy affects international transactions, domestic production and prices (Deniz Igan, 2013). Numerous researches have been conducted to investigate the traditional transmission channels of monetary policy such as interest rates, asset prices and foreign exchange rates (Boivin, Kiley and Mishkin, 2010). The theory of interest rate channels assumes that financial intermediaries do not play important role in the economy. Bernanke and Gertler (1995–1983) show that the traditional interest rate channel relies on one of three following assumptions: (i) For borrowers: loans and bonds are perfect substitutions for internal financing (ii) For lenders: loans and bonds are perfect substitutions for internal financing. (iii) Credit demand is not sensitive to interest rates.

However, the effect of monetary policy on the economy is greater than what can be explained by traditional interest rate channel, researchers have identified other monetary policy transmission channels such as lending channels, balance sheet channels, and risk-taking channels. Mishkin (1996) suggests that monetary policy affect banks, thereby affecting bank loans or lending channels. This lending channel is based on the notion that banks play significant role in financial system due to the fact that the lending channel is suitable for most businesses that need financial funds. Bernanke and Gertler (1983) argue that the balance sheet channels relate to the impact of monetary policy on loan demand. Higher interest rates increase interest payments while reduce the current value of assets and collateral. This will reduce the credibility of assets leading to higher disparity when using external finance. Consequently, credit growth will be slower, aggregate demand and supply will decrease. In addition, Bruno and Shin (2015) offers a risk-taking channel related to changes in the demand of financial resources caused by policy changes, leading to the changes of risk of the banks and financial intermediaries Low interest rates encourage financial intermediaries to take more risks to seek higher profits.

Meanwhile, Bernanke and Gertler (1983) argue that according to credit channel theory, the direct effect of monetary policy on interest rates is amplified by change in the cost of using external capital showing the difference between the cost of using external funds (loans or equity offerings) and internal funds (retained earnings). Cost of external funds represents the inefficiency of credit market when there is a big difference between the expected return of lender (returns from interest expenses) and cost of borrowers (interest expenses). Therefore, a change in monetary policy may increase or decrease interest rates, thereby increasing or decreasing the cost of external capital, affecting credit supply. Credit channel is not an alternative for traditional monetary policy transmission mechanism, but through credit channel, the impact of monetary policy on the economy will be significantly increased. This theory examines the impact of disadvantageous shock that raises the interest costs borrowers have to pay resulting in the increases the external fund costs, and the mechanism of these adverse effects. Financial shocks can affect corporate sector by cutting credit for valuable trading and investment opportunities (Kashyap et al, 1991).

Monetary policy is an important macro variable affecting corporate investment decisions. Jing et al. (2012) argue that expansionary monetary policy reduces financial constraints for private companies; however, this also leads to ineffective investments. In contrast, a good financial environment can help firms take advantage of investments, improve capital efficiency when they have better investment opportunities.

Monetary policy transmits to the economy through multiple channels such as currency channels (interest rates, foreign exchange, asset prices) and credit channels. According to classical economists, policy makers use leverage on short-term interest rates to influence capital costs, thus affecting commodity spending such as fixed assets, real estates, inventory. Changes in supply and demand affect the proportion of corporate investments. When the Central Bank raises interest rates, costs of debt financing as well as business financial constraints increase. Consequently, firm becomes more dependent on external funds and therefore reduces investment proportion due to the increase of capital costs.

Impact of monetary policy on corporate investment behavior through different channels gain great attention from researchers. Bernanke (1992) suggests the theory of monetary policy transmission channel considering the role of financial intermediaries in monetary policy transmission at micro level. Although theory of monetary policy transmission through credit channels is still controversial, this theory suggests that monetary policy affects the availability of financial resources by increasing or decreasing the supply of bank loans, thus affecting the supply of corporate investments. Bernanke (1995) argues that monetary policy influences long-term investments through interest rate and proportion of financial constraints. Chatelain and Tiomo (2003) argue that monetary policy affects corporate investments through interest rate channels, thus adjusting lending and borrowing costs, through credit channels this mechanism affects enterprise capital costs. Zulkhibri (2013) argues that monetary policy has a significant impact on the ability to access external funds when interest rates rise. Firms become more vulnerable, especially those with high leverage, internal capital becomes more important in the period of low liquidity.

Changes in monetary policy (for example, from loosening to tightening) will affect corporate investment opportunities and external financial constraints through credit and currency channels. Rational managers may consider adjusting their investment plans in response to these changes. From the view point of financial constraint of credit channel, in the period of tightening monetary policy, credit restriction and investment costs increase when credit supply decreases; according to NPV method and profit maximization theory, firms will reduce investment proportions in response to those changes. On the contrary, in the period of expansionary monetary policy, financial constraints and capital costs decrease due to the increase of credit supply, projects with negative NPV will become positive, firms will expand investments to maximize profits.

From the view point of monetary policy transmission, in the period of expansionary monetary policy, aggregate demand will increase when money base supply increases. Big companies expand productions and investment proportion due to increase of investment opportunities. In the period of tight monetary policy, total demand decline and capital costs increase due to money supply decrease. Firms reduce investment proportions leading to a decline in investment opportunities. Monetary policy affects the economy directly and indirectly. An increase in interest rates will weaken the balance sheet by reducing net cash flow of interest rates and real asset value. This amplifies the impact of monetary policy on borrowers' spending. This impact process can also take place indirectly. Suppose that tight monetary policy reduces spending, decrease in cash flow and asset value will correlate with spending decrease, thus the balance sheet is affected, then the value is reduced afterwards. This indirect channel has significant impact since it considers that the influence of financial factors may occur later when a change in monetary policy occurs.

Monetary policy affects financial factors. First, financial theories of business cycle emphasize the role of borrowers' balance sheet (Bernanke and Gertler (1989), Calomiris and Hubbard (1990), Gertler (1992), Greenwald and Stiglitz (1993), Kiyotaki and Moore (1993)). These theories begin with the idea that inefficient markets make borrowers' spending depend on their balance sheet status because of the link between net asset value and credit terms. This leads to a direct financial transmission mechanism: changes in balance sheet change amplify changes in spending.

3   Methdology

3.1   Data Description

The paper consists of 500 companies listed on two Vietnam stock exchanges: Ho Chi Minh City Stock Exchange (HOSE) and Hanoi Stock Exchange (HNX) in the period 2007 - 2016. The companies are selected as follows:

Excluding companies operating in banking sector, insurance sector, real estate sector, investment funds, securities companies. The sample only includes non-financial companies.

Excluding companies that do not have enough financial statements in 10 years from 2007–2016, assuming that firms that do not have enough financial statements are newly established, newly equitized or unsatisfactory to operate continuously, the financial statements of these companies are different and difficult to assess.

Finally, a data set of 200 non-financial balance sheets in 5 different sectors, from 2007 to 2016, represent observation of 2000 company-year is collected. Data are extracted from audited financial statements and audited consolidated financial statements of companies collected from www.stox.vn and www.vietstock.vn.

3.2 Models

First, to determine company investment behavior in Vietnam, we apply the empirical model of Bond et al. (2003) and Zulkhibri (2015) to estimate the technical method of error - correction, as follows:

$$I_{i,t}/K_{i,t-1} = \alpha_0 + \alpha_1 * \Delta S_{it} + \alpha_2 * \left(k_{i,t-2} - S_{i,t-2}\right) + \alpha_5 * DEBT_{it} + DEBT_{it} * T + DEBT_{it} * (1-T) + \varepsilon_{it} \quad (1)$$

where:

- I_{it} is company investment at time t, including capital expenditures on property, plant and equipment. It is measured by the proportion of firm capital expenditure on tangible fixed assets. This measurement is similar to the measurement in the researches of Kaplan and Zingales (1997), Chirinko, Fazzari and Meyer (1999), Bhagat, Moyen and Suh (2005), Love and Zicchino (2006), Moyen (2004), Odit et al. (2008), Nguyen Thi Ngoc Trang et al. (2013), Nguyen Ngoc Dinh (2015) and Zulkhibri (2015)
- $K_{i,t-1}$ is capital stock value. It is measured by book value of tangible fixed assets, this measurement is similar to the measurement of Odit et al. (2008), Nguyen Thi Ngoc Trang et al. (2013), Nguyen Ngoc Dinh (2015) and Zulkhibri (2015).
- S_{it} is calculated by natural logarithm of company revenue in which company revenue is measured by revenue/(1 + inflation)
- ΔS_{it} is the first difference of S_{it}, measuring the revenue growth rate. This measurement is similar to research of Zulkhibri (2015). $\Delta S_{i,t-1}$ is the first lag of Δs_{it}.
- k_{it} is natural logarithm of K_{it}.
- $DEBT_{it}$ is ratio of short-term debt/total debt. Besides, we replace the variable of external funding resources $DEBT_{it}$ by variable of cash flow CF_{it} to assess the effect of cash flow on investments as proxy for internal funding resources (Fazzari et al., 1988), where CF_{it} is measured by profit after tax plus depreciation plus interest expense.
- The existence of adjustment costs is presented by error-correction behavior $(k_{i,t-2} - s_{i,t-2})$ in the research model, where $k_{i,t-2}$ and $s_{i,t-2}$ are the second lag of k_{it} and s_{it} respectively. The adjustment costs reflect capital stock adjustment toward the target level. The error-correction behavior requires that the coefficient α_2 should be negative to be consistent with the presence of adjustment cost, if the current capital is lower (higher) than its target level, investment may be higher (lower) in the future (Guariglia, 2008; Zulkhibri, 2015).
- T is a dummy variable indicating tight monetary policy in the period of 2010 - 2011 due to economy stabilizing, inflation controlling (Nguyen Thi Hai Ha,

2012), and in the period of 2007 - 2008 due to the global financial crisis, so T = 1 if the time under consideration is 2007, 2008, 2010, 2011 and otherwise T = 0. In those years, the interest rates are higher; therefore, the selection of these years as proxies for period of tight monetary policy is reasonable.

- (1 − T) is proxy for period of loosening monetary policy in 2009, 2012 - 2014, 2015. We consider these two periods when interest rates change to examine the effect of monetary policy on corporate investments.

To estimate panel data, we can use pool ordinary least squares (OLS), fixe Effects (FE) and random effects (RE) model. In this paper, we apply Bayesian approach to test the statistical results due to the fact that this approach makes conclusions more reliable.

4 Results

4.1 Descriptive results

Table 1 shows the descriptive statistics for the main variables. The mean and standard deviation of INV is 0.029 and 0.1849, respectively. It shows that the fluctuation of company investment is not too wide with the minimum value is −0.7698 and the maximum value is 4,9482. Mean and standard deviation of company's size (SIZE) is 26.6 and 1.4429 respectively; the minimum value is 20.98 and the maximum value is 31.32. In comparison to the company which have the mean value of cash flow is 0.1282 and standard deviation value is 0.1023, company which have the mean value of debt ratio is 0.622 and standard deviation value is 0.3901 exhibit more volatility of debt ratio than that of the cash flow.

Table 1 Variables descriptive statistics

VARIABLE	MEAN	STD.DEV	MIN	MAX	OBSERVATION
INV	0.0298	0.1849	−0.7698	4.9482	1800
SIZE	26.6010	1.4429	20.9841	31.3219	1800
DENTASIZE	0.0999	0.3327	−2.8825	2.0376	1800
ERRORCORR	−0.0609	0.7483	−2.4543	3.3319	1600
DEBT	0.6233	0.3901	0.0000	1.0000	1800
CF	0.1282	0.1023	−0.4958	2.1018	1800

Source: Author's Calculation

4.2   The effect of external finance on corporate investment behavior by monetary policy transmission

Most coefficients of model 1 exhibit reasonable sign and statistical significance. The results show that external funding has a positive impact, and is statistically significant to investment capacity. This suggests that companies often rely on new loans to make new investments. The results are consistent with study of Harris et al. (1994) which suggests the relationship between external funds and investment behavior of large Indonesian companies. The estimated results of model 2 also show similar results. The most important finding of this study is the impact of monetary policy in different monetary periods. The effect of monetary policy on investment behavior in a tight monetary policy period is greater (coefficient of −0.0495) than that in the loose monetary policy period (coefficient −0.05). When Vietnamese companies depend on external funds (in particular short-term debts), limited access to external finance during tight liquidity period can lead to a significant reduction in investment proportion. This result is consistent with the evidence of capital market imperfections and liquidity constraint (Fazzari et al., 1988; Bernanke & Gertler, 1989; Hoshi et al., 1991).

In addition, the ERRORCORR coefficient of models 1 and 2 are both negative implying the existence of investment behavior adjustment costs. The results are similar to the study of Zulkhibri (2015).

Table 2 Results

	Without monetary policy interaction (Model 1)			With monetary policy interaction (Model 2)		
	Mean	Equal-tailed		Mean	Equal-tailed	
		[95% Cred. Interval]			[95% Cred. Interval]	
ΔS_{it}	0.072*	0.052	0.092	0.071*	0.052	0.091
$k_{i,t-2} - S_{i,t-2}$	−0.000098	−0.009	0.019	−0.0000561	−0.009	0.019
debt	0.024*	0.040	0.06	0.475*	0.093	0.692
Debt_T				−0.495	−0.826	−0.119
Debt_(1 − T)				−0.500	−0.833	−0.122
_cons	0.033	0.019	0.045	0.033	0.020	0.045

Source: Author's Calculation

4.3 Effect of external finance in monetary policy transmission mechanism to corporate investment behavior when company is financial constrained

In order to examine the impact of financial constraints on corporate investment behavior, we divided our data into three categories: (i) Size (*large* if the average value of company size which is calculated by natural logarithm of total assets is in the upper 25% of distribution while *small* if firm natural logarithm of total assets is in the lower 25% of distribution) (ii) *bank-dependent* if firm ratio of short-term borrowing to total assets falls in the upper of 25% of distribution while *non-bank-dependent* if firm ratio of short-term borrowing to total assets falls in the lower 25 quartile of the distribution; and (iii) leverage (*high* if firm ratio of total debts to total assets falls in the upper 25 quartile of the distribution while *low* if firm ratio of total debts to total assets falls in the lower 25 quartile of the distribution).

In Table 3, model 1 shows the results of company with size constraint; model 2 exhibits the results of company with leverage constraint and model 3 presents the results of company with or without bank-dependent constraint.

Table 3 shows that in terms of the size, companies tend to reduce investment proportion regardless of company size and monetary policy regimes.

In terms of the leverage, in tight monetary policy period, high leverage companies tend to reduce the proportion of investment when the bank loans increased at the significance level of 10%. In expansionary monetary policy period, when short-term bank loans increase, low-leverage companies will increase investment while high-leverage companies will reduce investment. As can be seen that high leveraged companies will reduce investment proportion in both expansionary and tightening monetary policy period when short-term bank loans increase. In expansionary monetary policy period, low-leverage companies tend to increase the proportion of investment when their bank debt ratio is high. Finally, in the tight monetary policy period, bank-dependent companies tend to reduce investment proportion when their short-term bank loans increase regardless of the extent to which company depend on the bank loans.

In the period of expansionary monetary policy, when firm short-term bank loans increase, non-bank-dependent company will increase investment proportion while bank-dependent company will reduce investment proportion.

5 Conclusion

This paper aims to evaluate the effect of short-term bank loans on corporate investment behavior in both tight monetary policy period and loose monetary policy period in 2007 - 2016. Based on the Bayesian method, our results show that companies

Table 3 Results of financially constrained firms

	Model 1			Model 2			Model 3		
	Mean	Equal-tailed [95% Cred. Interval]		Mean	Equal-tailed [95% Cred. Interval]		Mean	Equal-tailed [95% Cred. Interval]	
Inv									
ΔS_{it}	0.069*	0.050	0.088	0.069	0.050	0.088	0.071	0.050	0.091
$k_{i,t-2} - S_{i,t-2}$	−0.00033	−0.009	0.009	−0.0002	−0.009	0.009	−0.00009	−0.009	0.009
Debt	−0.067	−0.091	−0.041	0.076	−0.202	0.347	−0.072	−0.188	−0.011
Debt*small*T	0.057*	0.018	0.097						
Debt*small*(1 − T)	0.033*	0.009	0.051						
Debt*large*T	0.046*	0.025	0.069						
Debt*large*(1 − T)	0.043*	0.017	0.063						
Debt*low*T				−0.099	−0.377	0.199			
Debt*low*(1 − T)				−.090	−0.369	0.207			
Debt*high*T				−0.097	−0.370	0.184			
Debt*high*(1 − T)				−0.105	−0.379	0.185			
Debt*nonbank*T							0.038	−0.058	0.126
Debt*nonbank*(1 − T)							0.037	−0.024	0.100
Debt*bank*T							0.052	−0.008	0.099
Debt*bank*(1 − T)							0.047	−0.023	0.093
_cons	0.033	0.020	0.046	0.033	0.021	0.047	0.033	0.019	0.046

Source: Author's Calculation

with higher short-term bank loans tend to increase their investments. Simultaneously, when considering more monetary policy factors, the results show that monetary policy has more effect on corporate investment behavior in the tight monetary policy period compared to that in loose monetary policy period, implying that in the tight monetary policy period, company should limit its access to bank loans since increasing bank loans in this period makes the cost of capital higher resulting in ineffectiveness of investments. At the same time, it may increase the probability of company financial distress.

This study, however is subject to several limitations. Firstly, the data excludes non-financial companies. Secondly, research data categories are not divided by industry. Therefore, future studies may consider the effect of bank loans on corporate investment behavior in each period of monetary policy for financial firms since they are more vulnerable when the state changes monetary policies. At the same time, it is necessary to consider the effect of debt on investment behavior in the context of different monetary policies by firm industries. In addition, expanding research to consider the impact of financial crisis on the relationship between bank loans and corporate investment behavior is also interesting issue.

References

Aivazian, V. A., Ge, Y., & Qiu, J. (2005). The impact of leverage on firm investment: Canadian evidence. *Journal of Corporate Finance, 11*(1–2), 277–291. https://doi.org/10.1016/S0929-119 9(03)00062-2.

Bernanke, B., & Gertler, M. (1989). Agency Costs, Net Worth, and Business Fluctuations. *The American Economic Review, 79*(1), 14–31.

Bernanke, B. S. (1983). Non-monetary effects of the financial crisis in the propagation of the great depression. *National Bureau of Economic Research Working Paper Series, No. 1054.* doi: https://doi.org/10.3386/w1054.

Bernanke, B. S., & Gertler, M. (1995). Inside the black box: The credit channel of monetary policy transmission. *National Bureau of Economic Research Working Paper Series, No. 5146.* doi: https://doi.org/10.3386/w5146.

Besanko, D., & Thakor, A. V. (1987). Collateral and rationing: Sorting equilibria in monopolistic and competitive credit markets. *International Economic Review, 28*(3), 671–689. https://doi.org/10.2307/2526573.

Bhagat, S., Moyen, N., & Suh, I. (2005). Investment and internal funds of distressed firms. *Journal of Corporate Finance, 11*(3), 449–472. https://doi.org/10.1016/j.jcorpfin.2004.09.002.

Boivin, J., Kiley, M. T., & Mishkin, F. S. (2010). How has the monetary transmission mechanism evolved over time? *National Bureau of Economic Research Working Paper Series, No. 15879.* doi: https://doi.org/10.3386/w15879.

Bond, S., Elston, J. A., Mairesse, J., & Mulkay, B. (2003). Financial factors and investment in Belgium, France, Germany, and the United Kingdom: A Comparison using company panel data. *The Review of Economics and Statistics, 85*(1), 153–165.

Borensztein, E., & Lee, J.-W. (2002). Financial crisis and credit crunch in Korea: Evidence from firm-level data. *Journal of Monetary Economics, 49*(4), 853–875.

Bruno, V., & Shin, H. S. (2015). Capital flows and the risk-taking channel of monetary policy. *Journal of Monetary Economics, 71*(C), 119–132.

Carpenter, R. E., & Guariglia, A. (2008). Cash flow, investment, and investment opportunities: New tests using UK panel data. *Journal of Banking & Finance, 32*(9), 1894–1906. https://doi.org/10.1016/j.jbankfin.2007.12.014.

Chatelain, J.-B., & Tiomo, A. (2003). Monetary policy and corporate investment in France. *Monetary Policy Transmission in the Euro Area*, 187–197.

Chatelain, J. B., Generale, A., Hernando, I., von Kalckreuth, U., & Vermeulen, P. (2003). New findings on firm investment and monetary transmission in the euro area. *Oxford Review of Economic Policy, 19*(1), 73–83.

Chirinko, R. S., Fazzari, S. M., & Meyer, A. P. (1999). How responsive is business capital formation to its user cost?: An exploration with micro data. *Journal of Public Economics, 74*(1), 53–80. https://doi.org/10.1016/S0047-2727(99)00024-9.

Chittoo, M. P. O. H. B. (2008). Does financial leverage influence investment decisions? : The case of Mauritian firms. *Journal of business case studies, 4*(9, (9)), 49–60.

Fazzari, S., Hubbard, R. G., & Petersen, B. C. (1988). Financing constraints and corporate investment. *Brookings Papers on Economic Activity, 1988*(1), 141–206.

Firth, M., Lin, C., & Wong, S. M. L. (2008). Leverage and investment under a state-owned bank lending environment: Evidence from China. *Journal of Corporate Finance, 14*(5), 642–653. https://doi.org/10.1016/j.jcorpfin.2008.08.002.

Gertler, M. (1992). Financial capacity and output fluctuations in an economy with Multi-period financial relationships. *The Review of Economic Studies, 59*(3), 455–472. https://doi.org/10.2307/2297859.

Greenwald, B., & Stiglitz, J. (1993). New and Old Keynesians. *The Journal of Economic Perspectives, 7*(1), 23–44.

Guariglia, A. (2008). Internal financial constraints, external financial constraints, and investment choice: Evidence from a panel of UK firms. *Journal of Banking & Finance, 32*(9), 1795–1809. https://doi.org/10.1016/j.jbankfin.2007.12.008.

Haan, L., & d., & Sterken, E. . (2011). Bank-specific daily interest rate adjustment in the dutch mortgage market. *Journal of Financial Services Research, 39*(3), 145–159. https://doi.org/10.1007/s10693-010-0095-2.

Harris, J. R., Schiantarelli, F., & Siregar, M. G. (1994). The effect of financial liberalization on the capital structure and investment decisions of Indonesian manufacturing establishments. *The World Bank Economic Review, 8*(1), 17–47. https://doi.org/10.1093/wber/8.1.17.

Hitoshi, F., Ichiro, M., & Hiroshi, U. (2005). A historical evaluation of financial accelerator effects in Japan's economy: Bank of Japan.

Hoshi, T., Kashyap, A., & Scharfstein, D. (1991). Corporate structure, liquidity, and investment: Evidence from Japanese industrial groups. *The Quarterly Journal of Economics, 106*(1), 33–60. https://doi.org/10.2307/2937905.

Jing, Q. L., Kong, X., & Hou, Q. C. (2012). Monetary policy, investment efficiency and equity value. *Economic Research Journal, 47*(5), 96–106. (in Chinese).

Kaplan, S. N., & Zingales, L. (1997). Do investment-cash flow sensitivities provide useful measures of financing constraints? *The Quarterly Journal of Economics, 112*(1), 169–215. https://doi.org/10.1162/003355397555163.

Kashyap, A., K. & Stein, J., C. & Wilcox, D. W. 1991. "Monetary policy and credit conditions: evidence from the composition of external finance," Finance and Economics Discussion Series 154, Board of Governors of the Federal Reserve System (US).

Kiyotaki, N., & Moore, J. (1997). Credit Cycles. *Journal of Political Economy, 105*(2), 211–248. https://doi.org/10.1086/262072.

Lê Hà Diêm Chi (2016). Ảnh hưởng của dòng tiền đến đầu tư': nghiên cứu trong trường hợp hạn chế tài chính. *Tạp chí Công nghệ Ngân hàng, 118+119* 27–37.

Love, I., & Zicchino, L. (2006). Financial development and dynamic investment behavior: Evidence from panel VAR. *The Quarterly Review of Economics and Finance, 46*(2), 190–210. https://doi.org/10.1016/j.qref.2005.11.007.

Mishkin, F. S. (1996). Understanding financial crises: A developing country perspective. *National Bureau of Economic Research Working Paper Series, No. 5600.* doi: https://doi.org/10.3386/w5600.

Mizen, P., & Vermeulen, P. (2005). Corporate investment and cash flow sensitivity: what drives the relationship? : European Central Bank.

Moyen, S., & Sahuc, J.-G. (2005). Incorporating labour market frictions into an optimising-based monetary policy model. *Economic Modelling, 22*(1), 159–186. https://doi.org/10.1016/j.econmod.2004.06.001.

Myers, S. C. (1977). Determinants of corporate borrowing. *Journal of Financial Economics, 5*(2), 147–175. https://doi.org/10.1016/0304-405X(77)90015-0.

Myers, S. C., & Majluf, N. S. (1984). Corporate financing and investment decisions when firms have information that investors do not have. *Journal of Financial Economics, 13*(2), 187–221. https://doi.org/10.1016/0304-405X(84)90023-0.

Nguyên Thị Hải Hà. (2012). Hệ thống ngân hàng Việt Nam trước yêu cầu tái cấu trúc để phát triển bền vững. *Ngân Hàng Nhà Nước Việt Nam, 16,* 6–9.

Nguyên Ngọc Định. (2015). Hành vi của nhà quản lý và hoạt động đầu tư tại các doanh nghiệp Việt Nam. *Tạp Chí Phát Triển Và Hội Nhập, 21*(31), 51–56.

Trang, N. T. N., & Quyên, T. T. (2013). Mối quan hệ giữa sử dụng đòn bẩy tài chính và quyết định đầu tư. *Tạp Chí Phát Triển Và Hội Nhập, 9*(19), 10–15.

Uyên, N. T. U., Thoa, T. T. K., & Hoài, H. T. (2015). mối quan hệ động giữa tính bất ổn trong dòng tiền và quyết định đầu tư. *Tạp Chí Công Nghệ Ngân Hàng, 116,* 10–15.

Perotti, E. C., & Vesnaver, L. (2004). Enterprise finance and investment in listed Hungarian firms. *Journal of Comparative Economics, 32*(1), 73–87. https://doi.org/10.1016/j.jce.2003.09.009

Thông, P. L., Ninh, L. K., Nghiêm, L. T., Tú, P. A., & Khải, H. V. (2008). Phân tích các yếu tố ảnh hưởng đến quyết định đầu tư của các doanh nghiệp ngoài quốc doanh ở Kiên Giang. *Tạp Chí Khoa Học, Trường Đại Học Cần Thơ, 9,* 102–112.

Robert, E. C., Steven, M. F., & Bruce, C. P. (1998). Financing constraints and inventory investment: A comparative study with high-frequency panel data. *The Review of Economics and Statistics, 80*(4), 513–519.

Rungsomboon, S. (2005). Deterioration of firm balance sheet and investment behavior: Evidence from panel data on Thai firms*. *Asian Economic Journal, 19*(3), 335–356. https://doi.org/10.1111/j.1467-8381.2005.00216.x

Stiglitz, J. E., & Weiss, A. (1981). Credit rationing in markets with imperfect information. *The American Economic Review, 71*(3), 393–410.

Vijverberg, C.-P.C. (2004). An empirical financial accelerator model: Small firms' investment and credit rationing. *Journal of Macroeconomics, 26*(1), 101–129. https://doi.org/10.1016/j.jmacro.2002.09.004.

Zulkhibri. (2013). Corporate investment behaviour and monetary policy: Evidence from firm-level data for Malaysia. *Global Economic Review, 42*(3), 269–290.

Zulkhibri, M. (2015). Interest burden and external finance choices in emerging markets: Firm-level data evidence from Malaysia. *International Economics, 141,* 15–33. https://doi.org/10.1016/j.inteco.2014.11.002.

Non-interest Income and Competition: The Case of Vietnamese Commercial Banks

Nguyen Ngoc Thach, T. N. Nguyen Diep, and V. Doan Hung

Abstract This paper investigates the impact of competition and other macro and micro factors on non-interest income (NII) of 27 commercial banks in Vietnam in the period 2010–2017. The current study applies a variable selection technique is the Least Absolute Shrinkage and Selection Operator (LASSO), and the findings suggest that competition (proxied by COM and HHI variables) has a significant impact on NII of commercial banks. Our research results further show that five micro and macro factors, namely Loan loss provisions to total assets (RES), Cost to Income (COST), Voice and Accountability (VAE), Political Stability and Absence of Violence (PVE) and Regulatory Quality (RQE), also have an impact on NII of commercial banks. Based on the empirical findings, this study provides relevant implications for bank administrators as well as regulatory bodies in the banking field to enhance NII in particular and competitiveness in general of Vietnamese commercial banks

Keywords Competition · Non-interest income · Macro · Micro determinants · LASSO

1 Introduction

Commercial banks are important financial intermediaries which aid in the execution of socioeconomic activities undertaken by different stakeholders in the society. Banks serve primarily by bridging the gap between surplus and deficit in the economy. Bank's operating income is derived either from interest-related or non-interest

N. N. Thach
Banking University of Ho Chi Minh City, 39 Ham Nghi street, District 1, Ho Chi Minh City, Vietnam
e-mail: thachnn@buh.edu.vn

T. N. N. Diep (✉) · V. D. Hung
Lac Hong University, 10 Huynh Van Nghe street, Buu Long ward, Bien Hoa, Dong Nai, Vietnam
e-mail: ngocdiep1980.dhlh@gmail.com

V. D. Hung
e-mail: doanviethung2012@gmail.com

© The Editor(s) (if applicable) and The Author(s), under exclusive license to Springer Nature Switzerland AG 2021
N. Ngoc Thach et al. (eds.), *Data Science for Financial Econometrics*, Studies in Computational Intelligence 898,
https://doi.org/10.1007/978-3-030-48853-6_20

incomes, and the fundamental function that banks play as an intermediary generates interest income. This traditionally acts as a major source of income for banks, importantly determining bank performance. Recently, with relentless development in information and communication technologies, tough competition from fintech companies and increasingly diverse and complex consumer demands, banks find themselves compelled to diversify their fields of operation Chiorazzo et al. (2008). This supports the sustainability of their business in the competitive market, and opens up another source of income: non-interest income.

The new landscape has urged commercial banks to adapt and create more diverse products and services, apply advances in information technology to operate more efficiently the increasingly diverse customer market. For commercial banks, there are two main sources of income: (i) loans are invariably the main income stream with inherently high risk; (ii) non-interest incomes (NII) from other services such as entrusted service income, service fees, transaction fees, and other types of charges, etc. Which are meant to stabilize incomes and lower associated risks for the banks. Therefore, to maintain or even enhance competitive edge in the new environment, most commercial banks now increase NII, and this increasingly has made up a significant proportion of the bank total operating revenue.

This study analyzes the influence of competition and other macro and micro factors on NII of Vietnamese joint stock commercial banks. The structure of this study comprises of five parts. Part 2 summarizes the theoretical background; Part 3 depicts research methods and data; Part 4 presents and discusses research results; Part 5 gives conclusions and offers recommendations.

2 Theoretical Background

A bank's income comes from two main areas of operation, namely interest income and non-interest income. According to Stiroh (2004), non-interest income emanates from many different activities, which could be categorized into four main components fiduciary income, service charges, trading revenue, and other types of fee.

According to Porter (1985), the goal of competition is to win market share, and the nature of competition is to seek profit which is meant to be higher than the average profit that businesses have been earning. As a consequence, competition is regarded as a thrust that determines the success or failure of a business. The competitive strategy is the search for a favorable and sustainable position in its industry Porter (1985).

On the link between competition and non-interest income, Uchida and Tsutsui (2005) show that the competition among urban banks tend to be higher than that in other regions. Casu and Girardone (2009) empirically document a non-linear relationship between the level of competition, the level of concentration and the technical efficiency of Euro-zone commercial banks. Nguyen et al. (2012) opine that commercial banks with better competitive advantage are more likely to earn less

NII than other banks. Besides competition, other factors have been shown to exert considerable impact on bank NII, discussed as follows.

Size is a variable used in most studies on factors affecting the bank income, especially NII. Conventionally, larger banks are prone to gain more NII due to their better economic advantages (Al-Horani (2010), Nguyen et al. (2012), Hakimi et al. (2012), Damankah et al. (2014), Meng et al. (2018), Hamdi et al. (2017)). However, when a bank's size is larger than its management capability, the bank's profitability tends to decrease (Kosmidou et al. (2005), Atellu (2016), Trujillo-Ponce (2013)).

A commercial bank that has higher deposit to total assets could earn higher income due to better access to capital which aids the bank in meeting customer needs of loan. Nevertheless, the increase in revenue from traditional lending service can affect the likelihood of a bank joining activities that generate NII (Stiroh (2004), Shahimi et al. (2006), Aslam et al. (2015)). Net interest margin (NIM) is one of the ratios that are used to measure the performance of commercial banks. The significant impact of this factor is confirmed in Hahm (2008), Nguyen et al. (2012). NIM is calculated as the difference between interest income and interest expense, divided by the total assets of the bank Maudos and De Guevara (2004).

The ratio of loan to total assets indicates how much a bank focuses on its traditional lending activities. Stiroh (2004), Bailey-Tapper (2010), Craigwell and Maxwell (2006) have confirmed the impact of this ratio on NII.

Most studies document that the ratio of loan loss provision to total assets typically leads to negative impact on NII (Hahm (2008), Stiroh (2004), Shahimi et al. (2006)). With regard to ratio of cost to income, banks spend much money for inputs and activities that enhance the quality of their services, thus increasing NII (DeYoung and Roland (2001), Lepetit et al. (2008)). On the other hand, expenses could be considered a performance measure as banks could factor expenses and cost of capital in higher interest income or higher service charges Arora and Kaur (2009). Consequently, a higher ratio of cost to income implies higher NII.

In addition to micro factors, macro determinants could play a role as well in determining NII of banks. Hahm (2008) argues that banks in developed and developing countries with high GDP growth rates have lower NII compared to that in underdeveloped countries. Therefore, banks in underdeveloped markets are motivated to diversify their activities and generate more NII (Hakimi et al. (2012), Atellu (2016)). Controlling inflation is the basic goal of monetary policy that all central banks in the world are concerned about Orphanides and Wieland (2012). Stiroh (2004) argue that in countries with low inflation and high growth rates in capital markets, banks have better conditions to increase the supply of services and products, generating more NII. However, Atellu (2016) shows that inflation has a significant and negative impact on NII in Kenya. The significant impact of this factor suggests that inflation also plays a role in deciding NII of commercial banks in Kenya.

It can be said that the most important goal of central banks in the world is to stabilize the value of the national currency through the control of inflation. The basic interest rate of central banks tends to have impact on the operations of commercial banks, and is positively related to NII Damankah et al. (2014).

Finally, Stiroh (2004) show that NII has a negative correlation with governance quality and bank business strategy. In addition, the competition in the market also has a positive impact on NII. If a country is considered to be good in terms of overall governance, competition will be healthier, forcing banks to diversify their services and increase NII accordingly. According to Kaufmann et al. (2003), Nguyen Manh Hung (2018), Doan Anh Tuan (2018), the WGI national governance index from the World Bank database consists of 6 components, including: Voice and Accountability; Political Stability and Absence of Violence; Government Effectiveness; Regulatory Quality; Rule of Law; Control of Corruption.

3 Research Methodology

3.1 Research Model

Based on research models from competition-focused studies of Davis and Tuori (2000), Casu and Girardone (2009), Nguyen et al. (2012), Koetter et al. (2012), Maudos and Solís (2009), Vo Xuan and Duong Thi Anh (2017), the research model in the present study is as follows:

$$
\begin{aligned}
\text{NII}_{it} = {}& \beta_0 + \beta_1 \text{SIZE}_{it} + \beta_2 \text{DEP}_{it} + \beta_3 \text{NIM}_{it} + \beta_4 \text{LOAN}_{it} + \beta_5 \text{RES}_{it} + \beta_6 \text{COST}_{it} \\
& + \beta_7 \text{INF}_{it} + \beta_8 \text{GDP}_{it} + \beta_9 \text{VAE}_{it} + \beta_{10} \text{PVE}_{it} + \beta_{11} \text{GEE}_{it} + \beta_{12} \text{RQE}_{it} \\
& + \beta_{13} \text{RLE}_{it} + \beta_{14} \text{CCE}_{it} + \beta_{15} \text{COM}_{it} + \beta_{16} \text{HHI}_{it} + u_{it}
\end{aligned}
\tag{1}
$$

where the variables in model (1) are interpreted and measured as follows:

Dependent variable: Non-interest income (NII) is calculated by the ratio of the total non-interest income / Total Assets

Independent variables:

COM: Competition, calculated as follows: ((Interest income and equivalents + income from services) - (interest expenses and equivalents + expenses from services))/(Interest income and equivalents + Income from services)

HHI: Competition, measured using corrected Hirschman-Herfindahl Index (HHI) index (Nguyen Thi Lien and Nguyen Thi Kim (2018), Tibshirani (1996)). The data are taken from the financial reports of Vietnamese commercial banks and the index is calculated using the following formula:

$$
\text{HHI}_{it} = 1 - \left[\left(\frac{\text{INT}_{it}}{\text{TOR}_{it}} \right)^2 + \left(\frac{\text{COM}_{it}}{\text{TOR}_{it}} \right)^2 + \left(\frac{\text{TRA}_{it}}{\text{TOR}_{it}} \right)^2 + \left(\frac{\text{OTH}_{it}}{\text{TOR}_{it}} \right)^2 \right]
\tag{2}
$$

where it is bank i in year t; HHI_{it} is an indicator of income diversification; INT_{it} is the value of net income from interest and equivalents; COM_{it} is the value of net income from banking services; TRA_{it} is the value of net income from trading and investment activities; OTH_{it} is the value of net income from other activities; TOR_{it} is the value of total income from banking operations with:

$$TOR_{it} = INT_{it} + COM_{it} + TRA_{it} + OTH_{it} \tag{3}$$

$$NII_{it} = COM_{it} + TRA_{it} + OTH_{it} \tag{4}$$

Micro and macro variables include: SIZE: Ln (Total assets); DEP: Deposit/Total assets; NIM: (Interest income – Interest expense)/Total assets; LOAN: Total loans/ Total assets; RES: Loan loss provisions/Total assets; COST: Cost/Income; GDP: Annual GDP growth rate; INF: Inflation is measured by the consumer price index (CPI); VAE: Voice and Accountability; PVE: Political Stability and Absence of Violence; GEE: Government Effectiveness; RQE: Regulatory Quality; RLE: Rule of Law; CCE: Control of Corruption.

Data in the study were collected from the financial statements of 27 Vietnamese commercial banks in the period of 2010–2017, totaling 216 observations. The Software Stata 14.0 is used to perform the regression.

3.2 Research Methodology

Identifying the determinants that significantly affect the research subject is not an easy task for researchers. When shortlisting variables for linear models, studies often look at P-values to make a decision. This can be misleading (Kreinovich et al., 2019; Nguyen et al., 2019). For example, researchers could omit important variables that are highly correlated but also have high P-values. On the other hand, irrelevant variables can be included in the model, creating unnecessary complexity in handling the model. In addition, overfitting issue can emerge if the number of observations is smaller than the number of variables in the model.

LASSO, originally proposed by Tibshirani (1996), is an extension of OLS regression which performs both variable selection and regularization through a shrinkage factor. It is capable of enhancing the accuracy and the interpretability compared to classical regression methods Tibshirani (1996). In the same "spirit" of ridge regression, i.e., shrinkage estimation, LASSO is an estimation method for estimating parameters in linear regression models, but by shrinking the parameters with respect to another norm, namely L^1–norm, rather than L^2–norm, mean the L^1, L^2 norm on the space R^n. Where L^2–norm (euclidean distance from the origin) is $\|\beta\|_2 = \sqrt{\sum_{j=1}^{k} \beta_j^2}$ is also known as "L^2 regularization" (making parameters smaller, control parameters, using L^2 norm). Specifically, LASSO offers a solution to the minimization under constraint problem $\min_{\beta \in \mathbb{R}^k} \left[Y - X\beta_2^2 \right]$ subject to $\beta_1 \leq t$. Which the constraint $\beta_1 \leq t$ on the parameters is usually not part of the model (unless there is some prior knowledge on the parameter), but only a statistical device used to improve the MSE of the estimator. The geometry of LASSO explains why LASSO does the covariate selection while performing the estimation. First, the (OLS) objective function Q: $\mathbb{R}^k \to \mathbb{R}$ $Q(\beta) = E\left(Y - X\beta_2^2\right)$ is a quadratic form in β. As such, each level set, i.e., $L_c = \{\beta : Q(\beta) = c\}, c \in \mathbb{R}^+$ is an ellipsoid (this boundary and its interior form a convex set in $\mathbb{R}^k$, let $\tilde{\beta}$ be the point where $Q(\beta)$ is minimum (a global minimum since $Q(\beta)$ is convex in β), with L_c is the minimum value of the LASSO objective function, i.e., the OLS solution. As c gets larger and larger (than $Q(\tilde{\beta})$), the corresponding level sets (ellipsoids), indexed by c, get larger and larger.

Unlike the "sphere" in $\mathbb{R}^k$, using Euclidean distance L^2–norm, i.e., $\{\beta \in \mathbb{R}^k :$

$\|\beta\|_2^2 = t\}$ with $\|\beta\|_2 = \sqrt{\sum_{j=1}^{k} \beta_j^2}$, L^1 –"sphere" (boundary of the L^1 constraint) is

not a sphere (geometrically) but a diamond with "corners". As such, it is possible that a level set L_c could hit a corner first, i.e., the LASSO solution (as an estimate of β) could have some components (estimates of components of the model parameter β) equal to zero exactly. When the LASSO algorithm produces these zero estimates, say, for β_j, the corresponding covariates X_j should be left out, as there is no contribution (to Y) from their part. This is a covariate selection procedure, based on estimation. The other non zero estimates are used to explain the model as well as for prediction of Y on new X. LASSO regression is performed by trading off a small increase in bias for a large decrease in variance of the predictions, hence may improve the overall prediction accuracy.

This study applies LASSO method, because this approach performs well in the presence of multicollinearity problem, and it displays the ideal properties to minimize numerical instability that may occur due to overfitting problem. To improve the research accuracy, LASSO will minimize the parameter estimates to 0 and in some cases, equate the parameters close to zero and thus allow some variables to be excluded from the model.

The results indicate no significant effects of size (SIZE), Deposits/Total Assets (DEP), NIM, LOAN, INF. On the other hand, annual GDP growth rate, the elements of the WGI national governance index (Government Effectiveness (GEE), Rule of Law (RLE); Control of Corruption (CCE)) are significant determinants of NII of Vietnamese commercial banks.

4   Research Results

Research data were collected from the financial statements of 27 Vietnamese commercial banks in the period of from 2010 to 2017 with a total number of 216 observations. Table 1 shows the minimum, maximum, average value and standard deviation of these variables.

The regression results in Table 2 show that the LASSO method (LASSO regression) removes the variables with coefficients equal to or close to zero compared to the RIDGE Regression method. This method also identifies 7 variables affecting NII including Competition (COM), HHI, Loan loss provision/ Total assets

Table 1 Descriptive statistics of variables

Variable	Obs	Mean	Std. Dev.	Min	Max
COM	216	0.335713	0.120153	−0.1156	0.591161
SIZE	216	32.14166	1.087879	29.86479	34.723
DEP	216	0.628259	0.133083	0.25084	0.893717
NIM	216	0.025738	0.011734	−0.00641	0.074219
LOAN	216	0.525889	0.128379	0.147255	0.731251
RES	216	0.005693	0.004596	−0.00485	0.028806
COST	216	0.939057	5.837139	0.000574	86.30244
INF	216	0.068163	0.053266	0.006	0.187
GDP	216	0.061263	0.005218	0.0525	0.0681
VAE	216	−1.41189	0.044595	−1.49698	−1.35879
PVE	216	0.15309	0.0881	−0.02235	0.267359
GEE	216	−0.14568	0.125629	−0.26985	0.067519
RQE	216	−0.5785	0.069701	−0.66871	−0.45394
RLE	216	−0.40769	0.193897	−0.5914	0.048006
CCE	216	−0.49941	0.07789	−0.62358	−0.39632
HHI	216	0.226918	0.210341	−0.73043	0.499991

Source: Author's calculation using Stata 14

Table 2 Regression results using LASSO

Variable	LASSO Regression Coef.	RIDGE Regression Coef.
const	−0.0016165	0.0066218
COM	−0.0016274	−0.0097941
HHI	0.0144259	0.0152904
SIZE		0.0003
DEP		−0.0022993
NIM		0.0441655
LOAN		−0,003264
RES	0.2297043	0.3530947
COST	0.0001352	0.0001714
INF		−0.0234523
GDP		−0.0437131
VAE	−0.0009966	0.000865
PVE	−0.00207	−0.0153063
GEE		−0.0198674
RQE	−0.0031571	0.018926
RLE		0.0001857
CCE		−0.0065995

Source: Author's calculation using Stata 14

(RES), Cost/Income (COST), Voice and Accountability (VAE); Political Stability and Absence of Violence (PVE) and Regulatory Quality (RQE).

5 Conclusion

By performing LASSO algorithm, our study results show that the competition factor has an impact on NII, so in order to increase NII, commercial banks need to diversify products and services, for example, the supportive services for credit activities, to improve and generate more NII. It is necessary to apply modern internet technologies such as cloud computing, Big Data, Internet of Things to help commercial banks reshape their business, governance, and electronic payment systems, aiming at building smart digital banks in the future to increase income, especially value-added services for customers, because this is the NII that Vietnamese commercial banks have not utilized. It can be said that the development of the modern banking service market in Vietnam has had positive changes but is still quite fragmented and low in synchronicity, thus not being able to create sustainable edge for the banks. However, the diversification of product types is always accompanied by an increase in expenses, so managers of commercial banks should have policies to manage and balance costs for this diversification effort. Besides, managers of Vietnamese commercial banks need to improve their credit risk management practices, cut costs, and lower bad debt to increase capital flows, minimize losses and maintain public trust.

For policy makers, it is necessary to stabilize the business environment for commercial banks through regulations and institutions. Measures need to be taken to prevent the manipulation of interest groups in the banking system, to maintain stability and safety of the system, to improve legal frameworks to facilitate healthy competition among commercial banks, contributing to economic stability in general as well as national financial-monetary background in particular in the next periods.

References

Al-Horani, A. (2010) "Testing the relationship between abnormal returns and non-interest earnings: The case of Jordanian commercial banks". *International Research Journal of Finance and Economics*, 55(1), 108–117.

Arora, S., & Kaur, S. (2009) "Internal determinants for diversification in banks in India an empirical analysis". *International Research Journal of Finance and Economics*, 24(24), 177–185.

Aslam, F., Mehmood, B., & Ali, S. (2015). "Diversification in banking: Is noninterest income the answer for pakistan's case". *Science International (Lahore)*, 27(3), 2791–2794.

Atellu, A. R. (2016) "Determinants of non-interest income in Kenya's commercial banks". *Ghanaian Journal of Economics*, 4(1), 98–115.

Bailey-Tapper, S. A. (2010). "Non-interest Income, Financial Performance & the Macroeconomy: Evidence on Jamaican Panel Data". *Bank of Jamaica (BOJ) Working Paper*.

Carbo-Valverde, S., Rodriguez-Fernandez, F., & Udell, G. F. (2009) "Bank market power and SME financing constraints". *Review of Finance*, 13(2), 309–340.

Casu, B. & Girardone, C. (2009). "Testing the relationship between competition and efficiency in banking: A panel data analysis". *Economics Letters*, *105*(1), 134–137.

Chiorazzo, V., Milani, C., & Salvini, F. (2008). "Income diversification and bank performance: Evidence from Italian banks". *Journal of Financial Services Research*, *33*(3), 181–203.

Craigwell, R., & Maxwell, C. (2006). "Non-interest income and financial performance at commercial banks in Barbados". *Savings and Development*, 3, 309–328.

Damankah, B. S., Anku-Tsede, O., & Amankwaa, A. (2014). "Analysis of non-interest income of commercial banks in Ghana", International Journal of Academic Research in Accounting. *Finance and Management Sciences*, *4*(4), 263–271.

Davis, E. P., & Tuori, K. (2000). "The changing structure of banks' income-an empirical investigation". *Brunel University, Department of Economics and finance. Working Paper*, 2, 16–10.

De Young, R., & Roland, K. P. (2001) "Product mix and earnings volatility at commercial banks: Evidence from a degree of total leverage model". *Journal of Financial Intermediation*, *10*(1), 54–84.

Hahm, J.-H. (2008). "Determinants and consequences of non-interest income diversification of commercial banks in OECD countries". *East Asian Economic Review*, *12*(1), 3–31.

Hakimi, A., Hamdi, H., & Djelassi, M. (2012). "Modelling non-interest income at Tunisian banks", 2, 88–99.

Hamdi, H., Hakimi, A., & Zaghdoudi, K. (2017) "Diversification, bank performance and risk: have Tunisian banks adopted the new business model?". *Financial innovation*, *3*(1) 22.

Kaufmann, D., Kraay, A., & Mastruzzi, M. (2003). *Governance matters III: Governance indicators for 1996–2002*. The World Bank.

Koetter, M., Kolari, J. W., & Spierdijk, L. (2012). "Enjoying the quiet life under deregulation? Evidence from adjusted Lerner indices for US banks". *Review of Economics and Statistics*, *94*(2), 462–480.

Kosmidou, K., Tanna, S., & Pasiouras, F. (2005) *Determinants of profitability of domestic UK commercial banks: panel evidence from the period 1995–2002*. In *Money Macro and Finance (MMF) Research Group Conference*.

Kreinovich, V., Thach, N. N., Trung, N. D., & Thanh, D. V., Eds. (2019). *Beyond traditional probabilistic methods in economics*. Cham: Springer. https://doi.org/10.1007/978-3-030-04200-4.

Lepetit, L., et al. (2008) "Bank income structure and risk: An empirical analysis of European banks". *Journal of banking & finance*, *32*(8), 1452–1467.

Maudos, J., & Solís, L., (2009). "The determinants of net interest income in the Mexican banking system: An integrated model". *Journal of Banking & Finance*, *33*(10), 1920–1931.

Maudos, J. n., & De Guevara, J. F. (2004). "Factors explaining the interest margin in the banking sectors of the European Union". *Journal of Banking & Finance*, *28*(9), 2259–2281.

Meng, X., Cavoli, T., & Deng, X. (2018). "Determinants of income diversification: evidence from Chinese banks". *Applied Economics*, *50*(17), 1934–1951.

Nguyen, M. H. (2018). "Quan tri quoc gia va nhung goi mo cho tien trinh cai cach the che kinh te thi truong o Viet Nam". *Tap chi Khoa hoc DHQGHN: Kinh te va Kinh doanh*, *34*(27), 24–31.

Nguyen, H. T., Trung, N. D., & Thach, N. N. (2019). Beyond traditional probabilistic methods in econometrics. In V. Kreinovich, N. Thach, N. Trung & D. Van Thanh (Eds.), *Beyond traditional probabilistic methods in economics*. ECONVN 2019. Studies in Computational Intelligence, vol. 809. Cham: Springer. https://doi.org/10.1007/978-3-030-04200-4_1.

Nguyen Thi Lien Hoa, & Nguyen Thi Kim Oanh, (2018). "Da dang hoa thu nhap va rui ro cua he thong ngan hang thuong mai - bang chung thuc nghiem tai Viet Nam". *Ky yeu hoi thao khoa hoc An ninh tai chinh cua Viet Nam trong hoi nhap quoc te*, 213–229. ISBN:978-604-922-620-5.

Nguyen, M., Skully, M., & Perera, S. (2012). "Market power, revenue diversification and bank stability: Evidence from selected South Asian countries", Journal of International Financial Markets. *Institutions and Money*, *22*(4), 897–912.

Orphanides, A., & Wieland, V. (2012). "Complexity and monetary policy". *CEPR Discussion Paper*, 9, 167–204.

Porter, M. E. (1985). *The Competitive advantage: Creating and sustaining superior performance.* NY: Free Press.

Shahimi, S., et al. (2006). "A panel data analysis of fee income activities in Islamic banks". *Journal of King Abdulaziz University: Islamic Economics19*(2).

Stiroh, K. J. (2004). "Diversification in banking: Is noninterest income the answer?", Journal of Money. *Credit and Banking, 36*(5), 853–882.

Tibshirani, R. (1996). "Regression shrinkage and selection via the lasso". *Journal of the Royal Statistical Society: Series B (Methodological), 58*(1), 267–288.

Trujillo-Ponce, A. (2013). "What determines the profitability of banks? Evidence from Spain". *Accounting & Finance, 53*(2), 561–586.

Tuan, D. A. (2018) " Tac dong cua bat dinh chinh tri den hieu qua cua ngan hang thuong mai tai cac nen kinh te moi noi". *Tap chi khoa hoc dai hoc Da Lat, 8*(1S), 103–117.

Uchida, H. & Tsutsui, Y. (2005) "Has competition in the Japanese banking sector improved?". *Journal of Banking & Finance, 29*(2), 419–439.

Vinh, V. X., & Tien, D. T. A. (2017). "Cac yeu to anh huong den suc canh tranh cua cac ngan hang thuong mai Viet Nam". *Tap chi Khoa hoc DHQGHN: Kinh te ve Kinh doanh, 33*(1), 12–22.

Reconsidering Hofstede's Cultural Dimensions: A Different View on South and Southeast Asian Countries

Nguyen Ngoc Thach, Tran Hoang Ngan, Nguyen Tran Xuan Linh, and Ong Van Nam

Abstract This research aims to evaluate the impact of national wealth measured by real GDP per capita on the changes in cultural values in 12 selected countries in the South and Southeast Asian region. By applying the Metropolis-Hastings (MH) method, the findings show that national wealth has a nonlinear effect on cultural values expressed in Hofstede's Cultural Dimensions theory. Interestingly, in contrast to previous studies, this research finds that national wealth has a positive impact on Masculinity.

Keywords Bayesian analysis · Metropolis-Hasting method · Hofstede's cultural dimensions

1 Introduction

How does economic development affect cultural values and change cultural values? This issue has received a lot of attention not just from scientists but from several policymakers. Economic development not only creates the material foundation to

N. N. Thach (✉)
Institute for Research Science and Banking Technology, Banking University HCMC, 39 Ham Nghi Street, District 1, Ho Chi Minh City, Vietnam
e-mail: thachnn@buh.edu.vn

T. H. Ngan
HCMC Institute for Development Studies, 149 Pasteur, District 3, Ho Chi Minh City, Vietnam
e-mail: ngankdtt@hotmail.com

N. T. X. Linh (✉)
Ho Chi Minh Industry and Trade College, 20 Tang Nhon Phu, District 9, Ho Chi Minh City, Vietnam
e-mail: xuanlinh86@gmail.com

O. V. Nam
Banking University HCMC, 39 Ham Nghi Street, District 1, Ho Chi Minh City, Vietnam
e-mail: namov@buh.edu.vn

© The Editor(s) (if applicable) and The Author(s), under exclusive license to Springer Nature Switzerland AG 2021
N. Ngoc Thach et al. (eds.), *Data Science for Financial Econometrics*, Studies in Computational Intelligence 898,
https://doi.org/10.1007/978-3-030-48853-6_21

enhance the cultural values and build new values for a country but also acts as the spiritual foundation to foster cultural growth. Several studies looking into the effect of economic development on cultural changes confirm such an impact (McClelland 1961; Inglehart 1977; Altman 2001). However, some researchers believe that economic development has only a weak correlation with cultural values (Yeh and Lawrence 1995; Smith and Bond 1998). To sum up, the relationship between economic development and cultural values has not yet been established by research. In addition, since most studies were done in countries other than Asia, especially outside South and Southeast Asia, more studies in this geographic area are required to corroborate the research findings.

This study supports the hypothesis that the economic development of a country affects the cultural values there. It employs Hofstede's Cultural Dimensions theory. The data obtained from the World Bank's Gross Domestic Product (GDP) per capita were used to see if there was an impact of economic development on cultural changes in South and Southeast Asian countries. The results confirm that economic development has an impact on the cultural values in these Asian countries. Also, the GDP per capita in this region is positively correlated with the Masculinity Index and Indulgence versus Restraint Index; yet, after GDP per capita reaching a threshold, the result is contrary to the findings of some previous studies.

2 Theoretical Framework and Hypotheses

2.1 Hofstede's Cultural Dimensions Theory

Hofstede's Cultural Dimensions Theory is a framework for cross-cultural communication. It describes the influence of culture on community members and their behavior. Hofstede adopted this framework between 1967 and 1973 to analyze an IBM survey on human resources worldwide. In the study, he proposed six cultural dimensions such as Power distance—PDI, Individualism—IDV, Uncertainty avoidance—UAI, Masculinity—MAS, Long-term orientation—LTO, and Indulgence versus Restraint—IND.

Later on, several studies corroborated the value of this theoretical framework. Trompenaars (1993) and Smith et al. (2006) maintain that Hofstede's cultural dimensions are stronger indicators of cultural values than many others. According to Sojerd and Chris (2018), Hofstede's framework is a widely adopted one for cross-cultural comparisons. Together with Inglehart's, Hofstede's works have been cited more than 200,000 times, making them the most frequently cited frameworks in social sciences. The present study uses Hofstede's cultural dimensions as a basis for data analysis.

2.2 The Relationship Between Economic Performance and Cultural Values

The relationship between economic development and cultural values began to gain attention when Weber published his thesis entitled "Protestant ethics" in 1930; he claimed that Protestant ethics enhanced the development of capitalism as they emphasized savings, hard work, and accumulation of human capital (Andersen et al. 2013; Becker and Woessman 2009). Using the data obtained from 20 countries, Frake et al. (1991) confirmed that cultural values played an important role in economic performance. Contrary to these views, Marx's economic determinism postulated that economic forces defined and regulated all aspects of civilization including social, cultural, intellectual, political, and technological. Amariglio and Callari (1989) also conformed to this view of economic determinism where economic factors served as the foundation for all the other societal and political arrangements in society. Tang and Koveos (2008) also found evidence of the nonlinear relationship between national assets measured through GDP per capita and cultural values with internal factors as found in Hofstede's cultural dimensions.

2.3 Research Hypotheses

Based on the findings of Tang and Koveos (2008), the present study proposes the following hypotheses:

Hypothesis 1: Per capita economic growth increases the Power Distance Index, and after the income reaching a certain level, the Power Distance Index starts decreasing.

Hofstede (2001) argues that the middle class acts as a bridge between the powered and the powerless groups. According to Adelman and Morris (1967), the emergence of the middle class helps to reduce the power gap. However, if the income inequality is larger, the size of the middle class reduces in scope and the Power distance increases further.

The present paper tries to clarify how economic development changes the income gap. The Kuznets Curve theory (1955) is used to explain this and it suggests that various indicators of economic disparity tend to get worse with economic growth until the average income reaches a certain point. The impacts of savings accumulation of the upper class and urbanization increase the income gap in the early stages of economic development. However, the inequality diminishes as the economy grows, due to legal interventions such as tax policies, government subsidies or inflation (which reduces the value of savings), and technological development. It is also noted that economic development increases the Power distance until the GDP per capita reaches a certain threshold, after which the distance starts decreasing.

Hypothesis 2: Individualism first decreases and then increases with economic development.

A number of studies looking into the changes in Individualism agree that economic development is an important influential factor (Ogihara 2017). The economic environment significantly affects the behavior of the public. When there is a lack of resources needed for economic growth, it is difficult to implement any task on one's own and people tend to rely on others. The rejection by others greatly affects one's survival and, therefore, people often try to avoid such a situation. Thus, collectivism tends to dominate under difficult economic conditions. In contrast, having abundant resources allows individuals more freedom to pursue their own interests and reduces their need to rely on others. That is when the economy has not yet developed, individualism is not very visible. However, as the economy grows to a certain level, people tend to pursue their own interests more and Individualism appears to be more embraced.

Hypothesis 3: GDP growth per capita increases Uncertainty avoidance but after the income reaches a certain threshold, the Uncertainty avoidance starts decreasing.

According to Hofstede (1980), Uncertainty avoidance has three components: rule orientation, job stability, and stress. During the transition from an agricultural economy to an industrial economy, farmers had to give up their land to make room for factories, which in turn, urged them to earn a stable job for their survival. However, as national assets continue to increase and social structure gets more complex, the concerns about stability decrease. According to Inkeles (1993), as the quality of life has improved by industrialization and modernization, many in the developed countries tend to be less worried, but feel bored with their stable lives. The hypothesis is that after the income reaching a certain level, risk aversion decreases.

Hypothesis 4: GDP growth per capita reduces Masculinity.

Hofstede (2001) concluded that there was no correlation between Masculinity and economic development. However, several researchers have later on rejected this finding. The studies by Duflo (2012) and IMF (2013) confirm that feminism and economic growth have a two-way relationship. Economic growth helps enhance human rights and thus, women become more empowered. When more women join the workforce, they have a chance to realize their potential in the workplace. In the present study, the hypothesis is that per capita growth reduces the Masculinity Index.

Hypothesis 5: Long-term orientation (LTO) decreases at first and then increases with GDP per capita growth.

According to Read (1993), LTO has a strong correlation with national savings. In the present study, the attempt is to explain how economic development affects LTO through savings. According to the Life-cycle hypothesis (Read 1993), individuals from highly developed economies tend to increase current consumption and cut savings by expecting higher incomes in the future. In addition, the development of financial markets and fewer lending restrictions reduce the need for savings. In other words, higher-income individuals tend to have less savings. However, an individual's saving behavior is also determined by anthropological factors. When economic development improves the living conditions and increases the life expectancy in developed countries, personal savings increase too as people try to save more for retirement. That is, when the economy grows to a specific level, people tend to increase their savings.

Hypothesis 6: The higher the GDP per capita growth is, the higher the Index of Indulgence versus Restraint gets.

In 2010, Indulgence versus Restraint was added to Hofstede's Cultural Dimensions theory as the sixth dimension. Indulgence societies allow more freedom and satisfaction based on basic human needs. In contrast, restrained societies inhibit human satisfaction and enjoyment of life by imposing strict ethical standards on their individual members. People of indulgence societies tend to be more optimistic while people of restrained societies are more pessimistic and introverted. Using research data from 495,011 participants from 110 countries, Sojerd and Chris (2018) argued that people with high Indulgence versus Restraint (IVR) index were to choose big cities to reside in. It is more important for these people to live a meaningful life than to find a high paying job. So the hypothesis for the present study is that economic growth increases the IVR index.

3 Methodology and Data

3.1 Research Method

While using the p-value to test hypotheses has faced severe criticism (Briggs and Nguyen 2019); Owing to the fast development of computer science and informative technology for recent decades, Bayesian methods have been applied more and more widely (see, for example., Nguyen and Thach 2018; Anh et al. 2018; Nguyen and Thach 2019; Nguyen et al. 2019a, b; Sriboonchitta et al. 2019; Svítek et al. 2019; Kreinovich et al. 2019; Thach et al. 2019; Thach 2020a, b). Thach (2019a) also pointed that in a model which has many explanatory variable has more than one model capable interpreting the variables. Therefore, if only one model selected, it will lead to high risk that the model is not suitable. To overcome this drawback, Bayes statistic will evaluate all possible models in the model space and then summarize the statistical results based on the weight of model in the model space. With a small sample size, Bayesian statistics that combines observed data with prior information could increase the robustness of inferential results. This particular study uses Bayesian linear regression with the Metropolis-Hastings (MH) method.

The MH algorithm proceeds through three main loop stages:

Stage 1: Create the sample (proposal) values $a^{proposal}$ for each random variable. The proposal distribution for this sample is $q(.)$

$$a^0 \sim q(a)$$

where a^0 is the initial variable.

This sample from the proposal distribution $q(a^{(x)}|a^{(x-1)})$

$$a^{proposal} \sim q(a^{(x)}|a^{(x-1)})$$

Now, perform the iteration of the sample value with $x = 1, 2, \ldots$

A proposal distribution is a symmetric distribution if $q(a^{(x-1)}|a^{(x)}) = q(a^{(x)}|a^{(x-1)})$.

Stage 2: Compute the acceptance probability. Carry out the MH acceptance function to ensure a balance between the following two constraints:

$$\frac{\pi(a^{proposal})}{\pi(a^{(x-1)})} \tag{1}$$

This constraint aims that the sample is in a higher probability area under full joint density.

$$\frac{\pi(a^{(x-1)}|a^{proposal})}{\pi(a^{proposal}|a^{(x-1)})} \tag{2}$$

The second constraint makes the sample explore the space and avoid getting stuck at one site; as this sample can reverse its previous one move in the space.

MH acceptance function must have this particular form to guarantee that the MH algorithm satisfies the condition of detailed balance because the stationary distribution of the MH algorithm is, in fact, the target posterior, which is relevant here (Gilks et al. 1996).

The acceptance function in the case of symmetric proposals is:

$$\theta\left(a^{(x)}|a^{(x-1)}\right) = \min\left\{1, \frac{q\left(a^{(x-1)}|a^{(x)}\right)\pi\left(a^{(x)}\right)}{q\left(a^{(x)}|a^{(x-1)}\right)\pi\left(a^{(x-1)}\right)}\right\} \tag{3}$$

where θ is acceptance probability, $\pi(.)$ is the full joint density. The 'min' operator makes sure that the acceptance probability θ is never larger than 1.

When the proposal distribution is symmetric $q(a^{(x-1)}|a^{(x)}) = q(a^{(x)}|a^{(x-1)})$, rewrite the acceptance function as:

$$\theta\left(a^{(x)}|a^{(x-1)}\right) = \min\left\{1, \frac{\pi\left(a^{(x)}\right)}{\pi\left(a^{(x-1)}\right)}\right\} \tag{4}$$

Stage 3: During this final stage, rely on the acceptance probability θ, and decide to accept or reject a proposal. If a proposal number is smaller than θ, accept the value; else, reject it.

The Bayesian statistics relies on the assumption that all model parameters are random and can be in conjunction with prior information. This assumption is in contrast to the traditional approach, also known as frequentist inference, where all the parameters are considered unknown fixed quantities. The Bayesian analysis applies the Bayes rule to create formalism for incorporating the prior information, with evidence from available data. This creates a posterior distribution of model parameters by using the Bayes rule. The posterior distribution results are derived from the observed data of the model by updating the data with prior knowledge:

$$posterior \propto likelihood \times prior$$

One of the advantages of Bayesian statistics is that the results of the research do not significantly rely on the sample size.

Based on the hypotheses proposed in Sect. 2.3, there are six models:

$$Model\ 1 : PDI = \alpha_0 + \alpha_1 \log GDPer + \alpha_2 \log GDPer^2 + \varepsilon_1 \qquad (5)$$

$$Model\ 2 : IDV = \beta_0 + \beta_1 \log GDPer + \beta_2 \log GDPer^2 + \varepsilon_2 \qquad (6)$$

$$Model\ 3 : UAI = \gamma_0 + \gamma_1 \log GDPer + \gamma_2 \log GDPer^2 + \varepsilon_3 \qquad (7)$$

$$Model\ 4 : MAS = \delta_0 + \delta_1 \log GDPer + \delta_2 \log GDPer^2 + \varepsilon_4 \qquad (8)$$

$$Model\ 5 : LTO = \lambda_0 + \lambda_1 \log GDPer + \lambda_2 \log GDPer^2 + \varepsilon_5 \qquad (9)$$

$$Model\ 6 : LTO = \rho_0 + \rho_1 \log GDPer + \rho_2 \log GDPer^2 + \varepsilon_6 \qquad (10)$$

where PDI, IDV, UAI, MAS, LTO, and IVR are dependent variables to explain Hofstede's cultural dimensions, such as the Power distance, Individualism, Uncertainty avoidance, Masculinity, Long-term orientation, and the Indulgence versus Restraint respectively. The independent variable $logGDPer$ is the log of the average GDP per capita of the country. The average GDP per capita is converted into a logarithm to standardize the highly skewed economic variable to make it more amenable to least-square statistics, given that Hofstede's cultural scores have been normalized from 0 to 120. To explore the nonlinear relationship between national wealth and cultural values, the square of the log of the average GDP per capita of the country is implemented in each equation.

Because the prior studies were performed in the Frequentist framework, information about the model parameters from those studies is not available. Besides, as the data in these particular models are sufficiently large, prior information may have little impact on the robustness of the models. Therefore, in this particular study, three candidate simulations are built for each model above, such as completely non-informative, mildly informative, and more strongly informative priors. The first simulation applies flat prior, a completely non-informative prior for coefficient parameters; this is the most widely used non-informative prior in Bayesian analysis when a researcher has no prior knowledge. The second simulation incorporates Zellner's g-prior, a fairly informative prior. A normal prior is implemented in the third simulation for the strongly informative prior. After fitting the three regression models, a Bayesian factor test and a Bayes test model are used to select the most proper model.

Simulation 1 (for the first model presenting the relationship between PDI and GDP per capita) is as given below:

Likelihood model:

$$PDI \sim (\mu, \sigma^2)$$

Prior distributions:

$$\alpha_{LogGDP} \sim 1(flat)$$

$$\alpha_{Log2GDP} \sim 1(flat)$$

$$\alpha_{cons} \sim 1(flat)$$

$$\sigma^2 \sim jeffreys$$

where μ is the mean of the normal distribution of PDI, α is a vector of coefficients, σ^2 is the variance for the error term.

Simulation 2 is expressed as follows:

Likelihood model:

$$PDI \sim (\mu, \sigma^2)$$

Prior distributions:

$$\alpha | \sigma^2 \sim \text{zellnersg(dimension, df, priormean,} \sigma^2)$$

$$\sigma^2 \sim \text{Invgamma}\left(\frac{v_0}{2}, \frac{v_0 \sigma_0^2}{2}\right)$$

where v_0 is df (*prior degree of freedom*) and σ_0^2 is the residual of MS. The number of *dimensions, df, prior mean,* σ_0^2 and v_0 are obtained from the Ordinary least squares (OLS) regression results (Table 1).

Therefore, the prior distributions are as follows:

$$\alpha | \sigma^2 \sim \text{zellnersg}(3, 11, 131, -18, -157)$$

$$\sigma^2 \sim \text{Invgamma}(50.5, 7420.5)$$

Table 1 The results of OLS regression

Source	SS	df	MS	Number of obs = 12	
Model	366.71776	2	183.35888	F(2, 9) = 0.94	
Residual	1759.53224	9	195.503582	Prob > F = 0.4266	
Total	2126.25	11	193.295455	R-squared = 0.1725	
				Adj R-squared = −0.0114	
				Root MSE = 13.982	
PDI	Coef.	Std. err.	t	P > |t|	[95% Conf. interval]
LogGUP	131.0029	161.2791	0.81	0.438	−233.8356	495.8415
Log2GUP	−18.08308	20.8936	−0.87	0.409	−65.34768	29.18153
_cons	−156.5999	306.5653	−0.51	0.622	−850.0987	536.899

Source The authors' calculation

Simulation 3 is given as below:
Likelihood model:

$$PDI \sim (\mu, \sigma^2)$$

Prior distributions:

$$\alpha \sim N(1, 100)$$

$$\sigma^2 \sim \text{Invgamma}(20.5, 20.5)$$

The relationship between Hofstede's other five cultural dimensions and the GDP per capita can be modeled in the same way as above.

3.2 Data

The independent variables of the above equations are the logs of the average real GDP per capita of the country between 1990 and 1994 in US dollars (a constant as evaluated in 2011) for 12 South and Southeast Asian countries as in other studies, for example, Thach (2019b). These countries are Bangladesh, China, Hong Kong, India, Indonesia, Japan, Malaysia, Philippines, Korea Republic, Singapore, Thailand, and Vietnam. The data on GDP per capita of these countries were obtained from the World Development Indicators database (World Bank 2019). The data for the dependent variables for the period 1995–2004 were taken from Hofstede's work (2010). To evaluate the effects of real GDP per capita on cultural values, the data from the period 1990 to 1994 were used.

4 Empirical Results

4.1 Bayesian Simulations and Model Comparison

Table 2 summarizes the regression results of the first model describing the relationship between PDI and GDP per capita with three various priors.

After running regressions, the Bayesian factor test was performed, the result of which is demonstrated in Table 3.

According to Bayesian factor analysis, Simulation 1 with the highest Log (BF) and the highest Log marginal-likelihood (Log (ML)) was chosen (Table 3). In addition, the results of the Bayes model test also indicated that Simulation 1 had the highest posterior probability. Based on the comparison results, Simulation 1 was selected.

Table 2 The summary of regression results of model 1

	Mean	Std. Dev.	MCSE	Median	Equal-tailed [95% Cred. Interval]	
Simulation 1						
LogGDP	143.3986	183.4104	12.3899	147.9821	−230.5211	523.1816
Log2GDP	−19.71798	23.76702	1.60577	−19.94787	−68.29913	28.96438
cons	−179.4396	348.4736	23.5402	−192.2074	−887.7105	530.9829
σ^2	249.2538	153.4246	9.61718	209.5108	92.97592	639.7085
Simulation 2						
LogGDP	133.1992	137.6333	6.04367	129.7112	−147.3987	407.9753
Log2GDP	−18.38607	17.86245	0.789044	−17.78779	−53.54094	17.92186
cons	−160.5811	260.8665	11.3829	−157.4088	−689.6474	361.8393
σ^2	153.8164	49.35536	2.6781	144.0998	86.18076	278.4087
Simulation 3						
LogGDP	25.40806	7.649164	0.272748	26.07586	8.737797	39.00468
Log2GDP	−2.537541	1.844854	0.065687	−2.663258	−5.750307	1.436138
cons	13.96978	9.516499	0.395082	14.15636	−4.547108	33.06332
σ^2	194.2673	98.48264	2.67698	170.4605	78.40842	448.4162

Source The authors' calculation

Table 3 The summary of the Bayesian factor test and Bayes test model for model 1

	log(BF)	log(ML)	P(M—y)
Simulation 1		−36.1964	1
Simulation 2	−14.428	−50.6244	0
Simulation 3	−31.2398	−67.4362	0

Source The authors' calculation

The two important criteria to measure the efficient Markov chain Monte-Carlo (MCMC) are acceptance rate and extended auto-correlation (mixing). Convergence diagnostics of MCMC makes sure that Bayesian analysis based on an MCMC sample is valid. To detect convergence issues, the virtual inspection can be used. Graphical convergence diagnostics mainly consists of a trace plot, an auto-correlation plot, a histogram, and a kernel density estimate overlaid with densities approximated using the first and the second halves of the MCMC sample. All estimated parameters in these graphs are relatively valid. Trace plots and auto-correlation plots represent low auto-correlation; the shape of the graph is unimodal; the histograms and kernel density plots show normal distribution. Testing the convergence of PDI is a typical example (Fig. 1).

Figure 2 also demonstrates a fairly good mixing (low auto-correlation) of the MCMC chain of the variance. The graphical diagnostics looks reasonable with the chain traversing rather quickly and the auto-correlation dying off after about 25 lags; the histogram matches the density.

Fig. 1 Convergence test for the parameters of model 1. *Source* The authors' calculation

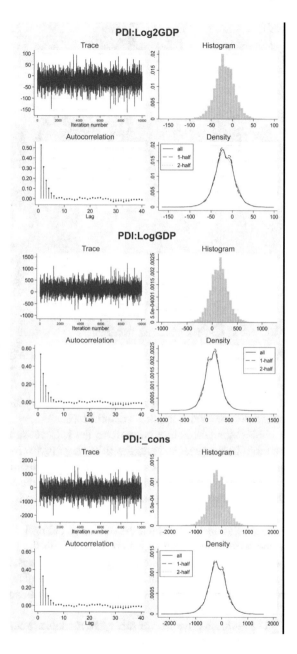

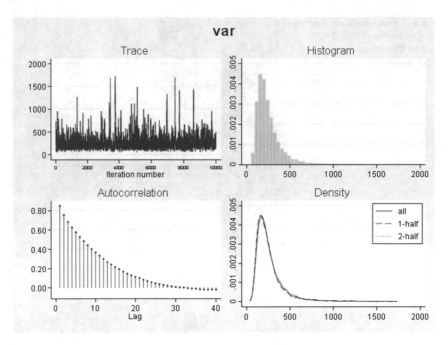

Fig. 2 Convergence test for the variance of model 1. *Source* The authors' calculation

Besides the visual diagnostics for the convergence of the MCMC chains, the effective sample size applies a helpful tool.

ESS estimates cannot be close to the MCMC sample size when using the MH method. In the present case, the ESS value is 5% higher than the MCMC sample size. Therefore, it can be said that low auto-correlation exists in the MCMC sample.

From the results of testing the convergence of the chains, it can also be concluded that the MCMC chains converge with the target distribution; hence, the Bayesian inference is valid.

The relationship between the other five cultural dimensions and GDP per capita is modeled in the same way as it is done with the linkage between PDI and GDP per capita. For Hofstede's other five cultural indices, the results of the Bayes factor test and Bayes model test also support the simulations which have flat prior for coefficients and *Jeffreys* for variance, in which the non-informative priors are used (Table 4).

The regression results for other five models are summarized in Table 5.

According to the results of the Bayesian analysis, the acceptance rate ranges from 0.248 to 0.341 and the values are reasonable. Besides, the results of testing the convergence of the models are supportive of Bayesian inference.

Table 4 The result of Bayesian factor test and Bayes model test of other models

	Prior	log(BF)	log(ML)	P(My)
Model 2				
Simulation 4	$\beta_i \sim 1(flate)$; $\sigma^2 \sim jeffreys$	.	-34.65941	1.0000
Simulation 5	$\beta_i \sim zellnersg(3, 11, -73, 9, 161, \sigma^2)$ $\sigma \sim igamma(5.5, 764.5)$	-24.07818	-58.73759	0.0000
Simulation 6	$\beta_i \sim N(1, 100)$ $\sigma^2 \sim igamma(2.5, 2.5)$	-24.49135	59.15076	0.0000
Model 3				
Simulation 7	$\gamma_i \sim 1(flat)$; $\sigma^2 \sim jeffreys$	.	-41.47065	1.0000
simulation 8	$\gamma_i \sim zellnersg(3, 11, 280, -35, -491, \sigma^2)$ $\sigma^2 \sim igamma(5.5, 3602.5)$	-16.72688	-58.19753	0.0000
Simulation 9	$\gamma_i \sim N(1, 100)$ $\sigma^2 \sim igamma(2.5, 2.5)$	-30.46374	-71.93439	0.0000
Model 4				
Simulation 10	$\delta_i \sim 1(flat)$; $\sigma^2 \sim jeffreys$	.	-38.23623	1.0000
Simulation 11	$\delta_i \sim zellnersg(3, 11, -216, 29, 452, \sigma^2)$ $\sigma^2 \sim igamma(5.5, 1474)$	-14.55465	-52.79088	0.0000
Simulation 12	$\delta_i \sim N(1, 100)$ $\sigma^2 \sim igamma(2.5, 2.5)$	-27.46756	-65.70379	0.0000
Model 5				
Simulation 13	$\lambda_i \sim 1(flat)$; $\sigma^2 \sim jeffreys$	.	-40.83935	1.0000
Simulation 14	$\lambda_i \sim zellnersg(3, 11, -246, 34, 501, \sigma^2)$ $\sigma^2 \sim igamma(5.5, 2849)$	-15.61166	-56.45101	0.0000
Simulation 15	$\lambda_i \sim N(1, 100)$ $\sigma^2 \sim igamma(2.5, 2.5)$	-29.16517	-70.00452	0.0000
Model 6				
Simulation 16	$\rho_i \sim 1(flat)$; $\sigma^2 \sim jeffreys$	.	-34.17845	1.0000
Simulation 17	$\rho_i \sim zellnersg(3, 11, 202, -25, -363, \sigma^2)$ $\sigma^2 \sim igamma(5.5, 676.5)$	-25.19841	-59.37686	0.0000
Simulation 18	$\rho_i \sim N(1, 100)$ $\sigma^2 \sim igamma(2.5, 2.5)$	-25.64125	-59.8197	0.0000

Where $i = 1, 2, 3$

Source The authors' calculation

Table 5 Regression results of the relationship between Hofstede's cultural dimensions (IDV, UAI, MAS, LTO, IVR) and GDP per capita

	Mean	Std. Dev.	MCSE	Median	Equal-tailed [95%Cred. Interval]	
Simulation 4						
LogGDP	−70.17924	151.004	2.39763	−68.59678	−377.1376	228.1945
Log2GDP	9.298309	19.54332	0.310523	9.050352	−29.34523	48.98223
cons	155.6332	287.3312	4.55867	152.6021	−414.5288	737.9677
σ^2	176.4907	96.36046	4.26546	151.872	66.80361	431.5422
Acceptance rate	0.341					
Efficiency: min	0.1483					
Simulation 7						
LogGDP	271.4475	331.5423	11.7844	276.1378	−381.317	915.5487
Log2GDP	−34.74463	43.03741	1.5203	−35.5799	−118.4917	50.27445
cons	−473.966	629.0669	22.3907	−485.4775	−1693.688	774.4687
σ^2	348.9132	213.6541	18.4505	286.0542	123.7431	945.4992
Acceptance rate	0.3038					
Efficiency: min	0.04564					
Simulation 10						
LogGDP	−214.8275	215.609	4.65206	−215.7055	−640.1963	228.492
Log2GDP	28.55694	27.8615	0.585953	28.58871	−28.19484	83.78117
cons	450.5367	410.8941	9.05473	450.7609	−395.6796	1266.812
σ^2	336.8769	190.8143	5.41454	288.0228	125.9904	863.2599
Acceptance rate	0.2927					
Efficiency: min	0.1242					
Simulation 13						
LogGDP	−227.3963	300.5612	5.689	−229.2711	−826.3232	385.9729
Log2GDP	31.2428	38.91391	0.743218	31.44537	−48.55419	108.9968
cons	465.2123	571.682	10.6735	464.0643	−698.6246	1601.426
σ^2	668.9798	416.2782	17.2837	556.7715	245.2968	1827.233
Acceptance rate	0.323					
Efficiency: min	0.09642					
Simulation 16						
LogGDP	203.325	147.1136	1.99131	202.8919	−86.62175	498.5326
Log2GDP	−25.33412	19.05639	0.258854	−25.3907	−63.65623	12.21917
cons	−365.3786	279.5425	3.78246	−364.0865	−925.8528	185.3921
σ^2	162.5102	103.4415	2.93446	134.8493	59.48839	444.4597
Acceptance rate	0.2665					
Efficiency: min	0.1243					

Source The authors' calculation

4.2 Discussion

The scatter plots of (log) average of GDP per capita in US dollars (a constant as evaluated in 2011) and Hofstede's cultural dimensions were examined in order to confirm the above results of the Bayesian simulation. Interestingly, these plots also show that GDP per capita has a nonlinear relationship with Hofstede's cultural scores (Figs. 3, 4, 5, 6, 7 and 8).

Figure 3 shows the relationship between the Power Distance Index and the log average of GDP per capita. According to Hypothesis 1, PDI first rises and then, after the income reaches a certain level, decreases with the log average of the GDP per capita. The graph in Fig. 3 demonstrates an upward trend first and then a downward one as expected. It can be explained that the rise in globalization and integration in the high-income countries has resulted in the interaction among their national cultures. The higher the income is, the more opportunities people get to come in contact with different cultures. People tend to absorb more external cultural values and have an increased balance in life and more satisfaction. This results in a lower Power Distance Index.

Figure 4 represents the relationship between the Individualism index and the log average of GDP per capita. It shows that Individualism first decreases and then increases with economic development, as discussed in Hypothesis 2. After overcoming a certain threshold, the economies further develop coupled with more open-door policies, which enable more people to expose themselves to western cultures, and personal freedom and self-actualization greatly are enhanced in the societies.

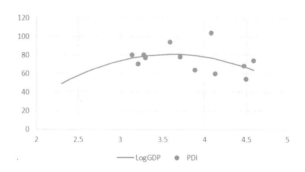

Fig. 3 Power distance and log average GDP per capital. *Source* The authors' calculation

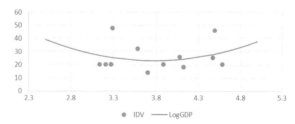

Fig. 4 Individualism and log average GDP per capital. *Source* The authors' calculation

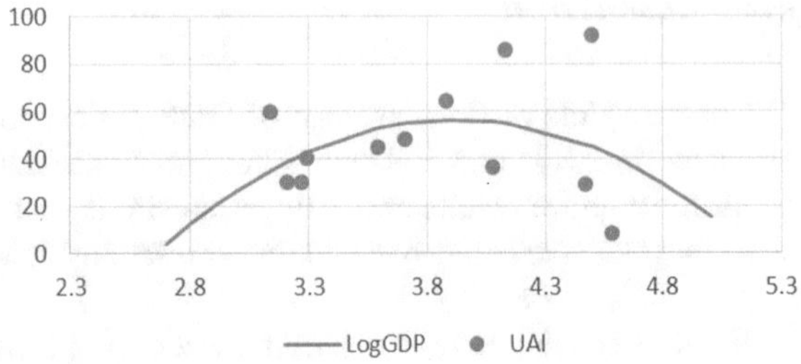

Fig. 5 Uncertainty avoidance and log average GDP per capital. *Source* The authors' calculation

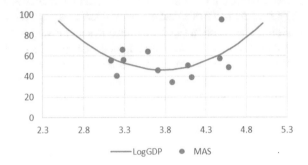

Fig. 6 Masculinity and log average GDP per capital. *Source* The authors' calculation

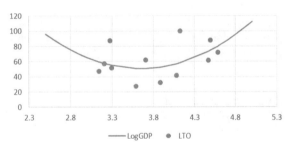

Fig. 7 Long-term orientation and log average GDP per capital. *Source* The authors' calculation

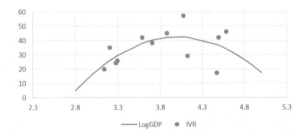

Fig. 8 Indulgence versus restraint and log average GDP per capita. *Source* The authors' calculation

Figure 5 demonstrates the relationship between Uncertainty Avoidance Index and the GDP per capita. As Hypothesis 3 has postulated, the level of Uncertainty avoidance first increases and then decreases. After the GDP per capita reaches a certain level, the Uncertainty Avoidance Index starts declining.

Figure 6 demonstrates the relationship between Masculinity and the log average of GDP per capita. It is found that the Masculinity Index is positively correlated with the log average of the GDP per capita after the index reaches a certain level. This is in contrast to Hypothesis 4. According to the hypothesis, the higher the average GDP per capita is, the higher is the Feminism Index, or, the lower is the Masculinity Index. The findings could reflect on the cultural and social values of the countries in this research. In the first stage, as the economy develops, women are encouraged to join the labor force. However, after reaching the threshold income, women are increasingly losing their roles at workplace, because Asian cultures tend to encourage women to take care of their families instead of joining the workforce. In the context of globalization, the influence of the western cultures that promote materialism, rivalry, and personal success results in fiercer competition among individuals, and women become less recognized in the workforce.

Figure 7 shows the relationship between Long-term orientation Index and national wealth. It depicts that the LTO first decreases and then increases as postulated in Hypothesis 5. That is, after the log average of the GDP per capita of these countries reach a certain level, the LTO index tends to increase.

Figure 8 demonstrates the relationship between Indulgence versus Restraint Index (IVR), and the log average of GDP per capita. According to Hypothesis 6, when GPD per capita increases, people tend to pursue their own happiness and have fewer social obligations and pressure. However, the results show that IVR decreases after reaching a certain threshold in the GDP per capita. Like in the case of the Masculinity Index, this could be attributed to the characteristics of the Asian people. In the first stage, when there is still much potential for development, income grows relatively fast, and the pressure of life decreases but after reaching the threshold, the competition gets fiercer. Besides, these countries have strict social rules and people appreciate academic qualifications a lot. Therefore, when income increases, people tend to invest more in the education of their children. The children in these countries tend to be under increased pressure to pursue education. Also, technological innovations require high-quality human resources, which forces workers to toil more and frequently upgrade their knowledge to enhance their work skills.

5 Conclusion

The paper examines the impact of national assets on cultural dimensions in 12 South and Southeast Asian countries by using the Metropolis-Hasting method. The growth rate of GDP per capita correlates positively with Individualism, Uncertainty, and Long-term Orientation Index and negatively with Power Distance, and Indulgence versus Restraint Index. The findings explain the paradoxical relationship between national property, and Masculinity and Indulgence versus Restraint. Also, these findings conflict with the results of some of the previous studies (Guiso et al. 2003; Tang and Koveos 2008). The contrast could be due to the strict social norms imposed on females in the region. Also, people in these societies are under constant pressure to be successful academically and financially, which leads them to have very little time for themselves. This could correlate with the fact that more people in highly developed economies such as Japan and South Korea tend to suffer from depression, stroke, and suicidal tendencies related to the workplace. Therefore, it can be suggested that policies to help remove social stereotyping of females could enable them maximize their potential. Also, work-related stress could be reduced to allow people to enjoy life more and thus, improve the quality of their lives during the course of economic development.

References

Adelman, I., & Morris, C. T. (1967). *Society, politics, and economic development: A quantitative approach*. Baltimore, MD: John Hopkins University.

Altman, M. C. (2001). Human agency, and economic theory: Culture as a determinant of material welfare. *Journal of Socio-Economics, 30*, 379–391.

Amariglio, J., & Callari, A. (1989). Marxian value theory and the problem of the subject: The role of commodity Fetishism, Rethinking Marxism. *A Journal of Economics, Culture & Society, 2*(3), 31–60.

Andersen, T. B., Bentzen, J., Dalgaard, C. J., & Sharp, P. (2013). Pre-reformation roots of the protestant ethic. Technical report, Competitive Advantage in the Global Economy (CAGE).

Anh, L. H., Le, S. D., Kreinovich, V., & Thach, N. N. (Eds.). (2018). *Econometrics for financial applications*. Cham: Springer.

Ata, C. B., Ljubica, D., & Can, S. (2018). Gender inequality and economic growth: Evidence from industry-level data. https://www.aeaweb.org/conference/2019/preliminary/paper/N8kTG95d.

Becker, S. O., & Woessmann, L. (2009). Was weber wrong? A human capital theory of protestant economic history. *The Quarterly Journal of Economics, 124*(2), 531–596.

Briggs, W. M., & Nguyen, H. T. (2019). Clarifying ASA's view on P-values in hypothesis testing. *Asian Journal of Economics and Banking, 3*(2) (to appear).

Cong, W., Wang T., Trafimow, D., & Jing, C.: Extending a priori procedure to two independent samples under skew normal settings. *Journal of Economics and Banking, 3*(2) (to appear).

Duflo, E. (2012). Women empowerment and economic development. *Journal of Economic Literature, 50*(4), 1051–1079.

Fernando, O. S., Gustavo, L. G., & Andrade, J. A. A. (2019). *A class of flat prior distributions for the Poisson-gamma hierarchical model*. Wiley Online Library

Forte, A., Garcia-Donato, G., & Steel, M. (2016). *Methods and tools for Bayesian variable selection and model averaging in univariate linear regression*.

Franke, R. H., Hofstede, G., & Bond, M. H. (1991). Cultural roots of economic performance: A research note. *Strategic Management Journal, 12*, 165–173.

Gelman, A., Gilks, W. R., & Roberts, G. O. (1997). Weak convergence and optimal scaling of random walk Metropolis algorithms. *Annals of Applied Probability, 7*, 110–120.

Gilks, W. R., Richardson, S., & Spiegelhalter, D. J. (1996). *Markov Chain Monte Carlo in practice*. London: Chapman and Hall.

Guiso, L., Sapienza, P., & Zingales, L. (2003). People's opium? Religion and economic attitudes. *Journal of Monetary Economics, 50*(1), 225–282.

Hofstede, G. (1980). *Culture's consequences: International differences in work-related values.* Beverly Hills, CA: Sage.

Hofstede, G. (2001). *Culture's consequences: Comparing values, behaviors, institutions, and organizations across nations.* Thousand Oaks, CA: Sage Publications.

Hofstede, G., Hofstede, G. J., & Minkov, M. (2010). *Cultures and organizations: Software of the mind: Intercultural cooperation and its importance for survival.* McGraw-Hill.

Inglehart, R. (1977). *The silent revolution: Changing values and political styles.* Princeton, NJ: Princeton University Press.

Inkeles, A. (1993). Industrialization, modernization and the quality of life. *International Journal of Comparative Sociology, 34*(1), 1–23.

International Monetary Fund. (2013). Women, Work and the Economy: Macroeconomic Gains from Gender Equity, IMF Staff Discussion Note 13/10 (2013).

Joseph, G. I., & Purushottam, W. L. (1991). On Bayesian analysis of generalized linear models using Jeffreys's Prior. *Journal of the American Statistical Association, 86*(416), 981–986.

Kreinovich, V., Thach, N. N., Trung, N. D., & Thanh, D.V. (Eds.). (2019). Beyond traditional probabilistic methods in econometrics. In: V. Kreinovich, N. Thach, N. Trung, & D. Van Thanh (Eds.) *Beyond traditional probabilistic methods in economics. ECONVN 2019.* Studies in computational intelligence (Vol. 809). Cham: Springer.

Kuznets, S. (1955). Economic growth and income inequality. *The American Economic Review, 45*(1), 1–28.

Liang, F., Paulo, R., Molina, G., Clyde, M. A., & Berger, J. O. (2008). Mixtures of g Priors for Bayesian variable selection. *Journal of the American Statistical Association, 103,* 410–423.

McClelland, D. C. (1961). *The achieving society.* New York: Free Press.

Ming-Hui, C., Joseph, G. I., & Sungduk, K. (2008). Properties and implementation of Jeffreys's Prior in binomial regression models. *Journal of the American Statistical Association, 103*(484), 1659–1664.

Nguyen, H. T., & Thach, N. N. (2018). A panorama of applied mathematical problems in economics. *Thai Journal of Mathematics. Special Issue*: Annual Meeting in Mathematics, 1–20.

Nguyen, H. T., & Thach, N. N. (2019). A closer look at the modeling of economics data. In: V. Kreinovich, N. Thach, N. Trung, & D. Van Thanh (Eds.) *Beyond traditional probabilistic methods in economics. ECONVN 2019.* Studies in computational intelligence (Vol. 809). Cham: Springer.

Nguyen, H. T., Sriboonchitta, S., & Thach, N. N. (2019a). On quantum probability calculus for modeling economic decisions. In: V. Kreinovich, & S. Sriboonchitta (Eds.) *Structural changes and their econometric modeling. TES 2019.* Studies in computational intelligence (Vol. 808, pp. 18–34). Cham: Springer.

Nguyen, H. T., Trung, N. D., & Thach, N. N. (2019b). Beyond traditional probabilistic methods in econometrics. In: V. Kreinovich, N. Thach, N. Trung, & D. Van Thanh (Eds.) *Beyond traditional probabilistic methods in economics. ECONVN 2019.* Studies in computational intelligence (Vol. 809). Cham: Springer.

Ogihara, Y. (2017). Temporal changes in individualism and their ramification in Japan: Rising individualism and conflicts with persisting collectivism. *Frontier in Psychology.*

Read, R. (1993). *Politics and policies of national economic growth.* Doctoral dissertation, Stanford University.

Sjoerd, B., & Chris, W. (2018). Dimensions and dynamics of national culture: Synthesizing Hofstede with Inglehart. *Journal of Cross-Cultural Psychology,* 1–37.

Smith, P., & Bond, M. H. (1998). *Social psychology across cultures.* London: Prentice Hall.

Smith, P. B., Peterson, M. F., & Schwartz, S. H. (2006). Cultural values, sources of guidance, and their relevance to managerial behavior: A 47-nation study. *Journal of Cross-Cultural Psychology, 33,* 188–208.

Sriboonchitta, S., Nguyen, H. T., Kosheleva, O., Kreinovich, V., & Nguyen T. N. (2019). Quantum approach explains the need for expert knowledge: On the example of econometrics. In: V. Kreinovich, & S. Sriboonchitta (Eds.) *Structural changes and their econometric modeling. TES 2019.* Studies in computational intelligence (Vol. 808). Cham: Springer.

Svítek, M., Kosheleva, O., Kreinovich, V., & Nguyen T. N. (2019). Why quantum (wave probability) models are a good description of many non-quantum complex systems, and how to go beyond quantum models. In: V. Kreinovich, N. Thach, N. Trung, & D. Van Thanh (Eds.) *Beyond traditional probabilistic methods in economics. ECONVN 2019*. Studies in computational intelligence (Vol. 809). Cham: Springer.

Tang, L., & Koveos, P. E. (2008). A framework to update Hofstede's cultural value indices: Economic dynamics and institutional stability. *Journal of International Business Studies, 39*, 1045–1063.

Thach, N. N. (2019a). Impact of the world oil price on the inflation on Vietnam: A structural vector autoregression approach. In V. Kreinovich, N. N. Thach, N. Trung & D. Van Thanh (Eds.), *Beyond traditional probabilistic methods in economics. ECONVN 2019*. Studies in Computational Intelligence (Vol. 809, pp. 694–708). Cham: Springer.

Thach, N. N. (2019b). How values influence economic progress? An evidence from south and southeast Asian countries. In N. Trung, N. Thach & V. Kreinovich (Eds.), *Data science for financial econometrics. ECONVN 2020*. Studies in Computational Intelligence. Cham: Springer (to appear).

Thach, N. N. (2020a). Endogenous economic growth: the Arrow-Romer theory and a test on Vietnamese economy. *WSEAS Transactions on Business and Economics, 17*, 374–386.

Thach, N. N. (2020b). How to explain when the ES is lower than one? A Bayesian nonlinear mixed-effects approach. *Journal of Risk and Financial Management, 13*(2), 21.

Thach, N. N., Anh, L. H., & An, P. T. H. (2019). The effects of public expenditure on economic growth in Asia countries: A Bayesian model averaging approach. *Asian Journal of Economics and Banking, 3*(1), 126–146.

Trompenaars, F. (1993). *Riding the waves of culture: Understanding cultural diversity in business*. London: Economist Books.

Weber, M. (1930). *The Protestant ethic and the spirit of capitalism* (Trans. by T. Parsons). New York: Scribner. (Originally published 1904).

World Bank. (2019). World Development Indicators. http://datatopics.worldbank.org/world-development-indicators/.

Yeh, R., & Lawrence, J. J. (1995). Individualism and Confucian dynamism: A note on Hofstede's cultural root to economic growth. *Journal of International Business Studies, 26*, 655–669.

Risk, Return, and Portfolio Optimization for Various Industries Based on Mixed Copula Approach

Sukrit Thongkairat and Woraphon Yamaka

Abstract This study aims to apply the concept of mixed copula to the problem of finding the risk, return, and portfolio diversification at the industry level in the stock markets of Thailand and Vietnam. Six industry indices are considered in this study. Prior to constructing the portfolio, we compare the mixed copula with the traditional copula to show the better performance of the mixed copula in terms of the lower AIC and BIC. The empirical results show that the mixed Student-t and Clayton copula model can capture the dependence structure of the portfolio returns much better than the traditional model. Then, we apply the best-fit model to do the Monte Carlo simulation for constructing the efficiency frontier and find the optimal investment combination from five portfolio optimization approaches including Uniform portfolio, Global Minimum Variance Portfolio (GMVP), Markowitz portfolio, Maximum Sharpe ratio portfolio, and Long-Short quintile. The findings suggest that, overall, the industry index of Vietnam and the consumer services index of Thailand should be given primary attention because they exhibit the highest performance compared to other industries in the stock markets. This suggestion is supported by the results of the Maximum Sharpe ratio portfolio (the best portfolio optimization approach) that assign the largest portfolio allocation to the industry sector for Vietnam and the consumer services sector for Thailand.

Keywords Mixed Copula · Portfolio optimization · Markowitz portfolio · Maximum Sharpe ratio portfolio · Long-Short quintile

S. Thongkairat · W. Yamaka (✉)
Faculty of Economics, Center of Excellence in Econometrics, Chiang Mai University,
Chiang Mai 50200, Thailand
e-mail: woraphon.econ@gmail.com

S. Thongkairat
e-mail: sukrit415@gmail.com

N. Ngoc Thach et al. (eds.), *Data Science for Financial Econometrics*,
Studies in Computational Intelligence 898,
https://doi.org/10.1007/978-3-030-48853-6_22

1 Introduction

Most investors in capital markets always seek to obtain high return and face low risk, but this ideal combination is hardly likely to happen. In reality, investors need to consider the trade-off between the risk and the return when making an investment decision or adjusting their portfolios. Among the risk and return measurement approaches, the portfolio selection theory has become a key to financial research. It was developed by Markowitz (1952, 1959) and it is generally used to suggest the asset allocation in a portfolio to achieve the best performance. In addition, this theory also helps assess risk and return at each combination of the assets.

In the computational aspect, there are many portfolio optimization approaches, such as Uniform portfolio, Global Minimum Variance portfolio, Markowitz portfolio, Maximum Sharpe ratio portfolio, and Long-Short quintile. These estimations are not difficult to compute if we consider similar assets in a portfolio. However, it becomes more difficult when there exists a complexity of the dependence structure (among various assets). Generally, the portfolio selection theory assumes that the dependence structure of the asset returns follows a multivariate normal distribution and there is linear correlation among the asset returns. In other words, the variance of the return on a risky asset portfolio depends on the variances of the individual assets and on the linear correlation between the assets in the portfolio Huang et al. (2009). However, in the real empirical study, the finance asset return distribution has fatter tails than normal distribution. Therefore, the multivariate normal distribution may not provide adequate results as the distribution assumption does not fit the real behavior of the financial data. In addition, it is rather exhibiting a nonlinear relationship between the assets in the portfolio.

In the literature, many studies confirmed that the assumption of normality of financial returns distribution could bring a negative effect on the risk measurement as well as distorting the portfolio optimization (see, Brooks et al. 2005; Jondeau and Rockinger 2003). Commonly, normal distribution is a symmetric distribution with low kurtosis. However, Jondeau and Rockinger (2003) suggested that the asset returns, recently, have a negative skewness and high kurtosis. In the other words, it has a higher probability to decrease than increase and this probability is high. Besides, Embrechts et al. (1999a, 1999b, 2002) and Forbes and Rigobon (2000) mentioned that the linear correlation coefficient does not perform well in capturing the dependence of the assets thus giving rise to misleading conclusions. As a consequence, we can say that the assumptions of normality and linear correlation in the traditional approach are not realistic for portfolio optimization objectives.

Over the past few years, the concepts related to the asymmetric dependence in financial returns have been introduced and presented. For examples, the extreme value theory and the concept of exceeding correlation (Beine et al. 2010; Hartmann et al. 2004; Longin and Solnik 2001; Poon et al. 2004). Nevertheless, those studies cannot provide an asymmetry in tail dependence. Note that tail dependence relates to the relationship that occurs within the extreme cases. To defeat this obstacle, this study resorts to the portfolio optimization based on the copula theory which enables us to establish a flexible multivariate distribution with different marginal distributions and different dependence structures. This approach enables us to construct the complex joint distribution of the portfolio which is not restricted by the normality and the linear correlation assumptions. Recently, various empirical studies (Maneejuk et al. 2018; Nguyen and Bhatti 2012; Tansuchat and Maneejuk 2018; Thongkairat et al. 2019) suggested employing mixed copula, which is more flexible than the traditionally used copulas, as it has an ability to capture both symmetric, asymmetric and different complicated dependence. Thus, in this study, we employ various models of portfolio optimizations based on the mixed-copula approach (uniform portfolio, Global Minimum Variance Portfolio, Markowitz portfolio, Maximum Sharpe ratio portfolio, and Long-Short quintile). We believe that our pursuance should be more reasonable and adequate in capturing the dependence among the assets in the portfolio.

Our work is on examining risk, return, and portfolio optimization at the industry level in the context of Thailand' and Vietnam' stock markets. The industry sectors in each country are increasingly integrated and thus it is interesting to quantify the industry risk and return in the Thai and Vietnamese stock markets. Risk and return at the industry level could provide an information that help bankers to establish the strategy for lending and borrowing to investors in each industry. Moreover, this information could support policymakers to achieve a potential policy for boosting an industrial growth.

In a nutshell, the main purposes of this study are to examine the dependence structure of assets in the portfolios at the industry level in the Thai and the Vietnamese stock markets using the mixed copula approach, and to determine the optimal portfolios. The mixed copulas comprise combinations between Elliptical copulas and Archimedean copulas, for instance, Gaussian-Frank copula, Gaussian-Joe copula, Student-t-Frank copula, Clayton-Joe copula, etc. The mixed copulas are then used in five portfolio optimization approaches namely, uniform portfolio, Global Minimum Variance Portfolio, Markowitz portfolio, Maximum Sharpe ratio portfolio, and Long-Short quintile, to arrive at the optimal investment in the portfolios at the industry level in the stock markets of Thailand and Vietnam

Following this Introduction, Sect. 2 briefly explains the methodology, the basic concept of copulas as well as the mixed copula approach. We describe the data in Sect. 3. Section 4 presents our empirical findings for Thailand and Vietnam. Finally, we conclude in Sect. 5.

2 Methods and Procedures

In this study, we combine Generalized Autoregressive Conditional Heteroscedasticity (GARCH) and mixed copula to fit the indices on industries for each country. This approach permits us to model the marginal distributions from GARCH and measure the relationship among different indices through mixed copula.

2.1 Generalized Autoregressive Conditional Heteroscedasticity (GARCH)

GARCH is applied to model the marginal distribution of the random variable. Let r_t be the return of the asset. The model is then written as follows:

$$r_t = \phi_0 + \varepsilon_t, \tag{1}$$

$$\varepsilon_t = h\eta_t, \tag{2}$$

$$h_t^2 = \alpha_0 + \sum_{i=1}^{I} \alpha_i \alpha_{t-i}^2 + \sum_{j=1}^{J} \beta_j h_{t-j}^2. \tag{3}$$

where Eq. (1) is the conditional mean equation while Eq. (5) is the conditional variance equation. ε_t is the residual, consisting of standard variance, h_t, and standardized residual, η_t. In this paper, the distribution of residual is permitted to be six different distributions as follows, normal distribution, student-t distribution, skew-t distribution, skewed normal distribution, generalized error distribution and/or skewed generalized error distribution. The standardized residual η_t will be granted by the best-fit GARCH (I, J), to be subsequently converted into a uniformly [0, 1]. Note that the marginal distributions do not need to be the same because the copula has an ability to join any complex dependence structure.

2.2 Copula Concept

The concept of copula has received a growing attention in finance and economics since Sklar (1959) proposed the concept and discussed the multidimensional distribution function. The function can be used to construct the complicated relationship among the marginal distribution of the random variables. In the other words, the copula can describe the dependence relationship among n marginal distributions $F_1(x_1), \ldots, F_n(x_n)$, so

$$F(x_1, \ldots, x_n) = C(F_1(x_1), \ldots, F_n(x_n)) = C(\mu_1, \ldots, \mu_n), \tag{4}$$

where x is random variable. C denotes the copula, $F(x_1, \ldots, x_n)$ is a joint distribution of $x_1, \ldots, x_n$. in uniform [0, 1], $\mu_1, \ldots, \mu_n$ is the marginal distribution in uniform [0, 1]. Then, we can obtain the copula density function c by

$$c(u_1, \ldots, u_n) = \frac{\partial^n C(F_1(x_1), \ldots, F_n(x_n))}{\partial u_1, \ldots, \partial u_n}. \tag{5}$$

As there are many dependence structures, this study will consider various commonly used copulas which are the Gaussian copula, the Student-t copula, and the Archimedean copula family such as the Clayton copula, Frank copula, Gumbel. Then, our mixed-copulas are constructed from the combination of these copulas. The copula density functions can be found in Joe (2005), McNeil and Nešlehová (2009), Schepsmeier and Stöber (2014).

2.3 Mixed Copulas

The convex combination is suggested by Nelsen in 2006 for combining different copula densities. Since the traditional copulas set some limitations on their dependence parameters, it is likely for a misspecification problem to arise. By using a mixed copula class, we will obtain several additional advantages. First, the dependence structures obtained from mixed copulas will not change even though the data is transformed into several types. Second, the model becomes more flexible in the wide range of dependence structures. A mixed copula can be constructed from the convex combination approach, thus the density function is defined by the following:

$$c_{mix}(u_1, \ldots, u_n \,|\theta) = w c_1(u_1, \ldots, u_n \,|\theta^1) + (1 - w) c_2(u_1, \ldots, u_n \,|\theta^2) . \quad (6)$$

where w is the weight parameter with values [0, 1]. θ^1 and θ^2 are the dependence parameters of the first and the second copula density functions, respectively.

2.4 Optimization Portfolio

Portfolio optimization is the process for selecting the most appropriate weight of the asset in the portfolio. The risk minimization is found with the additional constraint that the summation of asset's weights is equally to one (see, Table 1). In this study, we first stimulate one-step ahead return forecast $(\tilde{r}_{t+1})$ 1500 rounds from the best fitted copula-GARCH specification. Then, we can obtain the investment allocation under various portfolio optimization approaches as shown in the following table.

3 Data

This study uses a data set from Thomson Reuter DataStream, including daily indices on six industries for each country: consumer goods, consumer services, finance, health care, industry, and utilities. The research period covers 2015–2019. The simple statistics are reported in Table 2. It is clearly found that average return of financial index of Thailand is negative (-0.0001), while the rest are positive. We also observe

Table 1 Optimization portfolio approaches

Portfolio approach	Formula
Uniform portfolio	$\omega = \frac{1}{N}$
Global minimum variance portfolio (GMVP)	Minimize $\omega^T \sum \omega$
	Subject to $1^T \omega = 1 \quad ; \omega \geq 0$
Markowitz portfolio	Minimize $\mu^T \omega - \lambda \omega^T \sum \omega$
	Subject to $1^T \omega = 1 \quad ; \omega \geq 0$
Maximum Sharpe ratio portfolio (MAXSR)	Minimize $\tilde{\omega}^T \sum \tilde{\omega}$
	Subject to $\tilde{\omega}^T \mu = 1 \quad ; \tilde{\omega} \geq 0 \; and \; \omega = \tilde{\omega}/(1^T \tilde{\omega})$
Long-Short quintile portfolio (1) (Lquintile1)	μ
Long-Short quintile portfolio (2) (Lquintile2)	$\mu/diag(\sum)$
Long-Short quintile portfolio (3) (Lquintile3)	$\mu/\sqrt{diag(\sum)}$

Table 2 Descriptive statistics

	Thailand			Vietnam		
	Mean	Skewness	Kurtosis	Mean	Skewness	Kurtosis
Consumer goods	0.0001	0.1909	5.1092	0.0005	−0.0655	5.1789
Consumer services	0.0002	0.0473	4.9324	0.0009	0.0524	5.9947
Finance	−0.0001	0.3101	6.8772	0.0007	−0.2936	5.6713
Health care	0.0003	−0.3428	7.9171	0.0005	0.2274	5.0132
Industry	0.0003	0.0452	5.0882	0.0004	−0.4788	6.1798
Utilities	0.0002	−0.0478	6.2703	0.0002	−0.2672	5.7941

that the average returns of all industries in Vietnam are greater than Thailand. All series show very clear signs of non-normality with high kurtosis (>3) and skewness. The skewness of our indices ranges between -0.5 and 0.5.

Moreover, the Jarque-Bera test is also employed to test the normality of all returns. Following Goodman (1999) a Minimum Bayes factor between 1 and 1/3 is considered anecdotal evidence for the alternative hypothesis, from 1/3 to 1/10 moderate evidence, from 1/10 to 1/30 substantial evidence, from 1/30 to 1/100 strong evidence, from 1/100 to 1/300 very strong evidence and lower than 1/300 decisive evidence. As shown in Table 3, MBF values of the Jarque-Bera test are close to zero for all returns, meaning the decisive evidence supporting non-normality of the returns.

Table 3 The Jarque-Bera test

	Consumer goods	Consumer services	Health care
Thailand	220.5471	179.6684	1183.1160
MBF	0.0000	0.0000	0.0000
Vietnam	228.7176	431.0123	204.4798
MBF	0.0000	0.0000	0.0000
	Industry	Utilities	Finance
Thailand	209.7071	513.8071	740.0228
MBF	0.0000	0.0000	0.0000
Vietnam	529.3497	388.4319	359.0638
MBF	0.0000	0.0000	0.0000

Note MBF is calculated by $MBF(p) = \begin{cases} -\exp(1)p\log p & \text{for } p < 1/\exp(1) \\ 1 & \text{for } p \geq 1/\exp(1) \end{cases}$

4 Empirical Results

4.1 Model Selection

After having estimated the parameters of the marginal distributions from the GARCH process, we continue to estimate the copula parameters. Several classes of copula, including Elliptical copulas, Archimedean copulas, and mixed copulas, are considered. To find the best fitting copula function, the Akaike information criterion (AIC) and Bayesian information criterion (BIC) are used. Table 4 presents AIC and BIC values for each copula. From this table, the mixed Student-t and Clayton copula has the smallest value of AIC and BIC for both Thailand (-3088.745 and -3063.499) and Vietnam (-2137.196 and -2111.949). Hence, we can conclude that Student-t and Clayton copula is the best fitting function to describe the dependence structure of the multivariate return series.

4.2 Copula Parameter Estimates

Table 5 shows the estimated parameters from mixed Student-t and Clayton copula for Vietnam and Thailand, where w_1 is the weight of Student-t copula and w_2 is the weight of Clayton copula. The distribution of six industries' volatilities follows mixed Student-t and Clayton copula with the weight of student-t copula being 0.9890 and 0.7070 for Vietnam and Thailand respectively. Likewise, the weights of Clayton copula are 0.0528 and 0.0524, respectively, for Vietnam and Thailand. Therefore, the fitted mixed Student-t and Clayton copula with a greater weight for Student-t copula indicates that there is a high possibility of symmetric dependence among six indices in both countries. Six industries in both countries are likely to move in the

Table 4 Model selection using AIC and BIC

Copula	Thailand			Vietnam		
	AIC	BIC	LL	AIC	BIC	LL
Gaussian	−1703.2030	−1698.1540	852.6016	−1356.4470	−1351.3980	679.2237
Student-t	−1940.8320	−1930.7330	972.4158	−1566.5390	−1556.4410	785.2696
Gumbel	−1597.7270	−1592.6780	799.8635	−1148.4490	−1143.3990	575.2243
Clayton	−1320.1430	−1315.0940	661.0717	−1226.9860	−1221.9370	614.4931
Frank	−1307.6240	−1302.5740	654.8118	−920.2860	−915.2368	461.1430
Joe	−1166.6590	−1161.6100	584.3294	−711.0087	−705.9594	356.5043
Gaussian–Student-t	−2911.0950	−2885.8490	1460.5480	−1912.0430	−1886.7960	961.0214
Gaussian–Clayton	NaN	NaN	NaN	−1409.9990	−1389.8020	708.9995
Gaussian–Frank	−1893.2860	−1873.0890	950.6430	−1416.9670	−1396.7700	712.4834
Gaussian–Gumbel	−2946.1990	−2926.0020	1477.1000	−1983.5500	−1963.3530	995.7749
Gaussian–Joe	−1897.9920	−1877.7950	952.9959	−2000.1620	−1979.9650	1004.0810
Student-t–Clayton	**−3088.7450**	**−3063.4990**	**1549.3730**	**−2137.1960**	**−2111.9490**	**1073.5980**
Student-t–Frank	−2080.6370	−2055.3910	1045.3180	−1604.9940	−1579.7480	807.4970
Student-t–Gumbel	−3080.7930	−3055.5470	1545.3970	−2122.8830	−2097.6370	1066.4420
Student-t–Joe	−2094.8410	−2069.5940	1052.4200	−1605.1820	−1579.9360	807.5910
Clayton–Frank	−1645.8060	−1625.6090	826.9031	−1897.9920	−1877.7950	952.9959
Clayton–Gumbel	−1772.6180	−1752.4210	890.3088	−1387.4330	−1367.2360	697.7163
Clayton–Joe	−1769.8240	−1749.6270	888.9121	−1877.2750	−1857.0780	942.6375
Frank–Gumbel	−2534.2150	−2514.0180	1271.1070	−1453.6190	−1433.4220	730.8096
Frank–Joe	−2754.6180	−2734.4210	1381.3090	−1737.8860	−1717.6890	872.9432
Gumbel–Joe	−2425.0990	−2404.9020	1216.5490	−1399.3250	−1379.1280	703.6625

Note LL denotes log likelihood, the Bold number corresponds to the lowest values of AIC and BIC

Table 5 Parameter estimation under mixed Student-t–Clayton copula

Vietnam		Coef.	S.E.	t-stat	p-value
	θ_T	0.2999	0.0117	25.5934	0.0000
	θ_C	11.8791	1.2403	9.5776	0.0000
	df	24.9245	1.2380	20.1336	0.0000
	w_1	0.9890	4.9264	0.2008	0.8409
	w_2	0.0528	0.2627	0.2008	0.8408
Thailand		Coef.	S.E.	t-stat	p-value
	θ_T	0.3508	0.0115	30.3875	0.0000
	θ_C	13.0961	1.3952	9.3863	0.0000
	df	37.8568	1.5925	23.7724	0.0000
	w_1	0.7070	2.9825	0.2371	0.8126
	w_2	0.0524	0.2210	0.2371	0.8126

Table 6 Annualized performance, Vietnam

	Consumer goods	Consumer services	Financials	Health care	Industrials	Utilities
Return	0.1867	0.0995	0.1315	−0.195	0.2628	0.0252
Risk	0.1649	0.2352	0.1996	0.2204	0.1721	0.1598
Sharpe ratio	1.1322	0.4232	0.6587	−0.8847	1.5269	0.1577
VaR (95%)	−0.0141	−0.0210	−0.0173	−0.0203	−0.0151	−0.014
ES (95%)	−0.0168	−0.0246	−0.0207	−0.0241	−0.0176	−0.0169

same direction as the copula dependence θ_T are 0.2999 for Vietnam and 0.3508 for Thailand (Table 6).

Moreover, we illustrate the performance of each industry for both countries in Figs. 1 and 2. In the case of Vietnam, it is clear that the industry index shows the highest performance compared to other sectors along our study periods. For Thailand, we observe that the consumer services index has been increasing along our sample period and shows the highest performance in Thai stock market (Table 7).

4.3 Comparing Portfolio Performances

In the last section, we compare the performance of different portfolio optimization approaches on the basis of the Sharpe ratio. Note that the higher the Portfolio's Sharpe ratio, the better the risk-adjusted performance. For this reason, the highest Portfolio's Sharpe ratio is preferred in our study. As shown in Tables 8 and 9, the Maximum Sharpe ratio approach (MAXSR) shows the highest Sharpe ratio for both Vietnam and Thailand markets. In addition, we also plot the efficiency frontier which

Fig. 1 Buy and Hold performance in Vietnam

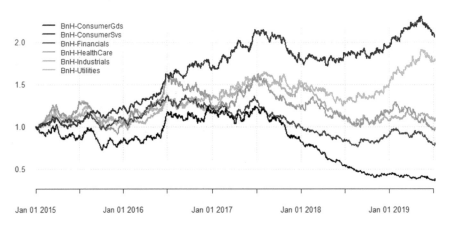

Fig. 2 Buy and Hold performance in Thailand

Table 7 Annualized performance, Thailand

	Consumer goods	Consumer services	Financials	Health care	Industrials	Utilities
Return	−0.0585	0.1373	−0.0137	0.0566	0.0736	0.0831
Risk	0.2055	0.1345	0.1434	0.1849	0.1381	0.1521
Sharpe ratio	−0.2847	1.0210	−0.0955	0.3061	0.5332	0.5465
VaR (95%)	−0.0188	−0.0114	−0.0127	−0.017	−0.0119	−0.0135
ES (95%)	−0.0222	−0.0136	−0.0151	−0.0198	−0.0141	−0.016

Table 8 The performance of different portfolio optimization approaches, Vietnam

	Uniform	GMVP	Markowitz	MAXSR	Lquintile1	Lquintile2	Lquintile3
Return	0.0819	0.1459	0.2628	0.243	0.2628	0.2628	0.2628
Risk	0.1543	0.1391	0.1721	0.1531	0.1721	0.1721	0.1721
Sharpe ratio	0.5309	1.049	1.5269	**1.5872**	1.5269	1.5269	1.5269

Note Bold number is the highest portfolio's Sharpe ratio

Table 9 The performance of different portfolios optimization approaches, Thailand

	Uniform	GMVP	Markowitz	MAXSR	Lquintile1	Lquintile2	Lquintile3
Return	0.0495	0.0753	0.1373	0.136	0.1373	0.1373	0.1373
Risk	0.1283	0.1143	0.1345	0.1332	0.1345	0.1345	0.1345
Sharpe ratio	0.3853	0.6592	1.021	**1.0214**	1.021	1.021	1.021

Note Bold number is the highest portfolio's Sharpe ratio

is the set of optimal portfolios that offers the highest expected return for a defined level of risk or the lowest risk for a given level of expected return. Portfolios that lie below the efficiency frontier are sub-optimal because they do not provide enough return for the level of risk. Portfolios that cluster to the right of the efficiency frontier are sub-optimal because they have a higher level of risk for the defined rate of return. According to Figs. 3 and 4, we can observe that MAXSR locates on the efficient frontier line in both m and Thailand cases. Thus, this result confirms the high performance of MAXSR in constructing the portfolio allocation as well as quantifying risk and return of portfolios. Note that Lquintile and Markowitz show the same performance in the Sharpe ratio for both Vietnam and Thailand cases.

Last but not least, we use the Maximum Sharpe ratio to find the optimal weight for the investment allocation purpose. In the case of the Vietnamese market, we should have a portfolio with the following weighs: 29.7648% in Consumer Goods sector, 0% in Consumer Services sector, 0.0003% in Financials sector, 0% in Health Care sector, 70.2345% in Industrials sector, 0.0002% in Utilities sector. For investment in the Thai stock market, an optimal portfolio is constituted of: Consumer Goods sector, 0%; Consumer Services sector, 97.2020%; Financial sector, 0%; Health Care sector, 0%; Industry sector, 0.4331%; and Utilities sector, 2.3646%.

5 Conclusion

The key factors that the investors generally consider in making their investment decision are risks and returns. In this study we quantify the risk, return, Sharpe ratio and compare the performance of different portfolio optimization approaches at the industry/sector level in Vietnam and Thailand over the period from 2015 to

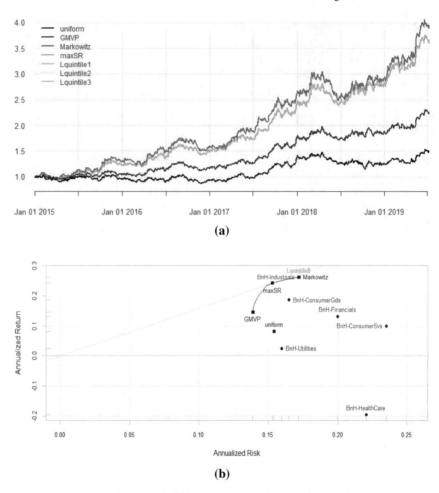

Fig. 3 Comparing portfolio optimization approaches in the case of Vietnam (**a** panel: Sharp ratio of different portfolios, **b** panel: efficiency frontier)

2019. In this study, we employed the mixed copula based GARCH to construct the complicated dependence structure of the assets in the portfolio. In order to make the reliable and acceptable estimation results, we firstly implement the GARCH to model the daily returns of six industries in both countries, then the dependence structure between variations of each industry's returns is captured by various classes of copula, such as Archimedean copula, Elliptical copula, and Mixed copula. To compare the portfolio optimization approaches, we consider the efficiency frontier and Sharpe ratio. Finally, the best performing approach is further used to construct the optimal set of asset allocation of the portfolio.

The findings from this study suggest that mixed Student-t and Clayton copula can better capture the dependence structure between industry and the performance

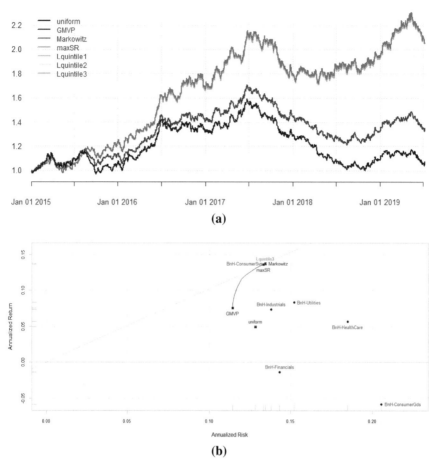

Fig. 4 Comparing portfolio optimization approaches in the case of Thailand (**a** panel: Sharp ratio of different portfolios, **b** panel: efficiency frontier)

of this copula is better than that of the traditional copula model. This result indicates that the advantage of our mixed copula can lead to better estimation of a portfolio's risk. The result of the mixed copula model shows a positive correlation among the industry indices.

Finally, the portfolio optimization result suggests that the investor should invest following Maximum Sharpe ratio approach. According to Maximum Sharpe ratio approach results, Vietnam has an expected return of 20.30% and a risk of 15.31% and the optimal weights for investment in each industry are as follows: 29.7648% in Consumer Goods sector, 0% in Consumer Services sector, 0.0003% in Financial sector, 0% in Health Care sector, 70.2345% in Industry sector, and 0.0002% in Utilities sector. In the case of Thailand, it has an expected return of 13.60% and a risk of 13.32%. The optimal weights for Consumer Goods sector, Consumer Services sector, Financial sector, Health Care sector, Industry sector, and Utilities sector are given as follows: 0%, 97.2020%, 0%, 0%, 0.4331%, and 2.3646% respectively.

Acknowledgements The authors are grateful to Puay Ungphakorn Centre of Excellence in Econometrics, Faculty of Economics, Chiang Mai University for the financial support.

References

Beine, M., Cosma, A., & Vermeulen, R. (2010). The dark side of global integration: Increasting tail dependence. *Journal of Banking and Finance*, *34*(1), 184–192.

Brooks, C., Burke, P., & Heravi, S., Persand, G. (2005). Autoregressive conditional kurtosis. *Journal of Financial Econometrics*, *3*(3), 399–421.

Embrechts, P., McNeil, A. J., & Straumann, D. (1999a). Correlation: Pitfalls and alternatives a short. *RISK Magazine*, 69–71.

Embrechts, P., Resnick, S. I., & Samorodnitsky, G. (1999b). Extreme value theory as a risk management tool. *North American Actuarial Journal*, *3*(2), 30–41.

Embrechts, P., McNeil, A. J., & Straumann, D. (2002). Correlation and dependence in risk management: Properties and Pitfalls. In M. Dempster (Ed.), *Risk management: Value at risk and beyond* (pp. 176–223). Cambridge: Cambridge University Press.

Forbes, K. J., & Rigobon, R. (2000). No contagion, only interdependence: Measuring stock market comovements. *Journal of Finance LVII*, *5*, 2223–2261.

Goodman, S. N. (1999). Toward evidence-based medical statistics. 1: The *p* value fallacy. *Annals of Internal Medicine*, *130*, 995–1004.

Hartmann, P., Straeman, S., & de Vries, C. G. (2004). Asset market linkages in crisis periods. *Review of Economics and Statistics*, *86*(1), 313–326.

Huang, J. J., Lee, K. J., Liang, H., & Lin, W. F. (2009). Estimating value at risk of portfolio by conditional copula-GARCH method. *Insurance: Mathematics and Economics*, *45*(3), 315–324.

Joe, H. (2005). Asymptotic efficiency of the two-stage estimation method for copula-based models. *Journal of Multivariate Analysis*, *94*(2), 401–419.

Jondeau, E., & Rockinger, M. (2003). Conditional volatility, skewness and kurtosis: Existence, persistence, and comovements. *Journal of Economic Dynamics and Control*, *27*(10), 1699–1737.

Longin, F., & Solnik, B. (2001). Extreme correlation of international equity markets. *Journal of Finance*, *56*(2), 649–676.

Markowitz, H. M. (1952). Portfolio selection. *Journal of Finance*, *7*(1), 77–91.

Markowitz, H. M. (1959). *Portfolio selection: Efficient diversification of investments*. Massachusetts: Yale University Press.

Maneejuk, P., Yamaka, W., & Sriboonchitta, S. (2018). Mixed-copulas approach in examining the relationship between oil prices and ASEAN's stock markets. In *International Econometric Conference of Vietnam*. Cham: Springer.

McNeil, A. J., & Neslehova, J. (2009). Multivariate Archimedean copulas, d-monotone functions and ℓ1-norm symmetric distributions. *The Annals of Statistics*, *37*(5B), 3059–3097.

Nguyen, C. C., & Bhatti, M. I. (2012). Copula model dependency between oil prices and stock markets: Evidence from China and Vietnam. *Journal of International Financial Markets, Institutions and Money*, *22*(4), 758–773.

Poon, S.-H., Rockinger, M., & Tawn, J. (2004). Modelling extreme-value dependence in international stock markets. *Statistica Sinica*, *13*, 929–953.

Schepsmeier, U., & Stöber, J. (2014). Derivatives and Fisher information of bivariate copulas. *Statistical Papers*, *55*(2), 525–542.

Sklar, M. (1959). Fonctions de repartition à n-dimensions et. leurs marges. Publications de l'Institut Statistique de l'Université de Paris (Vol. 8, pp. 229–231)

Tansuchat, R., & Maneejuk, P. (2018, March). Modeling dependence with copulas: Are real estates and tourism associated? In *International Symposium on Integrated Uncertainty in Knowledge Modelling and Decision Making* (pp. 302–311). Cham: Springer.

Thongkairat, S., Yamaka, W., & Chakpitak, N. (2019, January). Portfolio optimization of stock, oil and gold returns: A mixed copula-based approach. In *International Conference of the Thailand Econometrics Society* (pp. 474–487). Cham: Springer.

Herding Behavior Existence in MSCI Far East Ex Japan Index: A Markov Switching Approach

Woraphon Yamaka, Rungrapee Phadkantha, and Paravee Maneejuk

Abstract This paper examines the herding behavior in the MSCI Far East Ex Japan stock exchange (MSCI-FE) using the Markov switching approach. An analysis of daily data from December 26, 2012, to June 17, 2019, is considered to find evidence of herding behavior in MSCI-FE stock market under different market regimes. The market is divided into two regimes (upturn and downturn) to determine whether the herding exists in an upturn or downturn market. In this study we consider four models consisting of linear regression, linear regression with time-varying coefficient, Markov Switching regression (MS), and Markov Switching regression with a time-varying coefficient (MS-TV). To detect the herding behavior, we employ the Cross-Sectional Absolute Deviation (CSAD) method of Chang et al. (J Bank Finance 24(10):1651–1679, 2000) and Chiang and Zheng (J Bank Finance 34(8):1911–1921, 2010). The results show that herding behavior is present in MSCI-FE. From the results of MS and MS-TV models, the herding exists only in the market upturn regime, indicating that investors will neglect their information and choose to mimic larger investors during the market upturn. On the other hand, during the market downturn, we do not find any evidence of the herding behavior.

Keywords Herding behavior · CSAD · Markov switching · MSCI Far East Ex Japan

W. Yamaka · P. Maneejuk
Faculty of Economics, Center of Excellence in Econometrics, Chiang Mai University, Chiang Mai, Thailand
e-mail: woraphon.econ@gmail.com

P. Maneejuk
e-mail: mparavee@gmail.com

R. Phadkantha (✉)
Center of Excellence in Econometrics, Chiang Mai University,
Chiang Mai, Thailand
e-mail: rungrapee.ph@gmail.com

© The Editor(s) (if applicable) and The Author(s), under exclusive license
to Springer Nature Switzerland AG 2021
N. Ngoc Thach et al. (eds.), *Data Science for Financial Econometrics*,
Studies in Computational Intelligence 898,
https://doi.org/10.1007/978-3-030-48853-6_23

1 Introduction

In the financial investment theory, the nature of human is assumed to be rational and an investment decision is based on the available information and logic. But in fact, most investors often behave in deviation from logics and reasons, and they normally follow the decision and the action of other institutional investors by trading in the same direction over a period. This behavior is called herding behavior. The herding behavior can render both negative and positive contributions to the development of capital markets. It can be positive, if the herding behavior is based on the precise information of the investment, otherwise, it will be negative and thereby leading to the economic destruction or financial crisis; for examples, the capital market crash of Argentina during 2000– 2006 (Gavriilidis et al. 2007) and the Asian financial crisis in 1997 (Bonser-Neal et al. 2002; Bowe and Domuta 2004).

Babalos et al. (2015) mentioned that the asset prices driven by the herding behavior are not justified by their fundamental values. Also, Putra et al. (2017) believed that the herding behavior of investment leads to the distortion of asset prices in the market. The problems of the herding behavior have been explored in many works, such as Park (2011), Babalos et al. (2015), Phadkantha et al. (2019) and Brunnermeier (2001). They stated that the herding behavior could lead to the bubbles, systematic risk, and asymmetric volatility in financial markets during the market upturn. Moreover, during the periods of market turbulence, the herding behavior also poses a threat to financial stability and increases the market severity through a negative shock (Demirer and Kutan 2006; Pastpipatkul et al. 2016; Maneejuk et al. 2019; Economou et al. 2011). Specifically, Pastpipatkul et al. (2016) found that the dependence or the correlation behavior of the financial assets are different between the market upturn and downturn. Thus, the different market conditions might generate different herding behaviors. Chang et al. (2000), Chiang and Zheng (2010), Morelli (2010) also found that there is a high diversification during the market turbulence compared to the market upturn. The study of Lao and Singh (2011) provided an evidence for asymmetric effects of the herding behavior. They revealed that investors tend to herd more intensively during upward movement than during downward movement of the market. In addition, it has been suggested that the presence of herding behavior is most likely to occur during periods of extreme market, as investors would then be more triggered to follow the market consensus (Demirer et al. 2010). In this respect, the financial markets may involve a widespread irrationality in both market upturn and downturn. Although it has gained a lot of prominence among researchers, herding is still masked with an ambiguity in terms of herding behavior existence. Herding is a very interesting phenomenon to explore and it entails vital implications for investors and market regulators. In this study, we examine the herding behavior occurrence in an Asian stock market. Here, we consider the Morgan Stanley Capital International Far East Ex Japan index (MSCI-FE) as the scope of this study. This market index captures large and medium capitalization in both developed and emerging stock markets in Asian and the Far Eastern countries except Japan. The MSCI-FE index consists of two developed-market countries (Hong Kong and Singapore) and seven emerging-market countries (China, Indonesia, Korea, Malaysia, the Philippines, Taiwan, and Thailand). The market capitalization was valued at 4,230 billion dollars in 2018. The MSCI-FE market has high turnover and standard deviation but low Sharpe ratio when compared to MSCI All Country World Index. This indicates the importance of these nine stock markets.

To examine the presence of the herding behavior, we consider the most widely used method in the literature developed by Chang et al. (2000). This approach is based on the Capital Asset Pricing Model (CAPM) but the dependent variable is replaced by the Cross-Sectional Absolute Deviation (CSAD). $CSAD_t$ can be measured by

$$CSAD_t = \frac{1}{n} \sum_{i=1}^{n} |R_{i,t} - R_{m,t}|, \tag{1}$$

where $R_{i,t}$ is the return of a stock i and $R_{m,t}$ is the market return. The $CSAD_t$ is an indicator of the distribution of returns at time t. As suggested by Chang et al. (2000), the herding behavior test can be set as:

$$CSAD_t = a + \gamma_1 |R_{m,t}| + \gamma_2 R_{m,t}^2 + \varepsilon_t, \tag{2}$$

where $|.|$ is the absolute value. ε_t is the normally distributed error term. The existence of herding behavior of investors can be explained by coefficient γ_2. If the coefficient γ_2 is negative, this indicates that the herding behavior exists. This model can be estimated by regression approach and it has been applied in many markets, for examples, Amman Stock Exchange (Ramadan 2015), Indonesia and Singapore Stock Market (Putra et al. 2017), UK-listed Real Estate Investment Trusts (REITs) (Babalos et al. 2015), and the Stock Exchange of Thailand (Phadkantha et al. 2019). However, these studies may suffer from several deficiencies, including the inability to recognize that the herding pattern can change over time as market conditions change.

Babalos et al. (2015) mentioned that the nonlinearity in the relationship is likely to occur due to structural breaks in high-frequency financial data; thus, they suggested to use Markov Switching regression model for investigating the existence of the herding behavior. They compared the herding model based on linear regression and that on Markov Switching regression. They found that the existence of herding in the REITs market is rejected under the linear regression model, but their regime-switching model revealed substantial evidence of herding behavior under the crash regime for almost all sectors. This indicates that the conventional linear method may lack the ability in detecting the herding behavior when the market exhibits the structural change.

From this paper, we contribute to the literature in the following ways: Firstly, in the context of the present study, we trace the herding behavior through MSCI-FE index as a whole as opposed to previous studies that examined herding at the stock level. Secondly, although we also employ the Markov Switching model to investigate the presence of herding behavior, our examination is different from that by Babalos et al. (2015) in a methodological sense. We propose the Markov Switching time varying regression (MS-TV) herding model where the CSAD of MSCI-FE returns is allowed to follow multiple regime changes, and this allows us to capture nonlinear relationship as well as capture the full dynamics of the herding behavior occurring at each time point in two different market regimes, market upturn and market downturn.

The paper has four further sections. Section 2 presents the research methodology of the study and this section is further divided into four subsections. Data description is presented in Section 3. Section 4 provides the empirical results of the study. The conclusions of this study are presented in Sect. 5.

2 Methodology

In this section, we first introduce two regression equations for detecting the herding behavior. Then, the Markov Switching regression (MS) and the Markov Switching time varying regression (MS-TV) are explained.

2.1 Herding Behavior Detection Model

In addition to the herding detection equation of Chang et al. (2000) in Eq. (2), Chiang and Zheng (2010) modified Eq. (2) by adding $R_{m,t}$. This model permits the interpretation of asymmetric effects by estimating a single model, which is more streamlined than the initial regression of Chang et al. (2000). Therefore, we have

$$CSAD_t = a + \gamma_1 R_{m,t} + \gamma_2 |R_{m,t}| + \gamma_3 R_{m,t}^2 + \varepsilon_t \tag{3}$$

Again, to determine whether there exists herding behavior, we can observe the coefficient γ_3. If γ_3 is negative, it indicates that the market has a herding behavior.

2.2 Markov Switching in Herding Behavior

In this study, two-regime Markov Switching model is adopted to examine the herding behavior. Therefore, we can modify Eq. (2) and Eq. (3) as in the following:

$$
\begin{aligned}
CSAD_t &= a(s_t) + \gamma_1(s_t) |R_{m,t}| + \gamma_2(s_t) R_{m,t}^2 + \varepsilon_t(s_t), \quad s_t = 1 \\
CSAD_t &= a(s_t) + \gamma_1(s_t) |R_{m,t}| + \gamma_2(s_t) R_{m,t}^2 + \varepsilon_t(s_t), \quad s_t = 2
\end{aligned}
\tag{4}
$$

and

$$
\begin{aligned}
CSAD_t &= a(s_t) + \gamma_1(s_t) R_{m,t} + \gamma_2(s_t) |R_{m,t}| + \gamma_3(s_t) R_{m,t}^2 + \varepsilon_t(s_t), \quad s_t = 1 \\
CSAD_t &= a(s_t) + \gamma_1(s_t) R_{m,t} + \gamma_2(s_t) |R_{m,t}| + \gamma_3(s_t) R_{m,t}^2 + \varepsilon_t(s_t), \quad s_t = 2
\end{aligned}
\tag{5}
$$

where the disturbance $\varepsilon_t(s_t)$ is observable regime dependent errors which all of them are assumed to be normally distributed. $s_t = \{1, 2\}$ is an unobservable regime which is governed by the first order Markov chain, which is characterized by the transition probabilities $p_{ij} = \Pr(s_{t+1} = j \,|s_t = i), \sum_{j=1}^{k=2} p_{ij} = 1; \; i, j = 1, 2$. Thus, p_{ij} is the probability of switching from regime i to regime j and the transition probabilities can be formed in transition matrix P as follows,

$$P = \begin{bmatrix} p_{11} & p_{12} = 1 - p_{11} \\ p_{21} = 1 - p_{22} & p_{22} \end{bmatrix} \tag{6}$$

Similar to the one-regime herding detection model, we can investigate the herding existence from the regime dependent coefficient $\gamma_2(s_t)$ of Eq. (4) and $\gamma_3(s_t)$ of Eq. (5). Under this regime switching framework, the model can capture the different herding behavior occurring in the different period of market upturn ($s_t = 1$) or market downturn ($s_t = 1$).

2.3 Markov Switching in Herding Behavior with Time Varying Coefficients

Instead of assuming a constant coefficient within each regime, Markov switching model can relax this assumption by allowing the regime dependent coefficients to vary over time which is more likely to happen. Followings Kim (1994), the Markov Switching regression with time-varying coefficients for herding detection takes the form as first model

$$\begin{aligned} CSAD_t &= a_t(s_t) + \gamma_{1,t}(s_t) \left| R_{m,t} \right| + \gamma_{2,t}(s_t) R_{m,t}^2 + \varepsilon_t(s_t), & s_t = 1 \\ CSAD_t &= a_t(s_t) + \gamma_{1,t}(s_t) \left| R_{m,t} \right| + \gamma_{2,t}(s_t) R_{m,t}^2 + \varepsilon_t(s_t), & s_t = 2 \end{aligned} \tag{7}$$

and second model

$$\begin{aligned} CSAD_t &= a_t(s_t) + \gamma_{1,t}(s_t) R_{m,t} + \gamma_{2,t}(s_t) \left| R_{m,t} \right| + \gamma_{3,t}(s_t) R_{m,t}^2 + \varepsilon_t(s_t), & s_t = 1 \\ CSAD_t &= a_t(s_t) + \gamma_{1,t}(s_t) R_{m,t} + \gamma_{2,t}(s_t) \left| R_{m,t} \right| + \gamma_{3,t}(s_t) R_{m,t}^2 + \varepsilon_t(s_t), & s_t = 2 \end{aligned} \tag{8}$$

where $a_t(s_t)$, $\gamma_{1,t}(s_t)$, $\gamma_{2,t}(s_t)$ and $\gamma_{3,t}(s_t)$ are the regime dependent time varying coefficients.

$$\begin{aligned} a_t(s_t) &= \Pi_t(s_t)a_{t-1}(s_t) + u_{0t}(s_t) \\ \gamma_{1,t}(s_t) &= \Phi_t^1 \gamma_{1,t-1}(s_t) + u_{1t}(s_t) \\ \gamma_{2,t}(s_t) &= \Phi_t^2 \gamma_{2,t-1}(s_t) + u_{2t}(s_t) \\ \gamma_{3,t}(s_t) &= \Phi_t^3 \gamma_{3,t-1}(s_t) + u_{3t}(s_t) \end{aligned} \tag{9}$$

where $s_t = 1, 2$, Π_t is the coefficient of time-varying equation for $a_t(s_t)$ and Φ_t^k, $k = 1, 2, 3$ are the estimated coefficients of time-varying equation for $\gamma_{k,t}(s_t)$. $u_t(s_t)$ is the regime dependent errors of time-varying equation. Note that the time-varying equations are assumed to follow the AR(1) process and the variance of the time-varying equation cannot be constant over time. To predict the time varying coefficients and variances, we refer to Kalman Filter (Kim 1994). According to this model's properties, we can investigate the herding behavior over time.

3 Data Description

In this study, we can consider only one developed market country (Singapore) and six emerging market countries (China, Korea, Malaysia, the Philippines, Taiwan, and Thailand) to investigate whether there has been herding behavior in MSCI-FE index. Note that Hongkong stock market and Indonesia stock market are not included in our analysis because their data have a limited range. Table 1 presents the descriptive statistics of financial market returns of the countries under investigation. The daily data of market returns, collected from Bloomberg, span from December 26, 2014 to June 17, 2018. All data series are stationary as shown by the Augmented Dickey-Fuller test (ADF) and Minimum Bayes Factor Augmented Dickey-Fuller (MBF-ADF). The negative skewness indicates an asymmetric distribution in which the curve appears distorted or skewed to the left hence meaning that the MSCI-Far East Ex Japan index returns have a downward trend, or that there is a substantial probability of a big negative return. Also, considering the Jarque-Bera normality test and it is evident that stock returns reject the null hypothesis of normal distribution.

Table 2 shows the summary statistics of the cross-sectional absolute deviation (CSAD) (Column 2), market return (Column 3), absolute market return (Column 4), and market return squared (Column 5). This table provides the mean, median, maximum, minimum and standard deviation of all data series. The maximum of CSAD is 0.025 while the minimum is 0.000. This indicates that there is a small difference between MSCI-FE index return and all stock market index returns within this group. For the MSCI-FE index return, we find that the maximum market return is 0.035 while the minimum is -0.051, indicating the existence of both positive and negative returns in the MSCI-FE index where the negative return is larger than the positive return.

4 Empirical Result

4.1 Model Selection

Prior to examining the herding behavior, in this study, we consider four competing models which are linear regression model, linear regression with time varying parameter (TV), Markov Switching regression (MS) and Markov Switching time varying regression (MS-TV). The motivation of this study is to determine which of these four

Table 1 Data description

	Market	China	Korea	Malaysia	Philippines	Singapore	Taiwan	Thailand
Mean	0.001	0.001	0.001	0.000	0.002	0.000	0.002	0.001
Median	0.000	0.000	0.000	0.000	0.000	0.000	0.000	0.000
Maximum	0.035	0.059	0.037	0.039	0.063	0.028	0.045	0.052
Minimum	−0.051	−0.066	−0.045	−0.037	−0.072	−0.043	−0.067	−0.064
Std. Dev.	0.009	0.012	0.008	0.006	0.011	0.007	0.009	0.010
Skewness	−0.407	−0.221	−0.246	−0.323	−0.525	−0.245	−0.495	−0.173
Kurtosis	5.284	5.755	4.830	7.450	7.735	5.281	7.258	7.255
Jarque–Bera	413.483	547.397	252.454	1422.379	1654.121	382.854	1344.077	1281.748
MBFJarque-Bera	0.000	0.000	0.000	0.000	0.000	0.000	0.000	0.000
ADF test	−36.457***	−36.732***	−41.335***	−37.868***	−39.459***	−39.921***	−41.352***	−41.033***
MBF-ADF test	0.000	0.000	0.000	0.000	0.000	0.000	0.000	0.000

Table 2 Summary statistics

	$CSAD_t$	$R_{m,t}$	$|R_{m,t}|$	$R^2_{m,t}$
Mean	0.006	0.000	0.006	0.000
Median	0.005	0.000	0.005	0.000
Maximum	0.025	0.035	0.051	0.003
Minimum	0.000	−0.051	0.000	0.000
Std. Dev.	0.003	0.009	0.006	0.000

Table 3 Model selection

	Linear	Linear-TVTP	MS	MS-TVTP
First model	−16026.54	−3108.34	−16299.28	−15235.80
Second model	−16025.19	−3110.34	−16300.31	−16064.10

models is better in producing an analysis of herding behavior for MSCI-FE index. Hence, in this subsection, we construct the Akaike Information Criterion (AIC) to compare these four models. Then, we evaluate the four models based on the minimum AIC. The model comparison result is provided in Table 3, where we can see that MS outperforms other competing models. This indicates that there is a strong evidence in favor of the regime dependent coefficient models for all MSCI-FE index, confirming the existence of two market regimes for MSCI-FE index. In the case of the MS-TVTP, we find that it does not perform the best in our empirical examination, but its AIC is not far from that of the best fit model.

4.2 Estimation Results

The results of herding behavior examination of Eq. (1) and Eq. (3) are provided in Tables 4 and 5, respectively. Table 4 shows no herding behavior in MSCI-FE index. Three out of the four models show the positive value of the coefficient of R^2_{mt}, γ_2.

Table 4 shows the result of herding behavior under $CSAD_t = f(|R_{m,t}|, R^2_{m,t})$ specification. The parameter estimates are obtained from four models. It is important to note that the two regimes are easily distinguishable from the estimated levels of the constant parameter (a) for each state. In this case, the estimated constant term of regime 1 (0.0049) is higher than regime 2 (0.0035). This indicates that regime 1 is interpreted as the market upturn, while regime 2 is market downturn. Remind that the herding behavior is determined by the coefficient (γ_2). If the coefficient (γ_2) is negative, it means that there exists the herding behavior in the MSCI-FE market. However, the result obtained from the best fit MS model reveals the absence of herding behavior in MSCI-FE as indicated by the positive sign of $\gamma_2(s_t)$ for both market upturn and downturn. An explanation for this result lies in the cross-sectional diversity of

Table 4 The empirical tests examining the existence of herding effects in MSCI-FE index in the first model

Model		a	γ_1	γ_2	P
Model 1	Linear	0.0037	−0.0043	**0.2974**	
Model 2	Linear TV	0.0037	−0.0033	**0.2949**	
Model 3	MS (regime1)	0.0049	0.0159	**0.3745**	p_{11} 0.627
	MS (regime2)	0.0035	−0.0115	**0.2327**	p_{22} 0.877
Model 4	MS-TV (regime 1)	0.0041	0.2453	**−0.0029**	p_{11} 0.9999
	MS-TV (regime 2)	0.00004	0.00005	**0.00001**	p_{22} 0.9849

Note The parameter estimates of TV type models are calculated from the mean of the time varying parameter

Table 5 The empirical tests examining the existence of herding effects in MSCI-FE in the second model

Model		a	γ_1	γ_2	γ_3	P
Model 1	Linear	0.0037	−0.0053	0.3016	**−230.80**	–
Model 2	Linear TV	0.0037	−0.0027	0.2978	**−95.07**	–
Model 3	MS (regime1)	0.0052	0.0201	0.3142	**−1.8348**	0.6490
	MS (regime2)	0.0034	−0.0145	0.2709	**1.6016**	0.8860
Model 4	MS TV (regime 1)	0.0038	0.2824	−0.3138	**−0.0080**	0.9990
	MS TV (regime 2)	0.0000	0.0001	−0.0004	**0.0000**	0.9990

Note The parameter estimates of TV type models are calculated from the mean of the time varying parameter

MSCI-FE, resulting in a greater dispersion as every stock reacts differently to the market return movement. Our result corresponds to the results from linear and linear TV models. However, comparing the results of the MS model to MS-TV model, it is evident that the average time varying parameter of regime 1 shows a negative sign of $\gamma_2(s_t)$ in regime 1, but it is positive in regime 2. We infer that the MS-model may fail to capture herding effect under market upturn condition, although MS model outperforms MS-TVTP with respect to AIC. We plot the time varying coefficients in Fig. 1. The blue line presents the time varying coefficients under the market upturn regime. We can see that the time varying coefficients $\gamma_{2,t}(s_t)$ in the periods 2014–2016 are negative, which means the existence of intentional herding behavior during those periods. For the time varying coefficients $\gamma_{2,t}(s_t)$ of regime 2 or market downturn, the dynamic change is quite weak, and it is rather stable and positive along the sample period. This indicates that there is no herding behavior existing in market downturn.

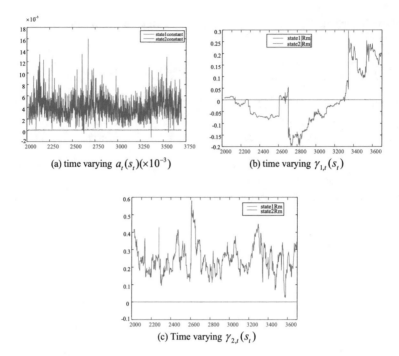

Fig. 1 The regime dependent time varying coefficient for Eq. (1): regime 1 (blue line) and regime 2 (red line)

Furthermore, the result of Table 4 also shows the probability of staying in regime 1 and regime 2. p_{11} is the probability of being in the market upturn while p_{22} is defined as the probability of persistence in the market downturn regime. According to the MS model, the transition probability of regime 1 and that of regime 2 are 0.627 and 0.877 respectively. This suggests that the market downturn regime is relatively more persistent.

Next, we turn our attention to the examination of herding from $CSAD_t = f(R_{m,t}, |R_{m,t}|, R_{m,t}^2)$ specification. Like the case of Table 4, we can examine the presence of the herding behavior from the coefficient of $R_{m,t}^2$. In this test, we have added the variable $R_{m,t}$ to permit the interpretation of asymmetric effects. Remind that the herding behavior of this model is determined by the coefficient (γ_3). If the coefficient (γ_3) is negative, it means that there exists the herding behavior in the MSCI-FE index. The results of four models provide a substantial evidence of the herding behavior of investors for the MSCI-FX as indicated by the negative sign of γ_3. In the case of linear and linear time-varying models, the estimated coefficients are -230.80 and -95.07, respectively. In the case of MS and MS-TV, the herding exists only in regime 1. We expect that if the $R_{m,t}$ is added into the herding equation, in the sense of econometrics, the omitted variable is solved, and thereby parameter estimate results. The regime dependent time varying coefficient $\gamma_{3,t}(s_t)$ for market

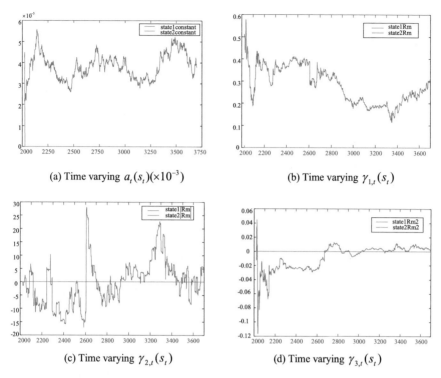

(a) Time varying $a_t(s_t)(\times 10^{-3})$

(b) Time varying $\gamma_{1,t}(s_t)$

(c) Time varying $\gamma_{2,t}(s_t)$

(d) Time varying $\gamma_{3,t}(s_t)$

Fig. 2 The regime dependent time varying coefficient for Eq. (3): Regime 1 (Blue line) and Regime 2 (Red line)

upturn (blue line) and market downturn (red line) are illustrated in Fig. 2d. We also observe that the time varying coefficient of regime 1 shows a negative sign during 2014–2016, which corresponds to the result from Table 4.

According to MS model, the probability of staying in regime 1 and that in regime 2 are 0.649 and 0.886, respectively. The market downturn regime is highly persistent for MSCI-FE index.

5 Conclusion

Herding behavior is formed when most of the individual investors follow the decisions of other institutional investors without any reason. The herding causes the prices of assets to deviate from their fundamental values and thereby leading to uncertainty and inefficiency in the market. Recently, various methods and models are proposed to examine the presence of herding behavior and its effect in the capital market; however, the results are mixed and not clear. Motivated by this reason, this study proposes a variant of the standard test model and employs a new herding behavior model for MSCI-FE index under different market regimes.

The main contribution of this paper is that we trace the herding behavior from the MSCI-FE index as a whole as opposed to previous studies that examined it at the stock level. Also, we introduce a new herding behavior model that exhibits the structural change in the market. The asymmetric nature of market upturn and downturn in the market cycles can be modeled by allowing the common factor dynamics to be subject to regime shifts according to a Markov process.

To detect the herding behavior, we employ the CSAD method of Chang et al. (2000) and Chiang and Zheng (2010). This study considers two herding equations. Then, four econometric models consisting of linear regression, linear regression with time varying coefficient, Markov Switching regression, and Markov Switching regression with time varying coefficient are used to estimate two herding equations.

The result of Eq. (1) provides a weak evidence supporting the existence of the herding behavior in MSCI-FE index as the coefficients $R^2_{m,t}$ are mostly present with the positive sign. We can find the presence of herding effect in market upturn regime during 2014–2016, according to the MS-TV result. In the case of the second herding equation, in contrast to the first equation result, we find that the coefficients $R^2_{m,t}$ show a negative value in most cases, except for market downturn.

From the above results, investors will neglect their own information and choose to mimic larger investors during the market upturn. On the other hand, in the downturn market, we do not find any evidence of herding behavior in MSCI-FE index. For future study, it should be possible to divide the period into two sub-periods, say 2014–2016 and 2017–2018, in order to confirm the result of this study. Also, if possible, the study should be pursued on the herding behavior to identify the potential trends in the future and forecast of the economic crisis based on herding behavior.

Acknowledgements The authors would like to thank Dr. Laxmi Worachai for her helpful comments on an earlier version of the paper. The authors are also grateful for the financial support offered by Center of Excellence in Econometrics, Chiang Mai University, Thailand.

References

Babalos, V., Balcilar, M., Gupta, R., & Philippas, N. (2015). Revisiting herding behavior in REITS: A regime-switching approach. *Journal of Behavioral and Experimental Finance, 8*(1), 40–43.

Babalos, V., Stavroyiannis, S., & Gupta, R. (2015). Do commodity investors herd? Evidence from a time-varying stochastic volatility model. *Resources Policy, 46*, 281–287.

Bonser-Neal, C., Jones, S. L., Linnan, D., & Neal, R. (2002). *Herding, feedback trading and foreign investors*. Indiana University.

Bowe, M., & Domuta, D. (2004). Investor herding during financial crisis: A clinical study of the Jakarta Stock Exchange. *Pacific-Basin Finance Journal, 12*(4), 387–418.

Brunnermeier, M. K. (2001). *Asset pricing under asymmetric information: Bubbles, crashes, technical analysis, and herding*. Oxford University Press on Demand.

Chang, E. C., Cheng, J. W., & Khorana, A. (2000). An examination of herd behavior in equity markets: An international perspective. *Journal of Banking and Finance, 24*(10), 1651–1679.

Chiang, T. C., & Zheng, D. (2010). An empirical analysis of herd behavior in global stock markets. *Journal of Banking and Finance, 34*(8), 1911–1921.

Demirer, R., & Kutan, A. M. (2006). Does herding behavior exist in Chinese stock markets? *Journal of international Financial Markets, Institutions and Money, 16*(2), 123–142.

Demirer, R., Kutan, A. M., & Chen, C. D. (2010). Do investors herd in emerging stock markets?: Evidence from the Taiwanese market. *Journal of Economic Behavior and Organization, 76*(2), 283–295.

Economou, F., Kostakis, A., & Philippas, N. (2011). Cross-country effects in herding behaviour: Evidence from four south European markets. *Journal of International Financial Markets, Institutions and Money, 21*(3), 443–460.

Gavriilidis, K., Kallinterakis, V., & Micciullo, P. (2007). The Argentine crisis: A case for herd behavior. *SSRN Electronic Journal.*

Kim, C. J. (1994). Dynamic linear models with Markov-switching. *Journal of Econometrics,* 144–165.

Lao, P., & Singh, H. (2011). Herding behaviour in the Chinese and Indian stock markets. *Journal of Asian Economics, 22*(6), 495–506.

Maneejuk, P., Yamaka, W., & Leeahtam, P. (2019). Modeling nonlinear dependence structure using logistic smooth transition Copula model. *Thai Journal of Mathematics,* 121–134.

Morelli, D. (2010). European capital market integration: An empirical study based on a European asset pricing model. *Journal of International Financial Markets, Institutions and Money, 20*(4), 363–375.

Park, B. J. (2011). Asymmetric herding as a source of asymmetric return volatility. *Journal of Banking and Finance, 35*(10), 2657–2665.

Pastpipatkul, P., Yamaka, W., & Sriboonchitta, S. (2016). Dependence structure of and co-movement between Thai currency and international currencies after introduction of quantitative easing. In *Causal inference in econometrics* (pp. 545–564). Cham: Springer.

Phadkantha, R., Yamaka, W., & Sriboonchitta, S. (2019). Analysis of herding behavior using Bayesian quantile regression. In *International Econometric Conference of Vietnam* (pp. 795–805). Cham: Springer.

Putra, A. A., Rizkianto, E., & Chalid, D. A. (2017). The analysis of herding behavior in Indonesia and Singapore stock market. In *International Conference on Business and Management Research (ICBMR 2017).* Atlantis Press.

Ramadan, I. Z. (2015). Cross-sectional absolute deviation approach for testing the herd behavior theory: The case of the ASE Index. *International Journal of Economics and Finance, 7*(3), 188–193.

Recovering from the Recession: A Bayesian Change-Point Analysis

Nguyen Ngoc Thach

Abstract After the 2008–2009 Great Recession, the world economy entered a long process of recovery. However, it was stated in a certain report that the world economy had decisively escaped from the recession. According to the author's viewpoint, an economy achieves real restoration only when the post-crisis growth rate of the economy regains the pre-crisis level. In order to confirm this, the present study was conducted using the Bayesian change-point method to specify the point of change in the economic growth of the world and in some of its regions. In addition, the average growth rate values between the pre-crisis and post-crisis (change-point) periods were compared, based on which, it could be determined whether the world and the regional economies have overcome the crisis. The empirical results revealed that Africa has been experiencing a sharp decline after the change-point, while the world as a whole and the rest of its regions, such as the developing and emerging countries including Vietnam, have been undergoing a decrease in their respective GDP growth. These findings confirmed that while the Asia-Pacific region has surely escaped from the recession, the other regional economies have not reached a complete retrieve so far.

Keywords Escape from the recession · Bayesian change-point analysis · Economic growth rate

1 Introduction

Despite the optimistic statements from certain organizations (such as National Financial Supervisory Commission 2014), low economic growth rates and a decline in the wealth of most developing and developed countries have been observed in the period post-2009. The experience of overcoming economic crises in the past rendered everyone to expect a rise in the world and regional economies after the 2008–2009 Great

N. N. Thach (✉)
Banking University HCMC, 39 Ham Nghi street, District 1, Ho Chi Minh City, Vietnam
e-mail: thachnn@buh.edu.vn

© The Editor(s) (if applicable) and The Author(s), under exclusive license to Springer Nature Switzerland AG 2021
N. Ngoc Thach et al. (eds.), *Data Science for Financial Econometrics*, Studies in Computational Intelligence 898, https://doi.org/10.1007/978-3-030-48853-6_24

Recession. However, the expected persistent regeneration of the economic indicators in the post-crisis economies has not occurred. Thach (2014) stated that in order to escape from the bottom of the crisis or to recover from the recession, economies require achieving full recovery of the key macroeconomic indicators, especially the real GDP growth.

Numerous studies on the impact of the 2008–2009 economic recession on the world and regional economies, using simple research applying descriptive statistics to complicated econometric models, have been published in the last ten years, in different languages. A common topic of research in these studies has been to analyze the effects of this crisis on the different countries and regions of the world, different industries and sectors, and different economic activities. In regard to such an influential phenomenon as that of the 2008–2009 Great Recession, there have been numerous studies thus it is difficult to conduct a perfect review of the literature. Indeed, performing Google search using the keywords "great recession" alone yields more than 88,700,000 results, not to mention the rest of the keywords associated with the 2008–2009 Great Recession such as "global financial crisis", "global economic crisis", "global financial and economic crisis", and so on. Therefore, concerning the effects of the 2008–2009 Great Recession, we introduce two research groups.

First, several studies have applied descriptive statistical analysis. Utilizing the UNCTAD database, Mareike and Kennan (2009) observed that the global economic crisis exerted a negative impact on the export volumes and prices of developing countries. In a study conducted on the recent global economic crisis, Boorman (2009) elaborated upon the influence of this crisis on the emerging market economies, with a focus on financial transmission mechanism. The research results revealed that emerging Europe, the CIS countries, and the newly industrialized Asian economies were the regions that experienced the worst effects. The findings of a study conducted by Dolphin and Chappell (2010) revealed that output, export, remittance, and capital flows decreased deeply in the emerging and developing countries during the period of the crisis. Lost GDP, lost export, lost remittance, etc., were estimated by comparing the figures over the period of 2003–2007 with those in the years 2008, 2009, and 2010. Gardo and Martin (2010) reviewed the financial and economic developments in central, eastern, and south-eastern regions of Europe since the occurrence of the global crisis, and concluded that the countries with the largest economic imbalances were the most affected ones. In a study conducted by Allen and Giovannetti (2011), the channels through which the financial and economic crisis was transmitted to Sub-Saharan Africa were analyzed. The afore-mentioned study revealed trade and remittance as the two main channels. Alcorta and Nixson (2011) analyzed the impact of the global economic crisis on the manufacturing sector in a sample of 11 low-income and middle-income countries and concluded that the countries which encountered major problems in their manufacturing sectors had poor infrastructure, high import dependence, and shortage of skilled labor. Barlett and Prica (2012), while investigating the impact of the global economic recession on 12 countries of south-east Europe, argued that the initial conditions at the beginning of the crisis determined the between-country differences in the growth impact. In an event analysis, Comert and Ugurlu (2015) studied the effects of the 2008 global economic crisis on the 15 most-affected developing economies, and the reasons underlying the bad performances of these countries. According to their results, trade channel was the most substantial mechanism in the transmission of the crisis from the advanced countries to the developing economies.

Second, in regard to the more complex models for the effects of the 2008–2009 Great Recession, Rose and Spiegel (2009) modeled the causes of the 2008–2009 economic crisis and its manifestations by applying a Multiple Indicators Causes (MIMIC) model on a cross-section of 107 countries. The authors explored over sixty potential causes of the crisis. Meanwhile, Berkmen et al. (2009) investigated the influence of the global economic crisis on 126 countries, by using both descriptive evidence and cross-country regression. The authors, therefore, were able to explore a broad set of explanatory variables generating differences in the growth impact across countries; among these variables, trade link and financial vulnerability were identified as the most important ones. McKibbin and Stoeckel (2009) modeled the global economic crisis as a combination of shocks to the global housing markets and the sharp increases in the risk premia of firms, households, and international investors, in a dynamic stochastic general equilibrium (DSGE) model. The model contained six production and trade sectors in 15 major economies and regions. With the application of panel data econometrics, Raz et al. (2012) compared the effects of the financial crises of 1997 and 2008 on the East Asian economies, and concluded that since these countries had managed to establish strong economic fundamentals, the magnitude of the impact of the 2008 crisis was relatively smaller in comparison to that of the 1997 crisis.

It is noteworthy that the above-stated studies applied either a simple analysis of data or econometric modeling using the frequentist approach. Moreover, these studies did not utilize the change-point method which is based on the Bayesian framework. The change-point method is appropriate for defining an abrupt change in a time series and for comparing the average values of an index (GDP growth rate, in case of the present study) between two periods split by the point of change. Therefore, in the present study, by performing a Bayesian change-point analysis utilizing the database on the economic growth of the world and that of a few emerging and developing regions, and comparing the economic growth in the period after the 2010–2018 crisis with that in the flourishing period prior to the Great Recession, the author verified the hypothesis on the real recovery of the studied economies from the recession.

2 Research Methodology and Data

2.1 Methodology

In recent years, the Bayesian approach and a few other non-frequentist probabilistic methods have become increasingly popular as powerful statistical techniques in social sciences (Nguyen et al. 2019; Briggs and Nguyen 2019; Kreinovich et al. 2019;

Tuan et al. 2019). For instance, Bokati and Kreinovich (2019) proposed a heuristic maximum entropy approach to portfolio optimization, or Thach et al. (2019) applied the Bayesian approach while conducting research on economic growth.

Previously, econometric models based on frequentist statistics have assisted researchers in achieving desired results. However, in several cases, especially in the ones predicting the future values of the variables of interest, prediction outcomes have been observed to be inaccurate. Owing to the updating of the prior information of all the model parameters with observed data, the Bayesian analysis enables reducing the dependence on the sample size. Therefore, the Bayes models are robust to sparse data. Prior information is defined as the expert knowledge available prior to the observed data. Moreover, the Bayesian approach provides straightforward probabilistic interpretations regarding the simulation results. For instance, in relation to the variable n in the present study, the probability that the estimate of this parameter was between 2001.284 and 2017.778 was observed to be 95%, or the threshold year ranged from 2001 to 2017 with 95% probability.

Change-point analysis has been conducted previously using traditional threshold regression (Tong 1983, 1990). Frequentist change-point investigations have also appeared in previous studies such as those conducted by Worsley (1986), Siegmund (1988), and several others. In particular, the change-point problem was investigated within the Bayesian framework by Carlin et al. (1992). In the Bayesian change-point models established by these authors, desired marginal posterior densities were obtained by utilizing an iterative Monte-Carlo method, an approach which avoids sophisticated analytical and numerical high-dimensional integration procedures.

A general formula of the threshold autoregressive (TAR) model is expressed as follows:

$$y_t = \beta_0^{(h)} + \sum_{i=1}^{p_h} \beta_1^{(h)} y_{t-i} + \varepsilon_t^{(h)} \quad \text{for } r_{h-1} \leq y_{t-1} < r_h \tag{1}$$

where h represents the number of regimes with $h = 1, ..., g$, and l represents the threshold lag of the model and a positive integer. In regard to each h, $\varepsilon_t^{(h)}$ is the sequence of independent and identically distributed normal random variables with mean = zero and variance = σ_h^2. The threshold values r_h satisfy $-\infty = r_0 < r_1 < \cdots < r_g = \infty$, and form a partition of the space of y_{t-l}.

In the case where $h = 2$, the Bayesian analysis of a TAR(2: p_1, p_2) process was proposed (p_1 and p_2 denote the first and second observations, respectively). The idea of the proposed procedure was to transform a TAR model into a change-point problem in linear regression via arranged autoregression. This resulted in obtaining a special model represented by Eq. (2):

$$y_t = \begin{cases} \beta_0^{(1)} + \sum_{i=1}^{p_1} \beta_1^{(1)} y_{t-i} + \varepsilon_t^{(1)}, & y_{t-l} \leq r \\ \beta_0^{(2)} + \sum_{i=1}^{p_2} \beta_1^{(2)} y_{t-i} + \varepsilon_t^{(2)}, & y_{t-l} > r \end{cases} \tag{2}$$

The change-time model obtained in the present study was fit as follows. In a sequence of counts $y_1, y_2, \ldots, y_N$, where the average of the counts has a certain value for the time steps 1 to n, and a different value for the time steps $n + 1$ to N, the counts at each time step i as a Poisson variable having the density function were modeled as follows:

$$Poisson\,(y; \mu) = e^{-\mu} \frac{\mu^y}{y!} = exp(y\log - \log(y!)) \tag{3}$$

where μ represented the mean of the distribution and was modeled as a flat distribution, which has the density of 1.

The initial mean μ_1 jumps to a new value μ_2 after a random time step n. Therefore, the generative model attains the following form:

$$y_i \sim \begin{cases} Poisson\,(y_i; \mu_1) \ 1 \le i \le n \\ Poisson\,(y_i; \mu_2) \ n < i \le N \end{cases}$$
$$\mu_i \sim 1(flat)$$
$$n \sim Uniform\,(1, 2, \ldots, N) \tag{4}$$

2.2 Bayesian Change-Point Model

Change-point modeling deals with the stochastic data, in general, time series, which change abruptly at a certain time point. The researchers are interested in specifying the point of change and the characteristics of the stochastic processes prior to and after the change.

In order to localize the point of change (n) in the real GDP growth rate of the world and that of certain selected regions, the following Bayesian model was fitted with the non-informative priors for all parameters used in the present study:

The likelihood model was as follows:

$$GDPr_t \sim Poisson\,(\mu_1), \ if \ year_t < n$$
$$GDPr_t \sim Poisson\,(\mu_2), \ if \ year_t > n \tag{5}$$

where $t = 1, \ldots, 18$.

The prior distributions were:

$$\mu_1 \sim 1$$
$$\mu_2 \sim 1 \tag{6}$$
$$n \sim Uniform\,(2000, \ 2018)$$

where variables μ_1, μ_2 have flat priors with a density equal to one, which is appropriate for the research, relative to which there have been no previous studies using the Bayesian approach, and n is assigned a uniform prior.

There were three parameters in the model established in the present study: μ_1, μ_2 and n. The mixture distribution of the GDPr variable was specified as the mean of a Poisson distribution by the following substitutable expression:

$$(\{\mu_1\} * sign\,(year\, <\, \{n\}) + \{\mu_2\} * sign\,(year\, >\, \{n\}))$$

Owing to the possible high autocorrelation in the MCMC chains, the MCMC size was increased to 40,000–80,000 in order to obtain higher precision in the estimates.

2.3 Data Description

The present study utilized real GDP, which is the most commonly used composite index standing for the main tendency of economic recovery. The time-series data on the real GDP annual growth rate of the world, Asia-Pacific region, Africa, the group of emerging and developing countries, and Vietnam, for the 2001–2018 period was obtained from the IMF database (IMF 2019a, b).

Figure 1 illustrates the GDP dynamics of the world along with some of the other regions and Vietnam, during the period between 2001 and 2018. In general, the average economic growth was higher for the 2001–2008 period compared to the 2009–2018 period. The Great Recession commenced with the US financial crisis in 2007 and spread rapidly throughout the world. The world's lowest economic growth was observed in the year 2009.

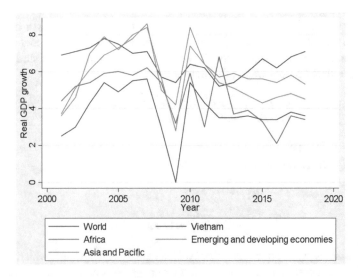

Fig. 1 Real GDP growth in the world during 2001–2018. *Source* IMF (2019a, b)

3 Empirical Results

3.1 Bayesian Simulation

The estimation results of the five models established in the present study (the world, emerging and developing economies, Asia-Pacific, Africa, and Vietnam) are listed in Table 1.

Similar to OLS regression, the posterior mean values are similar to the coefficient estimates, while standard deviation in Bayesian estimation is explained the same as standard error. However, credible intervals in Bayesian analysis are understood differently; the probability of the posterior mean value at the credible interval, for example, when of the World model was 2001.243–2017.653, was 95%. The precision of the posterior mean estimates was described by their Monte Carlo standard errors (MCSE). The smaller the MCSE, the more precise were the parameter estimates. When the MCMC sample size was 10,000, most of the MCSE estimates as calculated in the research models of the present study were accurate to one decimal position and sufficiently suitable for the MCMC algorithm (Table 1).

Table 1 Simulation results for five models

	Mean	Std. Dev	MCSE	Median	[95% Cred. Interval]	
World						
n	2008.98	5.26228	0.13613	2008.49	2001.24	2017.65
μ_1	3.56026	1.00413	0.02805	3.56324	1.33315	5.58033
μ_2	3.35771	0.8047	0.02023	3.31112	1.86955	5.15472
Emerging and developing economies						
n	2009.7	4.99265	0.15724	2010.19	2001.28	2017.62
μ_1	5.53072	1.30712	0.0289	5.62664	2.28535	7.88594
μ_2	4.76274	1.03304	0.03165	4.81356	2.63278	6.75119
Asia and Pacific						
n	2009	5.53599	0.15058	2008.79	2001.19	2017.73
μ_1	5.48237	1.32076	0.05335	5.62852	2.23951	7.8573
μ_2	5.57043	1.09058	0.03124	5.56388	3.48115	7.86871
Africa						
n	2010.38	4.16626	0.09068	2010.86	2001.65	2017.21
μ_1	4.91275	0.93875	0.02173	4.8432	3.33851	6.95378
μ_2	3.42555	0.84568	0.02025	3.43316	1.80058	5.06107
Vietnam						
n	2009.37	5.33305	0.13863	2008.88	2001.3	2017.72
μ_1	6.64897	1.1892	0.03767	6.49895	4.64275	9.45182
μ_2	6.36629	1.42632	0.03998	6.13573	4.38199	10.1683

Source The author's calculation

Furthermore, according to the estimation results of the present study presented in Table 1, the change occurred in 2008 for the world, in 2009 for the emerging and developing countries, Asia-Pacific, and Vietnam, and in 2010 for Africa. The decline in the economic growth rate, although significant, was slight, going from an estimated average of 3.56 to 3.36% for the world, from 5.53 to 4.76% for the group of emerging and developing economies, 6.65 to 6.37% for Vietnam, and 4.91 to 3.43% for Africa. The Asia-Pacific region, on the other hand, achieved a higher growth rate, going from 5.48% up to 5.57%, in comparison with the previous period.

3.2 Test for MCMC Convergence

MCMC convergence is required to be assessed prior to be able to trust the estimation results of a study, as the Bayesian inference is valid once the MCMC chain has converged to the target distribution. In order to achieve this validation, graphical diagnostics may be performed for all the parameters. Trace plots demonstrated that MCMC chains exhibited no trends and traversed the distribution rapidly. As observed in the autocorrelation plots, the MCMC chains were well mixed (low autocorrelation). The diagnostic plots did not indicate any signs of non-convergence for all the parameters. Therefore, it could be concluded that the MCMC chain was stationary. Figures 2 and 3 present the convergence diagnostics for the world economic growth model. The results were similar for the other four models.

Moreover, in addition to the convergence verification, the two major criteria which measured the efficiency of MCMC of the parameter estimates were the acceptance rate of the chains and the degree of autocorrelation. The model summaries in the present study indicated that the acceptance rate of these models ranged from 21 to 26%, while according to a report by Roberts and Rosenthal (2001), the range of 15–50% is acceptable.

A test on convergence was conducted for, in addition to all parameters, the functions of interest. The diagnostic plots for the ratio appeared satisfactory (Fig. 3). In the

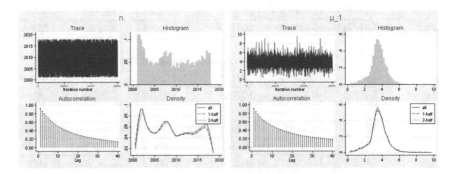

Fig. 2 Convergence test for parameters n and μ_1. *Source* The author's calculation

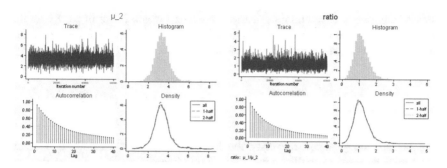

Fig. 3 Convergence test for parameter μ_2 and the ratio μ_1/μ_2. *Source* The author's calculation

model for the world economic growth, the ratio was as follows: $\mu_1/\mu_2 = 1.13401$. This implied that after 2008, the mean growth rate of the world economy decreased by a factor of approximately 1.13401, with a 95% credible range of [0.42, 2.16].

3.3 Discussion

The results of the present study indicated that during the period after the 2008–2009 Great Recession, the world and regional economies were moving slow on the road to regeneration. In other words, most of the economies investigated in the present study have not yet recovered from the recession. The only exception was the Asia-Pacific region with fast-growing Southeast Asian economies, which have fostered the economic recovery of the whole region, leading to a higher growth rate in the period of 2010–2017 compared to the pre–2009 period of economic boom.

Certain important explanations for the empirical results obtained in the present study are as follows. Global economic growth continues to be repressed. Risk remains high in several regions of the world, from sovereign crisis in certain Eurozone countries in the early 2010s to the recent US-China trade war, uncertainty related to the Brexit process in the UK, and the other geopolitical tensions that have increased the risk of inflating energy prices, the subdued demand for investment, and consumption in the developed and emerging market countries. These reasons are responsible for the sluggish recovery of the world economy to be conserved. IMF has recently forecasted the global economic growth to reach the value of 3.2% only in the year 2019 and 3.5% in the year 2020 (IMF 2019a, b).

Despite being located in a dynamic economic region of Southeast Asia, Vietnam has failed to regain a high growth rate during the period of 2001–2007. In spite of the efforts by the Government to improve the quality of institutional environment and support the domestic firms, Vietnam has not overcome the major barriers to growth, such as the slowly developing infrastructure system which is not in pace with the rapid economic growth, the low quantity of human resources, and the small and ineffective investment in education and R&D.

4 Conclusion

The present study aimed to verify the hypothesis that the world and regional economies have overcome from the 2008–2009 Great Recession. In order to meet this aim, the Bayesian change-point analysis was performed using the data on the economic growth rate of the world economy and that of the emerging and developing regions of the world on the basis of IMF's dataset, in order to identify a point of change in their economic growth and compare their average growth rates between the pre-crisis and post-crisis periods. According to the results of the present study, the point of change arrived approximately in 2008, 2009, or 2010, depending on the extent of impact generated by the global economic crisis on the countries and regions as well as on their capacity to cope with the crisis. It was observed that the most dramatic slowdown in the real GDP growth occurred in Africa after 2009, while a decline was observed in the other regions as well. Interestingly, contrary to what was, the Asia-Pacific region achieved fast recovery in the economic growth, owing to the strong economic recuperation in the countries of Southeast Asia.

Vietnam is located in the economically dynamic Asia-Pacific region. In the period after 2009, Vietnam's economy attained a rather rapid growth. Nevertheless, the average growth rate of this economy has not yet regained the level it had reached prior to the 2008–2009 Great Recession. Therefore, at present, it is necessary for Vietnam to actively implement institutional and structural reforms which are aimed at enhancing its economic growth.

References

Alcorta, L., & Nixson, F. (2011). *The global financial crisis and the developing world: Impact on and implications for the manufacturing sector*. UNIDO: Development policy and strategic research branch. Working paper 06/2010.

Allen, F., & Giovannetti, G. (2011). The effects of the financial crisis on Sub-Saharan Africa. *Review of Development Finance, 1*(1), 1–27.

Bartlett, W., & Prica, I. (2012). *The variable impact of the global economic crisis in South East Europe*. LSEE Research on South Eastern Europe.

Berkmen, P., Gelos, G., Rennhack, R., & Walsh J. P. (2009). *The global financial crisis: Explaining cross-country differences in the output impact*. IMF Working Paper WP/09/280.

Bokati, M., & Kreinovich, V. (2019). Maximum entropy approach to portfolio optimization: Economic justification of an intuitive diversity idea. *Asian Journal of Economics and Banking, 3*(2), 17–28.

Boorman, J. (2009). The impact of the financial crisis on emerging market economies: The transmission mechanism, policy response and the lessons'. In *Global Meeting of the Emerging Markets Forum 2009*, Mumbai.

Briggs, W. M., & Nguyen, H. T. (2019). Clarifying ASA's view on P-values in hypothesis testing. *Asian Journal of Economics and Banking, 3*(2), 1–16.

Carlin, B. P., Gelfand, A. E., Smith, A. F. M. (1992). Hierarchical Bayesian analysis of changepoint problems. *Journal of the Royal Statistical Society. Series C (Applied Statistics), 41*(2), 389–405.

Cömert, H., & Uğurlu, E. N. (2015). *The impacts of the 2008 global financial crisis on developing countries: The case of the 15 most affected countries.* ERC Working Papers in Economics 15/09

Dolphin, T., & Chappell, L. (2010). *The effect of the global financial crisis on emerging and developing economies.* Report of The Institute for Public Policy Research.

Gardo, S., & Martin, R. (2010). *The impact of the global economic and financial crisis on Central, Eastern and South-Eastern Europe: A stock-taking exercise.* Occasional paper series No 114.

IMF. (2019a). *Still sluggish global growth.* https://www.imf.org/en/Publications/WEO/Issues/2019/07/18/WEOupdateJuly2019

IMF. (2019b). *World economic development.* https://www.imf.org/external/datamapper/NGDP_RPCH@WEO/OEMDC/ADVEC/WEOWORLD?year=2019.

Kreinovich, V., Thach, N.N., Trung, N.D., & Thanh, D.V., eds. (2019). Beyond Traditional Probabilistic Methods in Economics. Cham: Springer, https://doi.org/10.1007/978-3-030-04200-4

Mareike, M., & Kennan, J. (2009). *The implications of the global financial crisis for developing countries' export volumes and values.* ODI Working Papers 305.

McKibbin, W. J., & Stoeckel, N. (2009). *The global financial crisis: Causes and consequences.* The Lowy Institute for International Policy. Working papers in international economics, 2(09)

National Financial Supervisory Commission. (2014). Macroeconomic report of 2013 and perspective for 2014.

Nguyen H. T., Trung N. D., & Thach N. N. (2019). Beyond traditional probabilistic methods in econometrics. In V. Kreinovich, N. Thach, N. Trung, & D. Van Thanh (Eds.), *Beyond traditional probabilistic methods in economics. ECONVN 2019.* Studies in Computational Intelligence (Vol. 809). Cham: Springer.

Raz, A. F., Indra, T. P. K., Artikasih, D. K., & Citra, S. (2012). Global financial crises and economic growth: Evidence from East Asian economies. *Bulletin of Monetary, Economics and Banking, 15*(2), 1–20.

Roberts, G. O., & Rosenthal, J. S. (2001). Optimal scaling for various Metropolis-Hastings algorithms. *Statistical Science, 16*(4), 351–367.

Rose, A. K., & Spiegel, M. M. (2009). *Cross-country causes and consequences of the 2008 crisis: Early warning.* The Federal Reserve Bank of San Francisco. Working Paper 2009–17.

Siegmund, D. (1988). Confidence sets in change point problems. *International Statistical Review, 56*(1), 31–48.

Thach, N. N. (2020). How values influence economic progress? An evidence from South and Southeast Asian countries. In N. Trung, N. Thach, & V. Kreinovich (Eds.), *Data science for financial econometrics. ECONVN 2020.* Studies in Computational Intelligence. Cham: Springer (to appear).

Thach, N. N., & Anh, L. H. (2014). Mot goc nhin khoa hoc ve su thoat day cua nen kinh te. Tap chi Cong nghe ngan hang [A View on Exit from Crisis Bottom. Banking Technology Review]. 99 (in Vietnamese).

Thach, N. N., Anh, L. H., & An, P. T. H. (2019). The effects of public expenditure on economic growth in Asia countries: A Bayesian model averaging approach. *Asian Journal of Economics and Banking, 3*(1), 126–149.

Tong, H. (1983). *Threshold models in non-linear time series analysis.* New York: Springer.

Tong, H. (1990). *Non-linear time series: A dynamical system approach.* New York: Oxford University Press.

Tuan, T. A., Kreinovich, V., & Nguyen, T. N. (2019). Decision Making Under Interval Uncertainty: Beyond Hurwicz Pessimism-Optimism Criterion. In Beyond Traditional Probabilistic Methods in

Economics, ECONVN 2019, Studies in Computational Intelligence. Edited by Vladik Kreinovich, Nguyen Ngoc Thach, Nguyen Duc Trung and Dang Van Thanh. Cham: Springer, vol. 809. https://doi.org/10.1007/978-3-030-04200-4_14

Worsley, K. J. (1986). Confidence regions and tests for a change-point in a sequence of exponential family random variables. *Biometrika, 73*(1), 91–104.

Ownership Structure and Firm Performance: Empirical Study in Vietnamese Stock Exchange

Tran Thi Kim Oanh, Dinh Thi Thu Hien, Hoang Thi Phuong Anh, and Dinh Thi Thu Ha

Abstract This paper aims to examine the relationship between ownership structure in terms of institutional and managerial ownership and firm performance. Using Bayesian approach and the data-set of companies listed in the VN30 basket of Vietnamese stock exchange in the period of 2012–2017, the results show a statistically significant negative relationship between institutional ownership and firm performance while a positive relationship between managerial ownership and firm performance at significant level.

Keywords Ownership structure · Managerial ownership · Institutional ownership · Agency cost

1 Introduction

Starting from the fact that there are companies managed by persons other than their owners, Berle and Means (1932) are among the first researchers to emphasize the separation between ownership and control exist in corporate environment, leading to negative impact on company value. In this context, managers benefit from exploiting the company's resources to serve their purposes. Most studies of the effect of owner-

T. T. Kim Oanh
University of Finance and Marketing, Ho Chi Minh City, Vietnam
e-mail: kimoanhtdnh@gmail.com

D. T. Thu Hien
Ho Chi Minh City Open University, Ho Chi Minh City, Vietnam
e-mail: hien.dtt@ou.edu.vn

H. T. Phuong Anh (✉)
University of Economics Ho Chi Minh City, Ho Chi Minh City, Vietnam
e-mail: anhtcdn@ueh.edu.vn

D. T. Thu Ha
Pacific Ocean University, Nha Trang, Vietnam
e-mail: dinhthuha.ueh@gmail.com

© The Editor(s) (if applicable) and The Author(s), under exclusive license
to Springer Nature Switzerland AG 2021
N. Ngoc Thach et al. (eds.), *Data Science for Financial Econometrics*,
Studies in Computational Intelligence 898,
https://doi.org/10.1007/978-3-030-48853-6_25

ship structure on company value are emphasized by agency cost theory (Jensen and Meckling 1976), which refers to the fact that in some cases, managers intend to make decisions that not aim at maximizing shareholder wealth. In addition, when board members own significant shareholdings, the agency costs will be reduced thanks to the adjustment of financial incentives between managers and owners (Fama and Jensen 1983).

Capital structure and ownership structure play an important role in the company performance and growth. Shareholders are the company real owners whose ownership structure and decisions affect company's operating strategy. Studies have shown that managerial ownership proportion have significant effect on company's performance and capital structure (Pirzada et al. 2015). Therefore, the proportion of managerial ownership is an important indicator of company performance and operating strategies to guarantee the stability and minimizing negative effects on company performance.

Topic about the relationship between ownership structure, capital structure and company performance has been researched in numerous countries. However, the research results are not consistent; besides, research on this topic in Vietnam is still limited.

This paper aims at 2 objectives:

Objective 1: examines whether the changes in the proportion of institutional ownership effect company performance.

Objective 2: examines whether the changes in the proportion of managerial ownership effect company performance.

2   Literature Review

2.1   Theory Background

Capital structure plays an important role for a company when it involves the ability to meet the needs of shareholders. Modigliani and Miller (1958) are the first to study the topic of capital structure and argue that capital structure does not play a decisive role in the value and performance of company. However, Lubatkin and Chatterjee (1994) as well as other studies prove the existence of the relationship between capital structure and the value of company. Modigliani and Miller (1963) show that their research model is ineffective when considering the impact of taxes since benefit from taxes on interest payments will increase company value when equity is replaced by debt.

However, in recent studies, the topic related to the influence of capital structure on company value no longer attracts the attention from researcher. Instead, they emphasize the relationship between capital structure and ownership structure in relation to the effect of strategic decision-makings by top managers (Hitt et al. 1991). These decisions will have an impact on company performance (Jensen 1986). Today, the main problem with the capital structure is to resolve the conflicts of corporate resource between managers and owners (Jensen 1989).

The fundamental theories related to ownership structure and firm performance, capital structure can be mentioned as follows.

2.1.1 Value and Corporate Performance of Firms

Capital structure is one of the key decisions of a company to maximize profits to the stakeholders. Moreover, a reasonable capital structure is also important for company to survive in competitive environment. Modigliani and Miller (1958) argue about the existence of an optimal capital structure when the risk of bankruptcy will be offset by tax savings due to debt using. Once this optimal capital structure is formed, company can maximize profits to stakeholders and these profits will often be higher than those of all equity firm.

Financial leverage is used to control managers but can also lead to bankruptcy of a company. Modigliani and Miller (1963) argue that a firm capital structure should consist entirely of debt because it will help to reduced tax on interest payments. However, Brigham and Gapenski (1996) argue that, in theory, the Modigliani–Miller (MM) model does not exist. But in fact, bankruptcy costs exist and correlate with the level of company debt. Thus, an increase in debt ratio may increase bankruptcy costs. Therefore, the authors argue that an optimal capital structure can only be obtained if the tax sheltering benefits from debt using be equal to bankruptcy cost. In this case, managers can identify when the optimal capital structure be achieved and try to maintain it. This is the only way that financial costs and weighted average cost of capital (WACC) are minimum so that the value and performance of company can increase.

2.1.2 The Agency Theory

Berle and Means (1932) were the first to develop Agency cost theory and they argued that there is an increase in the gap between ownership and control rights in large organizations, which is created by reduction in equity ownership. This situation will give managers the opportunity to pursue their own goals instead of increasing profits to shareholders.

Theoretically, shareholders are the owners of the company and the responsibility of the managers is to ensure the interests of shareholders are met. In other words, the responsibility of managers is to run the company in a way that profits to shareholders are the highest by increasing the profit and cash flow indicators (Elliot and Elliot 2002). However, Jensen and Meckling (1976) explained that managers do not always seek to maximize profits to shareholders. Agency theory is developed from these explanations and the agency problem is considered as a major factor determining company performance. The problem is that the interests of management and shareholders are not always the same, in this situation, managers—those responsible for running the company tend to achieve their own purpose instead of maximizing profit to shareholders. This means that managers will use excessive free cash flow available to achieve personal goals instead of increasing profits to shareholders (Jensen and Ruback 1983). Therefore, the main issue that shareholders must pay attention to is to ensure that managers do not use up the free cash flow by investing in unprofitable or negative NPV projects. Instead, free cash flow should be distributed to shareholders through dividend policy (Jensen 1986). The cost of managing directors to ensure they act in the interests of shareholders is treated as Agency costs. The greater the need to manage these directors, the higher the agency costs.

2.2   *Institutional Ownership and Firm Performance*

Most of studies on the relationship between ownership structure and firm performance emphasize the role of monitoring mechanism of this group of shareholders on management activities, creating both positive and negative impacts on firm performance and value.

Jensen and Meckling (1976) show that company performance will increase along with company ownership structure. However, Shleifer et al. (1988) showed a curvilinear correlation between the ownership structure and the company performance. The results show that at the beginning, company performance will increase, then gradually decline and eventually increase slightly when institutional ownership increases. Barnea, Haugen and Senbet (1985) found that under the impact of agency costs, ownership structures play an important role in company performance since it shows the separation between ownership and management. For example, high proportion of internal shareholders ownership may reduce agency costs, however, if this proportion reach a high level, it will lead to the opposite result. Similarly, outside investors or institutional investors tend to reduce agency costs by creating a relatively effective monitoring mechanism to company managers.

Shleifer and Vishny (1986), Bhojraj and Sengupta (2003) tested the hypothesis that institutional investors would be motivated to monitor company performance. Since institutional shareholders have more benefits than individual shareholders based on their voting rights, they can maintain some interventions to counter the board executive authority. Monks and Minow (2001) argue that this action is evidence that institutional shareholders take actions to protect their asset value.

Pound (1988) examines the positive correlation (efficient monitoring hypothesis) and negative correlation (conflict of interest hypothesis and strategic alignment hypothesis) between institutional ownership and company value. According to efficient monitoring hypothesis, institutional stakeholders own more information and ability to monitor company governance at a lower cost than minority stakeholders. Further, conflict of interest hypothesis emphasizes the fact that institutional stakeholders are able to develop activities that can bring current or potential profit in companies that they own stakes, thus they intend to be less interested in managerial discretion restriction. Finally, strategic alignment hypothesis highlights the fact that institutional stakeholders and managers define a common advantage in corporation. Besides, mitigation of monitoring function related to institutional stakeholders occur through this cooperation. McConnell and Servaes (1990) showed a positive relationship between the proportion of institutional shareholders ownership and Tobin's Q confirming the efficient monitoring hypothesis proposed by Pound (1988).

Brickley et al. (1988) classified institutional investors into two forms: institutional investors do not have business relationship with the company (pressure insensitive institutional investors) and institutional investors have business relationship with the company (pressure sensitive institutional investors). Almazan et al. (2005) identified a positive relationship between pressure insensitive institutional investors and a better director's remuneration discipline. In addition, an increase in the number of pressure insensitive institutional investors increases the effectiveness of management monitoring. Chen et al. (2007) concluded the positive relationship between independent long-term institutional shareholders and acquisition decisions. However, when monitoring benefits exceed costs, institutional shareholders will promote the monitoring process that could be detrimental to company, because through monitoring activities, they can collect information advantages to adjust their portfolio over time.

Morck et al. (1988) theoretical and empirically research the influence of ownership structure on company performance and found the similar research as Stulz (1988). They found that an increase in ownership structure and control authority will increase company value, however when this ownership proportion excessively increases, company value will decrease due to the impact of conservative management.

On the other hand, David and Kochhar (1996) mentioned that although institutional shareholders are able to mitigate the power of managers but there are some barriers that minimize this effect, of which: business relations with the companies they invest in, excessive government regulations are limiting their activities and limiting the information processing skills for monitoring companies. Leech (2000) argues that institutional investors are not always pursuing the control power in the companies they invest due to the fact that they own information that can cause damage to trading activities. Therefore, they tend to seek power in the form of influence rather than control.

Han and Suk (1998) confirmed the positive relationship between stock returns and institutional shareholder ownership, debating the role of institutional shareholders in the monitoring process against management. Similarly, Davis (2002) concluded a positive influence of institutional shareholder ownership on company performance. Cornett et al. (2007) determined a positively relationship between returns on assets and the ownership proportion of pressure insensitive institutional investors, as well as between returns on assets and number of institutional shareholders. However, Cornett et al. (2007) argued that institutional investors have potential business relationships with companies in which they invest will tend to compromise in monitoring activities to protect their business relations.

Qi et al. (2000) investigated the influence of ownership structure and firm performance of Chinese companies listed on Shanghai Stock Exchange in the period

1991–1996. Their results show that companies with high ownership proportion of state shareholders have negative correlation with firm performance.

Douma, George and Kabir (2003) find that both institutional and managerial ownership have significant influence company's performance. Their results show that ROA as proxy for company performance is positively correlated with managerial ownership but negative relation with the proportion of institutional ownership.

Earle et al. (2005) showed that the centralized ownership structure in a small number of shareholders could increase company performance through increasing monitoring, but this could also cause adverse effects when this group of shareholders exploit their power to gain their own benefit or to take advantage of small shareholders. Perrini et al. (2008) show that an increase in ownership structure may lead to a decline in costs generated from separation of ownership and management. However, by using their management power, major shareholders will ignore the interests of small shareholders.

Within company management mechanism, agency costs theory assumes that owners expect to maximize company profitability however in some case, manager's motivation might not be in consistent with the owner's expectation (Sánchez-Ballesta and García-Meca 2007). The proportion of owners ownership directly influence company performance, the main concern is that major shareholders tend to be proactive in monitoring the company helping to increase the company profitability, therefore shareholders concentration have a positive impact on company profitability (Sánchez-Ballesta and García-Meca 2007); however, according to agency cost theory, a high concentration of power in some shareholders groups may lead to inefficient decisions-making to maximize company value (Sánchez-Ballesta and García-Meca 2007). Morch et al. (1988), Claessens et al. (2002) shown that due to the benefits of effective monitoring, when the proportion of ownership increases, the value of company will also increase. However, when this proportion of ownership is mainly on the hand of some shareholders, the value of the company will be affected.

Studies have shown that the concentration of external shareholders can improve company performance by increasing monitoring activities (Grosfeld and Hashi 2007). However, major shareholders have the right to control managers to access to information sources and implement risk-taking governance activities, so a higher concentration proportion can lead to risk increasing in monitoring process (Grosfeld and Hashi 2007).

Pirzada et al. (2015) examined the effect of institutional ownership on the performance and capital structure of listed companies on Malaysian Stock Exchanges in the period 2001–2005. They found that institutional ownership has impact on company performance which is measured by Earning per Share (EPS) and Price/EPS (PE) ratios. However, the results do not support the hypothesis that there is a correlation between the ownership structure and capital structure. The authors argued that Malaysian companies often tend to finance their capital needs through internal funding resources rather than using external debt.

2.3 Managerial Ownership and Firm Performance

Relationship between managerial ownership and firm performance has attracted much attention from corporate finance researchers. Arguments claim that benefits between managers and shareholders are not entirely consistent. This conflict creates agency costs that reduce the value of the company; therefore, improving the proportion of managerial ownership not only helps to align the financial incentives of internal and external shareholders but also leads to better decisions that increase the company value. However, when managerial ownership reaches a certain level, managers are able to access to more freedom, leading to making decisions that only benefit managers and reduce company value. Thus, when managerial ownership reaches to a reasonable proportion, problem arise from agency cost can be reduced and company value can be maximized.

According to Jensen and Meckling (1976) as well as Fama and Jensen (1983), the internal ownership structure can lead to two types of corporate behavior, namely the convergence of interests between managers and shareholders, and the entrenchment effect. Jensen and Meckling (1976) argue that when the ownership proportion of internal stakeholder increases, the sensitivity of ineffective use of the company's resources decreases. Therefore, the contradiction between managers and shareholders will be reduced, being emphasized the convergence of interests between stakeholders. Han and Suk (1998) found a positive relationship between the proportion of internal ownership and stock's return, suggesting that when the ownership proportion of internal shareholders increases, their benefits tend to converge with the interests of outside shareholders. Similarly, Hrovatin and Urši (2002), using research samples of Slovenian companies, also showed a positive relation between the ownership of internal shareholders and the performance of the company.

On the other hand, if managers hold the majority of stocks with voting rights, they tend to take advantage in making decisions to achieve their own goals. In this context, the entrenchment effect is emphasized, which indicates a negative correlation between the internal ownership proportion and company value. Therefore, a centralized ownership structure can lead to major shareholders expropriate small shareholders. Han and Suk (1998) show that an excessive internal ownership might harm company value due to problem arise from manager's entrenchment.

Morck et al. (1988) shown that there are companies whose managerial ownership proportion are below the optimal level, and accordingly company performance can be enhanced by increasing this ownership. On the other hand, Demsetz and Lehn (1985) refer to those companies implementing the ownership proportion optimization, so there is no longer a relationship between ownership structure and company performance. Similarly, Morck et al. (1988), McConnell and Servaes (1990), Holderness et al. (1999) consider internal ownership proportion as an exogenous variable and found that changes in this proportion which resulted from capital structure changing are no longer effective and optimal. However, another research trend of Demsetz (1983), Demsetz and Lehn (1985), Cho (1998), Demsetz and Villalonga (2001) considered the internal ownership proportion as exogenous variable and found that changings in this proportion still bring certain effect. Cho (1998) argued that investments affect firm value, and in turn, firm value affect firm capital structure, however there are no reversed effects. The results emphasized that managerial ownership proportion cannot represent an effective incentive mechanism to maximize investment decisions.

Morck et al. (1988) found an average TobinQ increase when managerial ownership ratio increased from 0 to 5%, followed by a TobinQ decline when the managerial ownership proportion increased to 25%, however TobinQ continue to increase slightly when the managerial ownership ratio exceeds 25%. Evidences show that an increase in low managerial ownership companies can lead to convergence of interests between external shareholders and managers, helping to increase company value. However, an increase in high managerial ownership company can lead to a problem of governance centralization and reducing the value of company. McConnell and Servaes (1990) found that in 1976, for companies with low internal ownership rates, a 10% increase in this ratio could lead to an increase of TobinQ approximately 10%; Similarly, in 1986 with 10% increase in the internal shareholders ownership in companies with low internal ownership could increase TobinQ by 30%. However, for companies with high internal ownership, the correlation between TobinQ and the internal ownership ratio shows the opposite result. Holderness et al. (1999) retested the results of Morch et al. (1988) and found that there is only a positive relationship between internal ownership and firm value in companies with an internal ownership ratio of 0–5%. Short and Keasey (1999) emphasized the positive relationship between the interests of internal shareholders and the company goals in companies with internal ownership ratios below 12% and over 40%. Bhabra (2007) also confirmed the curve-shaped relationship between the internal ownership ratio and the company value, which shows that the positive relationship is below 14% and above 40%.

Core and Larcker (2002) tested the effect of establishing a target ownership proportion plan and found that before applying this plan, companies with a target ownership proportion plans showed lower stock prices as well as lower proportion of stocks held by management. However, after 2 years since applying this plan, these companies have higher performance, higher stock returns in 6 months of the plan announcement year. Therefore, raising the ownership proportion of managers can increase firm performance.

Chen et al. (2003) show that managerial ownership ratio has a linear relationship with the performance of Japanese companies. The authors found that institutional shareholders include State shareholders, executives, and board of directors are capable of manipulating company performance.

Li et al. (2007) studied the relationship between managerial ownership and firm performance of Chinese state-owned companies which were privatized in the period 1992–2000. The results showed that managerial ownership ratio has positive relationship with firm performance. Although the rates of ROA and ROS decline after the privatization, companies with high managerial rates, especially those with high CEO ownership, show less reduction in performance. The authors also found that the impact on company performance became less influential in companies with higher

CEO ownership ratios. The study supports the agency cost theory of Jensen and Meckling (1976) when arguing that a high ownership structure will motivate managers to act in the interests of shareholders.

3 Data and Methodology

In this paper, we use data set from 18 listed companies on VN30 index basket of Vietnamese stock exchange. This data set is used for the following reasons.

VN 30 is the first large-cap index in HOSE-Index which was implemented by Ho Chi Minh Stock Exchange on February 6, 2012.

VN30 index is calculated based on market capitalization including 30 component shares, representing 30 listed companies on HOSE, accounting for about 80% of total market capitalization and 60% of the total trading value of the market. Thus, companies listed on VN30 are HOSE listed companies with leading market capitalization and liquidity.

In addition, stocks in VN30 basket are the most liquid stocks, the price of stocks in VN30 basket will best reflect the relationship between stock supply and demand, thus limiting the speculation in poor liquidity stocks.

Thus, selecting data from companies in VN30 index guarantee market representative, meeting the requirements of data quality.

The research is based on quantitative research methods, including:

Descriptive statistics

Hypotheses testing about the relationship between company ownership structure including institutional shareholder ownership, managerial ownership and firm performance as measured by return on total assets (ROA). To test the effect of institutional and managerial ownership on the firm performance, this study is based on the research of Bhattacharya and Graham (2007), Gugler and Weigand (2003) and Chaganti and Damanpour (1991). The model is as follows:

$$FirmPerformance_{it} = \alpha + \beta_1\, ownership_{it} + \gamma X_{it} + \varepsilon_{it} \tag{1}$$

where:

Firm performance is dependent variable indicating performance of company. This variable can be measured by ROA, ROE.

Ownership is independent variable representing institutional and managerial ownership.

Institutional ownership is the amount of a company's available stock owned by mutual or pension funds, insurance companies, investment firms, private foundations, endowments or other large entities that manage funds on the behalf of others.

Managerial ownership, in which it means the percentage of equity owned by insiders and block holders, where insiders are defined as the officers and directors of a firm (Holderness 2003).

Table 1 Descriptive statistics

Variable	Observation	Mean	Std. Dev.	Min	Max
ROE	108	0.167	0.127	0.0034	0.912
ROA	108	0.075	0.096	0.0003	0.722
LEV	108	0.134	0.140	0	0.581
Institutional ownership	108	0.668	0.2419	0.2626	0.994
Managerial ownership	108	0.0827	0.122	0	0.426
SIZE	108	17.502	1.471	15.312	20.814
AGE	108	8	3.005	1	17

X are control variables including firm size (SIZE) calculated by natural logarithm of total assets; financial leverage (LEV) calculated based on total debt/total assets; business age (AGE) is number of years since listed on HOSE.

4 Results

4.1 Descriptive Statistics

Table 1 describes key variables in the study. In which, firm performance is measured by ROA, ROE, EPS. The average value of ROA is 7.55% and the fluctuation is quite large from 0.03 to 72.19%. This ratio is higher than previous studies such as those of Klapper, Laeven, and Love (2006) for four countries including the Czech Republic, Hungary, Finland and the Republic of Slovakia with an average ROA of about 3%; Gedajlovic and Shapiro (2002) for Japan with average ROA is 3.84%, Isik and Soykan (2013) in Turkey (3.03%) but lower than the average ROA of companies in Turkey in Önder (2003) research which is 11.9%. The average value of ROE is 16.798%, which is much higher than the study of Earle, Kucsera, and Telegdy (2005) in Budapest with ROE of 11.86%.

The average leverage ratio of 13.42%, which is smaller than that of Pirzada et al. (2015) for Malaysia at a rate of 19%. The average institutional ownership and managerial proportion are 66.83% and 8.27%, respectively.

4.2 Research Results

To estimate panel data, we can use pool ordinary least squares (OLS), fixed effects(FE) and random effects (RE) model. In this paper, we apply Bayesian

Table 2 Results

	ROA			ROE		
	Mean	Equal-tailed [95% Cred. interval]		Mean	Equal-tailed [95% Cred. interval]	
LEV	−0.1814	−0.38875	0.01792	−0.1444	−0.30689	0.02398
Institutional ownership	−0.0672	0.242658	0.11659	−0.11403	0.025416	0.24612
Managerial ownership	0.0294	0.36762	0.32398	0.38812	0.145738	0.62729
Size	−0.0055	−0.02734	0.01525	−0.0081	−0.02727	0.00946
Age	0.00367	−0.00884	0.01564	0.00425	−0.00517	0.01338
_cons	0.227	−0.22864	0.6936	0.1902	−0.15579	0.56682

Source Authors calculation

approach to test the statistical results due to the fact that this approach makes conclusions more reliable.

Table 2 shows that institutional ownership has a negative impact on firm performance in all models. These results are contrary to the research of Al-Najjar and Taylor (2008), Oak and Dalbor (2010), Tong and Ning (2004), Bokpin and Isshaq (2009) and Tornyeva and Wereko (2012). However, they are similar to those of Bhattacharya and Graham (2007), Gugler and Weigand (2003) and Chaganti and Damanpour (1991). According to Alfaraih et al. (2012) institutional ownership will have positive effect on firm performance when they have good corporate governance system and protect the interests of shareholders (Tornyeva and Wereko 2012; Hussain Tahir 2015; Chen et al. 2008; Cornett et al. 2007). However, institutional ownership can be detrimental to company when small shareholders are not protected or when institutions have relationship with company managers (Djankov 1999). Moreover, when the relationships between institutional investors and company managers are close, institutional stakeholders tend to allocate their voting rights to management decisions (McConnell and Servaes 1990). Djankov (1999) argues that when institutional stakeholders have significant power over company decisions, they can make decisions that benefit themselves instead of maximizing the value of company and the wealth of all shareholders. This may result in the case when the average ownership of institutional shareholders is more than 66%. In addition, according to Shin-Ping and Tsung-Hsien (2009) different types of institutional ownership will cause different impacts on firm performance. Some institutional stakeholders will help improve operation system while others might reduce firm efficiency. Institutional ownership divided into government and non-governmental organizations, domestic and foreign, ownership of financial institutions—mutual funds, insurance companies, venture capital funds, banks, securities, investment trust fund and other organizations. In particular, government organization ownership and large corporation ownership have significant negative relationship with firm performance while securities investment trust funds and firm performance witness positive relation. These results reflect the characteristics of firms in research sample, when most companies in the sample are mostly owned by the state or large corporations.

Results show that managerial ownership has a positive impact on firm performance in criteria. This result is consistent with studies by Bhagat and Bolton (2009), Kapopoulos and Lazaretou (2007), Marashdeh (2014). Jensen and Meckling (1976) propose that managerial ownership can help to reduce the agency problems by encouraging managers and owners to consider interests of managers, this gives them the motivation to increase company value. Horner (2010) argues that increase in managerial ownership will help managers achieve the best performance.

Financial leverage has a negative impact and is not statistically significant to company value. When company use its own capital, the greater the debt ratio, the greater the risk company incurs when suffering losses. At a certain time, debt increasing suggests that company financial position is not strong enough to finance company, thus high debt can negatively affect company value.

5 Conclusion

This study aims to examine the relationship between institutional and managerial ownership and firm performance. Using research sample of companies in VN30 basket over the period 2012–2017, we found a positive relationship between managerial ownership and firm performance. The results are similar to previous studies of Bhagat and Bolton (2009), Kapopoulos and Lazaretou (2007), Marashdeh (2014). In addition, we found a negative relationship between institutional ownership and firm performance, which are similar to research of Bhattacharya and Graham (2007), Gugler and Weigand (2003), Chaganti and Damanpour (1991), Shin-Ping and Tsung-Hsie (2009). The negative effect of Vietnamese institutional ownership on firm performance might be explained that high proportion of the state ownership could lead to low performance of company (Shin-Ping and Tsung-Hsien 2009). Accordingly, in order to improve firm performance, the state should gradually withdraw capital from these firms so that they can be self-reliant and able to make decisions.

References

Almazan, A., Hartzell, J. C., & Starks, L. T. (2005). Active institutional shareholders and cost of monitoring: evidence from managerial compensation. *Financial Management, 34*(4), 5–34.
Al-Najjar, B., & Taylor, P. (2008). The relationship between capital structure and ownership structure: New evidence from Jordanian panel data. *Managerial Finance, 34*(12), 919–933.
Alfaraih, M., Alanezi, F., & Almujamed, H. (2012). The influence of institutional and government ownership on firm performance: evidence from Kuwait. *International Business Research, 5*(10), 192.
Barnea, Haugen & Senbet (1985). Agency problems and financial contracting. *Journal of Banking and Finance, 1987, 11*(1), 172–175. https://econpapers.repec.org/article/eeejbfina/.

Berle, A., & Means, G. (1932). *The modern corporation and private property*. New York: Macmillan.

Bhattacharya, P. S., & Graham, M. (2007). Institutional ownership and firm performance: Evidence from Finland. Available at SSRN 1000092.

Bhagat, S., & Bolton, B. J. (2009). Sarbanes-Oxley, governance and performance. Available at SSRN 1361815.

Bhabra, G. S. (2007). Insider ownership and firm value in New Zealand. *Journal of Multinational Financial Management, 17*(2), 142–154.

Bhojraj, S., & Sengupta, P. (2003). Effect of corporate governance on bond ratings and yields: The role of institutional investors and the outside directors. *The Journal of Business, 76*(3), 455–475.

Bokpin, G. A., & Isshaq, Z. (2009). Corporate governance, disclosure and foreign share ownership on the Ghana Stock Exchange. *Managerial Auditing Journal, 24*(7), 688–703.

Brickley, J., Lease, R., & Smith, C. (1988). Ownership structure and voting on antitakeover amendments. *Journal of Financial Economics, 20*, 267–292.

Brigham, E., & Gapenski, L. (1996). *Financial management*. Dallas: The Dryden Press.

Chaganti, R., & Damanpour, F. (1991). Institutional ownership, capital structure, and firm performance. *Strategic Management Journal, 12*(7), 479–491.

Chen, J., Chen, D. H., & He, P. (2008). Corporate governance, control type and performance: The new Zealand story. *Corporate Ownership and Control, 5*(2), 24–35.

Chen, C. R., Guo, W., & Mande, V., (2003). Managerial ownership and firm valuation. *Pacific-Basin Finance Journal, 11*(3), 267–283. https://www.researchgate.net/publication/journal/0927-538X_Pacific-Basin_Finance_Journal.

Chen, X., Harford, J., & Li, K. (2007). Monitoring: Which institutions matter? *Journal of Financial Economics, 86*(2), 279–305.

Cho, M. H. (1998). Ownership structure, investment, and the corporate value: An empirical analysis. *Journal of Financial Economics, 47*(1), 103–121.

Claessens, S., & Djankov, S. (2002). Privatization benefits in Easter Europe. *Journal of Public Economics, 83*, 307–324.

Core, J. E., & Larcker, D. F. (2002). Performance consequences of mandatory increases in executive stock ownership. *Journal of Financial Economics, 64*(3), 317–340.

Cornett, M. M., Marcus, A. J., Saunders, A., & Tehranian, H. (2007). The impact of institutional ownership on corporate operating performance. *Journal of Banking and Finance, 31*(6), 1771–1794.

David, P., & Kochhar, R. (1996). Barriers to effective corporate governance by institutional investors: implications for theory and practice. *European Management Journal, 14*(5), 457–466.

Davis, P. (2002). Institutional investors, corporate governance and the performance of the corporate sector. *Economic Systems, 26*(3), 203–229.

Demsetz, H. (1983). The structure of ownership and the theory of the firm. *Journal of Law and Economics, 26*(2), 375–390.

Demsetz, H., & Lehn, K. (1985). The structure of corporate ownership: Causes and consequences. *The Journal of Political Economy, 93*(6), 1155–1177.

Demsetz, H., & Villalonga, B. (2001). Ownership structure and corporate performance. *Journal of Corporate Finance, 7*(3), 209–233.

Djankov, S. (1999). Ownership structure and enterprise restructuring in six newly independent states. *Comparative Economic Studies, 41*(1), 75–95.

Douma, George, & Kabir. (2003). Underperfomance and profit redistribution in business groups. Paper presented at the Business Economics Seminar, Antwerp, February.

Earle, J. S., Kucsera, C., & Telegdy, A. (2005). Ownership concentration and corporate performance on the Budapest stock exchange: Do too many cooks spoil the goulash. *Corporate Governance, 13*(2), 254–264.

Elliot, B., & Elliot, J. (2002). *Financial accounting and reporting* (12th ed.). London: Prentice Hall/Financial Times.

Fama, E. F., & Jensen, M. C. (1983). Separation of ownership and control. *Journal of Law and Economics, 26*(2), 301–325.

Gedajlovic, E., & Shapiro, D. M. (2002). Ownership structure and firm profitability in Japan. *Academy of Management Journal, 45*(3), 565–575.

Grosfeld, I., & Hashi, I. (2007). Changes in ownership concentration in mass privatised firms: Evidence from Poland and the Czech Republic. *Corporate Governance: An International Review, 15*(4), 520–534. https://econpapers.repec.org/article/blacorgov/.

Gugler, K., & Weigand, J. (2003). Is ownership really endogenous? *Applied Economics Letters, 10*(8), 483–486.

Han, K. C., & Suk, D. Y. (1998). The effect of ownership structure on firm performance: Additional evidence. *Review of Financial Economics, 7*(2), 143–155.

Hitt, M., Hoskisson, R., & Harrison, J. (1991). Strategic competitiveness in the 1990s: Challenges and opportunities for U.S. executives. *Academy of Management Executive, 5*(2), 7–22.

Holderness, C. G. (2003). A survey of blockholders and corporate control. *Economic policy review, 9*(1).

Holderness, C. G., Kroszner, R. S., & Sheehan, D. P. (1999). Were the good old days that good? Changes in managerial stock ownership since the great depression. *Journal of Finance, 54,* 435–469.

Horner, W. T. (2010). *Ohio's Kingmaker: Mark Hanna.* Man and Myth: Ohio University Press.

Hussain Tahir, S. (2015). Institutional ownership and corporate value: Evidence from Karachi stock exchange (KSE) 30-index Pakistan. *Praktini menadžment: struni asopis za teoriju i praksu menadžmenta, 6*(1), 41–49.

Hrovatin, N., & Urši, S. (2002). The determinants of firm performance after ownership transformation in Slovenia. *Communist and Post-Communist Studies, 35*(2), 169–190.

Isik, O., & Soykan, M. E. (2013). Large shareholders and firm performance: Evidence from Turkey. *European Scientific Journal, 9*(25).

Jensen & Meckling. (1976). The theory of the firm: Managerial behavior, agency costs, and ownership structure. *Journal of Financial Economics, 3,* 305–360; 40, 2, 36–52.

Jensen, M. (1986). Agency cost of free cash flow, corporate finance and takeovers. *American Economic Review Papers and Proceedings, 76,* 323–329.

Jensen, M. (1989). Eclipse of public corporation. *Harvard Business Review, 67*(5), 61–74.

Jensen, M., & Ruback, R. (1983). The market for corporate control: The Scientific Evidence. *Journal of Financial Economics, 11,* 5–50.

Kapopoulos, P., & Lazaretou, S. (2007). Corporate ownership structure and firm performance: Evidence from Greek firms. *Corporate Governance: An International Review, 15*(2), 144–158.

Klapper, L., Laeven, L., & Love, I. (2006). Corporate governance provisions and firm ownership: Firm-level evidence from Eastern Europe. *Journal of International Money and Finance, 25*(3), 429–444.

Leech, D. (2000). Shareholder power and corporate governance, University of Warwick, Warwick Economics Research Paper No. 564.

Lubatkin, M., & Chatterjee, S. (1994). Extending modern portfolio theory into the domain of corporate diversification: Does it apply? *Academy of Management Journal, 37,* 109–136.

Li, D., Moshirian, F., Nguyen, P., & Tan, L. (2007). Corporate governance or globalization: What determines CEO compensation in China? *Research in International Business and Finance, 21,* 32–49.

Li, D., Moshirian, F., Nguyen, P., & Tan, L.-W. (2007). Managerial ownership and firm performance: Evidence from China's privatizations. *Research in International Business and Finance, 21,* 396–413.

Marashdeh, Z. M. S. (2014). *The effect of corporate governance on firm performance in Jordan.* Doctoral dissertation, University of Central Lancashire.

McConnell, J. J., & Servaes, H. (1990). Additional evidence on equity ownership and corporate value. *Journal of Financial Economics, 27*(2), 595–612.

Modigliani, F., & Miller, M. (1958). The cost of capital, corporation finance, and the theory of investment. *American Economic Review, 48,* June, 261–197.

Modigliani, F., & Miller, M. (1963). Corporate income taxes and the cost of capital: A correction. *American Economic Review*, June, 433–443.

Monks, R., & Minow, N. (2001). *Corporate governance*. Oxford: Blackwell.

Morck, R., Shleifer, A., & Vishny, R. W. (1988). Management ownership and market valuation: An empirical analysis. *Journal of Financial Economics, 20*, 293–315.

Oak, S., & Dalbor, M. C. (2010). Do institutional investors favor firms with greater brand equity? An empirical investigation of investments in US lodging firms. *International Journal of Contemporary Hospitality Management, 22*(1), 24–40.

Önder, Z. (2003). Ownership concentration and firm performance: Evidence from Turkish firms. *METU studies in development, 30*(2), 181.

Perrini, F., Rossi, G., & Rovetta, B. (2008). Does ownership structure affect performance? Evidence from the Italian market. *Corporate Governance, 16*(4), 312–325.

Pirzada, K., Mustapha, M. Z. B., & Wickramasinghe, D. (2015). Firm performance, institutional ownership and capital structure: A case of Malaysia. *Procedia—Social and Behavioural Sciences, 311*, 170–176.

Pound, J. (1988). Proxy contests and the efficiency of shareholder oversight. *Journal of Financial Economics, 20*, 237–265.

Qi, D., Wu, W., & Zhang, H. (2000). Ownership structure and corporate performance of partially privatized Chinese SOE firms. *Pacific-Basin Finance Journal, 8*, 587–610.

Shin-Ping, L., & Tsung-Hsien, C. (2009). The determinants of corporate performance: A viewpoint from insider ownership and institutional ownership. *Managerial Auditing Journal, 24*(3), 233–247.

Sánchez-Ballesta, J.P., García-Meca, E. (2007). Ownership structure, discretionary accruals and the informativeness of earnings. *Corporate Governance: An International Review, 15*(4), 677–691.

Shleifer, A., & Vishny, R. W. (1986). Large shareholders and corporate control. *Journal of Political Economy, 94*(3), 461–489.

Shleifer, A., & Vishny, R. W. (1988). Management entrenchment: The case of manager-Specific Investment. *Journal of Financial Economics, 25*, 123–139.

Short, H., & Keasey, K. (1999), Managerial ownership and the performance of firms: Evidence from the U.K. *Journal of Corporate Finance, 5*(1), 79–101.

Stulz, R. (1988). Managerial control of voting rights: Financing policies and the market for corporate control. *Journal of Financial Economics, 20*(1–2), 25–54. https://www.researchgate.net/publication/journal/1879-2774_Journal_of_Financial_Economics.

Tong, S., & Ning, Y. (2004). Does capital structure affect institutional investor choices? *The Journal of Investing, 13*(4), 53–66.

Tornyeva, K., & Wereko, T. (2012). Corporate governance and firm performance: Evidence from the insurance sector of Ghana. *European Journal of Business and Management, 4*(13).

Support Vector Machine-Based GARCH-type Models: Evidence from ASEAN-5 Stock Markets

Woraphon Yamaka, Wilawan Srichaikul, and Paravee Maneejuk

Abstract Support Vector Machine (SVM) is a semiparametric tool for regression estimation. We will use this tool to estimate the parameters of GARCH models for predicting the conditional volatility of the ASEAN-5 stock market returns. In this study, we aim at comparing the forecasting performance between the Support Vector Machine-based GARCH model and the Maximum likelihood estimation based GARCH model. Four GARCH-type models are considered, namely ARCH, GARCH, EGARCH and GJR-GARCH. The comparison is based on the Mean Absolute Error (MAE), the Mean Squared Error (MSE), and the Root Mean Squared Error (RMSE). The results show that the stock market volatilities of Thailand and Singapore are well forecasted by Support Vector Machine-based-GJR-GARCH model. For the stock market of Malaysia, Indonesia and the Philippines, the Support Vector Machine-based-ARCH model beats all parametric models for all performance comparison criteria.

Keywords Support vector machine · GARCH · ASEAN-5 · Comparison

1 Introduction

Forecasting volatility of the financial assets returns is crucial in financial investment, and it has held the attention of academics and practitioners over the last few decades. The volatility of stock prices and market returns is viewed an uncertainty which has to be assessed for investment decision making and portfolios management. Therefore,

W. Yamaka · W. Srichaikul (✉) · P. Maneejuk
Faculty of Economics, Center of Excellence in Econometrics,
Chiang Mai University, Chiang Mai 50200, Thailand
e-mail: srichaikul.w@gmail.com

W. Yamaka
e-mail: woraphon.econ@gmail.com

P. Maneejuk
e-mail: mparavee@gmail.com

© The Editor(s) (if applicable) and The Author(s), under exclusive license
to Springer Nature Switzerland AG 2021
N. Ngoc Thach et al. (eds.), *Data Science for Financial Econometrics*,
Studies in Computational Intelligence 898,
https://doi.org/10.1007/978-3-030-48853-6_26

a reliable forecast of volatility of asset returns over an investment holding period is imperative for both investment risk taking and avoiding purposes. To be precise, it can help investors make the prudent investment decisions, discover investment opportunities, and frame the defensive strategies, etc. However, it is intuitive to think that most investors will opt for investing in stocks or markets that have lower volatility in order to get the optimal returns in the future.

Various models have been proposed for investigating volatility and comparing forecast performance. One of the predominant models is the Generalized Autoregressive Conditional Heteroscedasticity (GARCH) model of Bollerslev (1986). It is generalized from the ARCH of Engle (1982) in 1982 and has been widely used to capture the time-varying volatility of the data. However, as the model has intrinsically symmetric (the conditional variance responds symmetrically to positive and negative residuals), the forecasting results may be biased when the data exhibit a skewness (Franses and Van Dijk 1996). To address this problem, asymmetric GARCH-type models are proposed, for example GJR-GARCH model of Glosten, Jagannathan, and Runkle (1993) and EGARCH model of Nelson (1991). These models present potential improvements over the conventional GARCH models. The EGARCH model specifically earns its popularity in financial studies since it presents the asymmetric response of volatility to positive and negative returns.

Both symmetric and asymmetric GARCH models are generally estimated by the maximum likelihood estimation (MLE). This estimation requires specified distribution of innovations in order to estimate the parameters of the model. Nowadays, there are many distributions of innovations introduced as the functional form of the likelihood function, for example, Normal distribution, Student's t distribution, Generalized error distribution (GED). Thus, researchers have to find the best fit distribution of innovation thus making the estimation process time consuming, costly, and onerous. Fortunately, various semiparametric models are introduced to gain a better estimation for GARCH models as it does not require any assumptions on the functional form (Chen et al. 2010). One of the most used and efficient semiparametric models is support vector machine (SVM), which was developed by Vapnik (1995). It is a tool for linear and nonlinear input-output knowledge discovery. We can employ SVM to solve a pattern recognition as well as regression estimation problems.

Previous studies have applied support vector machines (SVM) for forecasting the stock market index movement. For instance, Hu et al. (2013) used SVM to predict the stock market. They revealed that SVM is a powerful predictive tool for stock price movement predictions in the financial market. Madge and Bhatt (2015) applied SVM to predict stock price direction and found this tool has little predictive ability in the short-run but definite predictive ability in the long-run. Upadhyay et al. (2016) were of opinion that the SVM is one of most efficient machine learning algorithms in modeling stock market prices and movements. Grigoryan (2017) studied and atudied of stock market trend using SVM and variable selection methods. Furthermore, Rungruang et al. (2019) made prediction of the direction of SET 50 index with in SVM.

In recent years, SVM has been further used for modeling the GARCH models. Chen et al. (2010) introduced the use of SVM to model ARMA process and apply it to forecast the conditional variance equation of the GARCH model in real data analysis. SVM-GARCH models significantly outperform the competing models in most situations of one-period-ahead volatility forecasting. Note that the conventional GARCH can perform better when the data is normal and large.

With numerous and strong suggestions to use SVM in GARCH model estimation, we therefore aim to confirm the superiority of the SVM by comparing with the MLE. To serve this purpose, we consider both symmetric and asymmetric GARCH models using MLE and SVM. Hence, four GARCH-type models, namely ARCH, GARCH, EGARCH, and GJR-GARCH, are considered in this study. We use the ASEAN stock markets including Indonesia, Malaysia, the Philippines, Singapore, and Thailand as the example data for comparing these two estimations. This is the first attempt for finding the best prediction method in predicting ASEAN economies, this makes perfect sense, and since SVM has been shown to be good.

The structure of the rest of this study is as follows. The next section provides the methodology used in this study. Section 3 presents ASEAN-5 stock markets data; Sect. 4 presents the comparison results. Finally, the conclusion is provided in Sect. 5.

2 Methodology

2.1 Support Vector Machine

For classification and regression problem, support vector machine (SVM) is one of the popular machine learning tools and was introduced by Vapnik (1995) in 1995. We are given the training data $\{(x_1, y_1), ..., (x_t, y_t)\}$ where $x_1 = (x_{11}, x_{12}, ..., x_{1m}), ..., x_t = (x_{t1}, x_{t2}, ..., x_{tm}), y \in R$. This model has the following form

$$y_t = \omega'\phi(x_t) + b \tag{1}$$

where $\omega \in R^m$, $b \in R$. $\phi(\cdot)$ is the nonlinear transfer function for projecting the inputs into the output space. In this case, we need to find the optimal value of weight ω. To do this, we have to minimize the norm of ω and additional penalization term.

$$\text{Minimize } \left\{ \tfrac{1}{2}||\omega||^2 + C \sum_{t=1}^{T} \left(v_t + v_t^*\right) \right\} \tag{2}$$

$$\text{subject to } \begin{cases} y_t - \omega'\phi(x_t) - b & \leq \varepsilon + v_t \\ \omega'\phi(x_t) + b - y_t & \leq \varepsilon + v_t^* \end{cases} \tag{3}$$

where v_t and v_t^* are positive slack variables, introduced to deal with the samples with prediction errors greater than ε. $C(\cdot)$ is the penalization. To solve Eqs. (2, 3), we can use the Lagrange multipliers as follows:

$$L = \frac{1}{2}\|\omega\|^2 + C\sum_{t=1}^{T}(v_t + v_t^*)$$

$$-\sum_{t=1}^{T}\lambda_{1t}(\varepsilon + v_t - y_t + \omega'\phi(x_i) + b) - \sum_{t=1}^{T}\lambda_{2t}v_t$$

$$-\sum_{t=1}^{T}\lambda_{3t}(\varepsilon + v_t^* + y_t + \omega'\phi(x_i) - b) - \sum_{t=1}^{T}\lambda_{4t}v_t^* \qquad (4)$$

We then minimize Eq. 4 with respect to ω, b, v_t and v_t^* but maximize it with respect to the Lagrange multipliers, λ_1, λ_2, λ_3 and λ_4. Then, the corresponding dual problem of the SVM can be derived from the primal problem by using the Karush-Kuhn-Tucker conditions as follows:

Minimize
$$L_k = \sum_{t=1}^{T}\sum_{t=1}^{T}(\lambda_{1t}-\lambda_{3t})(\lambda_{1t}-\lambda_{3t})\phi(x_t)\phi'(x_t) \\ +\varepsilon\sum_{t=1}^{T}(\lambda_{1t}+\lambda_{3t}) \qquad (5)$$

subject to

$$\sum_{t=1}^{T}(\lambda_{1t}+\lambda_{3t}) = 0 \qquad (6)$$

$$0 \le \lambda_{1t}, \lambda_{3t} \le C, \qquad (7)$$

where $\phi(x_t)\phi'(x_t)$ can be viewed as the inner-product kernel function $K(x_t, x_t)$ (Scholkopf and Smola 2001). So, we do not need to specify the explicit function of $\phi(x_t)$. In this study, we consider this kernel to be a quadratic function, thus

$$K(x_t, x_t) = \phi(x_t)\phi'(x_t) = x'_t x_t \qquad (8)$$

This procedure can be solved using quadratic programming (QP). Thus, we can obtain the optimal $\widehat{\omega}$ and $\widehat{b}$. Finally, we can predict the output as

$$\widehat{y} = \widehat{\omega}'\phi(x_t) + \widehat{b} \qquad (9)$$

2.2 GARCH-type Models

2.2.1 ARCH

The Autoregressive conditional heteroskedasticity (ARCH) model is introduced by Engle (1982). It is proposed to capture the heteroscedastic volatility existing in the time series. This model constructs the variance of a regression model's disturbances as a linear function of the lagged values of the squared regression disturbances. The model is given by

$$r_t = \mu + \varepsilon_t, \tag{10}$$

$$\sigma_t^2 = \alpha_0 + \alpha_1 \varepsilon_{t-1}^2 + \alpha_2 \varepsilon_{t-2}^2 + ... + \alpha_q \varepsilon_{t-q}^2, \tag{11}$$

where Eqs. 4, 5 are the conditional mean and variance. μ and ε_t are the intercept term and the error of the mean equation. $(\alpha_0, ..., \alpha_q)$ are the estimated parameters of the conditional variance equation which are constrained to be larger than zero. Note that ε_t is assumed to have normal distribution with mean zero and variance σ_t^2. However, this model has posed some problems, for instance it might need a large to capture all of the dependence and it cannot handle asymmetric effects.

2.2.2 GARCH

The GARCH-model is developed by Bollerslev (1986) to solve the problem in the ARCH-model. The GARCH-model becomes more parsimonious than ARCH as it does not require a high order to capture the volatility persistence. In this model, Bollerslev (1986) added the lag of the variance to capture the long-term influences. Thus, the variance of GARCH(q,p) model can be written as

$$\sigma_t^2 = \omega_0 + \sum_{j=1}^{q} \alpha_j \varepsilon_{t-j}^2 + \sum_{j=1}^{p} \beta_j \sigma_{t-j}^2, \tag{12}$$

where ω, α, β are restricted to be non-negative parameters and $\sum_{j=1}^{q} \alpha_j + \sum_{j=1}^{p} \beta_j \leq 1$ but should be close to unity for an accurate model specification.

2.2.3 EGARCH

Unlike ARCH and GARCH models, the exponential GARCH model (Nelson 1991) has an ability to capture the asymmetric effect and leverage effect in financial time series. The model can capture both negative shocks and positive shocks as well as the size effects. Also, the non-negative constraint is not required in this model. Thus, the variance of EGARCH(q,p) model can be written as

$$\log(\sigma_t^2) = \alpha_0 + \sum_{j=1}^{q} (\alpha_j \varepsilon_{t-j} + \gamma_j(|\varepsilon_{t-j}|)) + \sum_{j=1}^{p} \beta_j \log(\sigma_{t-j}^2), \qquad (13)$$

where α_j is the coefficient used for capturing the sign effects while γ_j is used for the size effect of the asymmetry. There is no restriction on α_0, α_j, and and γ; however, β must be positive and less than one to maintain stability.

2.2.4 GJR-GARCH

The GJR-GARCH model is suggested by Glosten, Jagannathan, and Runkle (1993) as the alternative model for EGARCH. Like EGARCH, the GJR-GARCH has also been used to capture the asymmetric effect of the shocks. The variance of this model can be written as

$$\sigma_t^2 = \omega + \sum_{j=1}^{q} (\alpha_j \varepsilon_{t-j}^2 + \gamma_j I_{t-j} \varepsilon_{t-j}^2) + \sum_{j=1}^{p} \beta_j \sigma_{t-j}^2, \qquad (14)$$

where I is the indicator function with the value 1 for $\varepsilon < 0$ and 0 otherwise. γ_j is the leverage term which uses to capture the asymmetry in the series. We note that the summation of ARCH effect α_j and GARCH effect β_j must be less than one; and γ must be greater than zero. We can notice that if $\gamma = 0.$, the GJR-GARCH converse to classical GARCH model.

2.3 GARCH Models Estimation Using SVM

As we mentioned above, there are two equations, namely mean and variance equations, in the GARCH-type model. To estimate this model, σ_t^2, ε_{t-j}^2 and σ_{t-j}^2 must be known in order to solve the primal problem of SVM in Eqs. (5–7). However, σ_t^2 and σ_{t-j}^2 are unobservable. Thus, we follow Chen et al. (2010) and compute σ_t^2 as the moving average of the contemporaneous and four lagged squared returns around each time point, that is

$$\sigma_t^2 = \frac{1}{5} \sum_{k=t-1}^{t} (y_t - \bar{y}_{5,t})^2, \qquad (15)$$

where $\bar{y}_{5,t} = \frac{1}{5} \sum_{k=t-1}^{t} y_k$.

2.4 Performance Criteria

As we compare various GARCH-type models using either MLE or SVM. The volatility prediction performance is evaluated using the following statistical measures, namely, the normalized mean absolute error (MAE), the mean squared error (MSE) and the root-mean-square error (RMSE). These measures can be computed as follows

1. Mean absolute error (MAE)

$$MAE = \frac{1}{T} \sum_{t=1}^{T} \left| \sigma_t^2 - \widehat{\sigma}_t^2 \right|. \tag{16}$$

2. Mean squared error (MSE)

$$MSE = \frac{1}{T} \sum_{t=1}^{T} (\sigma_t^2 - \widehat{\sigma}_t^2)^2. \tag{17}$$

3. Root-mean-square error (RMSE)

$$RMSE = \sqrt{\frac{1}{T} \sum_{t=1}^{T} \left(\sigma_t^2 - \widehat{\sigma}_t^2 \right)^2}, \tag{18}$$

where σ_t^2 is the actual volatility in real data and $\widehat{\sigma}_t^2$ is the forecasting values. As the actual σ_t^2 cannot be observed, we thus calculate the proxy of the actual volatility as

$$\sigma_t^2 = (y_t - \bar{y})^2, \tag{19}$$

where $\bar{y}$ is the mean of the observed real data.

3 ASEAN-5 Stock Markets Data Description

In this study, taken as data for the analysis are the daily closing price of ASEAN-5 stock markets consisting of the Stock Exchange of Thailand (SET), the Singapore Stock Exchange (SGX), the Philippines Stock Exchange (PSEi), the Kuala Lumpur Stock Exchange (KLSE), and the Jakarta Stock Exchange (JKSE). The daily time series data spans from March 23, 2012 to April 30, 2019 (totally 1552 observations). All data are collected from https://th.investing.com. All of these data are transformed into the log-return rate to ensure the stationary property. The description of our returns is provided in Table 1. In this study, we use Minimum Bayes factor (MBF) as the tool for checking the significant result. This MBF can be considered as an

Table 1 Descriptive statistics

Variable	Thailand (SET)	Singapore (SGX)	Philippines (PSEi)	Malaysia (KLSE)	Indonesia (JKSE)
Mean	0.0002	0.0000	−0.0004	0.0000	0.0003
Median	0.0006	0.0000	0.0000	0.0002	0.0005
Maximum	0.0448	0.0453	0.1284	0.0472	0.0578
Minimum	−0.0672	−0.0764	−0.0624	−0.0387	−0.0953
Std. Dev.	0.0093	0.0105	0.0136	0.006	0.0133
Skewness	−0.6762	−0.2735	1.1977	−0.3984	−0.4942
Kurtosis	8.3946	6.1998	13.757	9.0418	7.56
Jarque-Bera	2000.208 (0.0000)	681.4533 (0.0000)	7853.763 (0.0000)	2401.601 (0.0000)	1407.842 (0.0000)
ADF test	−38.2364 (0.0000)	−40.8687 (0.0000)	−44.249 (0.0000)	−36.1658 (0.0000)	−21.4926 (0.0000)

Note MBF is Minimum Bayes factor

alternative to the p-value (Held and Ott 2018). If $1 < MBF < 1/3$, $1/3 < MBF < 1=10$, $1=10 < MBF < 1/30$, $1/30 < MBF < 1/100$, $1/100 < MBF < 1/300$ and $MBF < 1/300$, there are, respectively, weak evidence, moderate evidence, substantial evidence, strong evidence, very strong evidence and decisive evidence for favoring the alternative hypothesis. According to Table 1, the Augmented Dickey-Fuller test (ADF) of unit root is conducted, and it shows a decisive evidence for stationarity in our returns. We also conduct the Jarque-Bera normality test as we observe that all returns exhibit a small skewness but high kurtosis. The result of Jarque-Bera shows a decisive evidence that stock returns reject the null hypothesis of a normal distribution. Thus, we can conclude that our returns are stationary and not normal. Therefore, we make the student-t distribution assumption in the MLE for estimating the GARCH models.

3.1 Comparison of the Forecasting Performance of the GARCH-Type Models

In this section, we present the results of in-sample volatility forecasting accuracy for each model by using real data as provided in Table 1. Various forecasting performance criteria (Eqs.16–19) are used in order to make the comparison. The comparison results are reported in Tables 2, 3, 4, 5 and 6. Each table shows the model comparison result for the individual stock market. Note the ARCH (1), GARCH (1,1), EGARCH (1,1) and GJR-GARCH(1,1) specifications are assumed in this study to make the comparison simple. We have left for future work the estimation of GARCH in which the estimates of future volatility will depend on higher order.

Table 2 Thailand (SET): performance measurement of SVM-GARCH and MLE-GARCH model

Model	SVM-GARCH		
	MAE	MSE	RMSE
ARCH (1)	0.00524	0.00005	0.00688
GARCH (1,1)	0.00522	0.00005	0.00681
GJR-GARCH (1,1)	**0.00519**	**0.00005**	**0.00678**
EGARCH (1,1)	0.00525	0.00005	0.00683
	norm-MLE-GARCH		
ARCH (1)	0.00642	0.00009	0.00925
GARCH (1,1)	0.00519	0.00006	0.00679
GJR-GARCH (1,1)	0.00642	0.00009	0.00923
EGARCH (1,1)	0.00642	0.00009	0.00924
	Std-MLE-GARCH		
ARCH (1)	0.00827	0.00018	0.01355
GARCH (1,1)	0.00900	0.00014	0.01186
GJR-GARCH (1,1)	0.00827	0.00018	0.01355
EGARCH (1,1)	0.00827	0.00018	0.01355

Table 3 Singapore (SGX): performance measurement of SVM-GARCH and MLE-GARCH model

Model	SVM-GARCH		
	MAE	MSE	RMSE
ARCH (1)	0.00598	0.00006	0.00741
GARCH (1,1)	0.00599	0.00006	0.00741
GJR-GARCH (1,1)	**0.00597**	**0.00006**	**0.00739**
EGARCH (1,1)	0.00600	0.00006	0.00741
	norm-MLE-GARCH		
ARCH (1)	0.00749	0.00011	0.01039
GARCH (1,1)	0.00601	0.00007	0.00743
GJR-GARCH (1,1)	0.00749	0.00011	0.01038
EGARCH (1,1)	0.00749	0.00011	0.01038
	Std-MLE-GARCH		
ARCH (1)	0.00826	0.00018	0.01354
GARCH (1,1)	0.00900	0.00014	0.01186
GJR-GARCH (1,1)	0.00825	0.00018	0.01354
EGARCH (1,1)	0.00826	0.00018	0.01354

Table 4 The Philippines (PSEi): performance measurement of SVM-GARCH and MLE-GARCH model

Model	SVM-GARCH		
	MAE	MSE	RMSE
ARCH (1)	**0.00826**	**0.00012**	**0.01115**
GARCH (1,1)	0.00876	0.00013	0.01152
GJR-GARCH (1,1)	0.00876	0.00013	0.01152
EGARCH (1,1)	0.00883	0.00014	0.01161
	norm-MLE-GARCH		
ARCH (1)	0.00843	0.00018	0.01355
GARCH (1,1)	0.00901	0.00014	0.01186
GJR-GARCH (1,1)	0.00826	0.00018	0.01354
EGARCH (1,1)	0.00821	0.00018	0.01348
	Std-MLE-GARCH		
ARCH (1)	0.00811	0.00018	0.01333
GARCH (1,1)	0.00901	0.00014	0.01186
GJR-GARCH (1,1)	0.00819	0.00018	0.01343
EGARCH (1,1)	0.00811	0.00017	0.01332

Table 5 Malaysia (KLSE): performance measurement of SVM-GARCH and MLE-GARCH model

Model	SVM-GARCH		
	MAE	MSE	RMSE
ARCH (1)	**0.00335**	**0.00002**	**0.00434**
GARCH (1,1)	0.00337	0.00002	0.00437
GJR-GARCH (1,1)	0.00336	0.00002	0.00436
EGARCH (1,1)	0.00339	0.00002	0.00439
	norm-MLE-GARCH		
ARCH (1)	0.00418	0.00004	0.00594
GARCH (1,1)	0.00339	0.00002	0.00442
GJR-GARCH (1,1)	0.00417	0.00004	0.00594
EGARCH (1,1)	0.00418	0.00004	0.00594
	Std- MLE- GARCH		
ARCH (1)	0.00811	0.00018	0.01333
GARCH (1,1)	0.00901	0.00014	0.01186
GJR-GARCH (1,1)	0.00831	0.00019	0.01359
EGARCH (1,1)	0.00831	0.00019	0.01359

Table 6 Indonesia (JKSE): performance measurement of SVM-GARCH and MLE-GARCH model

Model	SVM -GARCH		
	MAE	MSE	RMSE
ARCH (1)	0.00752	0.00009	0.00968
GARCH (1,1)	0.0077	0.0001	0.00975
GJR-GARCH (1,1)	0.00769	0.0001	0.00974
EGARCH (1,1)	0.00775	0.0001	0.00978
	norm-MLE-GARCH		
ARCH (1)	0.0093	0.00018	0.01326
GARCH (1,1)	0.00772	0.0001	0.00978
GJR-GARCH (1,1)	0.00916	0.00017	0.01313
EGARCH (1,1)	0.00917	0.00017	0.01314
	Std- MLE-GARCH		
ARCH (1)	0.00821	0.00018	0.01348
GARCH (1,1)	0.00901	0.00014	0.01186
GJR-GARCH (1,1)	0.00822	0.00018	0.01349
EGARCH (1,1)	0.00822	0.00018	0.0135

Tables 2, 3, 4, 5 and 6 present the results of the SVM-GARCH and the MLE-GARCH models. We compare the performance of eight models consisting of ARCH, GARCH, GJR-GARCH and EGARCH models under MLE with student-t distribution and ARCH, GARCH, GJR-GARCH and EGARCH models under SVM. The comparison is based on MAE, MSE, and RMSE. It can be seen that SVM-GARCH models outperform the traditional MLE-GARCH models for all cases. We observe that SVM-GJR-GARCH model is the best fit for Singapores and Thailands market returns, while SVM-ARCH model outperforms the traditional MLE-GARCH model for the stock market returns in the Philippines, Malaysia, and Indonesia. According to this empirical evidence based on real data, we can confirm a theoretical advantage of the SVM for estimating GARCH models.

The predictions made by the best fit SVM-GARCH model are plotted in Fig. 1 (black dash line). We also add the MLE-GARCH predictions in this figure for comparison. Note that we only show the MLE-GARCH followings the best specification of the SVM-GARCH, thus MLE-ARCH and MLE-GJR-GARCH predictions are illustrated in Fig. 1. Apparently in Fig. 1, the predictions by MLE-GJR-GARCH and SVM-GJR-GARCH are quite similar (see, Thailand and Singapore).

However, it is clear that the volatility predictions from MLE-ARCH are quite different from SVM-ARCH. From the plots, the forecast lines by SVM models capture more points than the MLE models do as the black dash line is so close to the blue points that the value is nearing zero (see, Malaysia and the Philippines). This indicates that SVM could improve forecasting performance.

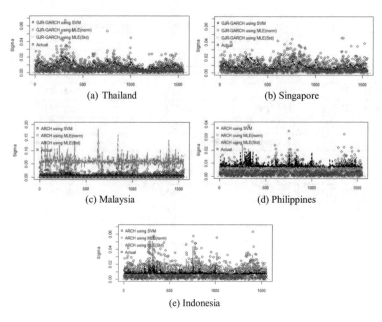

Fig. 1 Volatility Forecasts of ASEAN-5 Stock Markets. Note: Small dot line is the actual volatility; orange and yellow dash lines are forecasted by the parametric GARCH models (normal MLE and student-t MLE) while black dash line is obtained by SVM-GARCH

4 Conclusion

The study focuses on forecasting volatility using the GARCH model. We have employed the SVM to estimate the unknown parameters of GARCH model instead of using the MLE. To confirm the usefulness of SVM for estimating various GARCH-type models, i.e. ARCH, GARCH, GJR-GARCH, and EGARCH, we compare the prediction performance between GARCH model estimated by SVM (SVM-GARCH) and GARCH model estimated by MLE (MLE-GARCH). Empirical applications are made for forecasting the real data of daily closing price in ASEAN-5 stock markets.

According to the empirical results, we find that SVM based GARCH clearly outperforms the MLE based GARCH in terms of the lowest MAE, MSE, and RMSE. The SVM-GJR-GARCH model is the most appropriate to describe the volatility in Thailands and Singapores stock markets considering its lowest MAE, MSE and RMSE. Likewise, SVM-ARCH model is able to give more accurate predictions than regular the MLE on the stock market returns in the Philippines, Malaysia, and Indonesia.

References

Bollerslev, T. (1986). Generalized autoregressive conditional heteroskedasticity. *Journal of Econometrics, 31*(3), 307–327.

Chen, S., Hardle, W. K., & Jeong, K. (2010). Forecasting volatility with support vector machine based GARCH model. *Journal of Forecasting, 29*(4), 406–433.

Engle, R. F. (1982). Autoregressive conditional heteroscedasticity with estimates of the variance of United Kingdom inflation. *Econometrica: Journal of the Econometric Society*, 987–1007.

Franses, P. H., & Van Dijk, D. (1996). Forecasting stock market volatility using (nonlinear) Garch models. *Journal of Forecasting, 15*(3), 229–235.

Glosten, L., Jagannathan, R., & Runkle, D. (1993). On the relation between the expected value and the volatility nominal excess return on stocks. *Journal of Finance, 46*, 1779–1801.

Grigoryan, H. (2017). Stock Market Trend Prediction Using Support Vector Machines and Variable Selection Methods. In *2017 International Conference on Applied Mathematics, Modelling and Statistics Application (AMMSA 2017)*. Atlantis Press.

Held, L., & Ott, M. (2018).On p-values and Bayes factors. *Annual Review of Statistics and Its Application, 5*, 393–419.

Hu, Z., Zhu, J., & Tse, K. (2013). Stocks market prediction using support vector machine. In *2013 6th International Conference on Information Management, Innovation Management and Industrial Engineering* (Vol. 2, pp. 115–118). IEEE.

Madge, S., & Bhatt, S. (2015). Predicting stock price direction using support vector machines. Independent work report spring.

Nelson, D. B. (1991). Conditional heteroske dasticity in asset returns: A new approach. *Econometrica, 59*, 347–370.

Rungruang, C., Srichaikul, W., Chanaim, S., & Sriboonchitta, S. (2019). Prediction the direction of SET50 index using support vector machines. *Thai Journal of Mathematics, 1*(1), 153–165.

Scholkopf, B., & Smola, A. J. (2001). *Learning with kernels: Support vector machines, regularization, optimization, and beyond*. MIT Press.

Upadhyay, V. P., Panwar, S., Merugu, R., & Panchariya, R. (2016). Forecasting stock market movements using various kernel functions in support vector machine. In *Proceedings of the International Conference on Advances in Information Communication Technology & Computing* (p. 107). ACM.

Vapnik, V. (1995). *The nature of statistical learning theory*. New York: Springer.

Macroeconomic Determinants of Trade Openness: Empirical Investigation of Low, Middle and High-Income Countries

Wiranya Puntoon, Jirawan Suwannajak, and Woraphon Yamaka

Abstract This study aims to determine the effects of macroeconomic factors on trade openness. We use panel regression with heterogeneous time trends to explore the causal relationship between macroeconomic factors and trade openness. The analysis relies on a sample of 85 countries for the period 1990–2017. We segment the data set into three sub-panels according to per capita income classification that distinguishes countries as belonging to low-income, middle-income, and high-income groups. Various types of panel regression are also considered, and it is found that time fixed effects model is the best model for all income groups. The key finding of this study is that GDP per capita shows a decisive evidence for its positive effects on trade openness in all income groups.

Keywords Trade openness · Cross-country level · Macroeconomic determinants · Panel regression · Heterogeneous time trends

1 Introduction

Nowadays, no nation exists in economic isolation. Tahir et al. (2018) suggested that it is difficult for the nation to gain the international economic benefits without economic integration in these recent years. All national economies are linked together through various channels and forms such as the goods and services, labor, business

W. Puntoon
Center of Excellence in Econometrics, Chiang Mai University, Chiang Mai, Thailand
e-mail: wiranya.pun@gmail.com

J. Suwannajak (✉) · W. Yamaka
Faculty of Economics, Center of Excellence in Econometrics, Chiang Mai University, Chiang Mai, Thailand
e-mail: woraphon.econ@gmail.com

W. Yamaka
e-mail: suwannajak.j@gmail.com

© The Editor(s) (if applicable) and The Author(s), under exclusive license to Springer Nature Switzerland AG 2021
N. Ngoc Thach et al. (eds.), *Data Science for Financial Econometrics*, Studies in Computational Intelligence 898, https://doi.org/10.1007/978-3-030-48853-6_27

enterprise, investment funds, and technology. The linkages are also witnessed in non-economic elements such as culture and the environment. Many countries have pursued economic integration as a strategy to promote economic growth and reduce poverty and income inequality for more than half a century. The economic integration can enhance the economic competitiveness and industrialization resulting a better employment opportunity which subsequently leads to poverty reduction in the country.

One of the main channels driving an economic integration is trade openness. Several researchers have attempted to investigate the impact of trade openness on economic growth and found the impacts to be both positive and negative. Leamer (1988), Dollar (1992), Sachs and Warner (1995) and Herath (2010) found the positive effects of trade liberalization on the economic growth, while the negative impact was revealed by Dollar (1992), Barro and Sala-i-Martin (1995), Sachs and Warner (1995), Edwards (1998) and Greenaway et al. (1998). It is so surprising that trade could not improve the economy. However, Greenaway et al. (1998) clarified that the source of negative impact is the government intervention. They explained that government intervention could distort the trade competitiveness as it could reduce the ambition and confidence of the domestic workers.

As mentioned above, there are many evidences confirming the importance of trade openness in propelling the economic growth of countries. However, there is a paucity of scholarly works to investigate the factors affecting the trade openness, especially the macroeconomic factors. Tahir et al. (2018) tried to explain how macroeconomic variables affect trade openness for the South Asian Association for Regional Cooperation (SAARC) region. They found such macroeconomic determinants as investment both in physical and human capital and per capita gross domestic product (GDP) to have a positive effects on trade openness. However, the size of labor force and currency exchange rate show a negative impact, Mbogela (2019) examined the determinants of trade openness in the African countries. His study found that the population size, the income per capita and economic location are the most important factors boosting the trade openness. In the individual country point of view, Guttmann and Richards (2006) analyzed the empirical determinants of trade openness in Australia. They revealed that the low openness of Australia is due to the country's large geographic size and great distance to the rest of the world.

The available empirical literature has already paid attention to the determinants of the trade openness in various scopes, the general examination on the effects of macroeconomic variables on trade openness at cross-country level has been ignored. The question of what really determines trade openness or the volume of trade at a cross-country level is worth answering. As some previous works investigated this casual effects at a certain country level using time series data or sample of a few panel countries, the present study aims to address this problem and examine macroeconomic determinants of trade openness with a focus on the low, middle and high-income countries to see whether macroeconomic determinants really matter for trade openness. The reason for focusing on these low, middle and high-income groups is that these groups are different in economic characteristics; thus, the impact of macroeconomic variables on trade openness could be different. We believe that our empirical study could provide a very useful information and hence policy-makers are expected to benefit a lot from our findings.

To investigate the effects of macroeconomic variables on trade openness, we consider the panel regression model with heterogeneous time trends of Kneip et al. (2012). This model is more flexible than the traditional panel regression when cross-sectional dimension and time dimension are low. Kneip et al. (2012) mentioned that the structural assumption on the error term (the unobserved heterogeneity is assumed to remain constant over time within each cross-sectional unit) in the traditional panel regression becomes very implausible when cross-sectional dimension and time dimension are large. Thus, the unobserved heterogeneity is not estimated correctly. In this study, we model panel regression for 85 countries from 1990 to 2017. Thus, the panel regression model with heterogeneous time trends is employed in our work. This study makes at least two contributions to the literature. First, no paper to date has examined the impact of macroeconomic variables on trade openness different in economic characteristics, namely the low, middle and high-income countries. 85 countries of data is employed to investigate the macroeconomic impacts. Second, unlike previous studies that employed the convention panel regression, a more flexible panel regression model with heterogeneous time trends is employed in this study.

Paper is organized as in the following. Section 2 explains the methodology used in this study. Section 3 presents the data and our trade openness equation. Results and discussions are provided in Sect. 4. In the final section, conclusions are drawn and some policy implications are suggested.

2   Methodology

2.1   Panel Regression with Heterogeneous Time Trends

Panel regression is a tool for estimating the relationships between independent and dependent variables. The model is based on balance panel data, comprising both cross-sectional unit $i = 1, ..., N$ and time series, $t = 1, ..., T$. The appropriate estimation method for this model depends on the two error components. More formally, the most general formulation of a panel data regression model can be expressed by the following equation.

$$y_{it} = \sum_{k=1}^{K} \beta X_{itk} + \alpha_i(t) + \mu_{it}, \tag{1}$$

where y is an $NT \times 1$ vector of response variable, X is $NT \times K$ explanatory variables β is a $K \times 1$ vector of coefficient parameters and μ_{it} is a common stochastic error term or residual. The assumption about this error term is that $\mu_{i,t} \sim N(0, \sigma^2)$

and is usually not correlated with X and the individual-specific effects error component $\alpha_i(t)$. Note that our model allows $\alpha_i(t)$ to vary across individual countries and also over time. This time-varying individual effects $\alpha_i(t)$ are parametrized in terms of unobserved nonparametric functions $f_1(t), ..., f_d(t)$ According to Kneip et al. (2012), $\alpha_i(t)$ can be computed from

$$\alpha_i(t) = \sum_{l=1}^{d} \lambda_{it} f_l(t), \tag{2}$$

where λ_{it} are unobserved individual loadings parameters. If the $\lambda_{it} = 1$ and $f_l(t) = 1$ the model in Eq. (1) is conversed to the classical panel regression model with constant unobserved heterogeneity within each cross-sectional unit. In the estimation aspect, the two-step least squares estimation is employed. Firstly, the coefficient parameter β is estimated given the approximation of $\alpha_i(t)$. Secondly, the $\alpha_i(t)$ is computed by estimating the common factors $f_1(t), ..., f_d(t)$ and λ_{it} such that

$$f_l(t) = \sqrt{T} \gamma_{lt}, \tag{3}$$

$$\lambda_{il} = \frac{1}{T} f_l' \left(y_{it} - \sum_{k=1}^{K} \beta X_{itk} \right), \tag{4}$$

where γ_{lt} is the t element of the eigenvector γ_l that corresponds to the largest eigenvalue ρ_l $l = 1, ..., d$ of the empirical covariance matrix

$$\Sigma = \frac{1}{N} \sum_{i=1}^{N} \bar{\alpha}_i \bar{\alpha}_i', \tag{5}$$

where $\bar{\alpha}_i = Z_\kappa \left(y_{it} - \sum_{k=1}^{K} \beta X_{itk} \right), Z_\kappa = Z(Z'Z + \kappa R)^{-1} Z', Z$ has elements $\{z_s(t)\}_{s,t=1,...,T}$ and R elements $\left\{ \int z_s^m(t) z_k^m(t) dt \right\}_{s,k=1,...,T}$ is a preselected smoothing parameter to control derivative of the function of $\vartheta_i^m(t)$ (smoothness function). This function can be viewed as a spline function and in this study, we assume $m = 2$ which leads to cubic smoothing splines. For more estimation details, we refer to Bada and Liebl (2012).

2.2 Model Specifications of Panel Regression Model

As there are many specifications of the panel regression. In practice, the best specification of the model is important as it will lead to obtaining the accurate results. Therefore, in this study, four types of panel regression with heterogeneous time trends are considered, namely none of individual specific effects and time specific effects, fixed effects, time fixed effects and random effects which can be written as follows:

$$\text{None} \quad y_{it} = \sum_{k=1}^{K} \beta X_{itk} + \alpha_i(t) + \mu_{it}, \tag{6}$$

$$\text{Fixed effects} \quad y_{it} = \phi_i + \sum_{k=1}^{K} \beta X_{itk} + \alpha_i(t) + \mu_{it}, \tag{7}$$

$$\text{Time fixed effects} \quad Y_{i,t} = \theta_t + \sum_{k=1}^{K} \beta X_{itk} + \alpha_i(t) + \mu_{it}, \tag{8}$$

$$\text{Random effects} \quad Y_{i,t} = \phi_i + \theta_t + \sum_{k=1}^{K} \beta X_{itk} + \alpha_i(t) + \mu_{it}, \tag{9}$$

3 Data and Model Specification

The main objective of this paper is to examine the effects of macroeconomic determinants on trade openness for the low, middle and high-income countries. There are various macroeconomic determinants as discussed in the literature which matter for achieving higher growth of trade openness. In this study, we consider several macroeconomic variables consisting of economic characteristics (Real GDP per capita; USD), urban population (UP; people), gross capital formation (CF; USD), labor force (LF; people), and exchange rate (XRT; local currency to USD), Trade openness (TRADE) is simply defined as the summation of the export and import values as a percent of GDP. The data covers the years of 1990–2017 for 85 countries (low-income (15), middle-income (47), high income (23)). The data includes 2,380 observations in total. The selections of the countries are based on the availability of data for the level of income groups. All data are obtained from the database of World Bank and transformed into natural logarithm. Thus, we can present our empirical model for each group as follows:

$$\ln Trade_{it} = \beta_0 + \beta_1 \ln UP_{it} + \beta_2 \ln GDP_{it} + \beta_3 \ln CF_{it} + \beta_4 \ln LF_{it} + \beta_5 \ln XRT_{it} + \alpha_{it} + \mu_{it}, \tag{10}$$

Table 1 reports the descriptive statistics of low, middle and high-income countries. It shows the mean, median, maximum, minimum and standard deviation. We can observe that low-income countries show the lowest trade openness, while the highest trade openness is shown by high-income countries. The average of labor force and urban population of our sample are not much different in terms of mean. We see that

Table 1 Descriptive statistics (logarithm)

	Dependence variable	Independence variable				
	TRADE	UP	GDP	CF	LF	XRT
Low income						
Mean	−1.6493	3.3358	5.9771	20.3169	15.1262	5.5074
Median	−1.6518	3.4347	5.9918	20.382	15.2395	6.2034
Maximum	−0.4303	4.1043	6.9374	23.6579	17.0381	9.1147
Minimum	−2.6051	2.1809	4.798	16.9712	12.5985	−0.0736
Std. Dev.	0.4501	0.4085	0.4586	1.3371	1.0187	1.6926
Observations	420	420	420	420	420	420
Middle income						
Mean	−1.4558	3.8795	7.6312	22.7603	15.64	3.3975
Median	−1.3871	3.966	7.6608	22.6629	15.6819	3.2407
Maximum	−0.1169	4.5191	9.5882	29.3146	20.4874	10.0155
Minimum	−3.6201	2.7965	4.5559	17.8689	11.4355	−10.4292
Std. Dev.	0.612	0.3913	0.9077	2.0101	1.9246	2.7884
Observations	1316	1316	1316	1316	1316	1316
High income						
Mean	−0.4489	4.3893	10.0524	24.155	14.6772	1.6571
Median	−0.4259	4.4217	10.1775	24.4083	14.8685	1.2792
Maximum	1.1532	4.6052	11.5431	29.0201	18.9168	7.2453
Minimum	−2.1041	3.4391	7.8219	18.7399	11.1289	−0.9782
Std. Dev.	0.6497	0.2191	0.7557	2.2657	2.0693	2.0675
Observations	644	644	644	644	644	644

high-income countries are shown to be the highest urban population but the smallest workers with the mean value on 4.3893 and 14.6772, respectively.

Before estimation, we need to check the stationarity of the variables to make sure that they fluctuate around a constant mean, and their variances are constant. We used the panel unit root test based on the Levin, Lin and Chou (LLC) and Im, Pesaran and Shin Test (IPS), Augmented Dickey Fuller (ADF) and Phillips–Perron (PP). Table 1 shows the results of the tests at level for LLC, IPS, ADF and PP tests with no trend. We find there are many strong evidences supporting the stationarity of our panel data at the level as the MBF values are mostly less than 0.33. This indicates that we can further use the panel regression to analyze our panel data.

As we mentioned in Sect. 2.2, we estimate Eq. (10) using four types of panel regression. Thus, the best specification is obtained by comparing their performance. In addition, we also compare the result of the Panel regression with heterogeneous time trends and Panel regression with constant time trends. To choose the best fit of the model specification, we use Akaike Information criterion (AIC). AIC is the

Table 2 Panel unit root at the level with time trend

Variable	Method	Low income	Middle income	High income
Trade	LLC	−0.2333 (0.9732)	−4.0009 (0.0003)	−1.6705 (0.2478)
	IPS	0.4870 (0.8882)	−2.2442 (0.0806)	−1.2659 (0.4488)
	ADF	21.3943 (0.0000)	113.5120 (0.0000)	46.5146 (0.0000)
	PP	24.1750 (0.0000)	117.8600 (0.0000)	38.7909 (0.0000)
UP	LLC	1.3401 (0.4074)	−1.3631 (0.3950)	−14.4949 (0.0000)
	IPS	6.0088 (0.0000)	1.8748 (0.1725)	−97.7367 (0.0000)
	ADF	13.3691 (0.0000)	103.6590 (0.0000)	100.9190 (0.0000)
	PP	104.7640 (0.0000)	479.5360 (0.0000)	247.1930 (0.0000)
GDP	LLC	1.5621 (0.2952)	−1.0560 (0.5726)	−1.7699 (0.2088)
	IPS	3.3234 (0.0040)	5.2126 (0.0000)	2.2781 (0.0747)
	ADF	8.1478 (0.0000)	38.8147 (0.0000)	22.9910 (0.0000)
	PP	13.0219 (0.0000)	59.3108 (0.0000)	23.7999 (0.0000)
CF	LLC	0.9234 (0.6529)	−1.8612 (0.1769)	−1.0164 (0.5966)
	IPS	3.4732 (0.0024)	3.6775 (0.0012)	1.9818 (0.1403)
	ADF	8.0659 (0.0000)	51.8520 (0.0000)	23.8611 (0.0000)
	PP	9.5587 (0.0000)	74.2529 (0.0000)	21.7223 (0.0000)
LF	LLC	1.8486 (0.1811)	−2.2019 (0.0885)	−2.7849 (0.0207)
	IPS	6.9307 (0.0000)	2.9077 (0.0146)	2.7340 (0.0238)
	ADF	5.6580 (0.0000)	98.5143(0.0000)	36.9738 (0.0000)
	PP	42.5348 (0.0000)	208.9870 (0.0000)	75.6311 (0.0000)
XRT	LLC	−4.6909 (0.0000)	−9.2417 (0.0000)	−2.7077 (0.0256)
	IPS	−2.9688 (0.0122)	−8.5049 (0.0000)	−2.7962 (0.0201)
	ADF	54.0656 (0.0000)	249.5870 (0.0000)	52.6786 (0.0000)
	PP	73.3055 (0.0000)	396.9590 (0.0000)	56.2288 (0.0000)

Note () is the Minimum Bayes Factor (MBF), calculated by $MBF = \exp(t^2/2)$ (see, Maneejuk and Yamaka 2020)

tool for measuring the quality of statistical models for a given set of data. Given a collection of models, AIC estimates the performance of each model relative to each of the other models. Hence, AIC can be viewed as the model selection measure. In this study, best model can be chosen based on the smallest value of AIC. Table 2 shows AIC for each specification for the income subpanels which are low-income, middle-income and high-income. It is found that the smallest AIC is obtained from the time fixed effects for all three groups. We thus conclude that time fixed effects is the efficient specification for estimating our three income subpanels. Also this model specification provides more accuracy and reliable results when compared to the conventional model through AIC.

4 Estimation Results of the Panel Regression Model

Following the model selection result, Eq. (10) is estimated by time fixed effects model for all income groups to investigate the impacts of such macroeconomic factors as urban population, GDP per capita, gross capital formation, labor force and exchange rate on trade openness. The results are respectively provided in Tables 3, 4 and 5 for low-income countries, middle-income countries and high income countries.

Table 3 Model selection (logarithm)

		AIC		
		Low-income	Middle-income	High-income
Panel regression with constant time trends	None	73.32651	162.4754	128.3584
	Fixed effects	71.9293	159.575	128.0792
	Time fixed effects	31.96254	306.0489	70.95988
	Random effects	73.32651	162.4754	128.3584
Panel regression with heterogeneous time trends	None	25.51299	86.16482	1.741902
	Fixed effects	27.49016	88.50797	−1.502663
	Time fixed effects	24.60398	76.79495	−10.67905
	Random effects	26.27619	77.90326	−8.700712

Table 4 Estimation results of low-income countries

Parameter	Estimate	Std. error	z-value	MBF-test
β_0	6.7900	31.1000	0.2180	0.9765
UP	−1.1700	1.7700	−0.6610	0.8038
GDP	0.2050	0.0969	2.1200	0.1057*
CF	0.1630	0.0389	4.2000	0.0001***
LF	−0.6000	2.0700	−0.2900	0.9588
XRT	−0.0025	0.0950	−0.0261	0.9997

Note MBF denotes Minimum Bayes factor computed by e^{plogp} where p is p − value where * = decisive evidence, *** = moderate evidence (see Held and Ott 2018)

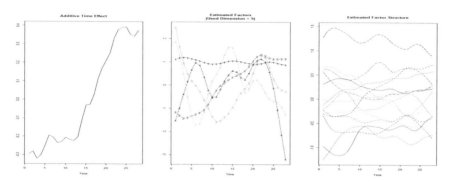

Fig. 1 Left panel: Estimated time-effects θ_t Middle panel: Estimated common factors $f_l(t)$, $l = 1, \ldots, 5$ Right panel: Estimated time-varying individual effects $\alpha_i(t)$

As for the low-income group, Table 4 shows that three macroeconomic factors have a negative effects on trade openness: urban population (-1.1700), labor force (-0.6000) and exchange rate (-0.0025), while GDP per capita and gross capital formation show a positive effects. According to MBF, our findings provide a weak evidence of urban population, labor force and exchange rate, moderate evidence of GDP per capita and decisive evidence of the gross capital formation effects on trade openness of low-income countries. The results reveal that the degree of trade openness of low-income countries would be influenced positively by higher per capita GDP and gross capital formation In addition, we plot the time-effects, common factors and time-varying individual effects. According to Fig. 1, the left panel shows that the different country have considerable different time-varying levels. This indicates that trade openness of low-income countries had substantially increased along 1990–2017. The middle panel shows the five estimated common factors. We can observe that the common factors had increased along 1990–2017, implying a higher variance of this common factors. Finally, the right panel illustrates the time-varying individual effects for 15 low-income countries. We can observe that the individual effects of each country seem to be stationary overtime.

Table 5 presents Panel fixed effects estimation results for middle-income countries. According to the results, the estimated coefficients of urban population and GDP per capita have, respectively, moderate and decisive evidences in the middle-income group. The coefficient of urban population is negative while the coefficient of GDP per capita is positive. This indicates that a 1% increase in urban population and in GDP per capita decreases and increases trade openness by a value of 2.71% and 0.347%, respectively. The possible explanation for the negative relationship effects of urban population on trade openness is that the resources are consumed by the population instead of being used in productive channels. For other variables, we obtain a weak evidence for the coefficients of exchange rate, gross capital formation and labor force.

We then plot the time-effects, common factors, and time-varying individual effects in Figure 2. The left panel shows that when the time has changed, the trade openness of

Table 5 Estimation results of middle-income countries

Parameter	Estimate	Std. error	z-value	MBF-test
β_0	4.7200	7.1000	0.6650	0.8016
UP	−2.7100	1.3300	−2.0300	0.1274*
GDP	0.3470	0.0450	7.7200	0.0000***
CF	0.0156	0.0265	0.5900	0.8403
LF	0.0918	0.3200	0.2870	0.9597
XRT	−0.0202	0.0356	−0.5680	0.8510

Note MBF denotes Minimum Bayes factor computed by $e^{p \log p}$, where p is where * = decisive evidence, *** = moderate evidence (see Held and Ott 2018)

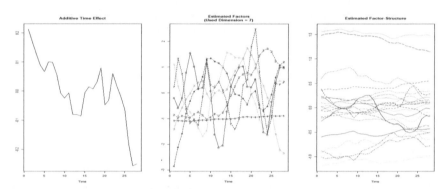

Fig. 2 Left panel: Estimated time-effects θ_t Middle panel: Estimated common factors $f_l(t)$, $l = 1, ..., 7$ Right panel: Estimated time-varying individual effects $\alpha_i(t)$

middle-income countries had decreased in 1990–2017. The seven estimated common factors are also provided in the middle panel. We can observe that the common factors $f_2(t)$ had increased along 1990–2017, implying a higher variance of this common factors. Finally, the right panel illustrates the time-varying individual effects for 47 middle-income countries. We can observe that the individual effects of each country are quite different.

Finally, Table 6 provides the estimated effects of macroeconomic factors on trade openness of high-income countries. The estimated coefficients of GDP per capita and exchange rate are, respectively, with decisive and moderate positive evidences, implying that a change in trade openness of a high-income country is strongly influenced by GDP per capita and exchange rate (depreciation). For urban population, labor force, and gross capital formation variables, we find a weak evidence supporting the impact of these variables on trade openness.

We further plot the time-effects, common factors, and time-varying individual effects in Fig. 3. The left panel shows that when the time has changed, the trade openness of high-income countries had decreased in 1990–2017. The six estimated common factors are also provided in the middle panel. We can observe that the common factors had obviously increased along 1990–2017, implying a higher variance

Table 6 Estimation results of high-income countries

Parameter	Estimate	Std. error	z-value	MBF-test
β_0	−2.5400	14.3000	−0.1770	0.9845
UP	−0.5610	3.1200	−0.1800	0.9839
GDP	0.4910	0.0519	9.4600	0.0000***
CF	0.0101	0.0264	0.3830	0.9293
LF	−0.0695	0.3380	−0.2060	0.9790
XRT	0.0696	0.0298	2.3300	0.0662*

Note MBF denotes Minimum Bayes factor computed by e^{plogp} where p is p − value where * = decisive evidence, *** = moderate evidence (see Held and Ott 2018)

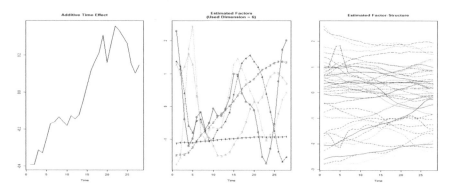

Fig. 3 Left panel: Estimated time-effects θ_t Middle panel: Estimated common factors $f_l(t)$, $l = 1, ..., 6$ Right panel: Estimated time-varying individual effects $\alpha_i(t)$

of this common factors. Finally, the right panel illustrates the time-varying individual effects for 23 high-income countries. We can observe a heterogeneous time-varying individual effects among high-income countries.

In sum, our panel time fixed effects model provides many evidences that GDP-per capital has a positive effects on trade openness for all three groups. The results reveal that the degree of trade openness would be influenced positively by higher per capita GDP. The results are consistent with the gravity model of international trade and also with the findings of Frankel and Romer (1999) and Tahir et al. (2018).

5   Conclusion

This paper focused on the low, middle- and high-income countries to investigate the effects of macroeconomic determinants on trade openness. The motivation for the current study stems from the beneficial impact that trade openness has on the economic growth of many countries. We consider annual panel dataset of 15 low-income countries, 47 middle-income countries and 23 high-income countries for the period 1990–2017. Several macroeconomic variables are investigated, namely urban population, GDP per capita, gross capital formation, labor force and exchange rate. Empirical analysis is carried out by using panel regression with heterogeneous time trends.

Prior to estimating our empirical openness model, four types of panel regression are employed, and AIC is used for selecting the best fit type. The result of model selection shows that time fixed effects panel regression is the best fit model for all three groups.

The results from our investigation indicate that the increase in gross capital formation decisively increases trade openness in low-income countries, while the increase in the urban population could decrease trade openness in middle-income countries. The exchange rate variable is found to produce strongly positive effects on trade openness of high-income counties. If exchange rate is increased (depreciated), it will lead to a higher trade competitiveness in the high-income countries. In addition, we also found that GDP per capita contributes a strong positive effects to trade openness. This finding confirms what the previous studies have established with regards to the influence of GDP per capita as a proxy for the level of economic development on the level of trade openness. GDP per capita has proved to be the most influential variable in this study, which is consistent with the traditional gravity model approach.

The study suggests that policymakers of each country in whatsoever income group take appropriate steps to liberalize more the international trade by first improving the national economic development.

References

Bada, O., & Liebl, D. (2012). phtt: Panel data analysis with heterogeneous time trends. *R package version, 3*(2).

Barro, R., & Sala-i-Martin, X. (1995). *Economic growth*. New-York: Mc Graw Hill.

Dollar, D. (1992). Outward-oriented developing economies really do grow more rapidly: Evidence from 95 LDCs, 1976–85. *Economic Development and Cultural Change, 40*(3), 523–544.

Edwards, S. (1998). Openness, productivity and growth: What do we really know? *The Economic Journal, 108*(477), 383–398.

Frankel, J., & Romer, D. (1999). Does trade cause growth? *American Economic Review, 89*(3), 379–399.

Greenaway, D., et al. (1998). Trade reform, adjustment and growth: What does the evidence tell us. *The Economic Journal, 108*, 1547–1561.

Guttmann, S., & Richards, A. (2006). Trade openness: An Australian perspective. *Australian Economic Papers, 45*(3), 188–203.

Held, L., & Ott, M. (2018). On p-values and Bayes factors. *Annual Review of Statistics and Its Application, 5*, 393–419.

Herath, H. M. S. P. (2010). Impact of trade liberalization on economic growth of Sri Lanka: An Econometric Investigation. In *Proceedings of the 1st, Internal Research Conference on Business and Information*.

Kneip, A., Sickles, R. C., & Song, W. (2012). A new panel data treatment for heterogeneity in time trends. *Econometric Theory, 28*(3), 590–628.

Leamer, E. (1988). Measures of openness. *Trade policy issues and empirical analysis* (pp. 145–204). Chicago: Chicago University Press.

Mbogela, C. S. (2019). An Empirical study on the determinants of trade openness in the African economies. *Advances in Management and Applied Economics*, *9*(3), 9–42.

Maneejuk, P., & Yamaka, W. (2020). Significance test for linear regression: how to test without P-values? *Journal of Applied Statistics*, 1–19.

Sachs, J. D., & Warner, A. (1995). Economic reform and the process of global integration. *Brookings Papers on Economic Activities*, *1*, 1–118.

Tahir, M., Hasnu, S. A. F., & Ruiz Estrada, M. (2018). Macroeconomic determinants of trade openness: Empirical investigation of SAARC region. *Journal of Asia Business Studies*, *12*(2), 151–161.

Determinants of Bank Profitability in Vietnam: An Empirical Lasso Study

Van Dung Ha and Hai Nam Pham

Abstract In this study, the factors that affected bank profitability in Vietnam over the period 2007–2018 are investigated. The net interest margin (NIM), the return on assets (ROA), and the return on equity (ROE) are considered proxies for bank profitability. Substantial predictive ability is analyzed by applying a variable selection technique known as the least absolute shrinkage and selection operator (LASSO). This analysis was based on considering a dataset of annual reports of 30 banks as well as macroeconomic factors. Finally, the results demonstrate that capital rate, equity, loan rate, bank size, deposit, and liquidity are used to determine bank profitability.

Keywords Bank profitability · LASSO · Vietnam banking system · Variable selection

1 Introduction

In the banking system of Vietnam, significant changes have been observed from 1988 to 2019. In the transforming process of Vietnam from the centrally planned economy to market economy, the development of banking system has played an important role. In order to follow up the market economy, Vietnam withdrew the centrally planned economy and applied the free market economy as well as joined international financial institutions such as the International Monetary Fund, World Bank, etc. The new banking era in the market economy receives much of capital in both human and capital ones from the inside economy and abroad. Only state-owned banks used to operate in Vietnam before 1988, whereas the banking system

V. D. Ha (✉)
Banking University Ho Chi Minh City (BUH), Ho Chi Minh City, Vietnam
e-mail: dunghv@buh.edu.vn

H. N. Pham
Ho Chi Minh City University of Technology (HUTECH), Ho Chi Minh City, Vietnam
e-mail: ph.nam@hutech.edu.vn

© The Editor(s) (if applicable) and The Author(s), under exclusive license
to Springer Nature Switzerland AG 2021
N. Ngoc Thach et al. (eds.), *Data Science for Financial Econometrics*,
Studies in Computational Intelligence 898,
https://doi.org/10.1007/978-3-030-48853-6_28

of Vietnam consists of 35 commercial banks, including state-controlled banks and private banks, in 2019.

The importance of maintaining a stable and strong banking system is proved by the impacts of the global financial crises on the banking system of Vietnam. In order to manage and supervise the banking system, which is a key channel of capital for the economy, identifying determinants of bank profitability in Vietnam is crucial.

In this study, methodological extensions that apply a variable selection approach, the least absolute shrinkage and selection operator (LASSO), pioneered by Tibshirani (1996), are used. Some recent studies determined that this selection approach can identify the most relevant predictors from an extensive set of candidate variables and can improve the predictive power (Fan and Li 2001; Tian et al. 2015). In the least absolute shrinkage and selection operator, strict assumptions such as the preselection of variables are not strictly required. Moreover, the results are still found to be statistically consistent as the number of observations approaches infinity (Van der Geer 2008). Importantly, the problem of multicollinearity can be significantly reduced by LASSO. In addition, LASSO is computationally efficient even when a large set of potential predictors is considered. Thus, the forecasting power can be improved by LASSO models.

This paper is organized as follows. In Sect. 2, the relevant literature is discussed. In Sect. 3, the data sources and methodology are presented. The empirical results are reported in Sect. 4 and Sect. 5 concludes the paper.

2 Literature Review

The studies on bank profitability are based on two main theories: market power (MP) theory and structural efficiency theory (efficient structure, ES).

The correlation between industry structure and performance is described in the market power hypothesis, which is also called the structure–conduct performance (SCP) hypothesis. The profit of a firm is dictated by its industry structure even when the firms attempt to differentiate themselves. Different industry structures have different dimensions such as the industry culture, regulatory environment, and concentration. The study of Karim et al. (2010) empirically supports the SCP hypothesis. The results demonstrate a significant positive relationship between industry structure and bank profitability. However, the relative-market-power (RMP) hypothesis is contradicted in some studies. According to this hypothesis, the banks receive higher yield as they become bigger and acquire more market shares. The profit of a bank can be earned from larger market shares as well as diversified sources of profits (Berger 1995).

Superior efficiency is considered to be a result of large industries, whereas market concentration is not considered a random event. In this hypothesis, efficient firms can increase their size and market share because they can generate higher profits, and consequently, these firms can increase their market concentration. In the previous studies, these hypotheses are often distinguished considering that market share is an independent variable that exhibits a positive coefficient supporting the ES hypothesis (Smirlock 1985). In the model of more than 2,700 units of state banks, bank profitability was displayed as a function of concentration, market share, as well as considered as an interaction term between market share and concentration. In conclusion, the difference depends on whether the market share is displayed as a proxy for the efficiency of a firm or for market power. It is concluded that the ES hypothesis can be true if and only if efficiency must be related to concentration and/or market share. With regard to the impacts of efficiency on the market structure, the ES hypothesis has been examined in some studies (Berger and Hannan 1992; Goldberg and Rai 1996). In the study of Berger (1995), two measures of efficiency, namely, X-efficiency and scale efficiency, were applied to test the structure–performance relationship. In order to explain the positive relationship between profits and concentration, lower cost resulting from superior management and production process can be used.

Another variable that is used in some studies of several banking sectors is called the scale of regulation. Jayaratne and Strahan (1998) in their study reported that operating costs and loan losses can be sharply decreased by the permission of statewide branching and interstate banking. Branching deregulation can bring improvement because interstate banks perform more efficiently than their rivals. Ownership characteristics can also influence bank profitability when management incentives can be differently modeled in the forms of bank ownership (Short 1979; Bourke 1989; Molyneux 1993).

Many researchers have tried to find the major determinants of bank profitability in terms of empirical study. In order to investigate the determinants of net interest margin (NIM) of banks in four Southeast Asian countries, Doliente (2005) employed the dealer model developed by Ho and Saunders (1981). The primary findings indicate that capital, operating expenses, collateral, and liquid assets, and loan quality are determinants of the NIM of a region. The secondary findings indicate that the noncompetitive structure of the banking systems of a region has large impacts on NIM. In addition, the banking and Asian currency crises have affected the profit size as reported in this study.

Kosmidou et al. (2007) investigated the determinants of bank profitability taking into consideration Greece banks, which operate abroad. During the period 1995–2001, they worked on an unbalanced panel consisting of 19 Greek bank subsidiaries located in 11 countries. They reported that both the operating experience and the level of profits of the parent bank significantly affected the profitability of subsidiary banks abroad, whereas the size of subsidiary banks demonstrated a negative impact on profitability. Similarly, some other factors such as liquidity, domestic stock market developments, concentration or market share, cost efficiency, and loan loss provisions in the host economies demonstrated no significant impacts on the profitability of subsidiary banks.

Athanasoglou et al. (2008) reported that bank specifics, industry specifics, and macroeconomic factors were determinants of bank performance by using an empirical framework that integrated the traditional structure conduct–performance hypothesis. In this paper, the GMM technique was used for panel data taken from Greek banks for the period 1985–2001. The findings suggest that bank profitability in Greek was moderate which explains that deviations from perfectly competitive market structures could not be large. Significant impacts are reported on bank profitability from all bank specifics except size. The results also indicate that the business cycle is positively related to bank profitability.

Berger and Bouwman (2013) investigated the impact of capital on the performance of banks. The results demonstrated that at all times, including normal times, market crises, and banking crises, capital helped banks with small sizes to increase their market share and the probability of survival. They also reported that capital increased the performance of large- and medium-sized banks during banking crises.

Apergis (2014) empirically examined the effects of nontraditional bank activities on the profitability of banks by using data from 1725 U.S. financial institutions for the period of 2000–2013. The empirical results reported that nontraditional bank activities showed a positive impact on the profitability of banks. From the literature, it can be concluded that most of the studies have been conducted in developed countries. However, a few studies have been carried out in developing economies. As an example of a single-country study, Sufian and Habibullah (2009) carried out their study in Bangladesh. The primary limitation of this study was that they ignored the endogeneity problem between profitability, risk, and capital. They applied data from 1997 to 2004 before the adaptation of BASEL II.

Batten and Xuan Vinh Vo (2019) studied the determinants of bank profitability in Vietnam from 2006 to 2014. The study indicated that bank size, capital adequacy, risk, expense, and productivity had strong impacts on the profitability of banks by using some of the econometric panel data methods. They also reported that bank-industry-specific environment influenced the profitability of banks. However, different profitability measures result in different directions of causality.

Truc and Thanh (2016) reported that the factors such as bank size, capital, ownership characteristics, inflation rate, GDP growth, and trading value of stock market influence the profitability of banks in Vietnam.

The previous studies with traditional statistical method show the consistent results of determinants of bank profitability. In addition to previous researches, this research using methodological extension—LASSO approach—may empirically contributes to the present literature and thus it is considered the novelty of the study.

3 Data and Methodology

3.1 Data Sources

In this study, the panel dataset is used, which is considered a relatively broad dataset of commercial banks in Vietnam. In the sample, 30 domestic Vietnamese commercial banks are included during the period 2007–2018. Moreover, different data sources such as the Vietnam Bureau of Statistics, Bankscope, the State Bank of Vietnam, and individual bank reports are followed in this study. Furthermore, the bank specifics are manually recorded from bank financial reports if these data are not found in other databases.

3.2 Methodology

LASSO, originally proposed by Tibshirani (1996), is applied in order to select the most appropriate predictors and to provide accurate forecasts. LASSO is an extended form of an OLS regression that performs both variable selection and regularization by using a shrinkage factor. Moreover, LASSO increases the accuracy and the interpretability of classical regression methods, which is an advantage (Tibshirani 1996).

In LASSO, both parameter estimation and variable selection can be simultaneous executed. LASSO reduces some coefficients in the linear regression and sets others to 0. Hence, LASSO attempts to retain the good features of both subset selection and ridge regression. In this approach, some coefficients are set to 0 by a small value of the threshold δ or a large value of the penalty term λ. Therefore, the LASSO performs a process of continuous subset selection. Moreover, correlated variables still have an opportunity to be selected. In addition, the LASSO linear regression can be generalized to other models, such as GLM, hazards model, etc. (Park and Hastie, 2007). In the beginning, when it was first developed, the LASSO techniques did not have a large diffusion because of the relatively complicated computational algorithms. However, this issue has been resolved by more recent proposals (Table 1).

Table 1 Variables summary

Variable		Formula	Notation
Dependent	Net interest margin	(interest income – interest expense)/total earnings assets	NIM
	Return on assets	Net income/total assets	ROA
	Return on equity	Net income/equity	ROE
Independent	Capital rate	Equity/total assets	CAP
	Equity	Log(equity)	EQT
	Loan rate	Net loan/total assets	NLTA
	Bank size	Log(total assets)	SIZE
	Deposit	Deposit/total assets	DEP
	Liquidity	Cash and securities/total assets	LIQ

4 Empirical Results

4.1 NIM

The difference between interest rates that banks charge for loans and those they pay on deposits is called the net interest margin (NIM). NIM is applied to indicate the profitability and growth of banks.

The parameter lambda that minimizes the mean-squared prediction error is indicated by an asterisk (*) and is selected in this study (Table 2).

The estimates of the determinants of the net interest margin are presented in Table 3. It can be seen from the table that capital rate (CAP), loan rate (NLTA), and deposit (DEP) are the three most important determinants of the profitability of banks. A positive relationship exists between the capital rate and the NIM. In the period 2007–2008, when the capital rate was increased by 1%, the NIM increased by 0.03858%. The loan rate is considered as the second determinant of the NIM. A positive relationship was found between the loan rate and the NIM for the period 2009–2018 excluding the period 2015–2016. In the period 2015–2016, when the loan rate was increased by 1%, the NIM decreased by 0.039%. The deposit rate is considered as the next important determinant of NIM. However, a positive relationship was observed between the deposit rate and the NIM for the period 2011–2014.

Table 2 The selected lamba

Year	Lambda	MSPE	Std. dev
07–08	0.1601*	0.00015	0.000038
09–10	0.0457*	0.00007	0.000013
11–12	0.0067*	0.00013	0.000017
13–14	0.0328*	0.00009	0.000014
15–16	0.0058*	0.00013	0.000035
17–18	0.0582*	0.00016	0.000082

Source Author's calculation

Table 3 Estimation results

Variables	07–08	09–10	11–12	13–14	15–16	17–18
CAP	0.0385	0.0688	0.1761	0.1027	0.0467	0.0335
EQT		0.0014		0.0016	0.0015	
SIZE			0.0045		0.0282	
NLTA		0.0159	0.0543	0.0313	−0.0390	0.0510
DEP	−0.0398		0.0395	0.0230	−0.0100	−0.0903
LIQ		−0.0028	0.0249			

Source Author's calculation

4.2 ROA

The other measurement of bank profitability is known as return on assets (ROA). The capability of a bank to generate profits from its asset management functions is reflected by this measurement criterion. Therefore, it is frequently used as the key ratio for evaluating the profitability of a bank (Table 4).

The estimates of the determinants of return on assets (ROA) are presented in Table 5. It can be seen from the table that capital rate (CAP), loan rate (NLTA), and deposit (DEP) are the three important determinants of the profitability of banks. A positive and significant relationship was observed between the capital rate and the ROA for the period 2007–2018. In the year 2007–2008, when the capital rate was increased by 1%, the ROA increased by 0.025%. However, a negative relationship was observed between the capital rate and ROA in the period 2017–2018. When the capital rate was increased by 1%, the ROA decreased by 0.166%. The loan rate is considered as the second determinant of ROA. A positive relationship was observed between the loan rate and the ROA for the period 2009–2018 excluding the period 2007–2008. In the period 2007–2008, when the loan rate was increased by 1%, the ROA decreased by 0.0001%. A negative relationship was observed between the deposit rate and the ROA for the period 2007–20,118, which indicates that banks with a high level of the deposit rate tend to have a low level of ROA.

Table 4 The selected lambda

Year	Lambda	MSPE	Std. dev
07–08	0.56495*	0.000087	0.000023
09–10	0.09266*	0.000060	0.000025
11–12	0.00514*	0.000121	0.000070
13–14	0.02223*	0.000015	2.976e−06
15–16	0.02911*	0.000015	2.664e−06
17–18	0.00040*	0.000026	7.691e−6

Source Author's calculation

Table 5 Estimation results

Variables	07–08	09–10	11–12	13–14	15–16	17–18
CAP	0.0246	0.0374	0.1353	0.0418		−0.1667
EQT		0.0006		0.0017	0.0014	0.0173
SIZE			0.0038			−0.0158
NLTA	−0.0001		0.0344	0.0049	0.0072	0.0023
DEP	−0.0375		−0.0119		−0.0011	−0.0031
LIQ	0.0118		0.0451			0.0066

Source Author's calculation

Table 6 The selected lambda

Year	Lambda	MSPE	Std. dev
07–08	0.11762*	0.00399	0.00122
09–10	1.09993*	0.00261	0.00061
11–12	1.52808*	0.01126	0.00696
13–14	0.29348*	0.00169	0.00027
15–16	0.01418*	0.00245	0.00035
17–18	0.00154*	0.00347	0.00082

Source Author's calculation

4.3 ROE

The return to the shareholders on their equity capital is defined as ROE. This indicates how skillfully banks invest their money. Higher ROE indicates that banks wisely invest and is likely profitable (Table 6).

The determinants of bank profitability such as capital rate (CAP), equity (EQT), bank size (SIZE), loan rate (NLTA), and deposit (DEP) are presented in Table 7. A negative and significant relationship was observed between the capital rate and the ROE for the period 2007–2018. In the period 2007–2008, when the capital rate was increased by 1%, the ROE decreased by 0.437%. However, the negative impact of the capital rate on ROE was observed to be relatively high in the period 2017–2018. When the capital rate was increased by 1%, the ROE decreased by 2.068%. It was observed that equity, bank size, and loan rate showed positive impacts on the ROE of banks, whereas the deposit rate showed a negative impact on the ROE in the research period.

The results of this study are observed to be relatively consistent with those of the previous studies. As observed by Athanasoglou et al. (2008), Berger and Bouwman (2013), Truc and Thanh (2016), the bank size demonstrates positive impacts on profitability. Among different financial determinants, deposit displays negative impacts on the profitability of banks (as shown by Batten and Vo (2019)), whereas other determinants are positively related to the performance of banks.

Table 7 Estimation results

Variables	07–08	09–10	11–12	13–14	15–16	17–18
CAP	−0.4374	−0.1290			−0.3955	−2.0681
EQT			0.0382	0.0149	0.0118	0.1695
SIZE	0.0150	0.0253		0.0073	0.0092	−0.1467
NLTA			0.0563	0.0550	0.2017	0.3057
DEP	−0.2696				−0.2960	−0.4058
LIQ	0.2051				0.1005	0.1130

Source Author's calculation

In summary, this research, which employs LASSO, has robust results compared to results of other studies using traditional statistical method. The similarity in determinants of bank profitability once confirms that such factor as capital rate, equity, loan rate, bank size, deposit, and liquidity determine net interest margin, return on assets, and return on equity of commercial banks in Vietnam.

5 Conclusion

The novelty of the paper is shown via the robustness of the determinants of bank profitability. Employing LASSO which can increase the prediction power and reduce the post-estimation problems (Fan and Li 2001; Tian et al. 2015; Van der Geer 2008), the paper finds the consistent results as those of previous researches that used traditional statistical method.

The existing literature suggests that variables that affect the profitability of banks are a function of the specific sector. Moreover, bank-specific variables, such as capital, bank size, or loan rate, play an important role in determining the profitability of banks.

Appropriate mechanisms for managing, screening, and monitoring should be properly put in place in order to improve the profitability of banks. In Vietnam, loan approvals and monitoring of troubled loans broadly depend on collateral, whereas, in fact, it should focus on the cash flow of the borrower which may result in low levels of nonperforming loans. Moreover, these mechanisms must take into consideration the specific circumstances of the banking sector when they are designed. In addition, the selection of managers and the organizational body as well as the procedures should be the responsibility of the boards and chief executives of banks. However, the internal decisions of managers are sometimes beyond their control. In order to further improve the quality of the banking sector, thus making it more profitable, the structure of assets and liabilities of banks should be revised as well as the bank size should be increased. The results of this study conform to most of the existing literature.

References

Apergis, N. (2014). The long-term role of non-traditional banking in profitability and risk profiles: Evidence from a panel of U.S. banking institutions. *Journal of International Money and Finance, 45*, 61–73.

Athanasoglou, P. P., Brissimis, S. N., & Delis, M. D. (2008). Bank-specific, industry-specific and macroeconomic determinants of bank profitability. *International Financial Markets, Institutions and Money, 18,* 121–136.

Batten, J., & Vo, X. V. (2019). Determinants of bank profitability—Evidence from Vietnam. *Emerging Markets Finance and Trade, 55*(1), 1–12.

Berger, A. N., & Bouwman, C. H. S. (2013). How does capital affect bank performance during financial crises? *Journal of Financial Economics, 109,* 146–176.

Berger, A. (1995). The relationship between capital and earnings in banking. *Journal of Money, Credit and Banking, 27*(2), 432–456.

Berger, A., & Hannan, T. (1992). The price-concentration relationship in banking: A reply. *The Review of Economics and Statistics, 74,* 376–379.

Bourke, P. (1989). Concentration and other determinants of bank profitability in Europe, North America and Australia. *Journal of Banking and Finance, 13,* 65–79.

Doliente, J. S. (2005). Determinants of bank net interest margins in Southeast Asia. *Applied Financial Economics Letters, 1*(1), 53–57.

Fan, J., & Li, R. (2001). Variable selection via nonconcave penalized likelihood and its oracle properties. *Journal of the American Statistical Association, 96*(456), 1348–1360.

Golbert, L., & Rai, A. (1996). The structure-performance relationship for European banking. *Journal of Banking and Finance, 20,* 745–771.

Ho, T. S. Y., & Saunders, A. (1981). The determinants of bank interest margin: Theory and empirical evidence. *Journal of Financial and Quantitative Analysis, 16*(4), 581–600.

Jayaratne, J., & Strahan, P. (1998). Entry restrictions, industry evolution, and dynamic efficiency: Evidence from commercial banking. *Journal of Law and Economics, XLI,* 239–270.

Karim, B. K., Sami, B. A. M., & Hichem, B. K. (2010). Bank-specific, industry-specific and macroeconomic determinants of African Islamic Banks' profitability. *International Journal of Business and Management Science.*

Kosmidou, K., Pasiouras, F., & Tsaklanganos, A. (2007). Domestic and multinational determinants of foreign bank profits: The case of Greek banks operating abroad. *Journal of Multinational Financial Management, 17,* 1–15.

Molyneux, P. (1993). Market structure and profitability in European banking. *Institute of European Finance, University College of North Wales, Research Paper 9.*

Park, M. Y., & Hastie, T. (2007). L -regularization path algorithm for generalized linear models. *Journal of Royal Statistical Society, 69*(4), 659–677.

Short, B. (1979). The relation between commercial bank profit rates and banking concentration in Canada, Western Europe, and Japan. *Journal of Banking and Finance, 3,* 209–219.

Smirlock, M. (1985). Evidence on the (non) relationship between concentration and profitability in banking. *Journal of Money, Credit, and Banking, 17*(1), 69–83.

Sufian, F., & Habibullah, M. S. (2009). Bank specific and macroeconomic determinants of bank profitability: Empirical evidence from the china banking sector.*Frontiers of Economics in China, 4,* 274–291.

Tian, S., Yu, Y., & Guo, H. (2015). Variable selection and corporate bankruptcy forecasts. *Journal of Banking & Finance, 52,* 89–100.

Tibshirani, R. (1996). Regression shrinkage ad selection via the lasso. *Journal of the Royal Statistical Society, Series B, 58*(1), 267–288.

Truc, N. P. N., & Thanh, N. P. T. (2016). Factors impact on profitability of banking system in Vietnam. *Journal of Development and Integration,* 52–59.

Van de Geer, S. (2008). High-dimensional generalized linear models and the lasso. *The Annals of Statistics, 36*(2), 614–645.

Financial Performance and Organizational Downsizing: Evidence from Smes in Vietnam

Van Dung Ha and Thi Hoang Yen Nguyen

Abstract In statistics, Lasso is a method that is used commonly for the correction of linear regression coefficients. In the present study, this method was applied to the problem of evaluating the impact of financial performance on organizational downsizing, on a dataset of over 2,500 small and medium enterprises in Vietnam, during the years 2013 and 2015. The results of the research demonstrated that certain financial factors, such as return on assets, firm equity, total sales during the year, asset efficiency, and profit margin, exerted an impact on firm downsizing (both capital and employment downsizing). The factors 'number of years since founded' and 'firm size' were also observed to exert significant effects on firm downsizing during these years.

Keywords Organizational downsizing · Financial performance · Lasso · Vietnam

1 Introduction

The stories of organizational downsizing of international companies, such as Hewlett-Packard, Cisco Systems, Lucent Technologies, General Motors, Fujitsu, American Airlines, Sun microsystems, and so on, are considered a development trend of strategic importance in the context of expansion in Europe (Filatotchev et al. 2000), USA (De Meuse et al. 2004), Australia (Farrell and Mavondo 2005), and Asian countries (Yu and Park 2006). The present study explored the effects of financial performance on organizational downsizing in the context of Vietnam. The present report has been organized according to the following structure: (i) theoretical bases

V. D. Ha (✉)
Banking University Ho Chi Minh City, Ho Chi Minh City, Vietnam
e-mail: dunghv@buh.edu.vn

T. H. Y. Nguyen
Thu Dau Mot University, Thu Dau Mot, Vietnam
e-mail: nthyen2011@gmail.com

N. Ngoc Thach et al. (eds.), *Data Science for Financial Econometrics*,
Studies in Computational Intelligence 898,
https://doi.org/10.1007/978-3-030-48853-6_29

and arguments to research issues; (ii) research data and models; (iii) research results; and (iv) conclusions.

2 Literature Review

There are several determinants of organizational downsizing, the financial aspect being one of them. Scaling up may concentrate on certain specific targets, such as (i) reducing operations costs, (ii) reducing management apparatus, (iii) rationalizing activities, (iv) increasing total efficiency, and (v) increasing competitiveness in the market (Jensen 1986; Neinstedt 1989; McKinley et al. 1995). In addition, several studies have reported downsizing to be associated with reduced profits and stock prices, as well as with delayed dividend growth, decreased work ethics, and increased workload of the employees (De Meuse and Tornow 1990; Worrell et al. 1991; Gombola and Tsetsekos 1992; Mishra and Spreitzer 1998).

While a few studies, such as the one conducted by Cascio and Young (2003), have reported that there is no statistical relationship between corporate financial performance and organizational downsizing, certain other studies have reported either negative or positive effects of financial performance on enterprise downsizing. Downsizing of the organization size is perceived in terms of various aspects that are significantly associated with the reduction in the number of employees and with the firm capital. The cohesion of an individual with an organization is based on, in addition to the constraints in the employment contract, the expectation that both the parties would fulfill their respective obligations (Robinson et al. 1994). In return, the organization creates a stable working environment for the employees, offers attractive salaries, and commits on providing personal development opportunities to the employees. If either party notices a possible violation of this agreement, it may lead to negative consequences. From the perspective of psychology, employees would view the downsizing of the organization as a consequence of the employers violating their responsibility toward the employees (De Meuse and Tornow 1990), leading to skepticism and reduced confidence in the senior management among the employees (Wiesenfeld et al. 2000; Lester et al. 2003; Feldheim 2007). This perception may exert a negative effect on the work performance of the individuals (Brockner et al. 1985), consequently affecting the maintenance of a positive working environment and negatively impacting the financial performance of the whole organization. In further negative cases, employees may even engage in vandalism (Buono 2003). However, a major benefit is received from the downsizing of a business when there is interference from the company management in issuing incentive policies to the remaining employees of the organization after the shrinking (Nienstedt 1989; Applebaum and Donia 2000). This allows a favorable working environment after the downsizing of the business, which may lead to high working morale (Hammer 1996). Most of the studies in the literature have researched the series of impacts on the financial performance caused by organizational downsizing, rather than studying the inverse cause-effect relationship. Therefore, an evaluation of the impacts of financial performance on firm downsizing, performed in the present study, would contribute to fulfilling this gap in the literature.

3 Data and Research Model

3.1 The Data

The present study employed the data obtained from a database of over 2,500 enterprises across nine provinces. The database used for this purpose, namely, the Viet Nam Small and Medium Enterprises data, is a collaborative effort of the Central Institute for Economic Management (CIEM), the Institute of Labour Science and Social Affairs (ILSSA), the Development Economics Research Group (DERG) of the University of Copenhagen, and the UNU-WIDER, and they have been collecting data biennially since the year 2005. The data was collected through a survey conducted in the following nine provinces of the country: Hanoi (including Ha Tay), Hai Phong, Ho Chi Minh City, Phu Tho, Nghe An, Quang Nam, Khanh Hoa, Lam Dong, and Long An. Because of the limitation of finance, the research can only access two time periods, 2013 and 2015. In the sample, enterprises were classified on the basis of the current definition of total capital provided by the World Bank, which defines small-scale enterprises like the ones with up to 50 employees and medium-sized enterprises as the ones with up to 300 employees. On the other hand, General Statistics Office defines small-scale enterprises like the ones having total assets of up to 20 billion and medium-sized enterprises as those having up to 100 billion of total assets. In the sample of year 2013, 314 SMEs were identified to have undergone downsizing, while this number reached 537 in the year 2015.

3.2 Research Model

A discontinuous data set provides the opportunity to estimate linear regressions. Owing to the unreliability of traditional Ordinary Least Squares (Wasserstein et al. 2019), the present study employed the Lasso method of estimation for the investigation of factors determining the downsizing of SMEs in Vietnam during the years 2013 and 2015.

In the present study, similar to previous studies, the following two dimensions of firm downsizing were considered: capital downsizing and employment downsizing (Filatotchev et al. 2000; Yu and Park 2006; Feldheim 2007; Carmeli and Scheaffer 2009).

Generally, research model is written as follows:

$$\text{EmpDZ}_i = \beta_0 + \beta_1 \text{ROA}_i + \beta_2 \text{Eqt}_i + \beta_3 \text{Size}_i + \beta_4 \text{AE}_i + \beta_5 \text{PM}_i + \beta_6 \text{Year}_i + \alpha_i$$

$$CapDZ_i = \beta_0 + \beta_1 ROA_i + \beta_2 Eqt_i + \beta_3 Size_i + \beta_4 AE_i + \beta_5 PM_i + \beta_6 Year_i + \alpha_i$$

where:

Measure of firm downsizing

Employment downsizing (EmpDZ): this is defined as the percentage reduction in employment (% in absolute value), and is calculated by subtracting the year-beginning employment from the year-end employment and then dividing the resultant value by year-beginning employment.

Capital downsizing (CapDZ): this is defined as the percentage reduction in total capital (% in absolute value), and is calculated by subtracting year-beginning total capital from the year-end total capital and then dividing the resultant value by year-beginning total capital.

Independent Variables

ROA   return on assets, calculated by dividing the profits by assets.
Eqt   firm equity, expressed in terms of the natural logarithm.
Size   total firm sales during the study year, natural logarithm.
AE   asset efficiency, calculated by dividing the sales by assets.
PM   profit margin, calculated by dividing the profits by sales.
Year   number of years since the organization was founded, natural logarithm.

3.3 The LASSO Linear Regression

The least squares estimates (OLS) for pairs (β_0, β) are based on minimizing the square error as follows:

$$minimize \left\{ \sum_{i=1}^{N} (y_i - \hat{y}_i)^2 \right\}$$

$$= \underset{\beta_0, \beta}{minimize} \left\{ \sum_{i=1}^{N} \left(y_i - \beta_0 - \sum_{j=1}^{p} x_{ij} \beta_j \right)^2 \right\} \qquad (1)$$

$$= \underset{\beta_0, \beta}{minimize} \| y - \beta_0 1 - X\beta \|^2$$

Like OLS, Lasso is also the estimation method for pairs (β_0, β) but by shrinking these parameters. Specifically, Lasso provides the solution to the minimization (1) with the condition that the parameter is tied:

$$\underset{\beta_0, \beta}{minimize} \left\{ \sum_{i=1}^{N} (y_i - \hat{y}_i)^2 \right\}$$

Conditions:

$$\|\beta_j\|_2 \leq t$$

Solving this problem is equivalent to solving the following minimization problem:

$$\underset{\beta_0, \beta}{\text{minimize}} \left\{ \sum_{i=0}^{N} (y_i - \hat{y}i)^2 + \lambda \|\beta_j\|_2 \right\}$$

where λ is a parameter greater than 0.

This paper employs the cross-validation method to select the optimal lambda value. In the cross-validation method, data is randomly divided into folds. In this research, we divide the data into 10 folds. The model is fit 10 times. When one fold is chosen, a linear regression is fit on the other nine folds using the variables in the model for that λ. These new coefficient estimates are used to compute a prediction for the data of the chosen fold. The mean squared error (MSE) of the prediction is computed. This process is repeated for the other nine folds. The 10 MSEs are then averaged to give the value of the cross validation error (CV MSE).

$$CV(\lambda) = \frac{1}{10} \sum_{i}^{10} MSE_i$$

The optimal lambda is selected such that of the smallest value of $CV(\lambda)$.

$$\hat{\lambda} = \underset{\lambda}{\text{argmin}} CV(\lambda)$$

The convention that has emerged following Hastie et al. (2015) is to consider a grid of 100 candidate values for λ. The default number of grid points is 100, and it can be changed by specifying option command grid(#) in stata software.

The largest value in the grid is the smallest value for which all the coefficients are zero, and we denote it by λ_{gmax}. The smallest value in the grid is λ_{gmin}, where $\lambda_{gmin} = r \times \lambda_{gmax}$ and r is set by the option command grid(#). The grid is logarithmic with the ith grid point given by:

$$ln\lambda_i = \frac{i-1}{n-1} \times lnr + ln\lambda_{gmax}$$

where n is the number of grid points and in this case $n = 100$.

4 Research Results

The present study selected a list of Lambda (λ) values in the experiment (capital downsizing and employment downsizing) and obtained the most optimal value. Test results corresponding to the Lambda correction parameter values of LASSO linear regression have been provided in the present report. In addition to offering the choice of selecting meaningful coefficients automatically, this method also reduces the variance, causing improvement in the generalization of the model.

4.1 In 2013

Table 1 presents a list of lambda (λ) values in descending order. Each lambda value is greater than 0. The λ value that resulted in the minimization of the mean-squared prediction error has been indicated by an asterisk (*). A hat (^) has been used to mark the largest λ value at which the MSPE remains within one standard error from the minimal MSPE value. The detailed results obtained for MSPE and the standard deviation of the LASSO linear regression method for the λ values used are provided in Table 1. As observed from the table, the best performance in terms of capital downsizing and employment downsizing was recorded for the λ values of 0.34162143 and 0.0011478, respectively.

The estimation results obtained with both Lasso and OLS are listed in Table 2. Signs of the coefficients remained the same, while the marginal effects did not. The coefficients obtained with Lasso were smaller than those obtained in the case of OLS estimations.

In 2013, ROA and profit margin were identified as two financial determinants in the case of both capital and employment downsizing. Both capital and employment downsizing were observed to be positively associated with ROA and negatively associated with profit margin, which implied that higher ROA led to a higher rate of downsizing of the SMEs in Vietnam. The other variables did not exert any impact on capital downsizing. Size, asset efficiency, and firm's age were observed to exert

Table 1 Parameter values for lambda correction

Capital downsizing				Employment downsizing			
	Lambda	MSPE	Standard deviation		Lambda	MSPE	Standard deviation
1	1.2566134	0.0001127	**0.0000506^**	95	0.0018276	0.0067632	0.0031678
5	0.8661348	0.0001079	0.00005004	97	0.0015173	0.0067631	0.0031677
13	0.4114842	0.0000994	0.00004516	99	0.0012597	0.0067631	0.0031676
15	**0.3416214**	0.0000983	**0.0000445***	100	**0.0011478**	0.0067631	**0.0031676***

Source: Author's calculation

Table 2 Lasso regression

	Capital downsizing		Employment downsizing	
	LASSO	Post-est OLS	LASSO	Post-est OLS
ROA	0.0385877	0.0495758	0.3581657	0.3616803
Eqt			−0.0974784	−0.0980694
Size			0.1180107	0.1185699
AE			0.2093819	0.2094055
PM	−0.0071742	−0.0169355	−0.0860155	−0.0857769
Year			0.2998590	0.2997078
_cont	0.0148873	0.0205473	−0.6072890	−0.6086249

Source: Author's calculation

positive impacts on employment downsizing, while firm equity affected employment downsizing negatively.

4.2 In 2015

In 2015, the λ values that demonstrated the best performance in the case of capital downsizing and employment downsizing were 0.0093028 and 0.2089690, respectively (Table 3).

Lasso estimations indicated that all the independent variables exerted impacts on firm downsizing. While the profit margin was observed to exert a negative effect on firm downsizing, the other variables were observed to affect firm downsizing positively (Table 4).

The financial determinants of firm downsizing for the year 2015 were as follows: ROA, equity, asset efficiency, and profit margin. These findings were similar to those obtained for employment downsizing in the year 2013. More significant determinants of capital downsizing were identified in the year 2015 in comparison to the year 2013.

Table 3 Parameter values for lambda correction

Capital downsizing				Employment downsizing			
	Lambda	MSPE	Standard deviation		Lambda	MSPE	Standard deviation
40	0.0112052	0.0000191	1.5E−06	41	0.2517039	0.0023635	0.0002059
42	**0.0093028**	0.0000191	**1.5E−06***	43	**0.2089690**	0.0023629	**0.000204***
44	0.0077233	0.0000191	1.5E−06	45	0.1734897	0.0023631	0.0002035
45	0.0070372	0.0000191	1.5E−06	47	0.1440342	0.0023634	0.0002027

Source: Author's calculation

Table 4 Lasso Regression

	Capital downsizing		Employment downsizing	
	LASSO	Post-est OLS	LASSO	Post-est OLS
ROA	0.0405621	0.0415886	0.0088346	0.0071827
Eqt	0.0020974	0.0021470	0.0178024	0.0194965
Size	0.0108078	0.0113413	0.1330224	0.1479233
AE	0.0108291	0.0111003	0.1410661	0.1542097
PM	−0.0078667	−0.0085775	−0.0800983	−0.0965458
Year	0.0405621	0.0415886	0.0088346	0.0071827
_cont	−0.0301596	−0.0312536	−0.3755614	−0.3812677

Source: Author's calculation

It is interesting that in both 2013 and 2015, all the financial determinants, with the only exception of profit margin, were observed to exert positive impacts on firm downsizing. This finding was similar to the results reported by Carmeli and Scheaffer (2009). However, better financial indicators could arise after the downsizing, and the results concerning the financial determinants were not as they were expected to be.

5   Conclusions

The present study evaluated the impacts of financial determinants on the downsizing of SMEs in Vietnam. Using the data of the years 2013 and 2015, the present study employed linear regression with the Lasso estimation method rather than the OLS ones. The following two dimensions of downsizing were studied: capital downsizing and employment downsizing. The results of the study demonstrated that the financial determinants of ROA, firm equity, and asset efficiency exerted positive impacts on firm's capital downsizing and employment downsizing, while the financial determinant of profit margin affected firm downsizing negatively. Moreover, it was observed that firm characteristics such as firm size and firm age played a significant role in determining firm downsizing. In addition to certain important theoretical and practical contributions, the present study provided a few pointers for future investigations in this area of research. Firstly, future research should be aimed to investigate the impact of these factors over a longer period of time. Secondly, future studies could employ a richer dataset for such an analysis.

References

Applebaum, S. H., & Donia, M. (2000). The realistic downsizing preview: A management intervention in the prevention of survivor syndrome (part 1). *Career Development International, 5*(7), 333–350.

Bouno, A. F. (2003). *The Hidden costs and benefits of organizational resizing activities*. InK. P. De Meuse & M. L. Marks (Eds.), *Resizing the organization: Managing layoffs, divestitures, and closings* (pp. 306–346). San Francisco, CA: Jossey-Bass.

Brockner, J., Davy, J., & Carter, C. (1985). Layoffs, self-esteem, and survivor guilt: Motivational, affective, and attitudinal consequences. *Organizational Behavior and Human Decision Processes, 36,* 229–244.

Carmeli, A., & Scheaffer, Z. (2009). How leadership characteristics affect organizational decline and downsizing. *Journal of Business Ethics, 86,* 363–378.

Cascio, W. F., & Young, C. E. (2003). Financial consequences of employment-change decisions in major US corporations. In K. P. De Meuse & M. L. Marks (Eds.), *Resizing the organization: Managing layoffs, divestitures and closings* (pp. 1982–2000). San Francisco, CA: Jossey Bass.

De Meuse, K. P., & Tornow, W. W. (1990). The tie that binds-has become very, very frayed. *Human Resource Planning, 13,* 203–213.

De Meuse, K. P., Bergmann, T. J., Vanderheiden, P. A., & Roraff, C. E. (2004). New evidence regarding organizational downsizing and a firm's financial performance: A long-term analysis. *Journal of Managerial Issues, 16*(2), 155–177.

Farrell, M. A., & Mavondo, F. (2005). The effect of downsizing-redesign strategies on business performance: Evidence from Australia. *Asia Pacific Journal of Human Resources, 43*(1), 98–116.

Feldheim, M. A. (2007). Public sector downsizing and employee trust. *International Journal of Public Administration, 30,* 249–271.

Filatotchev, I., Buck, T., & Zuckov, V. (2000). Downsizing in privatized firms in Russia, Ukraine, and Belarus. *Academy of Management Journal, 43,* 286–304.

Gombola, M. J., & Tsetsekos, G. P. (1992). The information content of plant closing announcements: Evidence from financial profiles and the stock price reaction. *Financial Management, 21*(2), 31–40.

Hammer, M. (1996). *Beyond reengineering*. London: HarperCollins.

Hastie, T., Tibshirani, R., & Wainright, M. (2015). *Statistical learning with sparsity: The lasso and generalizations*. Boca Raton: CRC Press.

Nienstedt, P. R. (1989). Effectively downsizing management structures. *Human Resource Planning, 12,* 155–165.

Jensen, M. (1986). Agency costs of free cash flow, corporate finance, and takeovers. *American Economic Review, 76,* 323–329.

Lester, S. W., Kickul, J., Bergmann, T. J., & De Meuse, K. P. (2003). *The effects of organizational resizing on the nature of the psychological contract and employee perceptions of contract fulfillment*. In K. P. De Meuse & M. L. Marks (Eds.), *Resizing the organization: Managing layoffs, divestitures, and closings* (pp. 78–107). San Francisco, CA: Jossey-Bass.

McKinley, W., Sanchez, C. M., & Schick, A. G. (1995). Organizational downsizing: Constraining, cloning, learning. *Academy of Management Executive, 9*(3), 121–145.

Mishra, A. I. L., & Spreitzer, G. M. (1998). Explaining how survivors respond to downsizing: The roles of trust, empowerment, justice, and work redesign. *Academy of Management Review, 23,* 567–588.

Robinson, S. L., Kraatz, M. S., & Rousseau, D. M. (1994). Changing the obligations and the psychological contract. *Academy of Management Journal, 37,* 437–452.

Wasserstein, R. L., Schirm, A. L., & Lazar, N. A. (2019). Moving to a world beyond "p < 0.05". *The American Statistician, 73*(1), 1–19.

Wiesenfeld, B. M., Brockner, J., & Thibault, V. (2000). Procedural fairness, managers' self-esteem, and managerial behaviors following a layoff. *Organizational Behavior and Human Decision Processes, 83,* 1–32.

Worrell, D. L., Davidson, W. N., & Sharma, B. M. (1991). Layoff announcements and stockholder wealth. *Academy of Management Journal, 34,* 662–678.

Yu, G., & Park, J. (2006). The effect of downsizing on the financial performance and employee productivity of Korean firms. *International Journal of Manpower, 27*(3), 230–250.

A New Hybrid Iterative Method for Solving a Mixed Equilibrium Problem and a Fixed Point Problem for Quasi-Bregman Strictly Pseudocontractive Mappings

**Kanikar Muangchoo, Poom Kumam, Yeol Je Cho,
and Sakulbuth Ekvittayaniphon**

Abstract In this paper, we introduce and study a new hybrid iterative method for finding a common solution of a mixed equilibrium problem and a fixed point problem for an infinitely countable family of closed quasi-Bregman strictly pseudocontractive mappings in reflexive Banach spaces. We prove that the sequences generated by the hybrid iterative algorithm converge strongly to a common solution of these problems.

1 Introduction

In this section, we discuss prior literature related to mixed equilibrium problem and fixed point problem. Since they require understanding of Banach space, Bregman distance, fixed point, and mapping, we encourage that readers who are not familiar with such topics read Sect. 2 (preliminaries) first before returning to the introduction.

K. Muangchoo · P. Kumam (✉)
KMUTTFixed Point Research Laboratory, Department of Mathematics,
Room SCL 802 Fixed Point Laboratory, Science Laboratory Building,
Faculty of Science, King Mongkut's University of Technology Thonburi (KMUTT),
126 Pracha-Uthit Road, Bang Mod, Thrung Khru, Bangkok 10140, Thailand
e-mail: poom.kumam@mail.kmutt.ac.th

K. Muangchoo
e-mail: ni_003@hotmail.com

Y. J. Cho
Department of Mathematics Education, Gyeongsang National University,
Jinju 52828, Korea
e-mail: yjchomath@gmail.com

School of Mathematical Sciences, University of Electronic Science
and Technology of China, Chengdu 611731, China

S. Ekvittayaniphon
Rajamangala University of Technology Phra Nakhon, 399 Samsen Rd.,
Vachira Phayaban, Dusit, Bangkok 10300, Thailand
e-mail: sakulbuth.e@rmutp.ac.th

© The Editor(s) (if applicable) and The Author(s), under exclusive license
to Springer Nature Switzerland AG 2021
N. Ngoc Thach et al. (eds.), *Data Science for Financial Econometrics*,
Studies in Computational Intelligence 898,
https://doi.org/10.1007/978-3-030-48853-6_30

417

Throughout this paper, let E be a real Banach space with the norm $\| \cdot \|$. If $\{x_n\}_{n \in \mathbb{N}}$ is a sequence in E, we denote the strong convergence and the weak convergence of $\{x_n\}_{n \in \mathbb{N}}$ to a point $x \in E$ by $x_n \to x$ and $x_n \rightharpoonup x$, respectively. Let C be a nonempty, closed and convex subset of E and $T : C \to C$ be a mapping. Then a point $x \in C$ is called a *fixed point* of T if $Tx = x$ and the set of all fixed points of T is denoted by $F(T)$. There are many iterative methods for approximation of fixed points of a nonexpansive mapping.

1.1 Iterative Algorithms for Finding Fixed Points

The concept of nonexpansivity plays an important role in the study of Mann-type iteration (1953) for finding fixed points of a mapping $T : C \to C$. The Mann-type iteration is given by the following formula:

$$x_{n+1} = \beta_n T x_n + (1 - \beta_n)x_n, \quad \forall x_1 \in C. \tag{1}$$

Here, $\{\beta_n\}_{n \in \mathbb{N}}$ is a sequence of real numbers satisfying some appropriate conditions. The construction of fixed points of nonexpansive mappings via the iteration has been extensively investigated recently in the current literature (see, for example, Reich (1979) and the references therein). In Reich (1979), Reich proved the following interesting result.

Theorem 1 *Let C be a closed and convex subset of a uniformly convex Banach space E with a Fréchet differentiable norm, let $T : C \to C$ be a nonexpansive mapping with a fixed point, and let β_n be a sequence of real numbers such that $\beta_n \in [0, 1]$ and $\sum_{n=1}^{\infty} \beta_n(1 - \beta_n) = \infty$. Then the sequence $\{x_n\}_{n \in \mathbb{N}}$ generated by (1) converges weakly to a fixed point of T.*

However, the sequence $\{x_n\}_{n \in \mathbb{N}}$ generated by Mann-type iteration (1) does not in general converge strongly. Some attempts to modify the Mann iteration method (1) so that strong convergence is guaranteed have recently been made. In 2008, Takahashi et al. (2008) studied a strong convergence theorem by the hybrid method for a family of nonexpansive mappings in Hilbert spaces. Let C be a nonempty closed convex subset of H. Then, for any $x \in H$, there exists a unique nearest point in C with respect to the norm, denoted by $P_C(x)$. Such a P_C is called the metric projection of H onto C, $x_0 \in H$, $C_1 = C$, $x_1 = P_{C_1}(x_0)$ and let

$$\begin{cases} y_n = \alpha_n x_n + (1 - \alpha_n)T_n x_n, \\ C_{n+1} = \{z \in C_n : \|y_n - z\| \le \|x_n - z\|\}, \\ x_{n+1} = P_{C_{n+1}}(x_0), \quad \forall n \ge 1, \end{cases} \tag{2}$$

where $P_{C_{n+1}}$ is the metric projection from C onto C_{n+1} and $\{\alpha_n\}$ is chosen so that $0 \le \alpha_n \le a < 1$ for some $a \in [0, 1)$ and $\{T_n\}$ is a sequence of nonexpansive mappings of C into itself such that $\bigcap_{n=1}^{\infty} F(T_n) = \emptyset$. They proved that if $\{T_n\}$ satisfies some appropriate conditions, then $\{x_n\}$ converges strongly to $P_{\bigcap_{n=1}^{\infty} F(T_n)}(x_0)$.

In 2010, Reich and Sabach (2010) proposed the following two iterative schemes for finding a common fixed point of finitely many Bregman strongly nonexpansive mappings (as defined in Definition 5) $T_i : C \to C$ $(i = 1, 2, \ldots, N)$ satisfying $\bigcap_{i=1}^{N} F(T_i) \neq \emptyset$ in a reflexive Banach space E:

$$
\begin{cases}
x_0 = x \in E, \quad \text{chosen arbitrarily,} \\
y_n^i = T_i(x_n + e_n^i), \\
C_n^i = \{z \in E : D_f(z, y_n^i) \leq D_f(z, x_n + e_n^i)\}, \\
C_n = \bigcap_{i=1}^{N} C_n^i, \\
Q_n = \{z \in E : \langle \nabla f(x_0) - \nabla f(x_n), z - x_n \rangle \leq 0\}, \\
x_{n+1} = proj_{C_n \cap Q_n}^f(x_0), \quad \forall n \geq 0,
\end{cases}
\tag{3}
$$

and

$$
\begin{cases}
x_0 \in E, C_0^i = E, i = 1, 2, \ldots, N, \\
y_n^i = T_i(v_n + e_n^i), \\
C_{n+1}^i = \{z \in C_n^i : D_f(z, y_n^i) \leq D_f(z, x_n + e_n^i)\}, \\
C_{n+1} = \bigcap_{i=1}^{N} C_{n+1}^i, \\
x_{n+1} = proj_{C_{n+1}}^f(x_0), \quad \forall n \geq 0,
\end{cases}
\tag{4}
$$

where $proj_C^f$ is the Bregman projection (as defined in Definition 4) with respect to f from E onto a closed and convex subset C of E. They proved that the sequence $\{x_n\}$ converges strongly to a common fixed point of $\{T_i\}_{i=1}^{N}$.

Furthermore, in 2010, Zhou and Gao (2009) introduced this definition of a quasi-strict pseudocontraction related to the function ϕ and proved the convergence of a hybrid projection algorithm to a fixed point of a closed and quasi-strict pseudocontraction in a smooth and uniformly convex Banach space. They studied the strong convergence of the following scheme:

$$
\begin{cases}
x_0 \in E, \\
C_1 = C, \\
x_1 = \Pi_{C_1}(x_0), \\
C_{n+1} = \{z \in C_n : \phi(x_n, Tx_n) \leq \frac{2}{1-k}\langle x_n - z, Jx_n - JTx_n \rangle\}, \\
x_{n+1} = \Pi_{C_{n+1}}(x_0), \forall n \geq 0,
\end{cases}
\tag{5}
$$

where $\Pi_{C_{n+1}}$ is the generalized projection from E onto C_{n+1}. They proved that the sequence $\{x_n\}$ converges strongly to $\Pi_{C_{F(T)}}(x_0)$.

1.2 Application for Solving Equilibrium Problems

In 2008, the equilibrium problem was generalized by Ceng and Yao (2008) to the mixed equilibrium problem: Let $\varphi : C \to \mathbb{R}$ be a real-valued function and $\Theta : C \times C \to \mathbb{R}$ be an equilibrium function. The mixed equilibrium problem (for short, MEP) is to find $x^* \in C$ such that

$$\Theta(x^*, y) + \varphi(y) \geq \varphi(x^*), \quad \forall y \in C. \tag{6}$$

The solution set of MEP (6) is denoted by Sol(MEP). In particular, if $\varphi \equiv 0$, this problem reduces to the equilibrium problem (for short, EP), which is to find $x^* \in C$ such that

$$\Theta(x^*, y) \geq 0, \quad \forall y \in C, \tag{7}$$

which is introduced and studied by Blum and Oettli (1994). The solution set of EP (7) is denoted by Sol(EP).

It is known that the equilibrium problems have a great impact and influence in the development of several topics of science and engineering. It turns out that many well-known problems could be fitted into the equilibrium problems. It has been shown that the theory of equilibrium problems provides a natural, novel and unified framework for several problems arising in nonlinear analysis, optimization, economics, finance, game theory, and engineering. The equilibrium problem includes many mathematical problems as particular cases, for example, mathematical programming problem, variational inclusion problem, variational inequality problem, complementary problem, saddle point problem, Nash equilibrium problem in noncooperative games, minimax inequality problem, minimization problem and fixed point problem, see Blum and Oettli (1994), Combettes and Hirstoaga (2005), Kumam et al. (2016), Muangchoo et al. (2019), Pardalos et al. (2010).

In 2014, Ugwunnadi et al. (2014) proved a strong convergence theorem of a common element in the set of fixed points of a finite family of closed quasi-Bregman strictly pseudocontractive mappings and common solutions to a system of equilibrium problems in a reflexive Banach space.

In 2015, Shehu and Ogbuisi (2015) introduced an iterative algorithm based on the hybrid method in mathematical programming for approximating a common fixed point of an infinite family of left Bregman strongly nonexpansive mappings which also solves a finite system of equilibrium problems in a reflexive real Banach space. Likewise, Xu and Su (2015) proved a new hybrid shrinking projection algorithm for common fixed point of a family of countable quasi-Bregman strictly pseudocontractive mappings with equilibrium, variational inequality and optimization problems. Wang (2015) proved a strong convergence theorem for Bregman quasi-strict pseudocontraction in a reflexive Banach space with applications.

In 2018, Biranvand and Darvish (2018) studied a new iterative method for a common fixed point of a finite family of Bregman strongly nonexpansive mappings in the frame work of reflexive real Banach spaces. They proved the strong convergence theorem for finding common fixed points with the solutions of a mixed equilibrium problem.

Motivated by the work given in Biranvand and Darvish (2018), Cholamjiak and Suantai (2010), Shehu and Ogbuisi (2015), Ugwunnadi et al. (2014), we prove that the sequences generated by the hybrid iterative algorithm converge strongly to a common solution of MEP (6) and a fixed point problem for an infinitely countable family of closed quasi-Bregman strictly pseudocontractive mappings in a reflexive Banach space.

2 Preliminaries

In this section, we introduce necessary definitions and results to be used in the main result. It is divided into five subsections: Banach spaces and fixed points, functions on Banach spaces, Bregman distances, known results on Bregman distances, and mixed equilibrium problems.

2.1 Banach Spaces and Fixed Points

We give the definition of Banach space and its properties, strong convergence and weak convergence of sequences in Banach space, and fixed points.

A *Banach space* is a vector space E over a scalar field K equipped with a norm $\|\cdot\|$ which is complete: every Cauchy sequence $\{x_n\}$ in E converges to $x \in E$. The dual space E^* of E is the set of continuous linear maps $x^* : E \to \mathbb{R}$ equipped with the new norm $\|\cdot\|^*$ defined by $\|x^*\|^* = \sup\{|x^*(x)| : x \in E, |x| \leq 1\}$.

Let E be a real Banach space (i.e. with $\mathbb{R}$ as the scalar field) with E^* as its dual space. We denote the value of $x^* \in E^*$ at $x \in E$ by $\langle x, x^* \rangle = x^*(x)$.

Let $\{x_n\}_{n \in \mathbb{N}}$ be a sequence in E. We say that the sequence *converges strongly* to $x \in E$ (denoted by $x_n \to x$) if $\lim_{n \to \infty} \|x_n - x\| = 0$. We also say that the sequence *converges weakly* to $x \in E$ (denoted by $x_n \rightharpoonup x$) if for any $x^* \in X^*$, we have $\lim_{n \to \infty} \langle x_n, x^* \rangle = \langle x, x^* \rangle$. It is well known that strong convergence implies weak convergence, but not the other way around.

Let $S = \{x \in E : \|x\| = 1\}$ be the unit sphere in E, and E^{**} be the dual space of E^*. A Banach space E is said to be

- *strictly convex* if $\left\|\frac{x+y}{2}\right\| < 1$ whenever $x, y \in S$ and $x \neq y$.
- *uniformly convex* if, for all $\epsilon \in (0, 2]$, there exists $\delta > 0$ such that $x, y \in S$ and $\|x - y\| \geq \epsilon$ implies $\left\|\frac{x+y}{2}\right\| \leq 1 - \delta$.
- *smooth* if the limit

$$\lim_{t \to 0} \frac{\|x + ty\| - \|x\|}{t} \tag{8}$$

exists for all $x, y \in S$.

- *uniformly smooth* if the limit (8) is attained uniformly in $x, y \in S$.
- *refexive* if the *evaluation map* $J : E \to E^{**}$ defined by the formula $J(x)(x^*) = x^*(x)$ for all $x \in E, x^* \in E^*$ is surjective.

We add that the following are well known:

(i) Every uniformly convex Banach space is strictly convex and reflexive.
(ii) A Banach space E is uniformly convex if and only if E^* is uniformly smooth.
(iii) If E is reflexive, then E is strictly convex if and only if E^* is smooth (see, for instance, Takahashi (2000) for more details).
(iv) If a Banach space E has a property that the modulus of convexity δ of E defined by

$$\delta(\varepsilon) = \inf \left\{ 1 - \frac{\|x + y\|}{2} : \|x\| \leq 1, \|y\| \leq 1, \|x - y\| \geq \varepsilon \right\}, 0 \leq \varepsilon \leq 2$$

is greater than 0 for every $\varepsilon > 0$, then E is uniformly convex.

Let C be a nonempty, closed and convex subset of E and $T : C \to C$ be a mapping.

- The mapping T is said to be *closed* if $x_n \to x$ and $T x_n \to y$ implies $T x = y$.
- A point $x \in C$ is called a *fixed point* of T if $T x = x$. The set of all fixed points of T is denoted by $F(T)$.
- Suppose that E is smooth. We said that a point $p \in C$ is an *asymptotic fixed point* of T if there exists a sequence $\{x_n\}$ in C such that $x_n \rightharpoonup p$ and $lim_{n \to \infty} \|x_n - T x_n\| = 0$. The set of all asymptotically fixed points of T is denoted by $\widehat{F}(T)$.

2.2 Functions on Banach Spaces

We give the notion of functions on Banach spaces, especially differentiability and convexity, and Legendre functions.

Let E be a real Banach space, and let $f : E \to (-\infty, +\infty]$ be a function. Define $B_r(x_0) = \{x \in E : \|x - x_0\| < r\}$ to be an open ball of radius r centered at $x_0 \in E$. Let dom $f = \{x \in E : f(x) < \infty\}$ be the domain of f, $int(dom \ f) = \{x \in dom \ f : \exists \varepsilon > 0, B(x, \epsilon) \subseteq E\}$ be the interior of the domain of f and f is said to be

- *proper* if the dom $f \neq \emptyset$.
- *lower semicontinuous* if the set $\{x \in E : f(x) \leq r\}$ is closed for all $r \in \mathbb{R}$.
- *convex* if

$$f(\alpha x + (1 - \alpha)y) \leq \alpha f(x) + (1 - \alpha) f(y), \quad \forall x, y \in E, \alpha \in (0, 1). \quad (9)$$

- *strictly convex* if the strict inequality holds in the inequality (9) for all $x, y \in dom \ f$ with $x \neq y$ and $\alpha \in (0, 1)$.

For any $x \in \text{int}(\text{dom} f)$, the *right-hand derivative* of f at x in the direction $y \in E$ is defined by

$$f'(x, y) := \lim_{t \to 0^+} \frac{f(x + ty) - f(x)}{t}. \tag{10}$$

The function f is said to be

- *Gâteaux differentiable at* x if the limit in (10) exists for all $y \in E$. In this case, $f'(x, y)$ coincides with the value of the *gradient* (∇f) of f at x.
- *Gâteaux differentiable* if f is Gâteaux differentiable for any $x \in \text{int}(\text{dom} f)$.
- *Fréchet differentiable at* x if the limit in (10) is attained uniformly for $\|y\| = 1$.
- *uniformly Fréchet differentiable* on a subset C of E if the limit in (10) is attained uniformly for $x \in C$ and $\|y\| = 1$.

We define the weak* topology on the dual space E^* to be the coarsest (or weakest) topology such that for any $x \in E$, the map $J(x) : E^* \to \mathbb{R}$ defined by $J(x)(x^*) = x^*(x)$ remains continuous. It is well known that if a continuous convex function $f : E \to \mathbb{R}$ Gâteaux differentiable, then ∇f is norm-to-weak* continuous (see, for example, the definition of norm-to-weak* in Butnariu and Iusem (2000) (Proposition 1.1.10)). Also, it is known that if f is Fréchet differentiable, then ∇f is norm-to-norm continuous (see Kohsaka and Takahashi 2005).

Definition 1 (Butnariu and Iusem 2000) Let $f : E \to (-\infty, +\infty]$ be a convex and Gâteaux differentiable function. The function f is said to be

(i) *totally convex* at $x \in \text{int}(\text{dom} f)$ if its modulus of total convexity at x, that is, the function $\nu_f : \text{int}(\text{dom} f) \times [0, +\infty) \to [0, +\infty)$ defined by

$$\nu_f(x, t) = \inf\{D_f(y, x) : y \in \text{dom} f, \|y - x\| = t\},$$

is positive whenever $t > 0$;
(ii) *totally convex* if it is totally convex at every point $x \in \text{int}(\text{dom} f)$;
(iii) *totally convex on bounded sets* if $\nu_f(B, t)$ is positive for any nonempty bounded subset B of E and $t > 0$, where the modulus of total convexity of the function f on the set B is the function $\nu_f : \text{int}(\text{dom} f) \times [0, +\infty) \to [0, +\infty)$ defined by

$$\nu_f(B, t) = \inf\{\nu_f(x, t) : x \in B \cap \text{dom} f\}.$$

From this point on, we shall denote by $\Gamma(E)$ the class of proper lower semicontinuous convex functions on E, and $\Gamma^*(E^*)$ the class of proper weak* lower semicontinuous convex function on E^*.

For each $f \in \Gamma(E)$, the *subdifferential* ∂f of f is defined by

$$\partial f(x) = \{x^* \in E^* : f(x) + \langle y - x, x^* \rangle \leq f(y), \forall y \in E\}, \quad \forall x \in E.$$

If $f \in \Gamma(E)$ and $g : E \to \mathbb{R}$ is a continuous convex function, then $\partial(f + g) = \partial f + \partial g$. Rockafellar's theorem (1970) ensures that $\partial f \subset E \times E^*$ is maximal monotone, meaning that we have

$$\langle \partial f(x) - \partial f(y), x - y \rangle \geq 0, \quad \forall x, y \in E, \tag{11}$$

and also that there is no monotone operator that properly contains it. For each $f \in \Gamma(E)$, the *(Fenchel) conjugate function* f^* of f is defined by

$$f^*(x^*) = \sup_{x \in E} \{ \langle x, x^* \rangle - f(x) \}, \quad \forall x^* \in E^*.$$

It is well known that

$$f(x) + f^*(x^*) \geq \langle x, x^* \rangle, \quad \forall (x, x^*) \in E \times E^*,$$

and $(x, x^*) \in \partial f$ is equivalent to

$$f(x) + f^*(x^*) = \langle x, x^* \rangle. \tag{12}$$

We also know that, if $f \in \Gamma(E)$, then $f^* : E^* \to (-\infty, +\infty]$ is a proper weak* lower semicontinuous convex function (see Phelps (1993) for more details on convex analysis).

A function $f : E \to \mathbb{R}$ is said to be *strongly coercive* if, for any sequence $\{x_n\}_{n \in \mathbb{N}}$ such that $\|x_n\|$ converges to ∞, we have

$$\lim_{n \to \infty} \frac{f(x_n)}{\|x_n\|} = \infty.$$

It is also said to be *bounded on bounded sets* if, for any closed sphere $S_r(x_0) = \{x \in E : \|x - x_0\| = r\}$ with $x_0 \in E$ and $r > 0$, $f(S_r(x_0))$ is bounded.

Definition 2 (Bauschke et al. 2001) The function $f : E \to (-\infty, +\infty]$ is said to be:

(a) *essentially smooth*, if ∂f is both locally bounded and single-valued on its domain;

(b) *essentially strictly convex*, if $(\partial f)^{-1}$ is locally bounded on its domain and f is strictly convex on every convex subset of dom ∂f;

(c) *Legendre*, if it is both essentially smooth and essentially strictly convex.

We can also define a Legendre function in a different way. Since E is reflexive, according to Bauschke et al. (2001), the function f is Legendre if it satisfies the following conditions:

(i) The interior of the domain of f, int(dom$f$), is nonempty;
(ii) f is Gâteaux differentiable on int(dom$f$);
(iii) dom $\nabla f = $ int(dom$f$).

We note that for a Legendre function f, the following conclusions hold Bauschke et al. (2001):

(a) f is Legendre function if and only if f^* is the Legendre function;
(b) $(\partial f)^{-1} = \partial f^*$;
(c) $\nabla f = (\nabla f^*)^{-1}$;
(d) $\mathrm{ran}\nabla f = \mathrm{dom}\nabla f^* = \mathrm{int}(\mathrm{dom} f^*)$;
(e) $\mathrm{ran}\nabla f^* = \mathrm{dom}\nabla f = \mathrm{int}(\mathrm{dom} f)$;
(f) The functions f and f^* are strictly convex on the interior of respective domains.

2.3 Bregman Distances

We give the definition of Bregman distance, Bregman projection, and some properties.

Definition 3 (Bregman 1967) Let $f : E \rightarrow (-\infty, +\infty]$ be a convex and Gâteaux differentiable function. The function $D_f : \mathrm{dom} f \times \mathrm{int}(\mathrm{dom} f) \rightarrow [0, +\infty)$ defined by

$$D_f(y, x) = f(y) - f(x) - \langle \nabla f(x), y - x \rangle \tag{13}$$

is called *Bregman distance* with respect to f.

It should be noted that D_f (see (13)) is not a distance in the usual sense of the term. In general, we have

- D_f is not necessarily symmetric (i.e. it's not guaranteed that $D_f(x, y) = D_f(y, x)$ for all $x, y \in E$).
- D_f does not satisfy the triangle inequality $D_f(x, z) \leq D_f(x, y) + D_f(y, z)$ for all $x, y, z \in E$.
- We have $D_f(x, x) = 0$, but $D_f(y, x) = 0$ may not imply $x = y$.

Instead, we have the following properties (cf. [Bauschke et al. (2001), Theorem 7.3 (vi), page 642]).

Proposition 1 (Property of Bregman distance) *Let $f : E \rightarrow (-\infty, +\infty]$ be a Legendre function. Then $D_f(y, x) = 0$ if and only if $y = x$.*
If f is a Gâteaux differentiable function, then Bregman distances have the following two important properties.

- *The three point identity: for any $x \in \mathrm{dom} f$ and $y, z \in \mathrm{int}(\mathrm{dom} f)$, we have (see Chen and Teboulle 1993).*

$$D_f(x, y) + D_f(y, z) - D_f(x, z) = \langle \nabla f(z) - \nabla f(y), x - y \rangle. \qquad (14)$$

- *The four point identity: for any* $y, w \in domf$ *and* $x, z \in domf$, *we have*

$$D_f(y, x) - D_f(y, z) - D_f(w, x) + D_f(w, z) = \langle \nabla f(z) - \nabla f(x), y - w \rangle. \qquad (15)$$

Let $f : E \to \mathbb{R}$ *a strictly convex and* Gâteaux *differentiable function. By* (13) *and* (14), *for any* $x, y, z \in E$, *the Bregman distance satisfies Chen et al.* (2011)

$$D_f(x, z) = D_f(x, y) + D_f(y, z) + \langle x - y, \nabla f(y) - \nabla f(z) \rangle. \qquad (16)$$

In particular,

$$D_f(x, y) = -D_f(y, x) + \langle y - x, \nabla f(y) - \nabla f(x) \rangle. \qquad (17)$$

Definition 4 The Bregman projection (1967) of $x \in domf$ onto the nonempty, closed, and convex set $C \subset domf$ is the necessarily unique vector $proj_C^f(x) \in C$ satisfying

$$D_f(proj_C^f(x), x) = \inf\{D_f(y, x) : y \in C\}.$$

Concerning the Bregman projection, the following are well known.

Lemma 1 (Butnariu and Resmerita 2006) *Let* E *be a Banach space and* C *be a nonempty closed convex subset of a reflexive Banach space* E. *Let* $f : E \to \mathbb{R}$ *be a totally convex and* Gâteaux *differentiable function and let* $x \in E$. *Then:*

(a) $z = proj_C^f(x)$ *if and only if* $\langle \nabla f(x) - \nabla f(z), y - z \rangle \leq 0$ *for all* $y \in C$;
(b) $D_f(y, proj_C^f(x)) + D_f(proj_C^f(x), x) \leq D_f(y, x)$ *for all* $y \in C$ *and* $x \in E$.

Let E be a Banach space with dual E^*. We denote by J the normalized duality mapping from E to E^* defined by

$$Jx = \{f \in E^* : \langle x, f \rangle = \|x\|^2 = \|f\|^2\},$$

where $\langle \cdot, \cdot \rangle$ denotes the generalized duality pairing. It is well known that if E is smooth, then J is single-valued.

If E is a smooth Banach space and set $f(x) = \|x\|^2$ for all $x \in E$, then it follows that $\nabla f(x) = 2Jx$ for all for all $x \in E$, where J is the normalized duality mapping from E into E^*. Hence, $D_f(x, y) = \phi(x, y)$ (Pang et al. 2014), where $\phi : E \times E \to \mathbb{R}$ is the Lyapunov functional defined by:

$$\phi(x, y) := \|x\|^2 - 2\langle x, Jy \rangle + \|y\|^2, \quad \forall (x, y) \in E \times E. \qquad (18)$$

Let C be a nonempty, closed and convex subset of E. The generalized projection Π_C from E onto C is defined and denoted by

$$\Pi_C(x) = \mathrm{argmin}_{y \in C} \phi(y, x).$$

This definition of $\Pi_C(x)$ was used in (5) in the introduction.

Definition 5 (Chen et al. 2011; Reich and Sabach 2010; Xu, and Su 2015) Let $T : C \to \mathrm{int}(\mathrm{dom} f)$ be a mapping. Recall that $F(T)$ denotes the set of fixed points of T, i.e., $F(T) = \{x \in C : Tx = x\}$, and $\widehat{F}(T)$ denotes the set of asymptotically fixed points of T. We provide the necessary notations of the nonlinear mapping related to Bregman distance as shown in the following. The mapping T is said to be

- *Bregman quasi-nonexpansive* if $F(T) \neq \emptyset$ and

$$D_f(p, Tx) \leq D_f(p, x), \ \forall x \in C, \ p \in F(T);$$

- *Bregman relatively nonexpansive* (Reich and Sabach 2010) if $\widehat{F}(T) = F(T)$ and

$$D_f(p, Tx) \leq D_f(p, x), \ \forall x \in C, \ p \in F(T);$$

- *Bregman strongly nonexpansive* (Bruck and Reich 1977; Reich and Sabach 2010) with respect to a nonempty $\widehat{F}(T)$ if

$$D_f(p, Tx) \leq D_f(p, x), \ \forall x \in C, \ p \in \widehat{F}(T)$$

and, if whenever $\{x_n\} \subset C$ is bounded, $p \in \widehat{F}(T)$, and

$$\lim_{n \to \infty} \left(D_f(p, x_n) - D_f(p, Tx_n) \right) = 0,$$

it follows that

$$\lim_{n \to \infty} D_f(Tx_n, x_n) = 0;$$

- *Bregman firmly nonexpansive* (Reich and Sabach 2011) if

$$\langle \nabla f(Tx) - \nabla f(Ty), Tx - Ty \rangle \leq \langle \nabla f(x) - \nabla f(y), Tx - Ty \rangle, \ \forall x, y \in C$$

or, equivalently

$$\begin{aligned} D_f(Tx, Ty) + D_f(Ty, Tx) + D_f(Tx, x) + D_f(Ty, y) \\ \leq D_f(Tx, y) + D_f(Ty, x), \ \forall x, y \in C; \end{aligned} \tag{19}$$

- *quasi-Bregman k-pseudocontractive* if there exists a constant $k \in [0, +\infty)$ and $F(T) \neq \emptyset$ such that

$$D_f(p, Tx) \leq D_f(p, x) + kD_f(x, Tx), \ \forall x \in C, \ p \in F(T);$$

- *quasi-Bregman pseudocontractive* if T is quasi-Bregman 1-pseudocontractive;
- *quasi-Bregman strictly pseudocontractive* if T is quasi-Bregman k-pseudocontractive with $k < 1$.

2.4 Known Results on Bregman Distances

We give known results on Bregman distances to be used in the proof of the main result.

Lemma 2 (Ugwunnadi et al. 2014) *Let* $f : E \to \mathbb{R}$ *be a Legendre function which is uniformly Fréchet differentiable and bounded on subsets of* E, *let* C *be a nonempty, closed, and convex subset of* E, *and let* $T : C \to C$ *be a quasi-Bregman strictly pseudocontractive mapping with respect to* f. *Then, for any* $x \in C$, $p \in F(T)$ *and* $k \in [0, 1)$ *the following holds:*

$$D_f(x, Tx) \leq \frac{1}{1-k} \langle \nabla f(x) - \nabla f(Tx), x - p \rangle. \tag{20}$$

Proof Let $x \in C$, $p \in F(T)$ and $k \in [0, 1)$, by definition of T, we have

$$D_f(p, Tx) \leq D_f(p, x) + kD_f(x, Tx)$$

and, from (13), we obtain

$$D_f(p, x) + D_f(x, Tx) + \langle \nabla f(x) - \nabla f(Tx), p - x \rangle \leq D_f(p, x) + kD_f(x, Tx),$$

which implies

$$D_f(x, Tx) \leq \frac{1}{1-k} \langle \nabla f(x) - \nabla f(Tx), x - p \rangle.$$

This completes the proof.

Lemma 3 (Ugwunnadi et al. 2014) *Let* $f : E \to \mathbb{R}$ *be a Legendre function which is uniformly Fréchet differentiable and bounded on subsets of* E, *let* C *be a nonempty, closed, and convex subset of* E, *and let* $T : C \to C$ *be a quasi-Bregman strictly pseudocontractive mapping with respect to* f. *Then* $F(T)$ *is closed and convex.*

Proof Let $F(T)$ be nonempty set. First we show that $F(T)$ is closed. Let $\{x_n\}$ be a sequence in $F(T)$ such that $x_n \to z$ as $n \to \infty$, we need to show that $z \in F(T)$. From Lemma 2, we obtain

$$D_f(z, Tz) \leq \frac{1}{1-k} \langle \nabla f(z) - \nabla f(Tz), z - x_n \rangle. \tag{21}$$

From (21), we have $D_f(z, Tz) \leq 0$, and from Bauschke et al. (2001), Lemma 7.3, it follows that $Tz = z$. Therefore $F(T)$ is closed.

Next, we show that $F(T)$ is convex. Let $z_1, z_2 \in F(T)$, for any $t \in (0, 1)$; putting $z = tz_1 + (1 - t)z_2$, we need to show that $z \in F(T)$. From Lemma 2, we obtain, respectively,

$$D_f(z, Tz) \leq \frac{1}{1-k}\langle \nabla f(z) - \nabla f(Tz), z - z_1\rangle \tag{22}$$

and

$$D_f(z, Tz) \leq \frac{1}{1-k}\langle \nabla f(z) - \nabla f(Tz), z - z_2\rangle. \tag{23}$$

Multiplying (22) by t and (23) by $(1 - t)$ and adding the results, we obtain

$$D_f(z, Tz) \leq \frac{1}{1-k}\langle \nabla f(z) - \nabla f(Tz), z - z\rangle, \tag{24}$$

which implies $D_f(z, Tz) \leq 0$, and from Bauschke et al. (2001), Lemma 7.3, it follows that $Tz = z$. Therefore $F(T)$ is also convex. This completes the proof.

Lemma 4 (Ugwunnadi et al. 2014) *Let $f : E \to \mathbb{R}$ be a convex, Legendre and Gâteaux differentiable function. In addition, if $f : E \to (-\infty, +\infty]$ is a proper lower semicontinuous function, then $f^* : E^* \to (-\infty, +\infty]$ is a proper weak* lower semicontinuous and convex function. Thus, for all $z \in E$,*

$$D_f\left(z, \nabla f^*\left(\sum_{i=1}^{N} t_i \nabla f(x_i)\right)\right) \leq \sum_{i=1}^{N} t_i D_f(z, x_i). \tag{25}$$

where $\{x_i\}_{i=1}^{N} \subset E$ and $\{t_i\}_{i=1}^{N} \subset (0, 1)$ with $\sum_{i=1}^{N} t_i = 1$.

Example 1 (Kassay et al. 2011) Let E is a real Banach space, $A : E \to 2^{E^*}$ be a maximal monotone mapping. If $A^{-1}(0) \neq \emptyset$ and the Legendre function $f : E \to (-\infty, +\infty]$ is uniformly Fréchet differentiable and bounded on bounded subsets of E, then the resolvent with respect to A,

$$Res_A^f(x) = (\nabla f + A)^{-1} \circ \nabla f(x)$$

is a single-valued, closed and Bregman relatively nonexpansive mapping from E onto $D(A)$ and $F(Res_A^f) = (A)^{-1}(0)$.

The following result was first proved in Kohsaka and Takahashi (2005) (see Lemma 3.1, pp. 511).

Lemma 5 *Let E be a Banach space and let $f : E \to \mathbb{R}$ be a Gâteaux differentiable function, which is uniformly convex on bounded sets. Let $\{x_n\}_{n\in\mathbb{N}}$ and $\{y_n\}_{n\in\mathbb{N}}$ be bounded sequences in E and $\lim_{n\to\infty} D_f(x_n, y_n) = 0$, then we have $\lim_{n\to\infty} \|x_n - y_n\| = 0$.*

The following lemma is slightly different from that in Kohsaka and Takahashi (2005) (see Lemma 3.2 and Lemma 3.3, pp. 511, 512):

Lemma 6 *Let E be a reflexive Banach space, let $f : E \to \mathbb{R}$ be a strongly coercive Bregman function and V be the function defined by*

$$V(x, x^*) = f(x) - \langle x, x^* \rangle + f^*(x^*), \quad \forall x \in E, \; x^* \in E^*.$$

The following assertions hold:

(i) $D_f(x, \nabla f^*(x^*)) = V(x, x^*)$ for all $x \in E$ and $x^* \in E^*$.
(ii) $V(x, x^*) + \langle \nabla f^*(x^*) - x, y^* \rangle \le V(x, x^* + y^*)$ for all $x \in E$ and $x^*, y^* \in E^*$.

It also follows from the definition that V is convex in the second variable x^* and

$$V(x, \nabla f(y)) = D_f(x, y).$$

The following result was first proved in Butnariu and Resmerita (2006) (see also Kohsaka and Takahashi 2005).

Lemma 7 (Reich and Sabach 2009) *If $f : E \to \mathbb{R}$ is uniformly Fréchet differentiable and bounded on bounded subsets of E, then ∇f^* is uniformly continuous on bounded subsets of E from the strong topology of E to the strong topology of E^*.*

Lemma 8 (Butnariu and Iusem 2000) *The function f is totally convex on bounded sets if and only if it is sequentially consistent: that is for any sequences $\{x_n\}_{n \in \mathbb{N}}$ in $dom(f)$ and $\{y_n\}_{n \in \mathbb{N}}$ in $Int(dom(f))$ such that $\{y_n\}$ is bounded and $\lim_{n \to \infty} f(x_n, y_n) = 0$, then we have $\lim_{n \to \infty} \|x_n - y_n\| = 0$.*

Lemma 9 (Reich and Sabach 2010) *Let $f : E \to \mathbb{R}$ be a totally convex and Gâteaux differentiable function. If $x_0 \in E$ and the sequence $\{D_f(x_n, x_0)\}$ is bounded, then the sequence $\{x_n\}$ is also bounded.*

Lemma 10 (Reich and Sabach 2010) *Let $f : E \to \mathbb{R}$ be a totally convex and Gâteaux differentiable function, $x_0 \in E$ and C be a nonempty closed convex subset of a reflexive Banach space E. Suppose that the sequence $\{x_n\}$ is bounded and any weak subsequential limit of $\{x_n\}$ belongs to C. If $D_f(x_n, x_0) \le D_f(proj_C^f x_0, x_0)$ for any $n \in N$, then$\{x_n\}$ strongly converges to $proj_C^f x_0$.*

Proposition 2 (Zalinescu 2002) *Let E be a reflexive Banach space and $f : E \to \mathbb{R}$ be a continuous convex function which is strongly coercive. Then the following assertions are equivalent:*

(1) f is bounded on bounded subsets and locally uniformly smooth on E.

(2) f^* is Fréchet differentiable and ∇f^* is uniformly norm-to-norm continuous on bounded subsets of $dom\, f^* = E^*$ and f^* is strongly coercive and uniformly convex on bounded subsets of E.

2.5 Mixed Equilibrium Problems

We recall the definition of mixed equilibrium problem as defined in (6) and give several known lemmas to be used in the proof of the main result.

Let $\varphi : C \rightarrow \mathbb{R}$ be a real-valued function and $\Theta : C \times C \rightarrow \mathbb{R}$ be an equilibrium function. The mixed equilibrium problem is to find $x^* \in C$ such that

$$\Theta(x^*, y) + \varphi(y) \geq \varphi(x^*), \quad \forall y \in C. \tag{26}$$

For solving the mixed equilibrium problem, let us make the following assumptions for a function Θ on the set C:

(A1) $\Theta(x, x) = 0$ for all $x \in C$;

(A2) Θ is monotone, i.e., $\Theta(x, y) + \Theta(y, x) \leq 0$ for all $x, y \in C$;

(A3) for each $y \in C$, the function $x \longmapsto \Theta(x, y)$ is weakly upper semicontinuous;

(A4) for each $x \in C$, the function $y \longmapsto \Theta(x, y)$ is convex and lower semicontinuous.

Definition 6 (Biranvand and Darvish 2018) Let C be a nonempty, closed and convex subsets of a real reflexive Banach space and let $\varphi : C \rightarrow \mathbb{R}$ be a lower semicontinuous and convex functional. Let $\Theta : C \times C \rightarrow \mathbb{R}$ be a functional satisfying $(A1) - (A4)$. The mixed resolvent of Θ is the operator $Res_{\Theta,\varphi}^f : E \rightarrow 2^C$

$$Res_{\Theta,\varphi}^f(x) = \{z \in C : \Theta(z, y) + \varphi(y) + \langle \nabla f(z) - \nabla f(x), y - z \rangle \geq \varphi(z), \ \forall y \in C\}. \tag{27}$$

The following results can be deduced from Lemma 1 and Lemma 2 due to Reich and Sabach (2010), but for reader's convenience we provide their proofs.

Lemma 11 (Biranvand and Darvish 2018) *Let E be a reflexive Banach space and $f : E \rightarrow \mathbb{R}$ be a coercive and Gâteaux differentiable function. Let C be a nonempty, closed and convex subset of E. Assume that $\varphi : C \rightarrow \mathbb{R}$ be a lower semicontinuous and convex functional and the functional $\Theta : C \times C \rightarrow \mathbb{R}$ satisfies conditions $(A1) - (A4)$, then $dom\left(Res_{\Theta,\varphi}^f\right) = E$.*

Proof Since f is a coercive function, the function $h : E \times E \rightarrow \mathbb{R}$ defined by

$$h(x, y) = f(y) - f(x) - \langle x^*, y - x \rangle$$

satisfies the following for all $x^* \in E^*$ and $y \in C$;

$$\lim_{\|x-y\| \rightarrow +\infty} \frac{h(x, y)}{\|x - y\|} = +\infty.$$

Then from Theorem 1 due to Blum and Oettli (1994), there exists $\hat{x} \in C$ such that

$$\Theta(\hat{x}, y) + \varphi(y) - \varphi(\hat{x}) + f(y) - f(\hat{x}) - \langle x^*, y - \hat{x} \rangle \geq 0$$

for any $y \in C$. This implies that

$$\Theta(\hat{x}, y) + \varphi(y) + f(y) - f(\hat{x}) - \langle x^*, y - \hat{x} \rangle \geq \varphi(\hat{x}). \tag{28}$$

We know that inequality (28) holds for $y = t\hat{x} + (1 - t)\hat{y}$ where $\hat{y} \in C$ and $t \in (0, 1)$. Therefore, we have

$$\begin{aligned}
\Theta(\hat{x}, t\hat{x} + (1 - t)\hat{y}) &+ \varphi(t\hat{x} + (1 - t)\hat{y}) + f(t\hat{x} + (1 - t)\hat{y}) - f(\hat{x}) \\
&- \langle x^*, t\hat{x} + (1 - t)\hat{y} - \hat{x} \rangle \\
&\geq \varphi(\hat{x})
\end{aligned} \tag{29}$$

for all $\hat{y} \in C$. By convexity of φ, we have

$$\begin{aligned}
\Theta(\hat{x}, t\hat{x} + (1 - t)\hat{y}) &+ (1 - t)\varphi(\hat{y}) + f(t\hat{x} + (1 - t)\hat{y}) - f(\hat{x}) \\
&- \langle x^*, t\hat{x} + (1 - t)\hat{y} - \hat{x} \rangle \\
&\geq (1 - t)\varphi(\hat{x}).
\end{aligned} \tag{30}$$

Since

$$f(t\hat{x} + (1 - t)\hat{y}) - f(\hat{x}) \leq \langle \nabla f(t\hat{x} + (1 - t)\hat{y}), t\hat{x} + (1 - t)\hat{y} - \hat{x} \rangle,$$

we can conclude from (30) and (A4) that

$$\begin{aligned}
t\Theta(\hat{x}, \hat{x}) + (1 - t)\Theta(\hat{x}, \hat{y}) &+ (1 - t)\varphi(\hat{y}) + \langle \nabla f(t\hat{x} + (1 - t)\hat{y}), t\hat{x} + (1 - t)\hat{y} - \hat{x} \rangle \\
&- \langle x^*, t\hat{x} + (1 - t)\hat{y} - \hat{x} \rangle \\
&\geq (1 - t)\varphi(\hat{x})
\end{aligned} \tag{31}$$

for all $\hat{y} \in C$. From (A1) we have

$$\begin{aligned}
(1 - t)\Theta(\hat{x}, \hat{y}) + (1 - t)\varphi(\hat{y}) &+ \langle \nabla f(t\hat{x} + (1 - t)\hat{y}), (1 - t)(\hat{y} - \hat{x}) \rangle \\
&- \langle x^*, (1 - t)(\hat{y} - \hat{x}) \rangle \\
&\geq (1 - t)\varphi(\hat{x})
\end{aligned} \tag{32}$$

or, equivalently,

$$(1 - t)[\Theta(\hat{x}, \hat{y}) + \varphi(\hat{y}) + \langle \nabla f(t\hat{x} + (1 - t)\hat{y}), \hat{y} - \hat{x} \rangle - \langle x^*, \hat{y} - \hat{x} \rangle] \geq (1 - t)\varphi(\hat{x}).$$

Thus, we have

$$\Theta(\hat{x}, \hat{y}) + \varphi(\hat{y}) + \langle \nabla f(t\hat{x} + (1 - t)\hat{y}), \hat{y} - \hat{x} \rangle - \langle x^*, \hat{y} - \hat{x} \rangle \geq \varphi(\hat{x}),$$

for all $\hat{y} \in C$. Since f is Gâteaux differentiable function, it follows that ∇f is norm-to-weak* continuous (see Phelps 1993, Proposition 2.8). Hence, letting $t \to 1^-$ we then get

$$\Theta(\hat{x}, \hat{y}) + \varphi(\hat{y}) + \langle \nabla f(\hat{x}), \hat{y} - \hat{x} \rangle - \langle x^*, \hat{y} - \hat{x} \rangle \geq \varphi(\hat{x}).$$

By taking $x^* = \nabla f(x)$ we obtain $\hat{x} \in C$ such that

$$\Theta(\hat{x}, \hat{y}) + \varphi(\hat{y}) + \langle \nabla f(\hat{x}) - \nabla f(x), \hat{y} - \hat{x} \rangle \geq \varphi(\hat{x}),$$

for all $\hat{y} \in C$, i.e., $\hat{x} \in Res^f_{\Theta,\varphi}(x)$. Hence, we conclude that $dom\left(Res^f_{\Theta,\varphi}\right) = E$. This completes the proof.

Lemma 12 (Biranvand and Darvish 2018) *Let $f : E \to \mathbb{R}$ be a Legendre function. Let C be a closed and convex subset of E. If the functional $\Theta : C \times C \to \mathbb{R}$ satisfies conditions $(A1) - (A4)$, then*

(1) $Res^f_{\Theta,\varphi}$ is single-valued;

(2) $Res^f_{\Theta,\varphi}$ is a Bregman firmly nonexpansive mapping Reich and Sabach (2010), i.e., for all $x, y \in E$,

$$\langle T_r x - T_r y, \nabla f(T_r x) - \nabla f(T_r y) \rangle \leq \langle T_r x - T_r y, \nabla f(x) - \nabla f(y) \rangle;$$

or, equivalently

$$D_f(Tx, Ty) + D_f(Ty, Tx) + D_f(Tx, x) + D_f(Ty, y) \leq D_f(Tx, y) + D_f(Ty, x);$$

(3) $F\left(Res^f_{\Theta,\varphi}\right) = Sol(MEP)$ is closed and convex;

(4) $D_f(q, Res^f_{\Theta,\varphi}(x)) + D_f(Res^f_{\Theta,\varphi}(x), x) \leq D_f(q, x), \quad \forall q \in F(Res^f_{\Theta,\varphi}),$ $x \in E$;

(5) $Res^f_{\Theta,\varphi}$ is a Bregman quasi-nonexpansive mapping.

Proof (1) Let $z_1, z_2 \in Res^f_{\Theta,\varphi}(x)$ then by definition of the resolvent we have

$$\Theta(z_1, z_2) + \varphi(z_2) + \langle \nabla f(z_1) - \nabla f(x), z_2 - z_1 \rangle \geq \varphi(z_1)$$

and

$$\Theta(z_2, z_1) + \varphi(z_1) + \langle \nabla f(z_2) - \nabla f(x), z_1 - z_2 \rangle \geq \varphi(z_2).$$

Adding these two inequalities, we obtain

$$\Theta(z_1, z_2) + \Theta(z_2, z_1) + \varphi(z_1) + \varphi(z_2) + \langle \nabla f(z_2) - \nabla f(z_1), z_1 - z_2 \rangle \geq \varphi(z_1) + \varphi(z_2).$$

So,

$$\Theta(z_1, z_2) + \Theta(z_2, z_1) + \langle \nabla f(z_2) - \nabla f(z_1), z_1 - z_2 \rangle \geq 0.$$

By $(A2)$, we have

$$\langle \nabla f(z_2) - \nabla f(z_1), z_1 - z_2 \rangle \geq 0.$$

Since f is Legendre it is strictly convex. So, ∇f is strictly monotone and hence $z_1 = z_2$. It follows that $Res^f_{\Theta,\varphi}$ is single-valued.

(2) Let $x, y \in E$, then we have

$$\Theta\left(Res^f_{\Theta,\varphi}(x), Res^f_{\Theta,\varphi}(y)\right) + \varphi\left(Res^f_{\Theta,\varphi}(y)\right)$$
$$+ \langle \nabla f\left(Res^f_{\Theta,\varphi}(x)\right) - \nabla f(x), Res^f_{\Theta,\varphi}(y) - Res^f_{\Theta,\varphi}(x) \rangle \quad (33)$$
$$\geq \varphi\left(Res^f_{\Theta,\varphi}(x)\right)$$

and

$$\Theta\left(Res^f_{\Theta,\varphi}(y), Res^f_{\Theta,\varphi}(x)\right) + \varphi\left(Res^f_{\Theta,\varphi}(x)\right)$$
$$+ \langle \nabla f\left(Res^f_{\Theta,\varphi}(y)\right) - \nabla f(y), Res^f_{\Theta,\varphi}(x) - Res^f_{\Theta,\varphi}(y) \rangle \quad (34)$$
$$\geq \varphi\left(Res^f_{\Theta,\varphi}(y)\right).$$

Adding the inequalities (33) and (34), we have

$$\Theta\left(Res^f_{\Theta,\varphi}(x), Res^f_{\Theta,\varphi}(y)\right) + \Theta\left(Res^f_{\Theta,\varphi}(y), Res^f_{\Theta,\varphi}(x)\right)$$
$$+ \langle \nabla f\left(Res^f_{\Theta,\varphi}(x)\right) - \nabla f(x) + \nabla f(y) - \nabla f\left(Res^f_{\Theta,\varphi}(y)\right), Res^f_{\Theta,\varphi}(y) - Res^f_{\Theta,\varphi}(x) \rangle$$
$$\geq 0.$$

By $(A2)$, we have

$$\langle \nabla f\left(Res^f_{\Theta,\varphi}(x)\right) - \nabla f\left(Res^f_{\Theta,\varphi}(y)\right), Res^f_{\Theta,\varphi}(x) - Res^f_{\Theta,\varphi}(y) \rangle$$
$$\leq \langle \nabla f(x) - \nabla f(y), Res^f_{\Theta,\varphi}(x) - Res^f_{\Theta,\varphi}(y) \rangle.$$

It means $Res^f_{\Theta,\varphi}$ is a Bregman firmly nonexpansive mapping.

(3)
$$x \in F\left(Res^f_{\Theta,\varphi}\right)$$
$$\Leftrightarrow x = Res^f_{\Theta,\varphi}(x)$$
$$\Leftrightarrow \Theta(x, y) + \varphi(y) + \langle \nabla f(x) - \nabla f(x), y - x \rangle \geq \varphi(x), \quad \forall y \in C$$
$$\Leftrightarrow \Theta(x, y) + \varphi(y) \geq \varphi(x), \quad \forall y \in C$$
$$\Leftrightarrow x \in Sol(MEP).$$

(4) Since $Res^f_{\Theta,\varphi}$ is a Bregman firmly nonexpansive mapping, it follows from Reich and Sabach (2010) that $F\left(Res^f_{\Theta,\varphi}\right)$ is a closed and convex subset of C. So, from (3) we have $F\left(Res^f_{\Theta,\varphi}\right) = Sol(MEP)$; is a closed and convex subset of C.

(5) Since $Res_{\Theta,\varphi}^f$ is a Bregman firmly nonexpansive mapping, we have from (19) that for all $x, y \in E$

$$
\begin{aligned}
D_f\big(Res_{\Theta,\varphi}^f(x), Res_{\Theta,\varphi}^f(y)\big) &+ D_f\big(Res_{\Theta,\varphi}^f(y), Res_{\Theta,\varphi}^f(x)\big) \\
&\leq D_f\big(Res_{\Theta,\varphi}^f(x), y\big) - D_f\big(Res_{\Theta,\varphi}^f(x), x\big) \\
&+ D_f\big(Res_{\Theta,\varphi}^f(y), x\big) - D_f\big(Res_{\Theta,\varphi}^f(y), y\big).
\end{aligned}
$$

Let $y = p \in F\big(Res_{\Theta,\varphi}^f\big)$, then we get

$$
\begin{aligned}
D_f\big(Res_{\Theta,\varphi}^f(x), p\big) + D_f\big(p, Res_{\Theta,\varphi}^f(x)\big) &\leq D_f\big(Res_{\Theta,\varphi}^f(x), p\big) \\
&- D_f\big(Res_{\Theta,\varphi}^f(x), x\big) + D_f(p, x) - D_f(p, p).
\end{aligned}
$$

Hence, we have

$$
D_f\big(p, Res_{\Theta,\varphi}^f(x)\big) + D_f\big(Res_{\Theta,\varphi}^f(x), x\big) \leq D_f(p, x).
$$

This completes the proof. ∎

3   Main Result

Motivated by the work of Shehu and Ogbuisi (2015), we prove the following strong convergence theorem for finding a common solution of the fixed point problem for an infinitely countable family of quasi-Bregman strictly pseudocontractive mappings and MEP (6) in a reflexive Banach space.

Theorem 2 *Let E be a Banach space and C be a nonempty, closed and convex subset of a reflexive Banach space E. Let $f : E \to \mathbb{R}$ be a strongly coercive Legendre function which is bounded, uniformly Fréchet differentiable and totally convex on bounded subsets of E. Let $\Theta_j : C \times C \to \mathbb{R}$, $j = 1, 2, \ldots, N$, be finite functions satisfying conditions $(A1) - (A4)$ and let $\varphi_j : C \to \mathbb{R}$, $j = 1, 2, \ldots, N$, be finite real-valued functions. Let $\{T_i\}_{i=1}^\infty$ be an infinitely countable family of quasi-Bregman strictly pseudocontractive mappings from C into itself with uniformly $k \in [0, 1)$. Assume that*

$$
\Omega := \bigcap_{i=1}^{\infty} F(T_i) \cap \bigcap_{j=1}^{N} Sol(MEP(\Theta_j, \varphi_j)) \neq \emptyset.
$$

Let $\{x_n\}$ be the sequence generated by the iterative schemes:

$$
\begin{cases}
x_1 = x_0 \in C_1 = C, \\
y_n^i = \nabla f^*[\alpha_n \nabla f(x_n) + (1 - \alpha_n)\nabla f(T_i x_n)], \\
u_n^i = Res_{\Theta_N,\varphi_N}^f \circ Res_{\Theta_{N-1},\varphi_{N-1}}^f \circ \cdots \circ Res_{\Theta_2,\varphi_2}^f \circ Res_{\Theta_1,\varphi_1}^f y_n^i, \\
C_{n+1} = \{z \in C_n : \sup_{i \geq 1} D_f(z, u_n^i) \leq D_f(z, x_n) + \frac{k}{1-k}\langle \nabla f(x_n) - \nabla f(T_i x_n), x_n - z \rangle\}, \\
x_{n+1} = proj_{C_{n+1}}^f (x_0), \ \forall n \geq 1,
\end{cases}
\tag{35}
$$

where

$$Sol(MEP(\Theta_j, \varphi_j)) = \{z \in C : \Theta_j(z, y) + \varphi_j(y) + \langle \nabla f(z) - \nabla f(x), y - z \rangle \geq \varphi_j(z), \forall y \in C\},$$

for $j = 1, 2, \ldots, N$, and $\{\alpha_n\}$ is a sequence in $(0, 1)$ such that $\lim sup_{n \to \infty} \alpha_n < 1$.
Then, $\{x_n\}$ converges strongly to $proj_{\Omega}^f(x_0)$ where $proj_{\Omega}^f(x_0)$ is the Bregman projection of C onto Ω.

Proof We divide the proof into several steps.

Step 1: We show that Ω is closed and convex. From Lemma 3, $\bigcap_{i=1}^{\infty} F(T_i)$ is closed and convex for all $i \geq 1$ and from (3) of Lemma 12, $\bigcap_{j=1}^{N} Sol(MEP(\Theta_j, \varphi_j))$ is a closed and convex for each $j = 1, 2, \ldots, N$. So, Ω is a nonempty, closed and convex subset of C. Therefore $proj_{\Omega}^f$ is well defined.

Step 2: We show that the sets C_n is closed and convex. For $n = 1$, $C_1 = C$ is closed and convex. Suppose that C_m is closed and convex for some $m \in \mathbb{N}$. For each $z \in C_m$ and $i \geq 1$, we see that

$$D_f(z, u_m^i) \leq D_f(z, x_m) + \frac{k}{1-k}$$
$$\langle \nabla f(x_m) - \nabla f(T_i x_m), x_m - z \rangle$$

$$D_f(z, u_m^i) - D_f(z, x_m) \leq \frac{k}{1-k} \langle \nabla f(x_m) - \nabla f(T_i x_m), x_m - z \rangle$$

$$\langle \nabla f(x_m), z - x_m \rangle - \langle \nabla f(u_m^i), z - u_m^i \rangle \leq f(u_m^i) - f(x_m) + \frac{k}{1-k} \langle \nabla f(x_m)$$
$$- \nabla f(T_i x_m), x_m - z \rangle$$

$$\langle \frac{1}{1-k} \nabla f(x_m) - f(u_m^i) - \frac{k}{1-k} \nabla f(T_i x_m), z \rangle \leq f(u_m^i) - f(x_m) + \langle \frac{1}{1-k} \nabla f(x_m), x_m \rangle$$
$$- \langle \nabla f(u_m^i), u_m^i \rangle - \langle \frac{k}{1-k} \nabla f(T_i x_m), x_m \rangle.$$

From the above expression, we know that C_m is closed and convex. By the construction of the set C_{m+1}, we see that

$$C_{m+1} = \{z \in C_m : \sup_{i \geq 1} D_f(z, u_m^i) \leq D_f(z, x_m) + \frac{k}{1-k} \langle \nabla f(x_m) - \nabla f(T_i x_m), x_m - z \rangle\}$$

$$= \bigcap_{i=1}^{\infty} \{z \in C_m : D_f(z, u_m^i) \leq D_f(z, x_m) + \frac{k}{1-k} \langle \nabla f(x_m) - \nabla f(T_i x_m), x_m - z \rangle\}.$$

$$(36)$$

Hence, C_{m+1} is also closed and convex.

Step 3: We show that $\Omega \subset C_n$ for all $n \in \mathbb{N}$. Note that $\Omega \subset C_1 = C$. Suppose $\Omega \subset C_m$ for some $m \in \mathbb{N}$ and let $p \in \Omega$. Then

$$D_f(p, u_m^i) = D_f\left(p, Res_{\Theta_N, \varphi_N}^f \circ Res_{\Theta_{N-1}, \varphi_{N-1}}^f \circ \cdots \circ Res_{\Theta_2, \varphi_2}^f \circ Res_{\Theta_1, \varphi_1}^f y_m^i\right)$$

$$\leq D_f\left(p, Res_{\Theta_{N-1}, \varphi_{N-1}}^f \circ \cdots \circ Res_{\Theta_2, \varphi_2}^f \circ Res_{\Theta_1, \varphi_1}^f y_m^i\right)$$

$$\vdots$$

$$\leq D_f\left(p, Res_{\Theta_1, \varphi_1}^f y_m^i\right)$$

$$\leq D_f\left(p, y_m^i\right)$$

$$= D_f\left(p, \nabla f^*[\alpha_m \nabla f(x_m) + (1 - \alpha_m)\nabla f(T_i x_m)]\right)$$

$$= V\left(p, \alpha_m \nabla f(x_m) + (1 - \alpha_m)\nabla f(T_i x_m)\right)$$

$$\leq \alpha_m V\left(p, \nabla f(x_m)\right) + (1 - \alpha_m)V\left(p, \nabla f(T_i x_m)\right)$$

$$= \alpha_m D_f\left(p, x_m\right) + (1 - \alpha_m)D_f\left(p, T_i x_m\right)$$

$$\leq \alpha_m D_f\left(p, x_m\right) + (1 - \alpha_m)\left[D_f\left(p, x_m\right) + k D_f\left(p, T_i x_m\right)\right]$$

$$\leq D_f\left(p, x_m\right) + k D_f\left(p, T_i x_m\right)$$

$$\leq D_f\left(p, x_m\right) + \frac{k}{1-k}\langle \nabla f(x_m) - \nabla f(T_i x_m), x_m - p\rangle.$$

$$(37)$$

That is $p \in C_{m+1}$. By induction, we conclude that $\Omega \subset C_n$ for all $n \in \mathbb{N}$.

Step 4: We show that $\lim_{n \to \infty} D_f(x_n, x_0)$ exists. Since $x_n = proj_{C_n}^f(x_0)$ which from (a) of Lemma 1 implies

$$\langle \nabla f(x_0) - \nabla f(x_n), y - x_n\rangle \leq 0, \quad \forall y \in C_n.$$

Since $\Omega \subset C_n$, we have

$$\langle \nabla f(x_0) - \nabla f(x_n), p - x_n\rangle \leq 0, \quad \forall p \in \Omega. \tag{38}$$

From (b) of Lemma 1, we have

$$D_f(x_n, x_0) = D_f\left(proj_{C_n}^f(x_0), x_0\right)$$

$$\leq D_f(p, x_0) - D_f\left(p, proj_{C_n}^f(x_0)\right) \leq D_f(p, x_0), \quad \forall p \in \Omega \subset C_n, \, n \geq 1. \tag{39}$$

This implies that the sequence $\{D_f(x_n, x_0)\}$ is bounded and hence, it follows from Lemma 9 that the sequence $\{x_n\}$ is bounded. By the construction of C_n, we have $x_m \in C_m \subset C_m$, and $x_n = proj_{C_n}^f(x_0)$ for any positive integer $m \geq n$. Then, we obtain

$$D_f(x_m, x_n) = D_f\left(x_m, proj_{C_n}^f(x_0)\right)$$

$$\leq D_f\left(x_m, x_0\right) - D_f\left(proj_{C_n}^f(x_0), x_0\right) \tag{40}$$

$$= D_f\left(x_m, x_0\right) - D_f\left(x_n, x_0\right).$$

In particular,

$$D_f(x_{n+1}, x_n) \leq D_f(x_{n+1}, x_0) - D_f(x_n, x_0).$$

Since $x_n = proj_{C_n}^f(x_0)$ and $x_{n+1} = proj_{C_{n+1}}^f(x_0) \in C_{n+1} \subset C_n$, we obtain $D_f(x_n, x_0) \leq D_f(x_{n+1}, x_0)$, $\forall n \geq 1$.
This shows that $\{D_f(x_n, x_0)\}$ is nondecreasing and hence the limit $\lim_{n \to \infty} D_f(x_n, x_0)$ exists. Thus, from (40), taking the limit as $m, n \to \infty$, we obtain $\lim_{n \to \infty} D_f(x_n, x_0) = 0$. Since f is totally convex on bounded subsets of E, f is sequentially consistent by Lemma 8. Therefore, it follows that $\|x_m - x_n\| \to 0$ as $m, n \to \infty$. Therefore, $\{x_n\}$ is a Cauchy sequence. Now by the completeness of the space E and the closedness of the set C, we can assume that $x_n \to p \in C$ as $n \to \infty$.
Moreover, we get that

$$\lim_{n \to \infty} D_f(x_{n+1}, x_n) = 0. \tag{41}$$

Since $x_{n+1} = proj_{C_{n+1}}^f(x_0) \in C_{n+1}$, we have for all $i \geq 1$ that

$$D_f(x_{n+1}, u_n^i) \leq D_f(x_{n+1}, x_n) \to 0, \quad n \to \infty. \tag{42}$$

It follows from Lemma 8 and (42) that

$$\lim_{n \to \infty} \|x_n - u_n^i\| = 0, \quad \forall i \geq 1. \tag{43}$$

This shows that $u_n^i \to p \in C$ as $n \to \infty$. for all $i \geq 1$. By Lemma 7, ∇f is uniformly continuous on bounded subsets of E, and we have that

$$\lim_{n \to \infty} \|\nabla f(x_n) - \nabla f(u_n^i)\| = 0, \quad \forall i \geq 1. \tag{44}$$

Step 5: We show that $p \in \cap_{i=1}^{\infty} F(T_i)$.
Denote $\Psi_j = Res_{\Theta_j, \varphi_j}^f \circ Res_{\Theta_{j-1}, \varphi_{j-1}}^f \circ \cdots \circ Res_{\Theta_2, \varphi_2}^f \circ Res_{\Theta_1, \varphi_1}^f$,
for $j = 1, 2, \ldots, N$, and $\Gamma_0 = I$. We note that $u_n^i = \Psi_N y_n^i$ for all $i \geq 1$. From (37) we observe that

$$D_f(p, \Psi_{N-1} y_n^i) \leq D_f(p, \Psi_{N-2} y_n^i) \leq \cdots \leq D_f(p, y_n^i)$$
$$\leq D_f(p, x_n) + \frac{k}{1-k} \langle \nabla f(x_n) - \nabla f(T_i x_n), x_n - p \rangle. \tag{45}$$

Since $p \in Sol(MEP(\Theta_N, \varphi_N)) = F(Res_{\Theta_N, \varphi_N}^f)$ for all $n \geq 1$, it follows from (45) and Lemma 12 (5) that

$$D_f(u_n^i, \Psi_{N-1} y_n^i) \leq D_f(p, \Psi_{N-1} y_n^i) - D_f(p, u_n^i)$$
$$\leq D_f(p, x_n) - D_f(p, u_n^i) + \frac{k}{1-k} \langle \nabla f(x_n) - \nabla f(T_i x_n), x_n - p \rangle. \tag{46}$$

From (43), (44), and (14), we get that $\lim\limits_{n\to\infty} D_f(u_n^i, \Psi_{N-1} y_n^i) = 0$ for all $i \geq 1$. Since f is totally convex on bounded subsets of E, f is sequentially consistent. From Lemma 8, we have

$$\lim_{n\to\infty} \|u_n^i - \Psi_{N-1} y_n^i\| = 0, \quad \forall i \geq 1. \tag{47}$$

Also, from (43) and (47), we have

$$\lim_{n\to\infty} \|x_n - \Psi_{N-1} y_n^i\| = 0, \quad \forall i \geq 1. \tag{48}$$

Hence,

$$\lim_{n\to\infty} \|\nabla f(x_n) - \nabla f(\Psi_{N-1} y_n^i)\| = 0, \quad \forall i \geq 1. \tag{49}$$

Again, since $p \in Sol(MEP(\Theta_{N-1}, \varphi_{N-1})) = F\left(Res^f_{\Theta_{N-1},\varphi_{N-1}}\right)$ for all $n \geq 1$, it follows from (45) and Lemma 12 (5) that

$$\begin{aligned}
D_f(\Psi_{N-1} y_n^i, \Psi_{N-2} y_n^i) &\leq D_f(p, \Psi_{N-2} y_n^i) - D_f(p, \Psi_{N-1} y_n^i) \\
&\leq D_f(p, x_n) - D_f(p, \Psi_{N-1} y_n^i) + \frac{k}{1-k} \langle \nabla f(x_n) - \nabla f(T_i x_n), x_n - p \rangle.
\end{aligned} \tag{50}$$

Similarly, we also have

$$\lim_{n\to\infty} \|\Psi_{N-1} y_n^i - \Psi_{N-2} y_n^i\| = 0, \quad \forall i \geq 1. \tag{51}$$

Hence, from (49) and (51), we get

$$\lim_{n\to\infty} \|x_n - \Psi_{N-2} y_n^i\| = 0, \quad \forall i \geq 1 \tag{52}$$

and

$$\lim_{n\to\infty} \|\nabla f(x_n) - \nabla f(\Psi_{N-2} y_n^i)\| = 0, \quad \forall i \geq 1. \tag{53}$$

In a similar way, we can verify that

$$\lim_{n\to\infty} \|\Psi_{N-2} y_n^i - \Psi_{N-3} y_n^i\| = \quad \cdots = \lim_{n\to\infty} \|\Psi_1 y_n^i - \Psi_{N-3} y_n^i\| = 0, \quad \forall i \geq 1.$$

$$\lim_{n\to\infty} \|x_n - \Psi_{N-3} y_n^i\| = \quad \cdots = \lim_{n\to\infty} \|x_n - y_n^i\| = 0, \quad \forall i \geq 1. \tag{54}$$

$$\lim_{n\to\infty} \|\nabla f(x_n) - \nabla f(\Psi_{N-3} y_n^i)\| = \quad \cdots = \lim_{n\to\infty} \|\nabla f(x_n) - \nabla f(y_n^i)\| = 0, \quad \forall i \geq 1.$$

Hence, we conclude that

$$\lim_{n\to\infty} \|\Psi_j y_n^i - \Psi_{j-1} y_n^i\| = 0, \quad \forall j = 1, 2, \cdots, N, \quad i \geq 1. \tag{55}$$

Since

$$y_n^i = \nabla f^*[\alpha_n \nabla f(x_n) + (1 - \alpha_n)\nabla f(T_i x_n)],$$

observe that

$$\begin{aligned}
\|\nabla f(y_n^i) - \nabla f(x_n)\| &= \|\nabla f\big(\nabla f^*[\alpha_n \nabla f(x_n) + (1 - \alpha_n)\nabla f(T_i x_n)]\big) - \nabla f(x_n)\| \\
&= \|\alpha_n \nabla f(x_n) + (1 - \alpha_n)\nabla f(T_i x_n) - \nabla f(x_n)\| \\
&= \|(1 - \alpha_n)\nabla f(T_i x_n) - (1 - \alpha_n)\nabla f(x_n)\| \\
&= \|(1 - \alpha_n)\big(\nabla f(T_i x_n) - \nabla f(x_n)\big)\| \\
&= (1 - \alpha_n)\|\nabla f(T_i x_n) - \nabla f(x_n)\|,
\end{aligned}$$

(56)

we obtain that

$$\lim_{n \to \infty} \|\nabla f(T_i x_n) - \nabla f(x_n)\| = \lim_{n \to \infty} \frac{1}{1 - \alpha_n}\|\nabla f(y_n^i) - \nabla f(x_n)\| = 0. \quad (57)$$

Since f is strongly coercive and uniformly convex on bounded subsets of E, f^* is uniformly Fréchet differentiable on bounded sets. Moreover, f^* is bounded on bounded sets, and from (57) we obtain

$$\lim_{n \to \infty} \|\nabla f(T_i x_n) - \nabla f(x_n)\| = 0, \quad \forall i \geq 1. \quad (58)$$

We have that $p \in F(T_i)$ for all $i \geq 1$. Thus $p \in \cap_{i=1}^{\infty} F(T_i)$.

Step 6: We show that $p \in \cap_{j=1}^{N} Sol(MEP(\Theta_j, \varphi_j))$. Since $u_n^i = \Psi_N y_n^i$ for all $i \geq 1$. From (6), we have that for each $j = 1, 2, \ldots, N$,

$$\Theta_j(\Psi_j y_n^i, z) + \varphi_j(z) + \langle \nabla f(\Psi_j y_n^i) - \nabla f(\Psi_{j-1} y_n^i), z - \Psi_j y_n^i \rangle \geq \varphi_j(\Psi_j y_n^i),$$

for all $z \in C$. From $(A2)$, we have

$$\begin{aligned}
\Theta_j(z, \Psi_j y_n^i) &\leq -\Theta_j(\Psi_j y_n^i, z) \\
&\leq \varphi_j(z) - \varphi_j(\Psi_j y_n^i) + \langle \nabla f(\Psi_j y_n^i) - \nabla f(\Psi_{j-1} y_n^i), z - \Psi_j y_n^i \rangle, \quad \forall z \in C.
\end{aligned}$$

(59)

And, we have

$$\begin{aligned}
\Theta_j(z, \Psi_j y_n^i) &\leq -\Theta_j(\Psi_j y_n^i, z) \\
&\leq \varphi_j(z) - \varphi_j(\Psi_j y_n^i) + \|\nabla f(\Psi_j y_n^i) - \nabla f(\Psi_{j-1} y_n^i)\| \, \|z - \Psi_j y_n^i\|, \quad \forall z \in C.
\end{aligned}$$

Hence,

$$\Theta_j(z, \Psi_j y_n^i) \leq \varphi_j(z) - \varphi_j(\Psi_j y_n^i) + \langle \nabla f(\Psi_j y_n^i) - \nabla f(\Psi_{j-1} y_n^i), z - \Psi_j y_n^i \rangle, \quad \forall z \in C.$$

Since $\Psi_j y_n^i \rightharpoonup p$ and Θ_j is lower semicontinuous in the second argument, φ_j is continuous then using (44), by taking $n \to \infty$, we have

$$\Theta_j(z, p) + \varphi_j(p) - \varphi_j(z) \leq 0, \quad \forall z \in C.$$

Setting $z_t = tz + (1-t)p$, $\forall t \in [0, 1]$. Since $z \in C$ and $p \in C$, we have $z_t \in C$ and hence

$$\Theta_j(z_t, p) + \varphi_j(p) - \varphi_j(z_t) \leq 0.$$

Now,

$$\begin{aligned}
0 &= \Theta_j(z_t, z_t) + \varphi_j(z_t) - \varphi_j(z_t) \\
&\leq t\Theta_j(z_t, z) + (1-t)\Theta_j(z_t, p) + t\varphi_j(z) + (1-t)\varphi_j(p) - \varphi(z_t) \\
&\leq t[\Theta_j(z_t, z) + \varphi_j(z) - \varphi_j(z_t)].
\end{aligned}$$

Since, $\Theta_j(z_t, z) + \varphi_j(z) - \varphi_j(z_t) \geq 0$. Then, we have

$$\Theta_j(p, z) + \varphi_j(z) - \varphi_j(p) \geq 0, \quad \forall z \in C.$$

Therefore, $p \in \bigcap_{j=1}^{N} Sol(MEP(\Theta_j, \varphi_j))$.

Step 7: Finally, we show that $x_n \to p = proj_\Omega^f(x_0)$. Let $\bar{x} = proj_\Omega^f(x_0)$. Since $\{x_n\}$ is weakly convergent, $x_{n+1} = proj_\Omega^f(x_0)$ and $proj_\Omega^f(x_0) \in \Omega \subset C_{n+1}$. It follows from (38) that

$$D_f(x_{n+1}, x_0) \leq D_f\big(proj_\Omega^f(x_0), x_0\big). \tag{60}$$

Now, by Lemma 10, $\{x_n\}$ strongly convergent to $\bar{x} = proj_\Omega^f(x_0)$. Therefore, by the uniqueness of the limit, we have that the sequence $\{x_n\}$ converges strongly to $p = proj_\Omega^f(x_0)$. This completes the proof.

Corollary 1 *Let E be a Banach space and C be a nonempty, closed and convex subset of a reflexive Banach space E. Let $f : E \to \mathbb{R}$ be a strongly coercive Legendre function which is bounded, uniformly Fréchet differentiable and totally convex on bounded subsets of E. Let $\{T_i\}_{i=1}^{\infty}$ be an infinitely countable family of quasi-Bregman strictly pseudocontractive mappings from C into itself with uniformly $k \in [0, 1)$. Assume $\Omega := \bigcap_{i=1}^{\infty} F(T_i) \neq \emptyset$. Let be the $\{x_n\}$ sequence generated by the iterative schemes:*

$$\begin{cases} x_1 = x_0 \in C_1 = C, \\ y_n^i = \nabla f^*[\alpha_n \nabla f(x_n) + (1 - \alpha_n)\nabla f(T_i x_n)], \\ C_{n+1} = \{z \in C_n : \sup_{i \geq 1} D_f(z, y_n^i) \leq D_f(z, x_n) + \dfrac{k}{1-k}\langle \nabla f(x_n) - \nabla f(T_i x_n), x_n - z\rangle\}, \\ x_{n+1} = proj_{C_{n+1}}^f(x_0), \ \forall n \geq 1, \end{cases}$$

$$(61)$$

where $\{\alpha_n\}$ is a sequence in $(0, 1)$ such that $\lim \sup_{n\to\infty}\alpha_n < 1$. Then, $\{x_n\}$ converges strongly to $proj_\Omega^f(x_0)$.

Corollary 2 *Let E be a Banach space and C be a nonempty, closed and convex subset of a reflexive Banach space E. Let $f : E \to \mathbb{R}$ be a strongly coercive Legendre function which is bounded, uniformly Fréchet differentiable and totally convex on bounded subsets of E. Let $\Theta_j : C \times C \to \mathbb{R}, \ j = 1, 2, \ldots, N$, be finite function satisfying conditions $(A1) - (A4)$ and let $\varphi_j : C \to \mathbb{R}, \ j = 1, 2, \ldots, N$, be finite real-valued function. Assume $\Omega := \bigcap_{j=1}^N Sol(MEP(\Theta_j, \varphi_j)) \neq \emptyset$. Let be the $\{x_n\}$ sequence generated by the iterative schemes:*

$$\begin{cases} x_1 = x_0 \in C_1 = C, \\ u_n^i = Res_{\Theta_N, \varphi_N}^f \circ Res_{\Theta_{N-1}, \varphi_{N-1}}^f \circ \cdots \circ Res_{\Theta_2, \varphi_2}^f \circ Res_{\Theta_1, \varphi_1}^f x_1, \\ C_{n+1} = \{z \in C_n : \sup_{i \geq 1} D_f(z, u_n^i) \leq D_f(z, x_n) + \dfrac{k}{1-k}\langle \nabla f(x_n) - \nabla f(T_i x_n), x_n - z\rangle\}, \\ x_{n+1} = proj_{C_{n+1}}^f(x_0), \ \forall n \geq 1, \end{cases}$$

$$(62)$$

where

$$Sol(MEP(\Theta_j, \varphi_j)) = \{z \in C : \Theta_j(z, y) + \varphi_j(y) + \langle \nabla f(z) - \nabla f(x), y - z\rangle \geq \varphi_j(z), \forall y \in C\},$$

for $j = 1, 2, \ldots, N$. Then, $\{x_n\}$ converges strongly to $proj_\Omega^f(x_0)$.

Acknowledgements This project was supported by the Center of Excellence in Theoretical and Computational Science (TaCS-CoE), KMUTT.

References

Bauschke, H. H., Borwein, J. M., & Combettes, P. L. (2001). Essential smoothness, essential strict convexity, and Legendre functions in Banach spaces. *Communications in Contemporary Mathematics*, 3(4), 615–647.

Biranvand, N., & Darvish, V. (2018). A new algorithm for solving mixed equilibrium problem and finding common fixed point of Bregman strongly nonexpansive mappings. *The Korean Journal of Mathematics*, 26(4), 777–798.

Blum, E., & Oettli, W. (1994). From optimization and variational inequalities to equilibrium problems. *The Mathematics Student*, 63, 123–145.

Brègman, L. M. (1967). The relaxation method of finding the common point of convex sets and its application to the solution of problems in convex programming. *USSR Computational Mathematics and Mathematical Physics*, 7, 200–217.

Bruck, R. E., & Reich, S. (1977). Nonexpansive projections and resolvents of accretive operators in Banach spaces. *Houston Journal of Mathematics*, 3(4), 459–470.

Butnariu, D., & Iusem, A. N. (2000). *Totally convex functions for fixed points computation and infinite dimensional optimization*. Dordrecht: Kluwer Academic Publishers.

Butnariu, D., & Resmerita, E. (2006). Bregman distances, totally convex functions and a method for solving operator equations in Banach spaces. *Abstract and Applied Analysis*, Article ID 84919, 1–39.

Ceng, L. C., & Yao, J. C. (2008). A hybrid iterative scheme for mixed equilibrium problems and fixed point problems. *Journal of Computational and Applied Mathematics*, *214*(1), 186–201.

Chen, G., & Teboulle, M. (1993). Convergence analysis of a proximal-like minimization algorithm using Bregman functions. *SIAM Journal on Control and Optimization*, *3*, 538–543.

Chen, J., Wan, Z., Yuan, L., Zheng, Y. (2011). Approximation of fixed points of weak Bregman relatively nonexpansive mappings in Banach spaces. *International Journal of Mathematics and Mathematical Sciences*, Article ID 420192, 23.

Cholamjiak, P., & Suantai, S. (2010). Convergence analysis for a system of equilibrium problems and a countable family of relatively quasi-nonexpansive mappings in Banach spaces. *Abstract and Applied Analysis*, Article ID 141376, 17.

Combettes, P. L., & Hirstoaga, S. A. (2005). Equilibrium programming in Hilbert spaces. *Journal of Nonlinear and Convex Analysis. An International Journal*, *6*(1), 117–136.

Kassay, G., Reich, S., & Sabach, S. (2011). Iterative methods for solving systems of variational inequalities in reflexive Banach spaces. *SIAM Journal on Optimization*, *21*, 1319–1344.

Kohsaka, F., & Takahashi, W. (2005). Proximal point algorithms with Bregman functions in Banach spaces. *Journal of Nonlinear and Convex Analysis*, *6*(3), 505–523.

Kumam, W., Witthayarat, U., Wattanawitoon, K., Suantai, S., & Kumam, P. (2016). Convergence theorem for equilibrium problem and Bregman strongly nonexpansive mappings in Banach spaces. *A Journal of Mathematical Programming and Operations Research*, *65*(2), 265–280.

Mann, W. R. (1953). Mean value methods in iteration. *Proceedings of the American Mathematical Society*, *4*, 506–510.

Muangchoo, K., Kumam, P., Cho, Y. J., Dhompongsa, S., & Ekvittayaniphon, S. (2019). Approximating fixed points of Bregman Generalized α-nonexpansive mappings. *Mathematics*, *7*(8), 709.

Pang, C. T., Naraghirad E., & Wen, C. F. (2014). Weak convergence theorems for Bregman relatively nonexpansive mappings in Banach spaces. *Journal of Applied Mathematics*, Article ID 573075.

Pardalos, P. M., Rassias, T. M., & Khan, A. A. (2010). Nonlinear analysis and variational problems. In: *Springer optimization and its applications* (Vol. 35). Berlin: Springer.

Phelps, R. P. (1993). Convex functions, monotone operators, and differentiability. In: *Lecture notes in mathematics* (2nd ed., Vol. 1364). Berlin: Springer.

Reich, S. (1979). Weak convergence theorems for nonexpansive mappings in Banch spaces. *Journal of Mathematical Analysis and Applications*, *67*, 274–276.

Reich, S., & Sabach, S. (2009). A strong convergence theorem for a proximal-type algorithm in reflexive Banach spaces. *Journal of Nonlinear and Convex Analysis. An International Journal*, *10*(3), 471–485.

Reich, S., & Sabach, S. (2010). Two strong convergence theorems for Bregman strongly nonexpansive operators in reflexive Banach spaces. *Nonlinear Analysis*, *73*, 122–135.

Reich, S., & Sabach, S. (2011). *Existence and approximation of fixed points of Bregman firmly nonexpansive mappings in reflexive Banach spaces: Fixed-point algorithms for inverse problems in science and engineering* (pp. 301–316). New York: Springer.

Rockafellar, R. T. (1970). On the maximal monotonicity of subdifferential mappings. *Pacific Journal of Mathematics*, *33*, 209–216.

Shehu, Y., & Ogbuisi, F. U. (2015). Approximation of common fixed points of left Bregman strongly nonexpansive mappings and solutions of equilibrium problems. *Journal of Applied Analysis*, *21*(2), 67–77.

Takahashi, W. (2000). *Nonlinear functional analysis. Fixed point theory and its applications*. Yokahama: Yokahama Publishers.

Takahashi, W., Takeuchi, Y., & Kubota, R. (2008). Strong convergence theorems by hybrid methods for families of nonexpansive mappings in Hilbert spaces. *Journal of Mathematical Analysis and Applications, 341*, 276–286.

Ugwunnadi, G., Ali, B., Idris, I., & Minjibir, M. (2014). Strong convergence theorem for quasi-Bregman strictly pseudocontractive mappings and equilibrium problems in Banach spaces. *Fixed Point Theory and Applications, 231*, 1–16.

Wang, Z. (2015). Strong convergence theorems for Bregman quasi-strict pseudo-contractions in reflexive Banach spaces with applications. *Fixed Point Theory and Applications, 91*, 1–17.

Xu, Y., & Su, Y. (2015). New hybrid shrinking projection algorithm for common fixed points of a family of countable quasi-Bregman strictly pseudocontractive mappings with equilibrium and variational inequality and optimization problems. *Fixed Point Theory and Applications, 95*, 1–22.

Zhou, H., & Gao, E. (2009). An iterative method of fixed points for closed and quasi-strict pseudo-contraction in Banach spaces. *Springer Science and Business Media LLC, 33*(1–2), 227–237.

Zălinescu, C. (2002). *Convex analysis in general vector spaces*. River Edge, NJ: World Scientific Publishing Co Inc.

Copula-Based Stochastic Frontier Quantile Model with Unknown Quantile

Paravee Maneejuk and Woraphon Yamaka

Abstract This study aims to improve the copula-based stochastic frontier quantile model by treating the quantile as the unknown parameter. This method can solve the problem of quantile selection bias as the quantile will be estimated simultaneously with other parameters in the model. We then evaluate the performance and accuracy of the proposed model by conducting two simulation studies and a real data analysis with two different data sets. The overall results reveal that our proposed model can beat the conventional stochastic frontier model and also the copula-based stochastic frontier model with a given quantile.

Keywords Stochastic Frontier · Frontier production function · Unknown quantile level · Copula

1 Introduction

The advantages of the stochastic frontier model (SFM) has been proved in various fields of study since the introduction by Aigner et al. (1977). The model has gained an increasing popularity as it can predict the effects of inputs on the output and then measure the technical efficiency of the production process, ranging from zero to one. In agriculture, the technical efficiency implies the ability of a farm or farmer to produce the maximum possible output from a given bundle of inputs or to produce a particular amount of output from the minimum amount of inputs. In other words, the efficiency is simply computed by the ratio of the observed output over the inputs

P. Maneejuk (✉)
Center of Excellence in Econometrics, Chiang Mai University, Chiang Mai, Thailand
e-mail: mparavee@gmail.com

W. Yamaka
Faculty of Economics, Center of Excellence in Econometrics, Chiang Mai University, Chiang Mai, Thailand
e-mail: woraphon.econ@gmail.com

N. Ngoc Thach et al. (eds.), *Data Science for Financial Econometrics*,
Studies in Computational Intelligence 898,
https://doi.org/10.1007/978-3-030-48853-6_31

445

of production. Practically, the model has two error components which are assumed to be independent of each other. The first is the noise error, U_i, which captures the exogenous variable shock effects; and the second error is actually the inefficiency term, V_i, which implies the loss of output occurred during the production process.

However, many recent studies have questioned the validation of the assumption of independence. For example, Das (2015) mentioned that inefficiency at time t might depend on the noise at time $t - 1$. Therefore, it is likely to have some relationship between the inefficiency term and the noise error. Smith (2008) and Wiboonpongse et al. (2015) later introduced a linkage between these two error terms constructed by Copulas. The Copulas is a joint distribution joining marginal distribution of the two error terms of the SFM. With that in mind, the Copula-based stochastic frontier model was introduced and appeared to be a promising field of study for researchers. (see, for example, Tibprasorn et al. 2015; Pipitpojanakarn et al. 2016; Maneejuk et al. 2017; Huang et al. 2018).

Nevertheless, both conventional SFM and the Copula-based SFM may not be robust to the heteroscedasticity and outliers as they still regress to the mean of the data. Kaditi and Nitsi (2010) mentioned that the traditional SFM still had some problems in the inefficiency distribution. It is not sensitive towards outliers and thereby leading to a misspecification problem. Therefore, they introduce an alternative model called quantile regression, which is firstly introduced by Koenker (2005), to estimate the production function in the agricultural sector. Also, the work of Duy (2015), and Gregg and Rolfe (2016) can confirm the usefulness of the model. They found that this model is suitable for efficiency estimation when dealing with heterogeneity in the country-level data. In other words, this model provides a better description of the production of efficient producers or firms at different quantile levels compared to the mean regression. Besides, Bernini et al. (2004) suggested that the quantile approach could improve the performance of the conventional SFM and provide new information. The SFM, along with quantile approach, allows us to observe asymmetric impacts of inputs on output and different efficiency levels across quantiles. Therefore, Bernini et al. (2004) introduced the stochastic frontier quantile model (SFQM). Later, this model was extended with Copulas as a copula-based SFQM by Pipitpojanakarn et al. (2016).

Although the copula-based SFQM model is useful for investigating the production function and its efficiency at the quantile of interest, we doubt that among all the quantile levels, which one is fit to the data? The traditional quantile regression model will be useful if we want to focus on some or a particular quantile level. In practice, one may consider many quintile levels to capture all possible effects of inputs on output; however, which quantile level is more worthwhile to be observed? When using the quantile model and extensions, we usually select a specific level of quantile, for instance $\tau = 0.1, 0.5, 0.9$, to represent each class/group. This, in turn, leads to a problem with quantile selection. Most of the existing studies use model selection criteria, like Akaike Information Criterion (AIC), to select the relatively best model at different quantile level or select the level of quantile based on experience and knowledge, which may, in turn, leads to a biased result. Instead of fitting the model with a specified quantile, this study, therefore, treats the quantile level as an unknown parameter and simultaneously estimates it with other parameters in the model. This idea is also considered by Jradi et al. (2019). They examined the optimal quantile level when estimating the technical efficiency. However, they still assumed that the two error components: the noise error, U_i, and the inefficiency term, V_i, are independent of each other. Therefore, in this study, we further develop the idea of unknown quantile SFM by incorporating it with the Copula. To summarize, this study develops the copula-based SFQM with unknown quantile. As we believe that regarding the quantile as an unknown parameter will enhance the performance of the model because the model is estimated at the most probable quantile value. In addition, a linkage between the two error terms is also constructed by Copulas.

The remainder of our paper unfolds as follows, Notation and a SFQM are defined in Sect. 2, with a brief review of technical efficiency that has been typically estimated in SFQM studies. In Sect. 3, we explain the model estimation followed by a simulation study shown in Sect. 4 to confirm the accuracy of the estimator. Then, in Sect. 5, we apply the model to analyze agricultural production in Asia, including the data set provided by Coelli (1996). Finally, Sect. 6 presents some concluding remarks.

2 Methodology

2.1 The Copula-Based Stochastic Frontier Quantile Model (SFQM) with Unknown Quantile

In this section, we explain the main idea and show the equation form of the stochastic frontier quantile model (SFQM) with unknown quantile. The model is similar to a simple linear regression model, but the error term is separated into two parts denoted by U_i and V_i, thus the model is written as

$$Y_i = X'_{ik}\beta(\tau) + E_i, \quad i = 1, \ldots, I, \tag{1}$$

$$E_i = U_i - V_i, \tag{2}$$

$$U \sim ALD(0, \sigma_U^2, \tau), \tag{3}$$

$$V \sim ALD^+(0, \sigma_V^2, \tau). \tag{4}$$

Y_i is the output variable and X'_{ik} is a $(I \times K)$ matrix of K input variables. The coefficient $\beta(\tau)$ presents the $(K \times I)$ matrix of estimated coefficients given a quantile level τ, where $\tau \in [0, 1]$. E_i is the composed error term, namely noise U_i, and inefficiency V_i. To include the quantile function properties in the model U_i is assumed to have the Asymmetric Laplace distribution (ALD) with mean zero and variance σ_U^2 while V_i is assumed to be truncated positive ALD with mean zero and variance σ_V^2. The functional form and explanation of this truncated distribution refer to the work of Horrace and Parmeter (2015). Generally, we can construct the quantile function and obtain different coefficients at different levels of the quantile. Also, the technical efficiency (TE) can vary across quantiles under the assumption of ALD. However, this study considers the quantile level as the unknown parameter, which may lead to a better estimation of variables and the more accurate results.

In the context of the SFQM, technical efficiency (TE) can be expressed as the ratio of the observed output (Y_i) to the frontier output (Y_i^*) conditional on the levels of inputs used by the country. The technical efficiency at a given quantile or $TE(\tau)$ is therefore

$$TE(\tau) = \exp(X'_{ik}\beta(\tau) + U_i - V_i)/\exp(X'_{ik}\beta(\tau) + U_i). \tag{5}$$

As the two error components are assumed to be correlated; therefore, Copula approach is conducted to join the marginal distributions of U_i and V_i.

2.2 Copula Functions

According to Sklars theorem, Copula approach becomes the most potent tool in joining the bivariate cumulative distribution function of the vector with marginals that are uniform. Copula contains all the information related to the dependence structure of its components. Thus, it is easier to construct the complicated dependence of the two error components in our model. Let the two random variables be x_1 and x_2, with margins F_1 and F_2, respectively. If F_1 and F_2 are continuous, the joint distribution function of the two-dimensional random vector x_1 and x_2 is given by

$$H(x_1, x_2) = P(X_1 \leq x_1, X_2 \leq x_2) = P(F_1(X_1) \leq F_1(x_1), F_2(X_2) \leq F_2(x_2)). \tag{6}$$

We assume that the marginal distributions of x_1 and x_2 are uniform on the interval [0, 1]. The term $H(x_1, x_2)$ is a joint distribution of x_1 and x_2 and there exists a unique copula $C(u, v)$, thus

$$H(x_1, x_2) = C(F_1(x_1), F_2(x_2)), \tag{7}$$

Note that u and v are the uniform distribution on [0, 1]. The corresponding bivariate density function is obtained by differentiating Eq. (7) concerning x_1 and x_2, thus we have

$$f(x_1, x_2; \theta) = c(F_1(x_1), F_2(x_2); \theta) f_1(x_1; \delta_1) f_2(x_2; \delta_2), \tag{8}$$

where θ is all parameters of the copula based model, θ is the dependence variable; and δ_1 and δ_2 are the estimated parameters of the marginal distributions.

3 Estimation of the Copula-Based SFQM

As the SFQM consists of two error components and we do not have observations on U_i and V_i. Therefore, the estimation of this model can be achieved by the simulated maximum likelihood (Smith 2008; Wiboonpongse et al. 2015). According to the joint copula density function Eq. (8), we can rewrite the density function of SFQM as

$$f(U, V; \theta) = c(u, v; \theta) f_1(U; \delta_1) f_2(V; \delta_2), \qquad (9)$$

where the terms $f_1(U; \delta_1)$ and $f_2(V; \delta_2)$ are the probability density function (pdf) of the error U_i and V_i, respectively. $c(u, v; \theta)$ is the copula density function. While the U_i and V_i cannot be observed directly in SFQM, but we can simulate V_i from the truncated positive ALD with mean zero, variance σ_V^2 and quantile τ. Then, we can obtain $U_i = V + E$ and rewrite the Eq. (9) as

$$f(U, V) = f(U, V + E) = c(v, v + e) f_1(U) f_2(V + E). \qquad (10)$$

To join the two error components, we consider the various bivariate copula functions from Elliptical or Archimedean class. Thus, we introduce six copula functions: Normal copula, Frank copula, Student-t copula, Gumbel copula, Clayton copula, and Joe copula as joint distribution function in copula part. Note that the marginal distribution in copula function is uniform [0, 1] so that the simulated $V_i + E_i$ and V_i are transformed by cumulative ALD and cumulative truncated ALD, respectively. To obtain the density function of the SFQM, we follow Smith (2008) and marginalize out v, and then yield

$$\begin{aligned} f(E \,|\, \theta) &= \int_0^M f(v, e) dv \\ &= E_V(f(V_i + E_i) c_\theta(F_1(V), F_2(V + E))). \end{aligned} \qquad (11)$$

As discussed in the works of Smith (2008) and Das (2015), the model is faced with multiple integrals in the likelihood. So, they suggested employing a simulated likelihood estimation function to attain the asymptotically unbiased simulators for the integrals. Thus, the exact likelihood function based on a sample $\{V_i^R\}_{r=1}^R$ of the size R, where $V_i^R = (V_1^R, \ldots, V_N^R)$, is expressed by

$$\begin{aligned} & L\left(\beta(\tau), \sigma_{(V+E)}, \sigma_V, \theta\right) \\ & = \sum_{i=1}^I \left(\frac{1}{R} \sum_{j=1}^R \log f(V_{ij} + E_{ij} \,|\, \delta) c\left(F_1(v_{ij}), F_2(V_{ij} + E_{ij}) \,|\, \theta\right)\right). \end{aligned} \qquad (12)$$

Then, we can maximize the log likelihood function, as shown in Eq. (12) using the BFGS algorithm to obtain the optimal parameter estimates.

4 Simulation Study

In this section, we conduct simulation studies to evaluate the performance and accuracy of the simulated maximum likelihood for fitting the proposed model. The data is generated from the following model.

$$Y_i = 1 + 2X_{1i} - 3X_{2i} + U_i - V_i, \tag{13}$$

where the two error components are generated from $U_i \sim ALD(0, \sigma_U, \rho)$ and $V_i \sim ALD^+(0, \sigma_V, \rho)$ distributions with mean zero and variance $\sigma_U = 1$ and $\sigma_V = 0.5$, respectively. Also, two simple examples of our model are considered in the simulation experiments.

- Simulation 1: We start with a simulation of uniform u and v by setting the Gaussian copula dependency $\theta = 0.5$. Then, the obtained u and v are transformed into two errors U_i and V_i through the quantile functions $F_{ALD}^{-1}(0, 1, \tau)$ and $F_{ALD^+}^{-1}(0, 0.5, \tau)$, respectively. The covariates X_{1i} and X_{2i} are simulated independently from the $uni(0, 1)$. Then, we can simulate Y_i from Eq. (13). In the first simulation, we set $\tau = (0.1, 0.2, 0.3, 0.4, 0.5, 0.6, 0.7, 0.8, 0.9)$ and generate 100 data sets with $N = 200$.
- Simulation 2: We consider six copula families to model the dependence structure of the two error components of the SFQM. We start with the simulation of uniform u and v by setting the true copula dependency θ equal to 0.5 for Gaussian, Students t, and Clayton copulas and equal to 1.5 for the remaining copulas, Gumbel, Joe, and Frank. For the case of Student's t copula, we set the true value of the additional

Table 1 Simulation result of Simulation 1

Parameter	TRUE	0.1	0.2	0.3	0.4	0.5	0.6	0.7	0.8	0.9
α	1	1.858	0.670	1.669	1.125	1.150	0.923	1.619	1.205	0.579
		(0.848)	(0.253)	(0.462)	(0.015)	(0.538)	(0.434)	(0.404)	(0.228)	(0.425)
β_1	2	1.561	1.853	1.741	1.912	1.827	1.831	1.865	1.747	2.699
		(0.914)	(0.192)	(0.235)	(0.117)	(0.155)	(0.188)	(0.250)	(0.025)	(0.055)
β_2	-3	-2.098	-3.143	-2.996	-3.335	-3.093	-3.160	-3.298	-3.067	-3.677
		(0.559)	(0.226)	(0.249)	(0.127)	(0.155)	(0.242)	(0.233)	(0.002)	(0.255)
σ_U	1	3.381	2.829	2.460	1.165	2.408	0.992	0.981	0.340	1.107
		(0.370)	(0.117)	(0.285)	(0.055)	(0.297)	(0.090)	(0.313)	(0.015)	(0.055)
σ_V	0.5	0.100	0.283	1.209	0.425	1.649	0.503	0.948	0.055	0.837
		(0.034)	(0.117)	(0.298)	(0.002)	(0.392)	(0.079)	(0.082)	(0.455)	(0.102)
θ	0.5	0.661	0.458	0.659	0.487	0.586	0.594	0.512	0.649	0.500
		(0.175)	(0.089)	(0.014)	(0.001)	(0.027)	(0.084)	(0.008)	(0.010)	(0.000)
τ		0.092	0.100	0.255	0.354	0.596	0.594	0.654	0.842	0.956
		(0.023)	-0.123)	(0.006)	(0.546)	(0.015)	(0.083)	(0.007)	(0.032)	(0.002)

( ) denotes standard deviation of the parameter estimates

Table 2 Simulation result of Simulation 2

$N=200$	Student-t		Clayton		Gumbel		Frank		Joe	
Parameter	TRUE	Est.	TRUE	Est.	TRUE	Est.	TRUE	Est.	TRUE	Est.
α	1	1.879	1	0.664	1	0.785	1	1.216	1	1.216
		(0.714)		(0.133)		(0.210)		(0.166)		(0.166)
β_1	2	1.860	2	1.799	2	2.415	2	1.836	2	1.836
		(0.146)		(0.656)		(0.046)		(0.148)		(0.148)
β_2	-3	-3.111	-3	-3.071	-3	-3.907	-3	-3.132	-3	-3.132
		(0.147)		(0.178)		(0.876)		(0.214)		(0.214)
σ_U	1	1.094	1	1.143	1	1.121	1	0.914	1	0.914
		(0.561)		(0.166)		(0.082)		(0.080)		(−0.080)
σ_V	0.5	0.395	0.5	0.476	0.5	0.411	0.5	0.607	0.5	0.607
		(0.040)		(0.199)		(0.100)		(0.043)		(0.043)
θ	0.5	0.482	0.5	0.581	1.5	2.125	1.5	1.48	1.5	2.48
		(0.007)		(0.096)		(0.386)		(0.385)		(0.385)
τ	0.5	0.607	0.5	0.583	0.5	0.518	0.5	0.475	0.5	0.475
		(0.019)		(0.096)		(0.013)		(0.029)		(0.029)
$N=400$	Student-t		Clayton		Gumbel		Frank		Joe	
Parameter	TRUE	Est.	TRUE	Est.	TRUE	Est.	TRUE	Est.	TRUE	Est.
α	1	1.438	1	0.815	1	0.910	1	0.934	1	1.014
		(0.415)		(0.064)		(0.091)		(0.092)		(0.135)
β_1	2	2.023	2	2.041	2	2.037	2	2.024	2	2.028
		(0.116)		(0.145)		(0.013)		(0.125)		(0.131)
β_2	-3	-3.132	-3	-3.110	-3	-3.146	-3	-3.117	-3	-3.056
		(0.114)		(0.102)		(0.109)		(0.101)		(0.110)
σ_U	1	0.953	1	1.123	1	1.164	1	0.975	1	0.555
		(0.368)		(0.147)		(0.055)		(0.070)		(0.047)
σ_V	0.5	0.453	0.5	0.205	0.5	0.572	0.5	0.572	0.5	0.584
		(0.035)		(0.136)		(0.076)		(0.047)		(0.040)
θ	0.5	0.430	0.5	0.641	1.5	1.285	1.5	1.553	1.5	2.061
		(0.003)		(0.052)		(0.327)		(0.072)		(0.113)
τ	0.5	0.558	0.5	0.600	0.5	0.608	0.5	0.442	0.5	0.565
		(0.032)		(0.077)		(0.025)		(0.026)		(0.026)

() denotes standard deviation of the parameter estimates

degree of freedom equal to 4. The data is generated similarly to the first simulation case, but the sample sizes are $N=200$ and $N=400$. Also, we fix $\tau=0.5$ for all simulation data.

The result of the simulation study is shown in Tables 1 and 2. These two tables show the mean and the standard deviation of parameter estimates from 100 datasets. The results from both simulation studies show a satisfactory performance of the model. The estimated mean parameters are entirely close to the true values for both cases. Besides, the standard deviations of parameter estimates, as shown in the braces, are acceptable. We also draw up the sample size from $N=200$ and $N=400$ and observe that the performance of the one step estimator is enhanced with increasing sample size, as reported in Table 2. To summarize, the overall simulation results indicate that the proposed model has acceptable performance, and the simulated maximum likelihood is suitable for fitting our model.

5 Application

In this section, the copula-based SFQM with unknown quantile is applied to analyze the agricultural production in Asia using the data set provided by Pipitpojanakarn et al. (2016). Another application is based on the data set of Coelli (1996) who early utilized the classical stochastic frontier model. The second application is conducted for comparison by using the same data set.

5.1 Application 1: Agricultural Production in Asia

We apply our proposed methods to analyze the data set about Asian agricultural production (see, Pipitpojanakarn et al. 2016). The data set consists of labor force (L), fertilizer (F), and agricultural area (A), as inputs for production, and agricultural product (Q) of 44 Asian countries in 2016. Thus, a specified model is as follows.

$$\ln Q_i = \alpha + \beta_1 \ln L_i + \beta_2 \ln F_i + \beta_3 \ln A_i + U_i - V_i, \tag{14}$$

where i corresponds to country. The best fitting copula for joining the two errors is selected first using the Akaike Information Criterion (AIC). In this study, we consider only six copula families consisting of Gaussian, Student-t, Gumbel, Frank, Clayton, and Joe. Besides, we also consider another two competing models: independent U_i and V_i based SFQM (denoted by Independence model), and the stochastic frontier (mean regression) model of Coelli (1996) (denoted by Tim Coellis Frontier model). The results are presented in Table 3. It is observed that the best model is the Student-t copula-based model where the value of AIC is minimum with 5.472.

Table 4 shows the estimated results of the Student-t copula-based SFQM with unknown quantile. Also, we conduct the Student-t copula-based SFQM with a given quantile level $\tau = 0.5$ and Coellis (Classical) stochastic frontier model and apply the same data set to these two models for making a comparison. For the proposed model, the unknown quantile is estimated to be 0.3910. The coefficient of each input variable appears to have the same direction, but the magnitude of the effects is quite different across models. For example, the coefficients of the labor force (L) are 0.0094 in the Student-t copula-based SFQM with $\tau = 0.391$, 0.0031 for Student-t copula-based SFQM with $\tau = 0.5$, and 0.0001 in the Coellis stochastic frontier, respectively. As the mixed results are obtained, we employ the AIC to select the best model given the data set, including the BIC. We find that our proposed model can outperform the other two models.

Table 3 AIC values of each Copula-based SFQM, independence model, and Tim Coellis Frontier model

Model	AIC
Gaussian	6.687
Student-t	5.472
Clayton	93.956
Gumbel	99.864
Frank	98.811
Joe	189.680
Independence model	282.190
Tim Coelli's Frontier model	14.055

Table 4 Estimation results

	Student-t-based SFQM with unknown quantile	Student-t based SFQM with quantile =0.5	Coelli's Frontier
α	4.473	4.880	4.794
	(0.093)	(0.015)	0.107
β_1	0.009	0.003	0.000
	(0.010)	(0.003)	(0.000)
β_2	0.072	0.159	0.001
	(0.014)	(0.005)	(0.002)
β_3	0.023	0.149	0.003
	(0.011)	(0.002)	(0.002)
σ_U	0.100	0.010	0.103
	(0.010)	(0.001)	(0.038)
σ_V	0.010	0.518	0.627
	(0.012)	(0.005)	(0.249)
θ	0.961	0.695	
	(0.026)	(0.003)	
df	4.969	4.984	
	(0.004)	(0.180)	
τ	0.391		
	(0.004)		
AIC	**5.472**	9.483	14.055
BIC	**21.529**	18.089	24.76

Note ( ) denotes the standard error and df is the degree of freedom

Table 5 AIC values of each Copula-based SFQM, independence model, and Tim Coellis Frontier model

	AIC
Gaussian	46.272
Student's t	47.459
Clayton	138.455
Gumbel	46.374
Frank	41.285
Joe	40.759
Independent model	48.858
Tim Coelli's Frontier model	44.054

5.2 Application 2: Production Analysis of Tim Coelli (1996)

In the second study, we use the data set of firm production collected and provided by Coelli (1996). The main objective of this study is to investigate the performance of our model and test whether our model can produce a better fit for the Coellis data set compared to his model (Coellis stochastic frontier). The data set contains cross-sectional data of 60 firms, consisting of firm output (Q) and inputs, namely capital (K) and labor (L). We use the same procedure as in Application 1 to perform the comparison. As a consequence, the results reported in Table 5 show that Joe copula-based SFQM with unknown quantile outperforms the other models as it has the smallest value of AIC.

As also reported in Table 6, the unknown quantile is estimated to be 0.3552. The AIC and BIC advocate that Joe copula-based SFQM with $\tau = 0.3552$ is the best model among all. On the other hand, the model with a given quantile level equal to 0.5 has the weakest performance in this study. The proposed model appears to be slightly better than the conventional model of Coelli (1996) in terms of AIC and BIC values. However, when considering the magnitude of some parameters, we find that the estimated values of σ_U and σ_V are quite different among the three models, but the results from Coellis Frontier model seem to have the most heterogeneity, indicating a bias of the technical efficiency in the model.

5.3 Estimates of Technical Efficiency Score

After estimating the production function model in Application 1 and 2, we then measure the score of technical efficiency using the Eq. (5). The idea of technical efficiency (TE) is simply the effectiveness of a firm to produce an output with a given set of inputs, such as labor, capital, and technology. A firm is considered a technically efficient firm if it can produce maximum output by using the minimum quantity of

Table 6 Estimation results

Model	Joe-based SFQM	Joe-based SFQM with quantile = 0.5	Coelli's Frontier
α	0.643	0.603	0.562
	(0.045)	(0.016)	(0.203)
β_1	0.273	0.279	0.281
	(0.027)	(0.013)	(0.048)
β_2	0.525	0.522	0.536
	(0.007)	(0.007)	(0.045)
σ_U	0.107	0.000	0.217
	(0.025)	(0.002)	(0.064)
σ_V	0.176	0.301	0.797
	(0.025)	(0.009)	(0.136)
θ	2.221	2.237	
	(0.050)	(0.014)	
τ	0.355		
	(0.014)		
AIC	40.759	210.099	44.055
BIC	50.249	213.589	54.526

Note () denotes the standard error

Table 7 Summary statistics of technical efficiency scores of application 1 and 2

Application 1	Student-t-based SFQM (black line)	Student-t-based SFQM with $\tau = 0.5$ (red line)	Coelli's Frontier (green line)
Min	0.623	0.145	0.468
Max	0.944	0.808	0.944
Mean	0.704	0.360	0.813
Application 2	Joe-based SFQM (black line)	Joe-based SFQM with $\tau = 0.5$ (red line)	Coelli's Frontier (green line)
Min	0.234	0.300	0.351
Max	0.879	0.905	0.937
Mean	0.714	0.750	0.741

inputs. In the context of the stochastic frontier analysis, technical efficiency can be defined as the ratio of the observed output to the corresponding frontier output conditional on the levels of inputs used by the firm, as written in Eq. (5). We use this concept to measure the TE score of both Applications and show the results in Fig. 1 and Table 7.

As presented in Fig. 1, we calculate the TE score from three different models: (1) copula-based SFQM with an estimated value of quantile = 0.391 (black line), (2) copula-based SFQM with a given quantile = 0.5 (red line), and 3) the Coellis

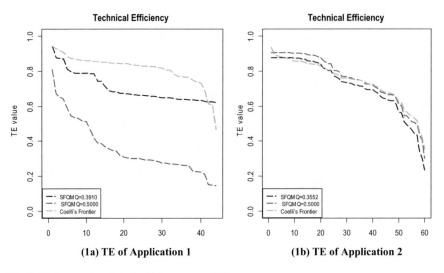

Fig. 1 Estimates of technical efficiency score (TE)

Frontier model (green line). Figure 1a illustrates the TE scores of Application 1, Asias agricultural production. The estimated TE scores are distinct across different models. However, Sect. 5.1 shows that the copula-based SFQM with unknown quantile is the best model with the lowest AIC. The TE scores obtained from the other two models somewhat deviate from that of the copula-based SFQM with unknown quantile. The Coellis Frontier model (green line) provides the overestimated TE score, while the copula-based SFQM with a given quantile provides the underestimated TE score.

Unlike Application 1, the TE scores of the three models are quite similar in Application 2. Table 7 shows that the average scores are approximately 0.7 for all models, and the maximum and minimum values are not much different. This result corresponds to the results of Sect. 5.2 that the performance of the models is not distinct in terms of AIC value. However, we can observe that the performance of our suggested model is slightly better than the other models as the value of AIC is smallest.

5.4 Conclusion

In this study, we extend the analysis of the copula-based stochastic frontier quantile model (SFQM) by considering the quantile level to be a parameter estimate. The model becomes more reliable and useful for investigating the causal effect of inputs on output as well as the technical efficiency (TE) measurement. Our parameter estimates and TE can reflect the data at the most suitable quantile value.

The simulation study is used to assess the performance of our proposed model. Two simulation studies are conducted, and the results confirm that the proposed model can perform well with accuracy since the means of almost all parameters are close to the true values. Finally, we point out the applicability of the model and methods through the two data sets. We compare the proposed model with the copula-based stochastic frontier at the median quantile and the conventional stochastic frontier model of Coelli (1996). The overall results from these applications show that the proposed model can outperform the rivals in terms of minimum AIC and BIC, indicating the gain from introducing the quantile level as a parameter estimate to the copula-based SFQM is substantial.

References

Aigner, D., Lovell, C. K., & Schmidt, P. (1977). Formulation and estimation of stochastic frontier production function models. *Journal of econometrics*, *6*(1), 21–37.

Bernini, C., Freo, M., & Gardini, A. (2004). Quantile estimation of frontier production function. *Empirical Economics*, *29*(2), 373–381.

Coelli, T. (1996). *A guide to FRONTIER Version 4.1: A computer program for stochastic Frontier production and cost function estimation.* CEPA Working Paper 96/08, http://www.uq.edu.au/economics/cepa/frontier.php. University of New England.

Das, A. (2015). Copula-based Stochastic Frontier model with autocorrelated inefficiency. *Central European Journal of Economic Modelling and Econometrics*, *7*(2), 111–126.

Duy, V. Q. (2015). Access to credit and rice production efficiency of rural households in the Mekong Delta.

Gregg, D., & Rolfe, J. (2016). The value of environment across efficiency quantiles: A conditional regression quantiles analysis of rangelands beef production in north Eastern Australia. *Ecological Economics*, *128*, 44–54.

Horrace, W. C., & Parmeter, C. F. (2015). A Laplace stochastic Frontier model. *Econometric Reviews*.

Huang, T. H., Chen, K. C., & Lin, C. I. (2018). An extension from network DEA to copula-based network SFA: Evidence from the US commercial banks in 2009. *The Quarterly Review of Economics and Finance*, *67*, 51–62.

Kaditi, E. A., & Nitsi, E.(2010). Applying regression quantiles to farm efficiency estimation. In *2010 Annual Meeting* (pp. 25–27).

Jradi, S., Parmeter, C. F., & Ruggiero, J. (2019). Quantile estimation of the stochastic frontier model. *Economics Letters*, *182*, 15–18.

Koenker, R. (2005). Quantile regression, no. 9780521845731 in Cambridge Books.

Maneejuk, P., Yamaka, W., & Sriboonchitta, S. (2017). Analysis of global competitiveness using copula-based stochastic frontier kink model. In *Robustness in econometrics* (pp. 543–559). Cham: Springer.

Pipitpojanakarn, V., Maneejuk, P., Yamaka, W., & Sriboonchitta, S. (2016). Analysis of agricultural production in Asia and measurement of technical efficiency using copula-based stochastic frontier quantile model. In *International Symposium on Integrated Uncertainty in Knowledge Modelling and Decision Making* (pp. 701–714). Cham: Springer.

Smith, M. D. (2008). Stochastic frontier models with dependent error components. *The Econometrics Journal*, *11*(1), 172–192.

Tibprasorn, P., Autchariyapanitkul, K., Chaniam, S., & Sriboonchitta, S. (2015). A copula-based stochastic frontier model for financial pricing. In *International Symposium on Integrated Uncertainty in Knowledge Modelling and Decision Making* (pp. 151–162). Cham: Springer.

Wiboonpongse, A., Liu, J., Sriboonchitta, S., & Denoeux, T. (2015). Modeling dependence between error components of the stochastic frontier model using copula: Application to intercrop coffee production in Northern Thailand. *International Journal of Approximate Reasoning*, *65*, 34–44.

Forecasting Volatility of Oil Prices via Google Trend: LASSO Approach

Payap Tarkhamtham, Woraphon Yamaka, and Paravee Maneejuk

Abstract Google search volume index has been widely used as a proxy of investor attention. In this study, we use Google search volume index to forecast energy return volatility. In order to find the keywords, we start with glossary of crude oil terms provided by the U.S. Energy Information Administration (EIA) and then add keywords based on Google Search's suggestions. Then, we arrive at a set of 75 Google keywords as a proxy of investor attention. As there are a large number of keywords to be considered, the conventional method may not be appropriate for the statistical inference. Thus, we propose using the LASSO to deal with this problem. Finally, we also compare the predictive power of LASSO with three types of stepwise method.

Keywords LASSO · Google search volume index · Forecasting

1 Introduction

Oil is a lifeblood of global economy and one of the most valuable commodities used in the world. According to the International Energy Agency (IEA 2018), the world oil consumption in 2016 was around 3,908 million tons. The world oil consumption at present predominantly about 64.5% is for transportation purpose, with 7.8% and 5.4% for industrial use and residential use respectively. In the recent years, oil plays an important role in our daily life as well as economy and social development. A fluctuation in world oil price generally leading to uncertainty would result in an economic recession within most countries and vice versa. While an increase in

P. Tarkhamtham · W. Yamaka (✉) · P. Maneejuk
Faculty of Economics, Center of Excellence in Econometrics,
Chiang Mai University, Chiang Mai, Thailand
e-mail: woraphon.econ@gmail.com

P. Tarkhamtham
e-mail: payap.tar@gmail.com

P. Maneejuk
e-mail: mparavee@gmail.com

© The Editor(s) (if applicable) and The Author(s), under exclusive license   459
to Springer Nature Switzerland AG 2021
N. Ngoc Thach et al. (eds.), *Data Science for Financial Econometrics*,
Studies in Computational Intelligence 898,
https://doi.org/10.1007/978-3-030-48853-6_32

oil price can contribute to a higher cost of the production process in oil-importing countries, a decrease in oil price can slow down the economic development in oil-producing countries. Thus, it is necessary to study the inherent mechanisms of oil price fluctuations in order to reduce the potential risks of oil price volatility (Wang et al. 2018). In the general economic approach, oil prices are determined by the movement of demand and supply of the oil. In addition, as the oil market is mainly influenced by regional activities which are born out of cooperation from various countries having different needs and environments, there are many more direct and indirect factors affecting oil price such as fundamental demand for oil, extreme weather, wars, OPEC's production capacity and policy, oil reserves of world's major consumers, and alternative energy. Guo and Ji (2013) investigated the relationship between oil price and market concerns which consist of oil demand, financial crisis, and Libya war and confirmed the relationship between oil price and these factors.

Moreover, numerous studies also reveal that the oil price volatility could be attributable to the oil-related events. For example, Ji and Guo (2015) examined the effects of oil-related events on oil price volatility and they concluded that oil price responded differently in each oil-related event. Zavadska et al. (2018) investigated the volatility patterns of Brent crude oil spot and futures prices during four major crises. They found that oil volatility reached its peak and was directly affected by the Gulf war and 2001 US terrorist attack period. However, the Asian and the Global Financial Crises seem to have an indirect impact on the oil market.

As we mentioned earlier, there are various factors affecting the oil price as well as its volatility. In this study, we forecast only the oil price volatility as it is central to asset pricing, asset allocation, and risk management (Zhang et al. 2019). The provision of accurate predictions of oil price volatility has always been a core activity for both investors and policymakers. The task has become even more important after the Great Economic Recession. However, it is difficult to obtain an accurately predicted result since the data of many factors are not up to date and most of them are often released with a delay of few months or quarters. Fortunately, in the recent years, the Google Search Volume Index (GSVI) has been proposed and widely used as the new predictor. Da et al. (2011) conducted a study which was the first ever to use the GSVI as a proxy of investor attention. They applied GSVI to predict stock price and the results showed that GSVI can capture the attention of retail investor and increase the accuracy of Russell 3000 stocks prediction. Then, many researchers have incorporated GSVI in the price volatility prediction models. Smith (2012) predicted volatility in the stock exchange market. Bijl et al. (2016) and Preis et al. (2013) used GSVI to forecast stock return and they confirmed that GSVI can predict the stock return, but the relationship between GSVI and stock return has changed overtime. Han et al. (2017) investigated the predictive power of GSVI in oil price movements and the results showed that GSVI exhibits statistically and economically significant in-sample and out-of-sample forecasting power to directly forecast oil prices. Wang et al. (2018) forecasted crude oil price using extreme learning machine and found that Google-based model outperforms the models based on the traditional data.

However, we find there are many keywords of GSVI and it is difficult to find the best keyword to enhance the prediction power. In the previous works, investigators typically estimated standard time series autoregressive models and augmented them by the use of GSVI as the exogenous variable. In their approach, some keyword was selected by the researchers and then they estimated the forecasting model based on one selected keyword. In the estimation aspect, it is difficult to include all the keywords when using the traditional estimations, says Ordinary Least Squares (OLS) or Maximum Likelihood Estimation (MLE). These methods are unable to produce sparse solutions (shrink insignificant coefficient to be zero) and are not computable when the number of coefficients is larger than that the number of observations. Thus, it will take quite an effort for researcher to get the best Google keyword.

To improve OLS and MLE, various techniques, such as subset selection and ridge estimation, are introduced. However, these two estimators are found to have some limitations. Although, the subset selection has an ability to select some related predictors by either retaining or dropping predictors from the model, it suffers from computational limitations, when the number of predictors is large. Ridge regression is a continuous process that shrinks coefficient; however, it does not set any coefficient to zero that makes the model hard to interpret. To deal with this problem, we use Least Absolute Shrinkage and Selection Operation (LASSO) which was introduced by Tibshirani (1996). LASSO is an estimation method that involves penalizing the absolute size of the regression coefficients and subset selection. It shrinks coefficient like a Ridge, and it can set some coefficient to be zero. From this advantage of the LASSO, many researchers have used LASSO in various fields like Gauthier et al. (2017) that tried to predict sound quality and the result showed that the LASSO algorithm is the most powerful candidate estimation for the predictive sound quality model. Wang et al. (2018) predicted ship fuel consumption by using LASSO regression, Support Vector Regression, and Gaussian Process, found that LASSO outperforms other traditional methods. Chen et al. (2018) applied LASSO to real time forecasts of endemic infectious diseases. Moreover, Panagiotidis et al. (2018) studied the determinants of Bitcoin return using LASSO.

Consequently, in this study, we forecast the oil return volatility based on LASSO estimation techniques. We consider 75 Google keywords as the exogenous variables in the volatility prediction model. To the best of our knowledge, this is the first attempt in applying the LASSO estimation for dealing with a large number of google keywords and also finding an appropriated keyword to predict the crude oil return volatility.

The rest of this paper is organized as follows: Sect. 2 explains the methodology. Section 3 describes the data. Section 4 shows the results of the study. Finally, Sect. 5 concludes.

2 Methodology

In this paper, we employ a Generalized Autoregressive Conditional Heteroskedasticity (GARCH) model to quantify the volatility of the oil return, then the LASSO regression model is used to investigate the prediction power of GSVI in forecasting the volatility of crude oil return. In this section, we firstly briefly review the GARCH model, then LASSO estimation is presented to fit our prediction model under the linear regression framework.

2.1 Generalized Autoregressive Conditional Heteroskedasticity (GARCH)

Generalized Autoregressive Conditional Heteroskedasticity (GARCH) model was proposed by Bollerslev (1986). The GARCH(p, q) is defined as follows

$$y_t = u + \sigma_t z_t \tag{1}$$

$$\sigma_t^2 = \omega + \alpha_1 \varepsilon_{t-1}^2 + \beta_1 \sigma_{t-1}^2 \tag{2}$$

where y_t is the observed variable and u is the intercept term of the mean Eq. (1) σ_t is the conditional volatility at time t, $z_t \sim N(0, 1)$ is a sequence of i.i.d. standardized residuals with mean equal to zero and variance equal to one. The restrictions are $\omega > 0, \alpha_1, \beta_1 > 0$ and $\alpha_1 + \beta_1 \leq 1$. The α_1 and β_1 are known as ARCH and GARCH parameters, respectively.

2.2 Least Absolute Shrinkage and Selection Operator (LASSO)

Prior to our describing the LASSO estimation in the regression framework, let us briefly explain the concept of the conventional OLS estimation. Considering the linear regression model:

$$y_t = x_t' \beta + \varepsilon = \beta_0 + \beta_1 x_1 + \cdots + \beta_p x_p + \varepsilon_t, \tag{3}$$

where y_t and x_t are the dependent and independent variables at time t. $\beta = (\beta_0, \beta_1, \ldots, \beta_p)$ is the estimated parameter. The parameter β_j, $j = 1, \ldots, p$, represents the effect size of independent variable j on the dependent variable. ε_t is referred to as the error at time t whereas the distribution is normal. The parameters of this model are estimated by loss function minimization. Thus, we can obtain the optimal parameters by using OLS

$$\left(\hat{\beta}_{ols} \right) = \underset{\beta}{\text{argmin}} \sum_{t=1}^{T} \left(y_t - \beta_0 - \sum_{j=1}^{P} \beta_j x_{tj} \right)^2, \tag{4}$$

Then, we turn our attention to LASSO which is the extension of the OLS. In this estimation, the loss function of OLS is minimized with the additional constraint so-called lasso penalty, which is $|\beta| \leq t$. Thus, this is an estimation method that involves penalizing the absolute size of the regression coefficients (Tibshirani 1996). In other words, LASSO minimizes the mean squared error subject to a penalty on the absolute size of coefficient estimates.

$$\left(\hat{\beta}_{Lasso}\right) = \underset{\beta}{\text{argmin}} \sum_{t=1}^{T} \left(y_t - \beta_0 - \sum_{j=1}^{P} \beta_j x_{tj} \right)^2, \tag{5}$$

$$\text{subject to } ||\beta||_1 \leq t,$$

where $t \geq 0$ is a tuning parameter. Tibshirani (1996) motivates the LASSO with two major advantages over OLS. First, due to the nature of the lasso-penalty, the lasso could force parameter estimates to be zero. In this sense, this estimator can remove irrelevant independent variables out from the regression model. Secondly, LASSO can outperform OLS in terms of prediction accuracy due to the bias-variance trade off.

In this study, we take these advantages of the LASSO and apply to our regression problem, where the dependent variable is the conditional variance σ_t^2, which is obtained from the GARCH process, and independent variable is Google keywords. Thus, we can rewrite the optimization in Eq. (5) as

$$\left(\hat{\beta}\right) = \underset{\beta}{\text{argmin}} \sum_{t=1}^{T} \left(\sigma_t^2 - \beta_0 - \sum_{j=1}^{P} \beta_j x_{tj} \right)^2, \tag{6}$$

$$\text{subject to } ||\boldsymbol{\beta}|| \leq t$$

where x_{tj} is the Google keywords j at time t. $||\boldsymbol{\beta}|| = |\beta_0| + |\beta_1| + |\beta_2| + \cdots + |\beta_p|$ is the l_1-norm or penalty function. This penalty function can condense some nonsignificant coefficients into zero, therefore, some irrelevant Google keywords are removed from the model. In practice, the constraint optimization in Eq. (6) can be rewritten as follows

$$\hat{\beta} = \underset{\beta}{\text{argmin}} \left\{ \frac{1}{2} \sum_{i=1}^{N} \left(\sigma_t^2 - \beta_0 - \sum_{j=1}^{P} \beta_j x_{ij} \right)^2 + \lambda ||\boldsymbol{\beta}|| \right\}, \tag{7}$$

where $\lambda > 0$ is a complexity parameter that controls the amount of shrinkage. For all t the solution for α is $\hat{\alpha} = \bar{y}$. We can assume without loss of generality that $\bar{y} = 0$ and hence omit α. The parameter $t \geq 0$ controls the amount of shrinkage that is applied to the estimates. Let $\hat{\beta}_j$ be the full least squares estimates and let $t_0 = \sum \left| \hat{\beta}_j \right|$. Values of $t < t_0$ will cause shrinkage of the solutions towards 0, and some coefficients may be exactly equal to 0.

Fig. 1 Time series
cross-validation (Hyndman
2016)

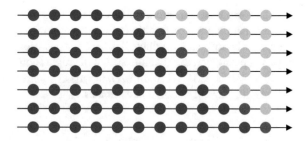

To estimate the regularization parameter λ, we adopt time-series cross-validation which introduced by Hyndman (2016). This procedure is sometimes known as "evaluation on a rolling forecasting origin". Figure 1 illustrates the series of training and test sets, where the blue is training sets, and the red is test sets. For example, we start with 5 observations of the training set and 1 observation of test set. We then perform run each set until the last observations. Thus, we can compute the forecast accuracy by averaging over the test sets. Finally, we can get a range of λ and then select the optimal λ value that corresponds to the lowest RMSE.

3 Data Description

In order to predict the crude oil return volatility, we use the weekly data covering the period of 2014–2018 collected from Thompson Reuter Data Stream. Then, we transform the crude oil price to rate of return by $r_t = \ln(p_t/p_{t-1})$ where p_t represents crude oil price at time t and p_{t-1} represents price at time $t-1$. GARCH(1, 1) model (Eq. 1) is then applied to forecast the volatility of the return of oil r_t. As our aim is to improve the volatility prediction of oil, we also collect the GSVI from the www.google.com. GSVI data is normalized by Google Trends with value between 0 and 100, where 100 is assigned to the date within the interval where the peak of search for that query is experienced, and zero is assigned to dates where search volume for the term has been below a certain threshold.

As there are many Google keywords related to the oil price movement, we reduce the number of keywords by considering the keywords in the glossary of crude oil terms provided by the U.S. Energy Information Administration (EIA). Then, some keywords are selected as follows: first we filter out the words for which Google Trends does not have enough data to generate time series. Second, we add keywords based on Google Search's suggestions and the literature review. Google search keywords are shown in Table 1.

Table 1 Google search keywords

Carbon tax	Extraction of petroleum	Oil company use
Clean energy	Fracking	Oil crisis
Climate change	Fuel oil	Oil export
Compressed natural gas	Gas well	Oil export ban
Crude oil	Gasoil	Oil opec
Crude oil consumption	Gasoline	Oil price
Crude oil news	Gasoline price	Oil prices news
Crude oil price	Global warming	Oil reserves
Crude oil production	Going green	Oil shale
Crude oil stocks	Greenhouse gas	Oil stocks
Diesel fuel	Hurricane	Oil supplies
Directional drilling	Iraq war	Oil well
Drilling	Kyoto protocol	Opec
Drilling rig	Libyan war	Opec conference
Energy crisis	Methane	Petroleum
Energy efficiency	Natural gas	Petroleum consumption
Energy independence	Natural gas price	Petroleum industry
Energy information administration	Natural gas production	Petroleum reservoir
Energy market	Natural resource	Pipeline
Energy sector	New York mercantile exchange	Renewable energy
Energy security	Offshore drilling	Russian ruble
Energy tax	Offshore fracking	Shale gas
Environment	Oil	Shale oil
Ethanol fuel	Oil and gas	Syrian conflict
Ethanol price	Oil and natural gas corporation	West texas intermediate

4 Empirical Results

In this section, we firstly present the result of GARCH model. The estimated result is shown in Table 2. We can observe that the estimated $\alpha_1 + \beta_1 = 0.9724$, indicating a high unconditional volatility of crude oil return and the stationary process of the GARCH is satisfied. Then, we extract the crude oil volatility return and illustrate in Fig. 2.

Table 2 GARCH model estimation

	μ	ω	α_1	β_1
GARCH(1, 1)	0.0004 (0.0027)	0.0001 (0.00006)	0.1777 (0.0531)	0.7947 (0.0589)

Fig. 2 Crude oil return and volatility

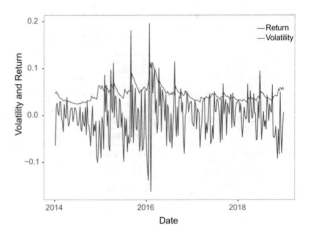

In Fig. 2, the volatility line (red) and the return line (blue) of the crude oil are plotted. We can observe that the volatility reaches the peak in 2016 corresponding to the large drop in crude oil price in the same period. We find that the resilient shale production and rising output from OPEC pile up the oil storage and thereby oppressing the price of the crude oil. As we can see in Fig. 2, crude oil return fell in early 2016. According to this plot, the volatility obtained from GARCH(1, 1) is quite reliable.

After that, we turn our attention to find the Google keywords that significantly explain the volatility of the crude oil return. In addition, we conduct in-sample and out-of-sample forecasts to investigate the performance of the LASSO estimation. To do this, we split the data into two sets which known as training set (in-sample) and test set (out-of-sample). We consider 80% of the full length of the data to be training set and the rest 20% is used as the test set. In this experiment, we also consider other variable selection methods, namely Stepwise method (Efroymson 1960). This method consists of three main approaches: Full Stepwise (Bidirectional), Forward Stepwise and Backward Stepwise methods. Note that these three methods have an advantage similar to that of the LASSO in terms of the ability to build the significant independent variable set for the regression model. However, the variable selection is based on the frequentist testing and the estimation is still based on the OLS.

To compare the in-sample and out-of-sample forecasts among the four competing estimations, we employ four loss functions of Root Mean Squared Error (RMSE), Median Absolute Error (MDAE), Mean Absolute Error (MAE) and Mean Absolute Percentage Error (MAPE), which can be expressed as

$$\text{RMSE} = \sqrt{\frac{1}{n} \sum_{i=1}^{n} \left(\sigma_i - \hat{\sigma}_i\right)^2}, \tag{8}$$

Table 3 In-sample LASSO and three conventional estimations

LASSO	Full-Stepwise	Forward stepwise	Backward stepwise
Carbon tax (1.836e−05)	Crude oil price (0.000801)	Crude oil price (0.000981)	Carbon tax (0.000181)
Crude oil news (1.989e−04)	Drilling (−0.000555)	Directional drilling (0.000144)	Crude oil price (0.000890)
Crude oil price (3.136e−04)	Energy tax (0.000142)	Drilling (−0.000475)	Directional drilling (0.000148)
Methane (2.189e−05)	Oil OPEC (0.000744)	Energy tax (0.000216)	Drilling (−0.000444)
Oil price news (1.414e−06)	Oil stocks (−0.000708)	Gasoline (0.000230)	Energy crisis (0.000131)
Syrian conflict (1.690e−05)	OPEC (−0.000709)	Methane (0.000113)	Energy tax (0.000190)
	Petroleum (0.000377)	Oil (−0.000279)	Natural resource (0.000192)
		Oil stocks (−0.000676)	Oil stocks (−0.000600)
		Pipeline (−0.000154)	Pipeline (−0.000143)
			Renewable energy (−0.000276)

Note The value in ( ) shows the coefficient of the model

$$MDAE = \text{median} \left| \sigma_i - \widehat{\sigma}_{ii} \right|, \tag{9}$$

$$MAE = \sqrt{\frac{1}{n} \sum_{i=1}^{n} \left| \sigma_i - \widehat{\sigma}_{ii} \right|}, \tag{10}$$

$$MAPE = \frac{1}{n} \sum_{i=1}^{n} \left| 100 \times \left(\frac{\sigma_i - \widehat{\sigma}_i}{\sigma_i} \right) \right| \tag{11}$$

where n is the number of observations, σ_i is the actual volatility value and $\widehat{\sigma}_i$ is the predicted volatility value.

The training data set is used to estimate our regression problem. Table 3 presents the in-sample estimation results for the different regression estimations. Table 4 shows the estimation performance.

The results of parameter estimates are shown in Table 4. We can see that LASSO selects only six Google keywords which are fewer than other technique. Backward Stepwise estimation selects ten Google keywords which is the largest number of keyword selection. We then look at the in-sample estimation performance, we can find that LASSO estimation is not superior to the other three competing estimations with respect to RMSE, MDAE, MAE and MAPE. Interestingly, it seems that all

Table 4 In-sample performance measures of LASSO and three conventional methods

Performance measures	LASSO	Forward stepwise	Backward stepwise	Full stepwise
RMSE	**0.010945**	0.008678	0.008508	0.008671
MDAE	**0.0062**	0.005	0.0051	0.0046
MAE	**0.007871**	0.006562	0.00643	0.00645
MAPE (%)	**16.25787**	13.66831	13.42196	13.50668

Table 5 Out-of-sample performance measures of LASSO and three conventional methods

Performance measures	LASSO	Forward stepwise	Backward stepwise	Full stepwise
RMSE	**0.008027**	0.009958	0.009012	0.009240
MDAE	**0.006300**	0.007600	0.007500	0.007200
MAE	**0.006862**	0.008482	0.007788	0.007856
MAPE (%)	**18.45003**	21.69907	21.26921	20.44368

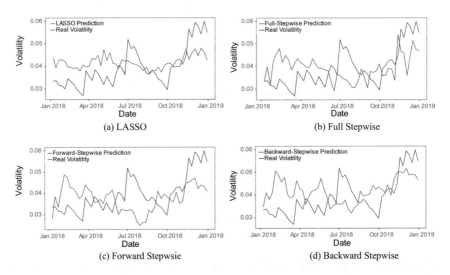

(a) LASSO

(b) Full Stepwise

(c) Forward Stepwsie

(d) Backward Stepwise

Fig. 3 Out-of-sample volatility forecast of oil return based on LASSO (**a**), full stepwise (**b**), forward stepwise (**c**), and backward stepwise (**d**)

Stepwise techniques are overfit since the in-sample estimation performance is greater than the out-sample estimation performance.

The result of the out-of-sample forecast for four methods are provided in Table 5. Apparently, LASSO has the lowest value for all error measures. This means that regression model estimated by LASSO is found to be superior to the three Stepwise techniques. Finally, the out-of-sample volatility forecasts are plotted in Figs. 3 and 4. The three stepwise estimations give very similar volatility forecasts, while the result

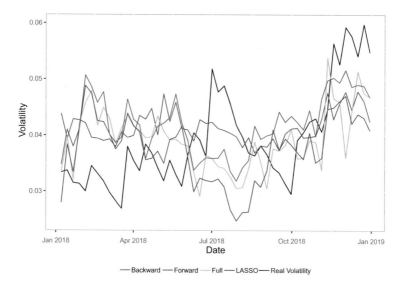

Fig. 4 Comparing out-of-sample volatility forecasts based on four estimations: realized volatility (black line), LASSO (purple line, full stepwise (green line), forward stepwise (blue line) and backward stepwise (red line)

of LASSO estimation shows a weak variation of the volatility forecast over this 20% out-of-sample set.

In essence, according to our in-sample and out-of-sample volatility forecasts of crude oil return, there are many evidences confirming the high forecasting power on crude oil return volatility under the regression framework. We find that six out of seventy-fifth Google keywords are the potential keywords which can improve the volatility forecasting.

5   Conclusion

In this study, we forecast the volatility of crude oil return under the linear regression context. To quantify the volatility, we apply GARCH model to forecast the conditional volatility and treat it as the dependent variable in the model. Then, we select 75 Google keywords to be the predictors in the forecasting model. However, it is quite difficult to incorporate all these predictors in the model as the conventional estimation like OLS may not provide the reliable result and it is sometime impractical when the number of parameter estimates is larger than the number of the observations. To deal with this limitation of the conventional method, we have suggested a LASSO estimation.

This method has been examined in both in-sample and out-of-sample forecast. We also consider three more conventional methods which are similar to LASSO in doing the variable selection. These methods are Full Stepwise, Forward Stepwise and Backward Stepwise methods. The results have shown that the forecasting model estimated by LASSO is superior to those by the traditional methods.

Acknowledgements This research work was partially supported by Chiang Mai University.

References

Afkhami, M., Cormack, L., & Ghoddusi, H. (2017). Google search keywords that best predict energy price volatility. *Energy Economics, 67*, 17–27.

Bijl, L., Kringhaug, G., Molnr, P., & Sandvik, E. (2016). Google searches and stock returns. *International Review of Financial Analysis, 45*, 150–156.

Bollerslev, T. (1986). Generalized autoregressive conditional heteroskedasticity. *Journal of Econometrics, 31*(3), 307–327.

Chen, Y., Chu, C. W., Chen, M. I., & Cook, A. R. (2018). The utility of LASSO-based models for real time forecasts of endemic infectious diseases: A cross country comparison. *Journal of Biomedical Informatics, 81*, 16–30.

Da, Z., Engelberg, J., & Gao, P. (2011). In search of attention. *The Journal of Finance, 66*(5), 1461–1499.

Efroymson, M. A. (1960). Multiple regression analysis. In A. Ralston & H. S. Wilf (Eds.), *Mathematical methods for digital computers*. Wiley.

Gauthier, P. A., Scullion, W., & Berry, A. (2017). Sound quality prediction based on systematic metric selection and shrinkage: Comparison of stepwise, LASSO, and elastic-net algorithms and clustering preprocessing. *Journal of Sound and Vibration, 400*, 134–153.

Guo, J. F., & Ji, Q. (2013). How does market concern derived from the Internet affect oil prices? *Applied Energy, 112*, 1536–1543.

Han, L., Lv, Q., & Yin, L. (2017). Can investor attention predict oil prices? *Energy Economics, 66*, 547–558.

Herrera, A. M., Hu, L., & Pastor, D. (2018). Forecasting crude oil price volatility. *International Journal of Forecasting, 34*(4), 622–635.

Hyndman, R. J. (2016). Measuring forecast accuracy. In M. Gilliland, L. Tashman, & U. Sglavo (Eds.), *Business forecasting: Practical problems and solutions* (pp. 177–183). Canada: Wiley.

International Energy Agency. (2018). Key world energy statistics.

Ji, Q., & Guo, J. F. (2015). Oil price volatility and oil-related events: An Internet concern study perspective. *Applied Energy, 137*, 256–264.

Lee, C. Y., & Cai, J. Y. (2018). LASSO variable selection in data envelopment analysis with small datasets. *Omega*. https://doi.org/10.1016/j.omega.2018.12.008.

Panagiotidis, T., Stengos, T., & Vravosinos, O. (2018). On the determinants of bitcoin returns: A LASSO approach. *Finance Research Letters, 27*, 235–240.

Preis, T., Moat, H. S., & Stanley, H. E. (2013). Quantifying trading behavior in financial markets using Google trends. *Scientific Reports, 3*(1684), 1–6.

Smith, G. P. (2012). Google internet search activity and volatility prediction in the market for foreign currency. *Finance Research Letters, 9*(2), 103–110.

Tibshirani, R. (1996). Regression shrinkage and selection via the LASSO. *Journal of the Royal Statistical Society: Series B (Methodological), 58*(1), 267–288.

Wang, J., Athanasopoulos, G., Hyndman, R. J., & Wang, S. (2018). Crude oil price forecasting based on internet concern using an extreme learning machine. *International Journal of Forecasting*, *34*(4), 665–677.

Wang, S., Ji, B., Zhao, J., Liu, W., & Xu, T. (2018). Predicting ship fuel consumption based on LASSO regression. *Transportation Research Part D: Transport and Environment*, *65*, 817–824.

Zavadska, M., Morales, L., & Coughlan, J. (2018). Brent crude oil prices volatility during major crises. *Finance Research Letters*. https://doi.org/10.1016/j.frl.2018.12.026.

Zhang, Y., Wei, Y., Zhang, Y., & Jin, D. (2019). Forecasting oil price volatility: Forecast combination versus shrinkage method. *Energy Economics*, *80*, 423–433.

Innovation and Earnings Quality: A Bayesian Analysis of Listed Firms in Vietnam

Minh Le, Tam Tran, and Thanh Ngo

Abstract Researchers have well documented the positive effects of innovation on growth and productivity. However, innovation might be associated with unexpected consequences, especially in term of earnings quality. In this paper, we employ the Bayesian linear regression to shed light on the relationship between innovation and earnings quality in the Vietnamese financial market. While other researches often use the Research and Development (R&D) expenditures, the number of patents and citations as measures of innovation, in this study, we use a new metric of innovation which is obtained from the Cobb-Douglas production function. Our research results indicate that firms with more innovations can lead to a decrease in earnings quality. Based on the findings, we recommend practitioners, especially investors and analysts should pay more attention to innovative firms when making investment decisions because these firms are more likely to participate in earnings manipulation.

Keywords Innovation · Earnings quality · Bayesian linear regression

1 Introduction

A great investment deal requires a lot of efforts. In order to earn a high rate of return, investors and analysts seek profitable firms with potential growth. One of the main factors that help firms achieve growth is innovation. Researchers have documented

M. Le (✉)
International Economics Faculty, Banking University of Ho Chi Minh City, HCMC, Vietnam
e-mail: minhlq@buh.edu.vn

T. Tran
Faculty of Finance, Banking University of Ho Chi Minh City, HCMC, Vietnam
e-mail: tamtm@buh.edu.vn

T. Ngo
Faculty of Finance and Banking, University of Economics and Law, VNU-HCM, Vietnam
e-mail: thanhnp@uel.edu.vn

N. Ngoc Thach et al. (eds.), *Data Science for Financial Econometrics*, Studies in Computational Intelligence 898, https://doi.org/10.1007/978-3-030-48853-6_33

that innovation can lead to economic growth (Bilbao-Osorio and Rodriguez-Pose 2004), and an increase in productivity (Hall et al. 2009). Innovation is also a predictor of firms with high future returns (Hirshleifer et al. 2013). Therefore, innovative firms might be ones that investors and analysts should put their eyes in.

Focusing on innovation alone is not enough to have a successful investment, however. In addition to search for innovative firms, investors and analysts must know at what price they should pay for the stocks. With regard to this manner, they need to perform the valuation process in which they employ earnings number as an important input to forecast the firms' future cash flows. The reliability of the forecast depends on how "good" an earnings number is or speaking differently, the quality of the earnings number.

As we have discussed, both innovation and earnings quality play critical roles in making investment decisions. But more interestingly, there might be a relationship between them. This relationship has been found in the literature. Lobo et al. (2018) show empirical evidence to point out that innovation can negatively impact financial reporting quality, meaning a firm with higher innovation can lead to greater earnings management. This happens because innovation creates opaque information environment which induces managers to engage in opportunistic acts. In another work, Kouaib and Jarboui (2016) document that the CEO's personal characteristics can result in real earnings management in the higher innovative firms. These papers illustrate that innovative environment can "trigger" managers to take part in earnings management.

Although the relationship between innovation and earnings quality has been explored, nevertheless, it is still underdeveloped. This especially becomes true when putting it in the context of the Vietnamese financial market. The Vietnamese stock market has been established since 2000, until now, it has been approximately 20 years in development. During this period, the market has witnessed the emergence of fraudulent financial reporting. In 2014, 80% of the Vietnamese listed firms have to restate their earnings number after audit (Vietstock 2014). This phenomenon has been continued and negatively affected the transparency of the market and eventually, the welfare of the investors. Despite the seriousness of the issue, the term "earnings quality" is quite new among Vietnamese analysts, investors, and researchers as well. Besides, because the Vietnamese stock market is still developing and there are many opportunities ahead, the market has become an ideal place to attract investment from innovative firms. However, as we have mentioned above, innovation, in turn, can be associated with greater earnings management and therefore, lower earnings quality. This interesting connection inspires us to dig deeply into the topic to answer whether the relationship between innovation and earnings quality exist in the Vietnamese stock market since both innovation and earnings quality are significant to the development of the market.

Our paper contributes to the literature in several important ways. First, we use a new measure to be a proxy for innovation. While prior studies often use the research and development (R&D) expenditures, the number of patents and citations for innovation measurement, in this paper, we try to gauge the technical efficiency by employing the Cobb-Douglas function and use it as the proxy for innovation. Second, our paper is the first one exploring the relationship between innovation and earnings quality in the Vietnamese financial market, a frontier market with a myriad of opportunities for development ahead. Third, we apply the Bayesian linear regression to investigate the mentioned relationship, which is a quite new approach in the field of earnings quality.

The rest of the paper is organized as follows: Section 2 reviews the literature and develops our hypothesis while Sect. 3 describes our research design. After that, we present data statistics and discuss the results in Sect. 4 before Sect. 5 concludes the paper.

2 Literature Review and Hypothesis Development

2.1 Earnings Quality Definition

In the accounting literature, there are two prominent terminologies associated with earnings manipulation: the first one is "earnings management" and the second one is "earnings quality". The definitions of these two terms are not the same, however, they are related to each other.

Healy and Wahlen (1999) stands on the perspective of standard setters to define the term "earnings management". According to their definition, "earnings management" occurs when managers use judgment in financial reporting and in structuring transactions to alter financial reports to either mislead some stakeholders about the underlying economic performance of the company or to influence contractual outcomes that depend on reported accounting numbers. (p. 368 Healy and Wahlen 1999). Hence, earnings management often involves the actions of manipulating financial reports with the aim of misleading stakeholders.

The second often-used term is earnings quality. Dechow and Schrand (2004) takes the point of view of the analyst to state that "a high-quality earnings number is one that accurately reflects the company's current operating performance, is a good indicator of future operating performance, and is a useful summary measure for assessing firm value." (p. 5). They also emphasize that an earnings number with high quality should be able to capture the intrinsic value of the firm.

Combining the definitions of the two terms, one can see that if earnings management occurs then that earnings number will have low quality and vice versa. A high-quality earnings number implies the absence of earnings management. In this paper, we mostly use the term earnings quality and when we use it, we refer to the definition by Dechow and Schrand (2004). In some occasions, we also use the term earnings management and in those cases, we understand it as the definition by Healy and Wahlen (1999).

2.2 Innovation Definition

There are numerous articles on innovation, both the determinants and the impacts of its. However, the term "innovation" is rarely defined clearly. Therefore, we would like to firstly mention about the definition of innovation to make our context become clearer.

According to Drucker (1993), "nnovation is the specific tool of entrepreneurs, the means by which they exploit change as an opportunity for a different business or a different service." (p. 19). In an effort to establish an international standard framework, the Organization for Economic Co-operation and Development (OECD) publishes the Oslo manual to provide guidelines on innovation statistics. In the latest version—the OECD Oslo Manual 2018 OECD (2018), innovation is defined as follows:

> An innovation is a new or improved product or process (or combination thereof) that differs significantly from the unit's previous products or processes and that has been made available to potential users (product) or brought into use by the unit (process). (p. 20)

The OECD also categorizes the innovation into two main types: the product innovations and the business process innovations. In this paper, we refer innovation to both two types of innovation and our measure for innovation, which will be discussed later in Sect. 3, reflects the combination of both the product innovations and the business process innovations.

2.3 The Relationship Between Innovation and Earnings Quality

Our idea to examine the impact of innovation on earnings quality originates from the debate in the literature about the accounting treatment of research and development (R&D) expenditures. The debate is about whether the R&D costs should be capitalized or expensed. The decision to capitalize or expense R&D investments should take into consideration "the trade-off between relevance and uncertainty of future benefits" from those investments (Kothari et al. 2002, p. 356). With regard to the uncertainty of future benefits of R&D expenditures, Kothari et al. (2002) complement the existing literature by providing evidence to show that R&D investments lead to more variability in future earnings than property, plant, and equipment (PP&E) investments. The results found by Kothari et al. (2002) imply an interesting connection between innovation and earnings quality. Because innovative firms often invest a significant amount on R&D and acquire more intangible assets than tangible assets and as what has been found by Kothari et al. (2002), one can suspect that a firm with high innovations can have low earnings quality. Later studies affirm this phenomenon. Srivastava (2014) conducts research to answer whether a decrease in earnings quality over the past 40 years is attributable to the economic changes or the changes in accounting rules. He points out that intangible intensity can negatively affect earnings quality. More specifically, his evidence shows that firms with higher intangible intensity exhibit more fluctuations in both revenues and cash flows, and therefore, lower earnings quality. He cites previous studies to explain that the accounting treatment requiring immediately expensing intangible assets can lead to more volatilities in expenses and eventually, in earnings number. Recently, Lobo et al. (2018) examine the impact of innovation on financial reporting quality more directly than previous studies. The authors show a negative connection between innovation and the quality of financial reports. This means firms with higher innovation are more likely to engage in earnings management, and hence, result in lower financial reporting quality. They argue that innovative firms often have more R&D investments as well as intangible assets. However, those assets are subject to information complexity, which create more difficulties for stakeholders to value and monitor those innovative firms. Therefore, firms with higher innovations often associate with poorer information environment. This opaque information environment, in turn, tempts managers to involve in greater earnings management which consequently reduces the financial reporting quality.

Motivated by prior studies, we are eager to examine whether there is a relationship between innovations and earnings quality in the Vietnamese financial market and in what direction that relationship occurs. Based on the reasoning of previous papers, we propose the following hypothesis for our study:

H1: Other things being equal, innovation negatively affect earnings quality.

3   Research Design

3.1   Earnings Quality Measures

The dependent variable in our research is Earnings Quality. In an excellent literature review, Dechow et al. (2010) summarize numerous ways to measure earnings quality. In our paper, we follow Kim et al. (2012) and use five different metrics to be proxies for earnings quality.

Discretionary accruals
Our first mean to gauge earnings quality is the modified Jones model which is firstly introduced by Dechow et al. (1995). There are some versions of the modified Jones model and we opt for the one proposed by Kothari et al. (2005). Specifically, the model that we use is:

$$\frac{ACC_{it}}{ASS_{it-1}} = \beta_0 \cdot \left(\frac{1}{ASS_{it-1}} \right) + \beta_1 \cdot \left(\frac{\Delta REV_{it} - \Delta REC_{it}}{ASS_{it-1}} \right)$$
$$+ \beta_2 \cdot \left(\frac{PPE_{it}}{ASS_{it-1}} \right) + \beta_3 \cdot ROA_{it-1} + \epsilon_{it} \tag{1}$$

In Eq. (1), the subscripts i and t denote firm i and year t, respectively. The detailed calculation of all variables in model (1) are as follows:

- ACC_{it}: Total Accruals of firm i in year t and equals to ΔCurrent assets$_{it}$ − ΔCurrent liabilities$_{it}$ − ΔCash$_{it}$ + ΔShort term debt included in current liabilities$_{it}$ − Depreciation & armotization costs$_{it}$. In which, ΔCurrent assets$_{it}$ equals to Current assets$_{it}$ − Current assets$_{it-1}$. The similar calculations apply for ΔCurrent liabilities$_{it}$, ΔCash$_{it}$, ΔShort term debt included in current liabilities$_{it}$,
- ASS_{it-1}: Total Assets of firm i in year $t - 1$,
- ΔREV_{it}: The change in revenues of firm i in year t and equals to REV_{it} − REV_{it-1},
- ΔREC_{it}: The change in net receivables of firm i in year t and is calculated by REC_{it} − REC_{it-1}',
- PPE_{it}: Gross property, plant, and equipment of firm i in year t,
- ROA_{it-1}: Return on Assets of firm i in year $t - 1$ and equals to Income before extraordinary items of firm i in year $t - 1$ divided by Total Assets of firm i in year $t - 1$.

The residual from model (1) is our first measure of earnings quality. The intuition behind model (1) is total accruals can be divided into two parts: the nondiscretionary accruals and the discretionary accruals. The residual in model (1) evaluates the discretionary accruals which, in turn, is a suitable measure of earnings management (Dechow et al. 2010). This means firms with higher residual will have lower earnings quality. We denote the first earnings quality measure as DA which is abbreviated for discretionary accruals.

Abnormal discretionary expenses

In addition to the modified Jones model, we also employ other measures of earnings quality to allow for the robustness check of our research results. Following Kim et al. (2012), our second proxy for earnings quality is calculated from the model by Roychowdhury (2006):

$$\frac{DE_{it}}{ASS_{it-1}} = \beta_0 + \beta_1 \cdot \left(\frac{1}{ASS_{it-1}}\right) + \beta_2 \cdot \left(\frac{REV_{it}}{ASS_{it-1}}\right) + \epsilon_{it} \qquad (2)$$

In model (2), DE_{it} is the discretionary expenses of firm i in year t which is equal to the sum of the advertising and the selling, general and administrative (SG&A) expenses. Roychowdhury (2006) and Kim et al. (2012) calculate the discretionary expenses as the sum of R&D, advertising and SG&A expenses. However, we are unable to extract the R&D expenses from the financial statements of the Vietnamese listed firms, therefore, we exclude the R&D expenses from the calculation of the discretionary expenses. Also in model (2), ASS_{it-1} is the total assets of firm i in year $t - 1$ and REV_{it} is the revenues of firm i in year t. The residual from model (2) capture the abnormal discretionary expenses and is our second metrics of earnings quality. We call the second measure as ADE which stands for abnormal discretionary expenses.

Abnormal production costs

The third measure is abnormal production costs. According to Roychowdhury (2006), productions costs equal to the cost of goods sold ($COGS$) plus the change in inventory. To calculate this measure, we follow the procedure by Roychowdhury (2006) and Kim et al. (2012) to use the following equations:

$$\frac{COGS_{it}}{ASS_{it-1}} = \beta_0 + \beta_1 \cdot \left(\frac{1}{ASS_{it-1}}\right) + \beta_2 \cdot \left(\frac{REV_{it}}{ASS_{it-1}}\right) + \epsilon_{it} \tag{3}$$

and:

$$\frac{\Delta INV_{it}}{ASS_{it-1}} = \beta_0 + \beta_1 \cdot \left(\frac{1}{ASS_{it-1}}\right) + \beta_2 \cdot \left(\frac{\Delta REV_{it}}{ASS_{it-1}}\right) + \beta_3 \cdot \left(\frac{\Delta REV_{it-1}}{ASS_{it-1}}\right) + \epsilon_{it} \tag{4}$$

In Eqs. (3) and (4), $COGS_{it}$ is the cost of goods sold of firm i in year t, ASS_{it-1} is the total assets of firm i in year $t - 1$, REV_{it} is the revenues of firm i in year t, ΔINV_{it} is the change in the inventory of firm i in year t, and ΔREV_{it-1} is the change in the revenues of firm i in year $t - 1$. Based on Eqs. (3) and (4), we then regress the following model:

$$\frac{PC_{it}}{ASS_{it-1}} = \beta_0 + \beta_1 \cdot \left(\frac{1}{ASS_{it-1}}\right) + \beta_2 \cdot \left(\frac{REV_{it}}{ASS_{it-1}}\right) + \beta_3 \cdot \left(\frac{\Delta REV_{it}}{ASS_{it-1}}\right)$$
$$+ \beta_4 \cdot \left(\frac{\Delta REV_{it-1}}{ASS_{it-1}}\right) + \epsilon_{it} \tag{5}$$

where PC_{it} is the production costs of firm i in year t and equal to $COGS_{it}$ plus ΔINV_{it} (Roychowdhury 2006; Kim et al. 2012). The residual from model (5) is our third measure of earnings quality. We name the third measure as APC which is the abbreviation of abnormal production costs.

Abnormal operating cash flows

The fourth metric is to capture the abnormal operating cash flows, which is estimated from the following equation (Roychowdhury 2006; Kim et al. 2012):

$$\frac{OCF_{it}}{ASS_{it-1}} = \beta_0 + \beta_1 \cdot \left(\frac{1}{ASS_{it-1}}\right) + \beta_2 \cdot \left(\frac{REV_{it}}{ASS_{it-1}}\right) + \beta_3 \cdot \left(\frac{\Delta REV_{it}}{ASS_{it-1}}\right) + \epsilon_{it} \tag{6}$$

In Eq. (6), OCF_{it} is the operating cash flows of firm i in year t. The residual from model (6) reflects the abnormal operating cash flows and is our fourth metric of earnings quality. We label it as OCF.

Combined measure
We follow Cohen et al. (2008) and Kim et al. (2012) to calculate the final metric of earnings quality which is the combination of the abnormal operating cash flows, the abnormal production cost, and the abnormal discretionary expenses. The formula to estimate this combined measure is:

$$Combine = OCFAPC + ADE \qquad (7)$$

3.2   Innovation Measure

Prior studies often use the R&D investment, the number of patens and the number of citations to be proxies for innovation. Separate from the other researches, in this paper, we utilize the Cobb-Douglas function to measure technical efficiency and use it as our metric for innovation.

For estimation purpose, we assume that the Cobb-Douglas technology can be represented by the following three-input, one output production function:

$$Y_{it} = f(K_{it}, L_{it}, M_{it}, t) \qquad (8)$$

where Y_{it} stands for firm's output, K_{it}, L_{it} and M_{it} represent production inputs: capital, labor, and materials. Subscripts i and t refer to firm and year, respectively. We define firm's output as total revenues. We measure capital by the equity capital, labor by the cost of employees and materials by the value of intermediate input. The production function itself is allowed to shift over time to account for technological change. We specify production function into a log-linear transformation of a Cobb Douglas form:

$$\ln Y = \alpha + \beta_K \cdot \ln K_{it} + \beta_L \cdot \ln L_{it} + \beta_M \cdot \ln M_{it} + \beta_t \cdot t + \epsilon_{it} \qquad (9)$$

The different between actual Y_{it} and predicted one is the Total Factor Productivity (TFP). TFP can be negative if the actual value is smaller than the predicted one and positive vice versa.

3.3   Model Specification

To achieve the research objective of the paper, we employ the following model to investigate the relationship between innovation and earnings quality:

$$EQ_{it} = \beta_0 + \beta_1 \cdot TFP_{it} + \beta_2 \cdot SIZE_{it} + \beta_3 \cdot \left(\frac{EQUITY}{SIZE}\right)_{it} + \beta_4 \cdot CPI_t + \epsilon_{it} \qquad (10)$$

In Eq. (10), the subscript i denotes firm i and the subscript t denotes year t. EQ is our earnings quality measures, which include DA, ADE, APC, OCF, and $Combine$. The calculation of these variables has been described earlier in Sect. 3.1. TFP is Total Factor Productivity which is our proxy for innovation. Except for TFP, all the other variables in the right-hand side of Eq. (10) are control variables. These control variables include: $SIZE$ is the natural logarithm of total assets, $EQUITY/SIZE$ is equal to equity capital divided by total assets and CPI is the consumer price index. Equation (10) is estimated within industry.

In order to estimate the parameters in Eq. (10), rather than using the classical linear regression (or frequentist linear regression) for panel data, we utilize the Bayesian linear regression instead. While the classical linear regression produces a single point estimate for the parameters, the Bayesian linear regression, however, offers us a posterior distribution of the parameters (Koehrsen 2018). The posterior distribution of the parameters is calculated as follows (Koehrsen 2018):

$$P\left(\beta \mid y, X\right) = \frac{P\left(y \mid \beta, X\right) \cdot P\left(\beta \mid X\right)}{P\left(y \mid X\right)} \tag{11}$$

In Eq. (11), $P\left(\beta \mid y, X\right)$ is the posterior probability distribution of the parameters, conditional upon the inputs and outputs, $P\left(y \mid \beta, X\right)$ is the likelihood of the data, $P\left(\beta \mid X\right)$ is the prior probability of the parameters and $P\left(y \mid X\right)$ is a normalization constant (Koehrsen 2018).

4 Results Discussion

4.1 Data Description

This paper uses data from Thompson Reuters. We collect data for listed firms in Hochiminh Stock Exchange and Hanoi Stock Exchange. We exclude firms in financial services and real estate developers, and firms without financial information to compute the earnings quality. The final data includes 591 unique firms.

We present the sample distribution by Thomson Reuters Business Classification (TRBC) industry name in Table 1. The construction and engineering industry has 542 firm-year observations, representing 16.29% of the total sample. The construction materials follow with 305 observations and 9.17% of the sample. The food processing is the third heavily represented industry with 220 observations and 6.61% of the sample.

Table 2 describes the statistics of the variables used in this paper. Our key dependent variable is earnings quality. We employ five proxies of the quality, including absolute value of discretionary accruals (DA), abnormal discretionary expenses (ADE), abnormal production cost (APC), abnormal levels of operating cash flows (OCF), and a combined proxy ($Combine$) which is defined as the expected direc-

Table 1 Descriptive statistics of variables

TRBC industry name	Number of observations	Percentage of sample	Cumulative percent
Construction and engineering	542	16.29	16.29
Construction materials	305	9.17	25.46
Food processing	220	6.61	32.07
Oil and gas refining and marketing	130	3.91	35.98
Pharmaceuticals	118	3.55	39.53
Iron and steel	99	2.98	42.50
Independent power producers	97	2.92	45.42
Construction supplies and fixtures	85	2.55	47.97
Agricultural chemicals	76	2.28	50.26
Consumer publishing	76	2.28	52.54
Non-paper containers and packaging	75	2.25	54.79
Coal	65	1.95	56.75
Electrical components and equipment	59	1.77	58.52
Business support services	57	1.71	60.23
Marine port services	55	1.65	61.89
Tires and rubber products	54	1.62	63.51
Oil and gas transportation services	51	1.53	65.04
Paper packaging	51	1.53	66.58
Specialty mining and metals	50	1.50	68.08
Miscellaneous specialty retailers	47	1.41	69.49
Apparel and accessories	43	1.29	70.78
Communications and networking	43	1.29	72.08
Commodity chemicals	42	1.26	73.34
Home furnishings	42	1.26	74.60
Passenger transportation, Ground & Sea	41	1.23	75.83
Hotels, motels and cruise lines	38	1.14	76.98
Marine freight and logistics	37	1.11	78.09
Heavy electrical equipment	35	1.05	79.14
Auto vehicles, parts and service retailer	34	1.02	80.16
Fishing and farming	33	0.99	81.15
Industrial conglomerates	31	0.93	82.09
Commercial printing services	29	0.87	82.96
Ground freight and logistics	28	0.84	83.80
Oil related services and equipment	27	0.81	84.61
Textiles and leather goods	27	0.81	85.42
Household products	26	0.78	86.20
Industrial machinery and equipment	25	0.75	86.96
Brewers	24	0.72	87.68

(continued)

Table 1 (continued)

TRBC industry name	Number of observations	Percentage of sample	Cumulative percent
Computer and electronics retailers	23	0.69	88.37
Electric utilities	21	0.63	89.00
Diversified industrial goods wholesaler	20	0.60	89.60
Household electronics	20	0.60	90.20
Auto and truck manufacturers	17	0.51	90.71
Water and related utilities	17	0.51	91.22
Food retail and distribution	16	0.48	91.70
Homebuilding	16	0.48	92.19
Highways and rail tracks	15	0.45	92.64
Paper products	14	0.42	93.06
Distillers and wineries	13	0.39	93.45
Airport operators and services	12	0.36	93.81
Aluminum	12	0.36	94.17
Leisure and recreation	12	0.36	94.53
Non-alcoholic beverages	11	0.33	94.86
Office equipment	11	0.33	95.19
Tobacco	11	0.33	95.52
Business support supplies	10	0.30	95.82
Diversified chemicals	10	0.30	96.12
Shipbuilding	9	0.27	96.39
Auto, truck and motorcycle parts	8	0.24	96.63
Aerospace and defense	7	0.21	96.84
Diversified mining	7	0.21	97.05
Employment services	7	0.21	97.26
IT services and consulting	7	0.21	97.48
Integrated oil and gas	7	0.21	97.69
Natural gas utilities	7	0.21	97.90
Oil and gas drilling	7	0.21	98.11
Semiconductors	7	0.21	98.32
Courier, postal, air freight & Land-bas	6	0.18	98.50
Drug retailers	6	0.18	98.68
Medical equipment, supplies and distribution	6	0.18	98.86
Renewable energy equipment and services	6	0.18	99.04
Heavy machinery and vehicles	5	0.15	99.19
Airlines	4	0.12	99.31
Personal products	4	0.12	99.43

(continued)

Table 1 (continued)

TRBC industry name	Number of observations	Percentage of sample	Cumulative percent
Phones and handheld devices	4	0.12	99.55
Casinos and gaming	3	0.09	99.64
Forest and wood products	3	0.09	99.73
Appliances, tools and housewares	2	0.06	99.79
Mining support services and equipment	2	0.06	99.85
Specialty chemicals	2	0.06	99.91
Advanced medical equipment and technology	1	0.03	99.94
Life and health insurance	1	0.03	99.97
Personal services	1	0.03	100
Total	**3,327**	**100**	
Author calculations			

Table 2 Descriptive statistics of variables

Variable	Obs.	Mean	Std. Dev.	Min	Max
DA	3,258	0.1241	0.1331	0.0020	1.1994
ADE	3,258	0.0684	0.0657	0.0014	0.4152
APC	3,260	0.0794	0.0740	0.0012	0.6107
OCF	3,259	0.1121	0.1082	0.0019	0.7603
Combine	3,326	0.1144	0.2506	−0.8400	8.1986
TFP	3,259	0.0002	0.0254	−0.1485	0.1325
SIZE	3,260	6.1300	1.3659	3.0749	10.0283
EQUITY/SIZE	3,261	0.5032	0.2180	0.0807	0.9532
CPI (%)	3,327	6.5992	5.8326	0.6320	23.0560

Source Author calculations

tions of the three variables: ADE, APC and OCF. The *Combine* is computed as $OCF - APC + ADE$. The computation of DA, ADE, APC, and OCF have been presented in Sect. 3.1. All continuous variables are winsorized at the top and bottom 1% of their distributions.

DA shows a mean value of 0.1241, representing the magnitude of the discretionary accruals account for 12.41% of the lagged total assets. The value is abnormally higher than in other studies in developed countries. For example, the U.S sample data has a mean DA of 0.5

The key independent variable is innovation. Studies in developed countries use the R&D expenditures, the number of patents or citations to represent innovation, however, these information are not publicly available in Vietnam. We use total factor productivity (TFP) as a proxy for innovation because TFP is defined as an increase in output caused by factors other than traditional inputs (such as capital, labour, and materials). We capture the annual TFP of each firm relying on Cobb-Douglas production function as presented in Eqs. (8) and (9). The value of TFP ranges from -14.85 to 13.25%. A firm which receives a negative value of TFP has a lower level of innovation and vice versa.

Our control variables include firm size (measured by natural logarithm value of total assets), capital structure (measured by the ratio of equity to total assets), and inflation (measured as CPI). The mean value of firm size is 6.13 (or VND1540 billion in local currency). This paper adds the firm size because we think that small firms may have different manipulation behavior from big ones. The mean value of the capital structure ($EQUITY/SIZE$) is 50.32%. We consider capital structure because we expect that more leveraged firms may be more prone to earning quality than less leveraged firms. Finally, the average value of CPI is 6.6%. This paper uses CPI as an environmental variable because a change in CPI will result in the same directional change in capital cost and thus expense manipulation.

4.2 Baseline Results

The goal of this paper is to find how innovation is related to earnings quality. We employ a Bayesian analysis. The procedure answers our research question by expressing uncertainty about unknown parameters using probability. In other words, a parameter is summarized by an entire distribution of values instead of one fixed value as in classical frequentist analysis.

The baseline results are presented in Table 3. Panel A shows the association between innovation (proxied by TFP) and the absolute value of discretionary accruals (DA). The mean value of TFP's coefficient is positive at 0.0528. From this mean value, we can conclude that on average innovation has a positive correlation to earnings manipulation. In other words, more innovated firms are more likely to engage in manipulation which in turn, results in lower earnings quality. This means firms with higher innovations have poorer earnings quality. This result is consistent with our main hypothesis that innovation negatively impacts on earnings quality and also concurs the findings by Lobo et al. (2018). However, the 95th percentile equal-tailed interval has values ranging from -0.1253 to 0.2329. The interval implies that a certain number of coefficients obtain a negative value. And one of the interesting questions is what the estimated probability is for a coefficient with a positive value. Using hypothesis interval testing, we have the answer that the estimated probability of this interval from 0 to 0.2359 is 72.34%. This result provides evidence that 72.34% of times the TFP obtains a positive coefficient. This finding is a superior characteristic that frequentist hypothesis testing does not have. Following Panel A, we also check the MCMC convergence as shown in Chart 1. The autocorrelation does not show any specific patterns. The posterior distribution of the relationship between TFP and DA resembles the normal distribution. Therefore, we have no reason to suspect nonconvergence (Fig. 1).

Table 3 Baseline analysis

	Dependent variable: absolute value of discretionary accruals (DA)											
	PANEL A				Equal-tailed		PANEL B				Equal-tailed	
	Mean	Std. Dev.	MCSE	Median	[95 % Cred.	Interval]	Mean	Std. Dev.	MCSE	Median	[95 % Cred.	Interval]
TFP	0.0528	0.0904	0.001	0.0524	−0.1253	0.2329	0.0305	0.0915	0.0009	0.0303	−0.1512	0.2100
SIZE							−0.0025	0.002	0.0000	−0.0025	−0.0065	0.0015
EQUITY/SIZE							0.0145	0.0127	0.0003	0.0145	−0.0102	0.0039
CPI (%)							0.0007	0.0004	0.0000	0.0007	−0.0001	0.0015
Constant	0.1255	0.0052	0.0005	0.1252	0.1156	0.1362	0.1234	0.0176	0.0006	0.1282	0.0941	0.1627
Obs.				3,193						3,193		
Acceptance rate				0.8042						0.259		
Clustered by				industry						Industry		
	Probability of a positive coefficients						Probability of a positive coefficients					
TFP				0.7234						0.6607		
SIZE										0.2708		
EQUITY/SIZE										0.9485		
CPI (%)										0.9700		

Source Author calculations

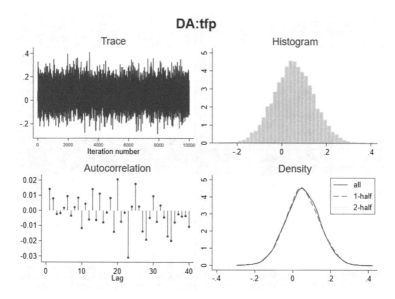

Fig. 1 Checking MCMC convergence

Panel B of Table 3 show results when we add control variables. The sign of TFP's coefficient remains positive at 0.0305. The variation of the coefficient is from -0.1512 to 0.2100 for 95th interval. The probability of getting a positive coefficient of TFP reduces moderately to 66.07% (from 72.34% in Panel A). $SIZE$'s coefficient obtains a negative mean value at -0.0025, suggesting that larger firms are less likely to manipulate earnings. However, the coefficient has a change of 27.08% to attain a positive value. Coefficient of capital structure is positive at 0.0145 in the 95th interval between -0.0102 to 0.0039. The positive mean value suggests firms with greater equity proportion are more likely to manipulate earnings. Finally, CPI's mean value coefficient is positive at 0.0007, suggesting that firms are more likely to engage in the manipulation during the time of high inflation. And the chance of being positive of the coefficient is 97%.

4.3   Robustness

In the robustness test, we employ four proxies, namely abnormal discretionary expenses (ADE), abnormal production cost (APC), abnormal levels of operating cash flows (OCF), and a combined proxy ($Combine$), of manipulation as dependent variables. Overall results show that the mean value of the coefficient of TFP remains positive. Table 4 provides results for robust checks. We remove information on standard deviation, MSCE, median, and equal-tailed interval for abbreviation (we can

Table 4 Robustness check

	(1)	(2)	(3)	(4)	(5)	(6)	(7)	(8)
	Dependent variables							
	ADE		**APC**		**OCF**		**Combine**	
TFP	0.1440	0.0311	0.1447	0.1223	0.2797	0.2454	0.6159	0.5055
$SIZE$		−0.0005		−0.0020		−0.0041		0.0011
$EQUITY/SIZE$		0.0131		−0.0121		0.0136		0.0096
CPI (%)		0.0002		0.0002		0.0008		0.0045
Constant	0.0824	0.0665	0.082	0.0947	0.1167	0.1304	0.1180	0.0775
Obs.	3,195	3,192	3,195	3,081	3,195	3,081	3,258	3,141
Acceptance rate	0.8072	0.8098	0.8051	0.8078	0.8095	0.8072	0.8061	0.8034
Clustered by	Industry							
	Probability of a positive coefficients							
TFP	0.8394	0.6619	0.9972	0.9878	0.9997	0.9994	0.9997	0.9983
$SIZE$		0.4135		0.0471		0.0092		0.6127
$EQUITY/SIZE$		0.9987		0.0545		0.8912		0.6550
CPI (%)		0.8177		1.000		0.9935		1.000

Source author calculations

provide upon requests). And we focus on the sign of coefficients and their probability of positive value to check if the results are similar to our baseline analysis.

Column 1 and 2 present the results when we use ADE as the dependent variable. The mean coefficients of TFP are positive at 0.1440 and 0.0311, respectively. The probability that the coefficients are positive is 83.94% and 66.19%, respectively. Column 3 and 4 display the results when APC is the dependent variable. The mean coefficients of TFP are positive at 0.1447 and 0.1223, respectively. The likelihood that the coefficients are positive is 99.72% and 98.78%, respectively. Column 5 and 6 show the results when OCF is used as the dependent variable. The mean values of TFP's coefficients are positive at 0.2797 and 0.2454 with the likelihood of being positive at 99.97 and 99.94%. Finally, Column 7 and 8 represent the results when and a combined proxy ($Combine$) is employed as the dependent variable. The mean values of TFP's coefficients are 0.6159 and 0.5055 with the likelihood of being positive at 99.97 and 99.83%. Among control variables, only CPI remains stable with its positive sign.

When using the Bayesian linear regression, the prior distribution of the parameters is an important concern. We anticipate some of the model parameters could have large scale and longer adaptation may be critical for the MCMC algorithm to reach optimal sample for these parameters. Therefore, we fit Bayesian regression model using Bayes prefix with default priors and increase the burn-in period from the default value of 2,500 to 5,000. Theoretically, the default prior for the coefficients, normal (0, 10,000) is expected to have a strong shrinkage effect on the constant term. However, we do not find much difference between the constant term obtained from xtreg command (0.1260) and those from Bayesian regressions, ranging from 0.0665 to 0.1304 as shown in Table 4. The selection of default prior does not have much effect on the coefficient of the constant term due to the absolute values of earning quality and innovation are quite small. For example, innovation proxied by TFP has a mean value of 0.0002. And the mean value of five measures of earning quality ranges from 0.0684 to 0.1241 (as shown in Table 2).

Overall, the positive correlation between innovation and real activity manipulation is supported in the robustness analysis. This result is similar to the one in the baseline analysis. We can say that we find robust evidence on a positive relationship between innovation and earning manipulation or in other words, a negative relationship between innovation and earnings quality across 5 proxied of manipulation using Bayesian analysis. This finding is analogous to the ones founded by Kothari et al. (2002), Srivastava (2014) and recently by Lobo et al. (2018). Kothari et al. (2002) and Srivastava (2014) explain that the nature of accounting treatment of innovation-related items such as the R&D investments and intangible assets might lead to a deficiency of earnings quality. In another work, Lobo et al. (2018) argue that innovation-related assets are often associated with information complexity and therefore, poor information environment. This increases the likelihood that managers in innovative firms will involve in earnings management. In addition to these arguments, we think that the nature of uncertainty of innovation might be another reason that causes a decrease in earnings quality. In some cases, innovation might create differentiated products and firms can significantly boost revenues with these new products and therefore, their earnings number increase. However, on another side, innovation might fail and in those cases, firms have to suffer from substantial investments in R&D and hence, their net income decrease. This nature of uncertainty in innovation leads to high fluctuations in earnings number, and therefore, cause a decline in the quality of earnings number.

5 Conclusion

Prior studies have documented a negative relationship between the R&D expenditures, intangible intensity and earings quality (Kothari et al. 2002; Srivastava 2014) as well as the negative relationship between innovation and financial reporting quality (Lobo et al. 2018). Motivated by prior studies, in this paper, we examine the relationship between innovation and earnings quality in the Vietnamese financial market. Using Bayesian analysis approach, we find evidence to support a negative impact of innovation on earnings quality. The result is consistent with previous researches.

The result might be explained by several reasons. First, it might be due to the nature of the accounting treatment of the R&D investments and intangible assets that can result in a decline in earnings quality. This has been mentioned by researchers, for example, Kothari et al. (2002) and Srivastava (2014). Second, innovative firms are often surrounded by opaque information environment which might induce managers to engage in earnings manipulation (Lobo et al. 2018). In this research, we add the third reason that the nature of uncertainty of innovation can lead to fluctuations in earnings number and hence, its quality decreases.

Based on what we have found in this paper, we would like to notice the practitioners in the Vietnamese financial market about the issue. Our research results suggest that investors and analysts should put more attention to innovative firms before making investment decisions because these firms are at a higher chance that will take part in earnings management.

References

Bilbao-Osorio, B., & Rodriguez-Pose, A. (2004). From R&D to innovation and economic growth in the EU. *Growth and Change A Journal of Urban and Regional Policy, 35*(4), 434–455.

Cohen, D., Dey, A., & Lys, T. (2008). Real and accrual-based earnings management in the pre- and post-Sarbanes-Oxley periods. *The Accounting Review, 83*(3), 757787.

Dechow, P. M., & Schrand, C. M. (2004). *Earnings quality*. USA: The Research Foundation of CFA Institute.

Dechow, P. M., Ge, W., & Schrand, C. M. (2010). Understanding earnings quality: a review of the proxies, their determinants and their consequences. *Journal of Accounting and Economics, 50*(2–3), 344–401.

Dechow, P., Sloan, R., & Sweeney, A. (1995). Detecting earnings management. *The Accounting Review, 70*, 193–225.

Drucker, P. F. (1993). *Innovation and entrepreneurship practice and principles, 19*. Publishers Inc: Harper & Row.

Hall, B. H., Francesca, L., & Mairesse, J. (2009). Innovation and productivity in SEMs: empirical evidence for Italy. *Small Business Economics, 33*(1), 13–33.

Healy, P. M., & Wahlen, J. M. (1999). A review of the earnings management literature and its implications for standard setting. *Accounting Horizons, 13*(4), 365383.

Hirshleifer, D., Hsu, P. H., & Li, D. (2013). Innovative efficiency and stock returns. *Journal of Financial Economics, 107*(3), 632–654.

Kim, Y., Park, M. S., & Wier, B. (2012). Is earnings quality associated with corporate social responsibility? *The Accounting Review, 87*(3), 761–796.

Koehrsen, W. (2018). *Introduction to Bayesian linear regression—An explanation of the Bayesian approach to linear modeling.* https://towardsdatascience.com/introduction-to-bayesian-linear-regression-e66e60791ea7. Accessed 06 Aug 2019.

Kothari, S. P., Leone, A., & Wasley, C. (2005). Performance matched discretionary accrual measures. *Journal of Accounting and Economics, 39*(1), 163197.

Kothari, S. P., Laguerre, T. E., & Leone, A. J. (2002). Capitalization versus expensing: evidence on the uncertainty of future earnings from capital expenditures versus R&D outlays. *Review of Accounting Studies, 7*, 355382.

Kouaib, A., & Jarboui, A. (2016). Real earnings management in innovative firms: does CEO profile make a difference? *Journal of Behavioral and Experimental Finance, 12*, 40–54.

Lobo, G. J., Xie, Y., & Zhang, J. H. (2018). Innovation, financial reporting quality, and audit quality. *Review of Quantitative Finance and Accounting, 51*(3), 719–749.

Roychowdhury, S. (2006). Earnings management through real activities manipulation. *Journal of Accounting and Economics, 42*(3), 335370.

Srivastava, A. (2014). Why have measures of earnings quality changed over time? *Journal of Accounting and Economics, 57*, 196217.

The Organization for Economic Co-operation and Development (OECD). (2018). *Oslo Manual 2018—Guidelines for collecting, reporting and using data on innovation* (4th ed.). https://www.oecd.org/science/oslo-manual-2018-9789264304604-en.htm. Accessed 04 Aug 2019.

Vietstock. (2014). *After auditing, 80 % of enterprises must adjust earnings after taxes.* https://vietstock.vn/2014/04/sau-kiem-toan-80-doanh-nghiep-phai-dieu-chinh-lai-sau-thue-737-341507.htm. Accessed 02 Aug 2019.

Impact of Macroeconomic Factors on Bank Stock Returns in Vietnam

My T. T. Bui and Yen T. Nguyen

Abstract The paper is conducted to identify the trend and the amount of influence of macroeconomic factors such as interest rates, exchange rates, money supply, industrial production index, domestic gold prices, and the stock market index on stock return of Vietnamese listed commercial banks. The authors collect a sample of monthly secondary data over the period from January 2012 to June 2018 to analyze on panel data. The research method is the combination of two approaches: LASSO (The Least Absolute Shrinkage and Selection Operator)—a method of machine learning and Bayesian analysis. The result shows that the interest rates and the growth of industrial production index have significant impacts on the stock return of Vietnamese bank stocks with the probability 1 and 0.9, respectively. Whereas the interest rates have negative effect, the growth of industrial production index has positive effect. Besides, this paper shows no impact of the other macroeconomic factors, such as exchange rates, money supply M2, domestic gold prices, and VN-Index.

Keywords Bank stock return · Macroeconomic factors · LASSO · Bayesian analysis · Panel data

1 Introduction

According to data from State Securities Commission, in the past five years, Vietnamese bank stocks have achieved price growth of 154% while VN-Index have increased by 96%. Compared to the world, the returns of Vietnamese bank stocks are very high. In September 2018, Vietcombank, Techcombank, BIDV, and HDBank shares increased by 20%, 20%, 37%, and 5%, respectively compared to the same

M. T. T. Bui (✉) · Y. T. Nguyen
Department of Mathematical Economics, Banking University of Ho Chi Minh City,
Ho Chi Minh City, Vietnam
e-mail: mybtt@buh.edu.vn

Y. T. Nguyen
e-mail: yennt@buh.edu.vn

© The Editor(s) (if applicable) and The Author(s), under exclusive license
to Springer Nature Switzerland AG 2021
N. Ngoc Thach et al. (eds.), *Data Science for Financial Econometrics*,
Studies in Computational Intelligence 898,
https://doi.org/10.1007/978-3-030-48853-6_34

point of time last year. Meanwhile according to Nasdaq Global (the Global Systemically Important Banks, G-SIBs), the price movements of 30 important banks worldwide decreased by 10% in value within a year. Therefore, Vietnamese bank stocks have received much attention and expectation of investors. Knowledge of bank stock return is even more essential for investors as well as market makers and macroeconomic managers. In the context of fluctuation in macroeconomics in Vietnam and the world, it is necessary to study the stock return of Vietnamese listed bank under the impact of macroeconomic factors. Previously, this issue has been also attracted many authors such as Koskei (2017), Mouna and Anis (2016), Mouna and Anis (2013), Nurazi and Usman (2016), Paul and Mallik (2003), Saeed and Akhter (2012). Most of the above results used multi-variable regression models on time series data or panel data by applying Least Square (LS) methods. In these researches, based on the probability value (p-value) of sample statistical values, the results were drawn. Before the tendency without p-value in statistics and econometrics, the paper "Impact of Macroeconomic Factors On Bank Stock Return in Vietnam" introduces another method by combination of LASSO (Least Absolute Shrinkage and Selection Operator) and Bayesian method to analyze panel data. The study hopes to provide a reliable empirical evidence for policymakers and investors on the issue of bank stock return under the influence of fundamental macroeconomic factors such as interest rates, exchange rates, industrial production index, money supply, gold prices and VN-Index. The paper consists of four sections. Section 1 presents the introduction of the research. Section 2 shows the theoretical framework and previous empirical studies. Section 3 is for quantitative analysis and research results, and Sect. 4 is the conclusion.

2 Theoretical Background and an Overview of Empirical Studies

2.1 The Return of Stock

Return of stock is the ratio that measures the profit on a stock in a period of time, calculated by the formula:

$$R = \frac{(P_t + D) - P_{t-1}}{P_{t-1}}, \tag{1}$$

where R is the return of stock over the period from the point $t - 1$ to point t of time; P_{t-1} and P_t are the stock market prices at point of time $t - 1$ and t; D is the value of dividend of the stock during that period.

In the short term, the dividend can be ignored, the stock return is following the formula:

$$R = \frac{P_t - P_{t-1}}{P_{t-1}}. \tag{2}$$

The approximate term of formula (2) is:

$$R = \ln\left(\frac{P_t}{P_{t-1}}\right),$$

where ln is the natural logarithm function.

2.2 Theories on the Impact of Macroeconomic Factors on Stock Returns

The return of stock price is usually explained by two theories, the Capital Asset Pricing Model (CAPM) and the Arbitrage Pricing Theory (APT). Both of these theories explain the relationship between risk and expected stock return based on the assumption that capital market is efficient market and investors can invest in risk-free securities and invest in a portfolio of many common stocks in the market.

- CAPM was introduced independently by many authors upon the previous research of Markowitz about modern portfolio diversification. CAPM can be interpreted as the expected return of a capital asset that equals the profit from a risk-free asset adding a risk premium based on the systematic risk.
- APT was proposed by Ross (1976), can be considered as a complement to CAPM of capital asset valuation. APT is described through the general formula:

$$E(R) = R_f + \beta_1 R_1 + \beta_2 R_2 + \cdots + \beta_k R_k,$$

where $E(R)$ is expected profit of securities; R_f is a risk-free profit; $R_1, \ldots, R_k$ are the relative amount of changes in value of factor j in a unit of time; $\beta_1, \ldots, \beta_k$ are the sensitivity of securities to factor j.

Both CAPM and APT base on the assumption that through diversification of portfolio, non-systematic risk can be decreased at the least level. Therefore, the risk factors can be considered from systematic risks. However, APT refers to many factors that may cause systematic risk whereas CAPM does not consider. The systematic risk factors mentioned in APT are: (1) Inflation changes; (2) Changes in GNP are indicated by industrial production index; (3) Changes in the bond interest rate curve; (4) Short-term interest rates; (5) Differences between short-term interest rates and long-term interest rates; (6) A diversified stock index; (7) Changes in gold prices or other precious metal prices; (8) Changes in exchange rates; (9) Oil prices, etc. However, APT does not specify exactly the number of macroeconomic factors that are sufficient to consider the impact on stock return of securities.

Chen et al. (1986), examined the impact of some macroeconomic factors to the stock return of companies in the United States from 1953 to 1983. Macroeconomic factors were considered such as industrial output factors, expected and unexpected inflation, the difference between the return on bonds rate Baa and under and the return on a long-term government bond, the difference between long-term bond and treasury-bill rate, stock indexes of the market, and oil prices. The research result indicated that the difference between long and short interest rate, expected and unexpected inflation, industrial production and the difference between high and low grade bonds are significant impacted factors on the stock returns of companies.

APT and the empirical research of Chen, Roll, and Ross have become important theoretical frameworks for studies of the impact of macroeconomic factors on stock returns.

2.3 Empirical Studies on the Impact of Macroeconomic Factors on Stock Return

Impact of macroeconomic factors on bank stock returns have been the interested issue of many scholars. Some recent research results can be mentioned below.

Koskei, in the paper Koskei (2017), examined the impact of exchange rates on the listed stock return in Kenya. The author applied a regression model with random effects for panel data. The sample was collected during 2008–2014. While the dependent variable was the stock return of listed banks on Nairobi Securities Exchange and the independent variable was the exchange rate, research also set control variables such as treasury bond, interest rates, inflation, market capitalization values and foreign portfolio equity sales, purchases, and turnovers. The result showed that the exchange rate and inflation had negative impact on the bank stock return in Kenya while the effect of interest rate was not found.

Nurazi and Usman studied the impact of basic financial information and macroeconomic factors on bank stock returns in Indonesia from 2002 to 2011; see Nurazi and Usman (2016). By employing the Least Square method, the result indicated that interest rates, exchange rates, inflation had negative impact on the bank stock returns.

Mouna and Anis studied the sensitivity of the stock returns of financial sector (financial services, banking, and insurance) to factors, including market returns, interest rates and exchange rates in eight countries (in Europe, United States and China) during the period of the world financial crisis 2006–2009; see Mouna and Anis (2016). Applying of four models—variate GARCH-in-mean model and volatility spillovers, research indicated some results for banking sector during the crisis. That showed the positive effects of exchange rates on stock returns in Germany, the USA, and Italy, but the negative effect in The United Kingdom. For short-term interest rates, it could be seen the positive effects in USA and Spain while the negative in Greek and France. For long-term interest rates, the positive and negative effects were found in France and Italy, respectively. For market returns, they was seen the positive effects on bank stock returns in most countries, except Italy.

In Saeed and Akhter (2012), Saeed and Akhter analyzed time series data with case studies in Pakistan from June 2000 to June 2010. Dependent variable was banking index which was calculated by taking return of weighted average of prices of 29 listed banks in Karachi Stock Exchange. The independent variables were the relative changes of macroeconomic factors of money supply, exchange rates, industrial production, short-term interest rates and oil prices. The result indicated that oil prices had positive impact on banking index while money supply, exchange rates, industrial production, and short term interest rates affected the banking index negatively.

Paul and Mallik with a similar case study in Australia in the period 1980–1999, analyzed macroeconomic factors such as inflation rates, interest rates and real GDP; see Paul and Mallik (2003). The research methodology consisted of cointegration tests and estimating an error correction model for examining the long-run relationship between the return of bank and finance stock prices and macroeconomic variables. The study revealed that the return of bank and finance stock prices were cointegrated with all three macroeconomic factors. The interest rates had negative effect, GDP growth had positive effect while inflation had no significant influence on stock returns.

3 Research Method

3.1 Research Model and Research Hypothesis

Based on APT and empirical evidences in different stock markets, the paper "Impact of Macroeconomic Factors on Bank Stock Return in Vietnam" focuses on basic macroeconomic factors including interest rates, exchange rates, money supply, industrial production index, domestic gold prices, and VN-Index. The research model is described in Fig. 1.

Based on the fluctuations of the macroeconomic factors and stock returns of Vietnamese commercial listed banks from 2012 to 2018, the research hypotheses are proposed as:

- H1: Interest rate has negative impact on stock returns of Vietnamese listed commercial banks.
- H2: Exchange rate has positive impact on stock returns of Vietnamese listed commercial banks.
- H3: Money supply M2 has negative impact on stock returns of Vietnamese listed commercial banks.
- H4: Industrial production index has positive impact on stock returns of Vietnamese listed commercial banks.
- H5: Domestic gold price has negative impact on stock returns of Vietnamese listed commercial banks.
- H6: VN-Index has positive impact on stock returns of Vietnamese listed commercial banks.

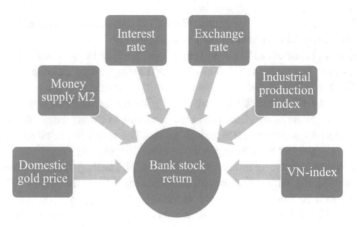

Fig. 1 The research model

3.2   Data and Variables

The paper employs regression analysis in panel data. In this panel data, the time dimension consists of 78 periods corresponding to 78 months from January 2012 to June 2018, and the object dimension includes eight Vietnamese listed commercial banks. The total number of observations is 624.

The stock returns are calculated based on the monthly average stock price data. The paper selects eight joint-stock commercial banks that are listed on Hanoi Stock Exchange (HNX) and Hochiminh Stock Exchange (HOSE). Banks with stock codes VPB, HDB, and BID are ignored because these stock price data are quite short. Data of explanatory variables are also calculated as monthly average. The average values instead of the value at end of month could minimize outlier values which often occur at the end of months. With average value, it can more accurately reflect the volatility trend of the data in reality. Data of explanatory variables are collected from reliable sources such as International Financial Statistics (IFS), State Bank of Vietnam (SBV), and Asian Development Bank (ADB). Information on the data of variables is summarized in Tables 1 and 2.

3.3   Research Method

Most previous empirical studies have used probability value (p-value) to test hypothesis for the effect of explanatory variables on explained variable. However, lately, researchers have noticed the defects of p-value. P-value is often misinterpreted because its meaning is misunderstood. A few common misunderstanding about p-value are that: (i) P-value is the probability of a scientific hypothesis, (ii) P-value is the probability of false detection, and (iii) The lower the p-value is, the higher the

Table 1 Definiton, calculation, and source of sample data

Variables	Symbols	Unit	Calculation and Definition	Source
Explained variable	RETURN	%/month	The bank stock return in one month	HNX, HOSE
Explaintory variables	INT	%/year	One-month term deposit interest rate	IFS
	LNEXR		Natural logarithm of exchange rate of VND/USD	IFS
	M2	%	Money supply M2 growth rate compared to the end of last year	SBV
	IIP	%	Growth rate of industrial production index over the same period last year	ADB
	LNGOLD		Natural logarithm of 37.5gram SJC gold	SJC
	LNVNI		Natural logarithm of the index of Ho Chi Minh Stock Exchange	HOSE

Table 2 Descriptive statistics

	RETURN	INT	LNEXR	M2	IIP	LNGOLD	LNVNI
Mean	−1.0222	6.2107	9.9781	7.6997	8.2208	8.2266	6.3862
Max	0.6069	14.004	10.0256	18.85	22.0559	8.4651	7.0442
Min	−643.4535	4.68	9.9441	0.25	−10.14	8.0977	5.8772
Std.Dev.	25.7592	2.1851	0.0284	4.6318	4.6373	0.0958	0.2636
Obs.	624	624	624	624	624	624	624

Source Authors' analysis

level of influence is; see Goodman (2008). Therefore, the study does not apply frequentist analysis. Instead, LASSO method and the Bayes analysis on panel data will be combined to clarify the impact of macroeconomic factors on bank stock returns. Firstly, LASSO is used to remove factors with little or no impact on the returns. Then, Bayesian analysis is applied to find out the trend and the amount of impact of the explanatory variables on the explained variable.

3.3.1   The Least Absolute Shrinkage and Selection Operator (LASSO)

Consider the population regression function:

$$Y_i = \beta_0 + \beta_1 X_{1i} + \beta_2 X_{2i} + \cdots + \beta_k X_{ki} + u_i.$$

LASSO method can be summarized as following: Finding the sample regression function $\hat{Y}_i = \hat{\beta}_0 + \hat{\beta}_1 X_{1i} + \hat{\beta}_2 X_{2i} + \cdots + \hat{\beta}_k X_{ki}$ such that

$$\min \left\{ RSS = \sum_{i=1}^{n} \left(Y_i - \hat{Y}_i \right)^2 = \sum_{i=1}^{n} \left(Y_i - \hat{\beta}_0 - \hat{\beta}_1 X_{1i} - \cdots - \hat{\beta}_k X_{ki} \right)^2 \right\}$$

$$\text{subject to } \sum_{j=1}^{k} |\hat{\beta}_j| \leq t, \tag{3}$$

where $\hat{\beta}_j, j = 1, \ldots, k$ and $\hat{Y}$ are the estimation value of the parameters $\beta_j, j = 1, \ldots, k$ and explained variable (or response) Y, respectively; RSS (Residual Sum of Squares) is the sum of squares of the difference between Y and $\hat{Y}$; and t is a tuning parameter.

The idea of LASSO is based on the traditional OLS estimation method with the constraint condition of the parameters. If tuning parameter t is small, the value of $\hat{\beta}_j, j = 1, \ldots, k$ will be small, the solution of LASSO will specify the value of $\hat{\beta}_j, j = 1, \ldots, k$ some exactly zero. When t is extremely large, the constraint is not very restrictive, and so the $\hat{\beta}_j, j = 1, \ldots, k$ can be large. In fact, if t is large enough, the solution of (3) will simply yield the OLS solution.

An equivalent formulation of optimization in (3) is:

$$\min \left\{ \sum_{i=1}^{n} \left(Y_i - \hat{\beta}_0 - \hat{\beta}_1 X_{1i} - \hat{\beta}_2 X_{2i} - \cdots - \hat{\beta}_k X_{ki} \right)^2 + \lambda \sum_{j=1}^{k} |\hat{\beta}_j| \right\}, \tag{4}$$

where $\lambda \geq 0$ is called penalty level.

When $\lambda = 0$, LASSO gives solution as same as OLS. When λ becomes sufficiently large, LASSO gives the null model in which all coefficient estimators equal zero. In other words, the penalty λ controls the complexity of the model. For suitable λ between these two extremes, some coefficients are zero, hence it is able to remove some predictors from the model. The precise mathematical proofs show that when λ increases, LASSO solution reduces the variance and increases the bias as the trade-off; see Hastie et al. (2009). The best λ is the value that gives the most accurate model for predicted value of the response on the test data set (the out-of-sample data). One of the criteria AIC (Akaike's information criterion), BIC (Bayesian's information criterion), or MSE (Mean Square Error) on the test set is chosen to determine the best λ; see Ahrens et al. (2019), Hastie et al. (2015).

A procedure usually applied to estimate the best λ and $\hat{\boldsymbol{\beta}} = (\hat{\beta}_0, \ldots, \hat{\beta}_k)$ is *Cross-validation*. In more details, the data set is randomly divided into $K > 1$ groups, K is usually chosen as 5 or 10. The first group is fixed as the test set and the remaining $K - 1$ groups are considered as the training set. For each value of λ, LASSO is applied on the training set to get the fitted model. Then, the fitted model is applied on test set and criteria MSE is concerned. Still this value of λ, the role of the test set is swapped for the second group, the third group, ..., and Kth group. We will get K values of MSE corresponding to K test sets. The average of these MSE is called Cross-validation Mean Square Errors (CVMSE). When λ changes in a certain range, the best $(\lambda, \hat{\boldsymbol{\beta}})$ are the values that produce the smallest CVMSE; see Hastie et al. (2015).

Compared to OLS, LASSO shows a larger MSE on sample data. However, on out-of-sample, there are some different results. Tibshirani, in Tibshirani (1996), examined four examples on simulated data to compare the MSE of LASSO with OLS on test data set. Three of four these examples give smaller MSE of LASSO. Therefore, by making sacrifice bias for variance of solutions $\hat{\beta}$, LASSO may outperform OLS in predictability; see Ahrens et al. (2019), Tibshirani (1996). Besides, by removing some predictors that have very little or no impact on the response, LASSO not only provides a simpler model than OLS, it can also help to overcome the well-known limitations of linear regression (possible learning, over interpretation of p-value,) because the parameters are not associated with standard errors and p-value.

The most disadvantage of LASSO is that LASSO estimators are not unique in some cases; see Hastie et al. (2015), Tibshirani (1996). Therefore, for the purpose of interpretation, it is necessary to combine LASSO with other methods. Unlike OLS, LASSO is not invariant to linear transformations, hence there is a need to standardize the data, i.e., all of the series of explanatory variables have zero mean and unit variance; see Tibshirani (1996).

3.3.2 Bayesian Model

By Bayesian analysis approach, all parameters are supposed to be random variables. Posterior distribution of parameter θ with given data D can be inferred by Bayesian rule:

$$P_{posterior}(\theta|D) \propto P_{likelihood}(D|\theta) P_{prior}(\theta).$$

where $P_{likelihood}(D|\theta)$ is likelihood function and $P_{prior}(\theta)$ is prior probability distribution.

In frequentist analysis, fixed effects and random effects are often used to analyze panel data. However, in Bayesian analysis, random effects model is applied:

$$Y_{it} = \beta_0 + \beta_1 X_{1it} + \cdots + \beta_k X_{kit} + u_i + \epsilon_{it} \tag{5}$$

where u_i is the random effect for ith objects and t expresses the point of time.
Selecting prior distributions

Bayesian inferences from two models will produce the same results if the likelihood functions of two models are proportional; see Berger and Wolpert (1988). Moreover, Bernstein von Mises theorem states that the larger data samples are, the less influence the prior distribution has on a the posterior distribution, therefore, Bayesian inferences based on likelihood function will yield the same results. For these reasons, in this research, choosing priors are more focused.

There are some choices of prior distributions: uninformative priors that is"flat" relative to the likelihood function; improper priors for example uniform prior; informative priors that is not dominated by the likelihood function and has an high impact on the posterior distribution; and hierarchical priors using hyper-parameters. Uninformative priors may result in improper posterior and strong informative priors may be subjective and bias, whereas hierarchical priors provide a compromise between these choices by using informative prior family of distributions and uninformative hyper-priors for the hyper-parameters; see Balov (2016).

Sampling method

Posterior distribution is used to learn about the parameters including point estimates such as posterior means, medians, percentiles, and interval estimates such as credible intervals and interval hypothesis tests for model parameters. Moreover, all statistical tests about parameters can be expressed as probability statements based on the estimated posterior distribution.

If the posterior distribution combines with the prior distribution, the inference of Bayesian analysis can proceed directly. However, except for some special models, posterior distribution is rarely precise and needs to be estimated through simulations. Markov Chain Monte Carlo (MCMC) sampling, such as Metropolis-Hasting methods or Gibbs method or sometimes their combination, can be used to simulate potentially very complex posterior models with an arbitrary level of precision. Besides, some Bayesian estimators, such as posterior mean and posterior standard deviation, involve integration. If the integration cannot be performed analytically to obtain a closed-form solution, sampling techniques such as Monte Carlo integration and MCMC and numerical integration are commonly used.

Gibbs sampler, actually a special case of Metropolis-Hasting algorithm updates each model parameter once according to its full conditional distribution. Some advantages of Gibbs sampling are efficient, because all proposals are automatically accepted, and unnecessary to add any tuning for proposal distributions. However, for most posterior distributions in practice, full conditionals are either not available or are very difficult to sample from, in that case adaptive Metropolis-Hasting is required. In another case, for some model parameters or groups of parameters, full conditionals are available and are easy to generate samples from. A hybrid Metropolis-Hasting algorithm implemented by Gibbs algorithm updates for only some blocks of parameters can greatly improve efficiency of MCMC sampling.

Testing of simulation performance

Based on an MCMC algorithm, inference through Bayesian analysis is efficient if the chains must be representative, and the estimators are accurate and stable; see

Kruschke (2014). Therefore, the convergence of MCMC must be verified before any inference.

Current practice for checking representativeness of MCMC often focuses on some methods: visual inspection of trajectory and considering a numerical description of convergence. The first visual method is a visual examination of the chain trajectory through trace plot. If the chain is representative, they should overlap and mix well. The other visual method is to check how well the density plots of three chains do overlap after the burn-in period through density plots. Yu and Mykland proposed a graphical procedure for assessing the convergence of individual parameters based on cumulative sums, also known as a cusum plot in his 1998 paper Yu and Mykland (1998). By definition, any cusum plot starts at 0 and ends at 0. For a chain without trend, the cusum plots should cross the x-axis. For example, early drifts may indicate dependence on starting values. If an early drift is detected, an initial part of the chain should be discard and run it longer. Cusum plots can be also used for assessing how fast the chain is mixing. The slower the mixing of the chain is, the smoother the cusum plots are. Conversely, the faster the mixing of the chain is, the more jagged the cusum plots are.

The correlation of the chain in k steps ahead can be used to measure the stable and accuracy of MCMC. The highly autocorrelation chains reveal a problem for the stable and accuracy of MCMC. Kass, Carlin, and Gelman provided a measure of independent information of autocorrelation chains, called effective sample size, denoted by ESS:

$$ESS = \frac{N}{1 + 2 \sum_{k=1}^{\infty} ACF(k)},$$

where N is denoted the actual number of steps in the chains, $ACF(k)$ is the autocorreltion at lag k of MCMC; see Kass et al. (1998). The another criteria that reflects effective accuracy of the chain is Monte Carlo standard error (MCSE):

$$MCSE = \frac{SD}{\sqrt{ESS}},$$

where SD is the standard deviation of the chain. MCSE is interpreted on an accuracy in the estimate of the posterior mean. Besides, the chain has a reasonable stable and accurate if a ESS of (at least) is desirable. In general, averaged efficiencies over all chains above 10% is good, sometimes even above 8% for the Metropolis-Hasting algorithm. The another important criteria measuring the efficiency of MCMC is reported for acceptance rate. An acceptance rate specifies the proportion of proposed parameter values that was accepted by the algorithm. When the acceptance rate is close to 0, then most of the proposals are rejected, which means that the chain failed to explore regions of appreciable posterior probability. The other extreme is close to 1, in which case the chain stays in a small region and fails to explore the whole posterior domain. An efficient MCMC has an acceptance rate that is neither too small nor too large and also has small autocorrelation. In the paper Roberts et al. (1997),

Table 3 Model comparison rule based on Bayes factor

$\log_{10}(BF_{12})$	BF_{12}	Evidence against M_1
0 to 1/2	1 to 3.2	Bare mention
1/2 to 1	3.2 to 10	Substantial
1 to 2	10 to 100	Strong
>2	>100	Decisive

Roberts, Gelman, and Gilks showed that in the case of a univariate posterior like the models in this paper, the optimal value is 0.45.

Comparing Bayesian models

After checking convergence, the next step of Bayesian analysis is model comparison based on deviance information criterion (DIC), log marginal likelihood, the posterior probabilities of model, and Bayes factor (BF). DIC is given by the formula: $D(\bar{\theta}) + 2p_D$, where $D(\bar{\theta})$ is the deviance of sample mean and p_D is the effective complexity, a quantity equivalent to the number of parameters in the model. Models with smaller DIC, larger log marginal likelihood, larger posterior probability are preferred.

The most decisive criteria in comparison models M_1 and M_2 with given data D is Bayes factor. $P_{Likelihood}(D|M_1) = m_1(D)$ and $P_{Likelihood}(D|M_2) = m_2(D)$ are the marginal likelihood of $M1$ and $M2$ with respect to D, respectively. Besides, by applying Bayes rule, posterior odds ratio is following the formula:

$$\frac{P_{Posterior}(M_1|D)}{P_{Posterior}(M_2|D)} = \frac{P_{Likelihood}(D|M_1)}{P_{Likelihood}D|M_2)} \frac{P_{Prior}(M_1)}{P_{Prior}(M_2)}.$$

If all priors are equally probability, that is, $P_{Prior}(M_1) = P_{Prior}(M_2)$, the posterior odds ratio is simplified to the marginal likelihood ratio, called the Bayes factors (BF). In paper Jeffreys (1961), Jeffreys recommended an interpretation of Bayes factor based on half-unit of the log scale. The following table provides the rule of thumb (Table 3).

3.4 Regression Model

Regression models for panel data are usually set as pooled regression model, fixed effects model (FEM) and random effects model (REM). According to Baltagi (1999), Greene and Zhang (2008) and data analysis in the research sample, some of following comments are concluded:

- The pooled regression model is inconsistent with the research situation, because there are significant differences of bank stock returns between the eight listed commercial banks.

- With the structure sample data, the number of periods $T = 78$ is quite large compared to the number of objects $N = 8$, so the difference in regression results by FEM and REM is negligible. This study uses REM effects for analysis by LASSO and Bayesian method.

Thus, REM is selected to examine the impact of macroeconomic factors on the stock returns of Vietnamese listed commercial banks from January 2012 to June 2018. The population regression function of quantitative analysis is:

$$RETURN_{it} = \beta_0 + \beta_1 LS_t + \beta_2 LNEXR_t + \beta_3 M2_t + \beta_4 IIP_t +$$
$$+ \beta_5 LNGOLD_t + \beta_6 LNVNI_t + u_i + \epsilon_{it},$$

where β_0 is the intercept coefficient; $\beta_1, \ldots, \beta_6$ are the parameters of the regression model, showing the effect of each macroeconomic factor on the bank stock return; $u_i, i = 1, \ldots, 8$ show the impacts of each bank's own characteristics on its return; ϵ_{it} is random error of the regression model.

3.5 Results and Discussions

3.5.1 The Result of LASSO

From the original LASSO which was introduced by Frank and Tibshirani, later studies have presented some choices of tuning parameter/penalty level to overcome possible phenomena in model; see Frank and Friedman (1993), Tibshirani (1996).

- Square-Root LASSO supplies the optimal penalty level, which is independent of the unknown error variance under homoskedasticity and valid for both Gaussian and non-Gaussian errors; see Belloni et al. (2011).
- Rigorous LASSO sets the penalty level for the presence of heteroskedasticity, non-Gaussian and cluster-dependent errors in balance or unbalance panel data; see Belloni et al. (2016). The penalty level λ is estimated by using iterative algorithms; see Ahrens et al. (2019).

The research uses Rigorous LASSO for squareroot LASSO to control the heteroskedasticity and cluster-dependent errors.

Before performing Rigorous LASSO estimation, the variables are standardized according to the formula:

$$\tilde{X}_{ij} = \frac{X_{ij} - \overline{X}_j}{\sqrt{\dfrac{1}{n} \sum_{i=1}^{n} \left(X_{ij} - \overline{X}_j \right)^2}},$$

where $\overline{X}_j$ is the sample mean of X_j (Table 4).

Table 4 Names of variables after standardizing

Original variable	Standardized variable
RETURN	STD_{RETURN}
INT	STD_{INT}
LNEXR	STD_{LNEXR}
M2	STD_{M2}
IIP	STD_{IIP}
LNGOLD	STD_{LNGOLD}
LNVNI	STD_{LNVNI}

Table 5 The estimation result by Rigorous LASSO

Selected	Rigorous LASSO	Post-est OLS
STD_{INT}	−0.0517192	−0.1200104
STD_{IIP}	0.0244747	0.0927659
Constant	−0.0321475	−0.0321475

Source Authors' analysis

The estimation result by Rigorous LASSO is shown in Table 5. Interest rate STD_{INT} and industrial production index STD_{IIP} have impact on Vietnamese bank stock returns in the period from January 2012 to June 2018. Thus, in the next analysis, the model retains two explanatory variables INT and IIP.

3.5.2 The Result of Bayesian Analysis

The two level random effects model is considered as following:

$$RETURN_{it} = \beta_0 + \beta_1 INT_t + \beta_2 IIP_t + u_t + \epsilon_{it} = \beta_1 INT_t + \beta_2 IIP_t + \tau_t + \epsilon_{it}, \quad (6)$$

where u_i is the random effect for ith bank, ϵ_{it} is the random error for ith bank at point of time t, and $t = 1, \ldots, 78$ are identified the months from January 2012 to June 2018. Based on the selection of prior distribution mentioned in Sect. 3.3.2, normal priors are set for the regression coefficients and inverse-gamma priors for the variance parameters.

$$\epsilon_{it} \sim \text{i.i.d. } N(0, \sigma_0^2)$$
$$\tau_t \sim \text{i.i.d. } N(\beta_0, \sigma_{bank}^2)$$
$$\beta_0 \sim N(0, 100)$$
$$\beta_1 \sim N(0, 100)$$
$$\beta_2 \sim N(0, 100)$$
$$\sigma_0^2 \sim InvGamma(0.001, 0.001)$$
$$\sigma_{bank}^2 \sim InvGamma(0.001, 0.001)$$

From the aforementioned sampling method in Sect. 3.3.2, to fit this model, six types of Monte Carlo simulations are employed to chose the reasonable efficiency result.

- Gibbs algorithm is used in the first method.
- Adaptive Metropolis—Hasting algorithm is applied in the second simulation.
- Separating into blocks for parameter in which random effects parameters are in the same block to be sampled can improve efficiency of the Metropolis-Hasting algorithm.
- In the fourth simulation, MCMC procedure can be more efficient by employing a Gibbs algorithm for a block of parameters.
- Sampling random-effects parameters in separate blocks from the another parameters, in case full Gibbs sampling is not available.
- In the last simulation, the fifth simulation is changed by grouping the random effects parameters in one block.

It is necessary to increase the number of iterations for the burnin process of MCMC simulation and to use thinning interval to increase the efficiency and decrease the autocorrelation of MCMC of six mentioned algorithms for model (5). By observing trace plot, density plot, cusum plot of each MCMC algorithm, the representativeness of each chain is not a problem. The coefficient estimators of variables INT and IIP of six simulations are presented in Table 6.

It can be seen that the coefficient of the variable INT ranges from -1.45 to -1.69 and the coefficient of the variable IIP ranges from 0.39 to 0.42. All interval hypothesis tests for model parameters show negative coefficients of INT and positive coefficients of IIP with a approximately probability level of 1 and 0.9, respectively.

Looking at Table 7, the simulation having the most suitable acceptance rate is the third one. Besides, relying on average efficiencies, the sixth is the most likely except the first simulation. However, Table 8 reports that the fourth simulation is preferred with the smallest DIC value, the largest log marginal likelihood, and the largest the posterior probability. Specially, the most decisive criteria between the six simulations is Bayes factor. Against the first simulation, the value of logarithm of BF corresponding to the fourth simulation is 6.34, which implies very sustainable evidence in favor of the fourth simulation. Moreover, the MCSE of two coefficient estimators of INT and IIP of this simulation are quite small among the others. Therefore, the result of the fourth simulation can be more significant.

Table 6 Point estimators for coefficients and their MCSE

Method	Coefficient of INT		Coefficient of IIP	
	Point estimator	MCSE	Point estimator	MCSE
The first simulation	−1.5481	0.0048	0.3896	0.0023
The second simulation	−1.486	0.0089	0.4058	0.1692
The third simulation	−1.6941	0.0398	0.3383	0.0078
The fourth simulation	−1.5002	0.029	0.4049	0.009
The fifth simulation	−1.4900	0.0504	0.4076	0.0162
The sixth simulation	−1.4512	0.0658	0.4186	0.0222

Source Authors' analysis

Table 7 The acceptance rate and the average efficiency of the algorithms for model

Simulation	Acceptance rate	Average efficiency
The first simulation	0.8719	82.08%
The second simulation	0.18	1.2%
The third simulation	0.4956	7.5%
The fourth simulation	0.9439	8.6%
The fifth simulation	0.5738	7.8%
The sixth simulation	0.877	17.03%

Source Authors' analysis

3.5.3   Discussions

The combination of two methods, Rigorous LASSO and Bayesian analysis, have clarified the effects of macroeconomic factors on listed commercial bank stock returns in Vietnam from 2012 to June 2018. The result can be expressed in the following formula:

$$\widehat{RETURN}_{it} = -1.5002 INT_t + 0.4049 IIP_t + \hat{\tau}_t.$$

Interest rate has a negative effect on the bank stock returns. This even happens with the probability nearly 1. The result indicates that when the other macro factors do not change, an 1% increase in interest rate will cause the decrease of the bank stock returns by 1.5002%. The hypothesis H1 is accepted. The trend of influence of interest rate is similar to results in these paper Nurazi and Usman (2016), Paul and Mallik (2003), Saeed and Akhter (2012). This result is also consistent with qualitative analysis. Interest rates represent the prices of the right to use the currency in a unit of time. When interest rates on the currency market rise rapidly, investors have more

Table 8 Model comparison

Simulation	DIC	log(ML)	log(BF)	P(M|Y)
The first simulation	5816.986	−2904.536	.	0.0017
The second simulation	5809.619	−2968.821	−64.28567	0.0000
The third simulation	5814.705	−2928.852	−24.31617	0.0000
The fourth simulation	5815.993	−2898.189	6.346405	0.9954
The fifth simulation	5816.33	−2904.034	.5015122	0.0029
The sixth simulation	5816.1	−2912.107	−7.571281	0.0000

Source Authors' analysis

choices to distribute their money. Therefore, the cash flows of investment in bank stocks will definitely be shared by other investment channels. This leads to a decline in bank stock returns.

Industrial production index has a positive effect on the bank stock returns. This trend of effect occurs with the probability about 0.9. The result shows that when the other macro factors do not change, if the change of industrial production index rise by 1%, the bank stock returns will go up by 0.4049%. The hypothesis H4 is accepted. This co-movement trend is similar to Paul and Mallik (2003), and contrary with Saeed and Akhter (2012). In Vietnam, it can be concluded that, during the period from January 2012 to June 2018, the growth of the real production sector will stimulate the rise of the stock market in general and the bank stock market in particular.

Exchange rate VND/USD, money supply M2, domestic gold prices, and VN-Index have no impact or very little impact on Vietnamese listed commercial bank stock returns in the period from January 2012 to June 2018. The hypothesis H2, H3, H5, and H6 are rejected. Therefore, these factors have been excluded from the model from the first step of analysis by LASSO.

4 Conclusions

The research results by quantitative analysis reveals that the interest rate and the growth of industrial production index have significant impacts on the bank stock returns with the probability over 90%. Interest rate performs the negative effect while the growth of industrial production index shows the positive effect. Besides, this research shows no impact of the other macroeconomic factors, such as exchange rates, money supply M2, domestic gold prices, and VN-Index.

The research results show that changes in macro management such as interest rate policies, industrial production development policies, will have certain impact on the stock market, especially banking stock market. The results imply some issue. Firstly, managing interest rates is not only a macro issue of State Bank of Vietnam, it also requires compliance with regulations of commercial banks. The race of interest rates among commercial banks may adversely affect the bank stock returns. Secondly, at the present, the development of industrial production in Vietnam really needs support from commercial banking system. In the opposite direction, according to the research results, the positive growth of industrial production lead to the positive changes in bank stock returns. The relationship between industrial production and bank stock returns is truly close. The study hopes to provide a reliable empirical evidence of macroeconomic factors to the profitability of bank stocks that may be a premise for bank managers and macroeconomic managers to come up with suitable policies for the sustainable development of the stock market of Vietnamese banking industry.

References

Ahrens, A., Hansen, C., & Schaffer, M. (2019). Lassopack: model selection and prediction with regularized regression in stata. *Institute of Labor Economics*, IZA DP No. 12081.

Asia Development Bank (2019). *Statistics*. Retrieved from http://www.adb.org/countries/viet-nam/main.

Balov, N. (2016). Bayesian hierarchical models in Stata. In *Stata Conference*. Stata Users Group, No. 30.

Baltagi, B. H. (1999). *Econometric analysis of cross section and panel data*. Cambridge, Mass: MIT Press.

Belloni, A., Chernozhukov, V., Hansen, C., & Kozbur, D. (2016). Inference in high dimensional panel models with an application to gun control. *Journal of Business & Economic Statistics*, *34*(4), 590–605.

Belloni, A., Chernozhukov, V., & Wang, L. (2011). Square-root lasso: pivotal recovery of sparse signals via conic programming. *Biometrika*, *98*(4), 791–806.

Berger, J. O., & Wolpert, R. L. (1988). *The likelihood principle*. IMS.

Chen, N. F., Roll, R., & Ross, S. A. (1986). Economic forces and the stock market. *Journal of Business*, *59*, 383–403.

Frank, I. E., & Friedman, J. H. (1993). A statistical view of some chemometrics regression tools. *Technometrics*, *35*(2), 109–135.

Gelman, A., & Rubin, D. (1992). Inference from iterative simulation using multiple sequences. *Statistical Science*, *7*, 457–472.

Goodman, S. (2008). A dirty dozen: twelve p-value misconceptions. *Seminars in Hematology*, *45*(3), 135–140.

Greene, W. H., & Zhang, C. (2003). *Econometric analysis*. Upper Saddle River, NJ: PrenticeHall.

Hanoi Stock Exchange. (2019). *Statistics*. Retrieved from http://hnx.vn/web/guest/ket-qua.

Hastie, T., Tibshirani, R., & Friedman, J. (2009). *The elements of statistical learning: data mining, inference, and prediction*. Springer.

Hastie, T., Tibshirani, R., & Wainwright, M. (2015). *Statistical learning with sparsity: the lasso and generalization*. Chapman and Hall/CRC Press.

Hochiminh Stock Exchange. (2019). *Statistics*. Retrieved from https://www.hsx.vn/.

International Financial Statistics. (2019). *Statistics*. Retrieved from http://www.data.imf.org.

Jeffreys, H. (1961). *Theory of probability*. Oxford, UK: Oxford University Press.

Kass, R., Carlin, B., Gelman, A., & Neal, R. (1998). Markov chain Monte Carlo in practice: a roundtable discussion. *The American Statistician, 52*(2), 93–100.

Koskei, L. (2017). The effect of exchange rate risk on stock returns in Kenyas listed financial institutions. *Research Journal of Finance and Accounting, 8*(3).

Kruschke, J. (2014). *Doing Bayesian data analysis: a tutorial with R, JAGS, and Stan.* Academic Press.

Mouna, A., & Anis, M. J. (2016). Market, interest rate, and exchange rate risk effects on financial stock returns during the financial crisis: AGARCH-M approach. *Cogent Economics & Finance, 4*(1), 1125332.

Mouna, A., & Anis, M. J. (2013). The impact of interest rate and exchange rate volatility on banks returns and volatility: evidence from Tunisian. *The Journal of Commerce, 5*(3), 01–19.

Nurazi, R., Usman, B. (2016). Bank stock returns in responding the contribution of fundamental and macroeconomic effects. *JEJAK: Journal Ekonomi Dan Kebijakan, 9*(1), 131–146.

Paul, S., & Mallik, G. (2003). Macroeconomic factors and bank and finance stock prices: the Australian experience. *Economic Analysis and Policy, 3*(1), 23–30.

Roberts, G. O., Gelman, A., & Gilks, W. (1997). Weak convergence and optimal scaling of random walk Metropolis Algorithm. *The Annals of Applied Probability, 7*(1), 110–120.

Ross, S. (1976). The arbitrage theory of capital asset pricing. *Journal of Economic Theory, 13*(3), 341–360.

Saeed, S., & Akhter, N. (2012). Impact of macroeconomic factors on banking index in Pakistan. *Interdisciplinary Journal of Contemporary Research in Business, 4*(6), 1–19.

State Bank of Vietnam. (2019). *Statistics.* Retrieved from http://www.sbv.gov.vn.

Tibshirani, R. (1996). Regression shrinkage and selection via the Lasso. *Journal of the Royal Statistical Society. Series B, 58*(1), 267–288.

Yu, B., & Mykland, P. (1998). Looking at Markov samplers through cusum path plots: a simple diagnostic idea. *Statistics and Computing, 52*(2), 93–100.

The Impact of Auditor Size and Auditor Tenure on Banks' Income Smoothing in Developing Countries: Evidence from Vietnam

Ha T. T. Le

Abstract Income smoothing is intentional alteration of a company's earnings by its management in order to reduce the fluctuations in the company's reported profit, resulting in a biased presentation of the related financial statements. Previous studies in Vietnam and other countries have found evidence of banks smoothing their income by adjusting provisions for loan loss. The purpose of this paper is to examine the impact of auditor size and auditor tenure on income smoothing practice through loan loss provisions by Vietnamese commercial banks. Based on a sample of 21 Vietnamese banks during the selected period from 2008 to 2017, the findings show that Vietnamese banks smooth their earnings through loan loss provisions. However, Big 4 auditors do not show a significant impact on reducing this income smoothing, while there is moderate evidence that long auditor tenure has a mitigating effect on this kind of earnings management. The results of the study are consistent with prior empirical evidence that find longer auditor tenure having an impact on reducing earnings management, whereas big audit firms do not always make a difference from other audit firms in relation to constraining earnings management, especially in the context of a developing country with a relatively weak audit environment. The study results have significant implications for the regulators and standard setters in developing a strong environment for the audit profession.

1 Introduction

The purpose of this study is to examine the impact of auditor size and auditor tenure on income smoothing through loan loss provisions by Vietnamese commercial banks. Loan loss provision (LLP) is normally a large expense in a bank's financial statements (DeBoskey and Jiang 2012). As an accounting estimate, LLP is subjective to

H. T. T. Le (✉)
Faculty of Accounting and Auditing, Banking Academy 12 Chua Boc,
Dong Da Hanoi, Vietnam
e-mail: haltt@hvnh.edu.vn

© The Editor(s) (if applicable) and The Author(s), under exclusive license
to Springer Nature Switzerland AG 2021
N. Ngoc Thach et al. (eds.), *Data Science for Financial Econometrics*,
Studies in Computational Intelligence 898,
https://doi.org/10.1007/978-3-030-48853-6_35

513

management manipulation to achieve a desired reported profit. Income smoothing through LLP is a form of earnings management in which a bank tries to reduce the fluctuations in its reported earnings by purposefully recording a lower LLP when its profit is low and a higher LLP when it has a higher profit. In prior literature, banks are considered as smoothing their earnings when the banks' LLP is positively correlated with earnings before tax and provisions (EBTP).

Income smoothing results in a biased presentation of a company's performance. Nevertheless, it is important to distinguish the different between income smoothing and fraudulent accounting. Fraud is illegal and is a violation of accounting standards. Whereas, income smoothing, as a form of earnings management, could be based on accounting choices that are allowed under generally accepted accounting principles (Dechow and Skinner 2000; Kitiwong 2019). For example, income smoothing could be achieved by aggressive recognition of provisions or understatement of bad debts, which are based on exercising judgment on accounting estimates (Dechow and Skinner 2000). Grey areas and loopholes in accounting standards give rooms for income smoothing practices.

A number of studies have found evidence that banks smooth their reported earnings through LLP, such as Dao (2017), DeBoskey and Jiang (2012), Fonseca and González (2008), Kanagaretnam et al. (2003), Wahlen (1994). Auditors are expected to limit income smoothing practice to improve the quality of reported earnings. Auditor size and auditor tenure are two aspects of audit quality measures that have received considerable attention of researchers recently. The question is whether auditor size and auditor tenure have a significant impact on restricting earnings management by banks. There have been many studies investigating the impact of audit quality on earnings management; however, research on this topic in the banking industry is still limited. Therefore this research is conducted to examine the impact of auditor size and auditor tenure on income smoothing by Vietnamese commercial banks.

Based on a sample of 21 Vietnamese banks in the selected period from 2008 to 2017, the findings show that Vietnamese banks smooth their earnings through loan loss provisions. However, Big 4 auditors do not show a significant impact on reducing this income smoothing, while there is moderate evidence that long auditor tenure has a mitigating effect on this kind of earnings management.

The results of the study are consistent with prior empirical evidence that find longer auditor tenure having an impact on reducing earnings management, whereas big audit firms do not always make a difference from other audit firms in relation to constraining earnings management, especially in the context of a developing country with a low risk of litigation against auditors, insufficient enforcement of observance with accounting and auditing standards, and limitations of the accounting regime. The study contributes to the debate on the role of auditor size and auditor tenure on earnings management in the banking industry.

The rest of the paper is organized as follows. Research hypotheses are developed in Sect. 2. In Sect. 3, the author describes the sample selection and research design. The empirical results of the study are reported and discussed in Sects. 4 and 5. Finally, Sect. 6 summaries major findings, addresses the limitation and offers suggestions for future studies.

2 Hypotheses Development

2.1 Auditor Size

The impact of audit firm size on the quality of a client's financial reporting is one of the most controversial topics in auditing research. Big audit firms are generally considered as possessing more resources to invest in building audit methodology and staff training (Craswell et al. 1995), and desire to protect their reputation and defend themselves from the risk of litigation (Francis et al. 1999).

A majority of studies in the topic have investigated the relationship between auditor size and discretionary accruals—a proxy for earnings management. Many studies have found evidence that big audit firms are better at constraining earnings management in their clients' financial statements. For example, Becker et al. (1998) find that US firms audited by non-Big 6 auditors report significantly higher level of discretionary accruals than those with Big 6 auditors. Tendeloo and Vanstraelen (2008) and Alzoubi (2018) also find a lower level of earnings management associated with Big 4 auditors in European and Jordan.

On the other hand, a number of other studies do not find evidence that big auditors are superior to other firms at limiting earnings management. For example, Piot and Janin (2007) document that there is no difference between big audit firms and smaller firms in relation to earnings management in France, which is argued to be caused by a lower risk of litigation against auditors in the country. Maijoor and Vanstraelen (2006) investigate the impact of national audit environment and audit quality in France, Germany and the UK. The results show no evidence of an international Big 4 audit quality in Europe, while a stricter audit environment constrains the level of earnings management, irrespective of the auditor size. Some studies even find that big audit firms are associated with a higher level of earnings management. For example, Alves (2013) and Antle et al. (2006) find evidence of positive relationship between big audit firms and discretionary accruals for Portugal and UK listed companies, while Lin et al. (2006) document a positive association between Big 5 auditors and earnings restatement in the US. Similarly, Jeong and Rho (2004) find no significant difference in audit quality between Big 6 and non-Big 6 auditors under weak regulatory institutional regimes in Korea.

In the banking industry, there are fewer studies addressing the impact of audit quality on earnings management. Kanagaretnam et al. (2010) study the impact of auditor size and auditor specialization on earning management in banks from 29 countries. The results show that Big 5 auditors have a significant impact on reducing income-increasing discretionary loan loss provisions, but insignificant influence on income-decreasing earnings management. Regarding the impact of audit quality on income smoothing, DeBoskey and Jiang (2012) find that auditor industry specialization has a significant influence on reducing income smoothing through loan loss provisions in US banks. In particular, the interaction term between auditor specialization and profit before tax and provisions (EBTP) has a mitigating effect on the positive relationship between LLP and EBTP. However, the study has not investigated the impact of auditor size on earnings smoothing by the banks.

Ozili (2017) investigates the role of Big 4 auditors on income smoothing through LLP on a sample of African banks. In contrast to the results reported by Kanagaretnam et al. (2010), the study finds that the interaction term between Big 4 firms and the banks' EBTP are positively correlated with LLP, which indicates that Big 4 audit firms do not have an impact on reducing income smoothing in the study sample.

From prior literature, if auditor size helps reduce income smoothing, we should observe a less positive correlation between EBTP and LLP for banks audited by big auditors. Even though there is conflicting evidence as to the impact of auditor size on earnings management, the general expectation is that big audit firms mitigate income smoothing by banks. Therefore, the first hypothesis is stated as follows:

H1. The relation between loan loss provisions and earnings is less positive for a bank audited by a Big 4 firm than that for a bank whose auditor is not a Big 4 firm.

2.2   Auditor Tenure

Auditor tenure is defined as the number of years an auditor is maintained by a client (Myers et al. 2003), or the length of the relationship between a company and an audit firm (Johnson et al. 2002). Auditor tenure is an aspect of the relationship between an auditor and his client.

It is generally believed that in the early year of the relationship between an auditor and a client, audit risk is higher, as the auditor may not have sufficient knowledge about the client. This situation can be mitigated with the understanding and knowledge of the client developed over the length of the auditor-client relationship.

A majority of previous studies have found a significant impact of long auditor tenure on the quality of clients' financial statements. Johnson et al. (2002) investigate the relationship between auditor tenure and the quality of financial reporting, as proxied by the magnitude of discretionary accruals and the persistence of current accruals in the US. The results show that short-term audit tenure (two to three years) is associated with a lower reporting quality than medium tenure (four to eight years), while there is no evidence that longer tenure (from nine years) is related to lower reporting quality relative to medium tenure. Carcello and Nagy (2004) also discover that financial statement fraud is more likely to happen in the first three year of an audit engagement, while no evidence to prove that financial statement fraud is more related to long auditor tenure. Similarly, Alzoubi (2018) suggests that firms with auditor tenure from five years record lower level of discretionary accruals. In the banking industry, Dantas and Medeiros (2015) also find that a long-term auditor-client relationship reduces the level of earnings management in a sample of Brazilian banks.

However, several research find that a lengthy client-auditor association negatively affect earnings quality, which could be due to auditor independence and skepticism being impaired in a familiar, long lasting relationship. Chi and Huang (2005) find Taiwanese evidence that even though auditor tenure is negative correlated with discretionary accruals, the quadratic form of auditor tenure is positively associated with discretionary accruals. The result means that while long auditor tenure increases the auditor ability to detect earnings management in the initial period of an audit engagement, excessive familiarity is associated with a lower earnings quality for lengthy auditor tenure. The study also finds that optimal cutoff point for auditor tenure is around five years for the lowest level of discretionary accruals (Chi and Huang 2005). Similarly, a study by González-Díaz et al. (2015) in a sample of state-owned organizations in Spain reveals that audit quality is improved over the initial years of an engagement, but then diminishes as the length of auditor tenure increases.

In Vietnam, according to Circular 39/2011/TT-NHNN dated 15 December 2011 by the State Bank of Vietnam (SBV), an audit firm is allowed to audit a particular bank for a maximum of five consecutive years. Thus the maximum auditor tenure for banks is five years. The question of interest is whether an audit with longer tenure (four or five years) is related to a higher reporting quality than a short-term tenure (three years or less). In particular, the research objective is to find out whether longer auditor tenure has a mitigating impact on income smoothing practice of banks. According to the common expectation that longer auditor tenure improves audit quality, the second hypothesis is developed as follows:

H2. The relation between loan loss provisions and earnings is less positive for a bank with longer auditor tenure.

3 Research Design

3.1 Sample Selection

The data used in the study are obtained from the audited financial statements of commercial banks in Vietnam for the period from 2008 to 2017. The financial statements are available on finance websites including S&P Capital IQ platform and Vietstock.vn. After eliminating banks that lack detailed financial accounting information and auditor's report, the final sample comprises of 21 banks with 210 bank-year observations. The annual GDP growth rates from 2008 to 2017 of the country are also obtained from S&P Capital IQ platform.

3.2   Model Specification

The research investigates the relationship between audit firm size and auditor tenure on Vietnamese banks' income smoothing through loan loss provision based on the following models:

$$
\begin{aligned}
LLP = {} & \alpha_0 + \alpha_1 \cdot EBTP + \alpha_2 \cdot BIG4 \cdot EBTP + \alpha_3 \cdot BIG4 + \alpha_4 \cdot \Delta NPL \\
& + \alpha_5 \cdot BegNPL + \alpha_6 \cdot \Delta LOAN + \alpha_7 \cdot BegLOAN + \alpha_8 \cdot LCO + \alpha_9 \cdot BegLLA \quad (1) \\
& + \alpha_{10} \cdot LnASSET + \alpha_{11} \cdot GROWTH + \alpha_{12} \cdot GDP + YEARDUMMIES + \epsilon
\end{aligned}
$$

$$
\begin{aligned}
LLP = {} & \beta_0 + \beta_1 \cdot EBTP + \beta_2 \cdot TENURE \cdot EBTP + \beta_3 \cdot TENURE + \beta_4 \cdot \Delta NPL \\
& + \beta_5 \cdot BegNPL + \beta_6 \cdot \Delta LOAN + \beta_7 \cdot BegLOAN + \beta_8 \cdot LCO + \beta_9 \cdot BegLLA \quad (2) \\
& + \beta_{10} \cdot LnASSET + \beta_{11} \cdot GROWTH + \beta_{12} \cdot GDP + YEARDUMMIES + \epsilon
\end{aligned}
$$

where:

LLP	Loan loss provisions divided by beginning total assets;
EBTP	Earnings before tax and provisions, divided by beginning total assets;
BIG4	Dummy variable that equals 1 if the auditor is a Big 4 firm and 0 otherwise;
BIG4.EBTP	Interaction term between BIG4 and EBTP
TENURE	Dummy variable that equals 1 if the auditor tenure is more than three years and 0 otherwise;
TENURE.EBTP	Interaction term between TENURE and EBTP
Δ NPL	Change in the value of non-performing loans divided by beginning total assets;
BegNPL	Beginning of the year non-performing loans divided by beginning total assets;
Δ LOAN	Change in total loans divided by beginning total assets;
BegLOAN	Beginning of the year total loans divided by beginning total assets;
LCO	Net loan write off in the year divided by beginning total assets;
BegLLA	Beginning of the year loan loss allowance divided by beginning total assets;
LnASSET	Natural logarithm of total assets at the end of the year;
GROWTH	The growth of total assets from the beginning to the end of the year;
GDP	Rate of change in Gross domestic product in the year;
YEARDUMMIES	Dummy variable for fiscal year.

Independent variables

The variables of interest are *EBTP* and the interaction terms *BIG4.EBTP* and *TENURE.EBTP*. From prior studies, a positive relationship between *EBTP* and *LLP* is an indication of banks' income smoothing via *LLP*. The interaction terms reveal whether a Big 4 auditor and longer auditor tenure mitigate income smoothing practice. In particular, a negative and significant coefficient on the interaction terms is an indicator of auditor size/auditor tenure weakening the positive relationship between *LLP* and *EBTP*. In this case, auditor size and auditor tenure help reduce banks' income smoothing through *LLP*. The study does not include both interaction terms in the same regression equation because using the two interaction terms of *EBTP* simultaneously may result in multicollinearity problem.

Besides the coefficient on the interaction variables, we are also interested in the coefficients on *BIG4* and *TENURE*. A positive and significant relationship between *LLP* and *BIG4* shows that banks audited by Big 4 auditors have higher *LLP*. Similarly, a positive and significant relationship between *LLP* and *TENURE* shows that a bank with longer auditor tenure records higher *LLP*. Even though higher *LLP* does not directly mean higher reporting quality, financial statement users and auditors are normally more worried about income-increasing earnings management and asset overstatements than with income-decreasing earnings management and asset understatements, thus a conservative *LLP* estimation could be more desirable.

Following prior literature, the study uses a dummy variable to measure the auditor size, 1 is given to a bank if it is audited by one of the Big 4 firms (PWC, Deloitte, KPMG and Ernst & Young) and 0 if otherwise. Regarding auditor tenure, prior studies use different measures for auditor tenure. Some studies such as Myers et al. (2003), Chi and Huang (2005) use the cumulative number of years that the audit firm has been engaged by the client as the measure for auditor tenure. However, this study employs a dummy variable approach following Johnson et al. (2002), Lim and Tan (2010), taking into account the fact that the relationship between auditor tenure and earnings quality may not be linear. Therefore *TENURE* is given 1 when the audit firm is employed for more than three years and 0 if otherwise.

Control variables
Control variables are included in both equations based on prior studies and the characteristics of loan loss provisioning in Vietnamese commercial banks. According to accounting regulations for Vietnamese banks issued by the State Bank of Vietnam (SBV), *LLP* is made based on both non-performing loans and total loan balances. Therefore beginning non-performing loans *(BegNPL)*, the change in *NPL* (ΔNPL), beginning total loans *(BegLOAN)* and the change in total loans ($\Delta LOAN$) are included in the models. The inclusion of *BegNPL* and ΔNPL is consistent with most prior studies on LLP such as Beaver and Engel (1996), Dantas et al. (2013), Kanagaretnam et al. (2010). The presence of *BegLOAN* and $\Delta LOAN$ are similar to prior literature such as Dantas et al. (2013), DeBoskey and Jiang (2012), Kanagaretnam et al. (2010), except for that some studies uses current loans and change in loans instead of beginning loans and change in loans. It is important to note that in order to calculate total loans and non-performing loans, the author includes the balances of non-performing loans that Vietnamese banks sold to Vietnam Asset Management Company, a state-owned enterprise established in 2013 by the Government of Vietnam to help Vietnamese banks to deal with bad debts. These loans are sold in exchange for the special bonds issued by Vietnam Asset Management Company according to guidance from the

State Bank of Vietnam. This inclusion is based on the fact that the banks retain the risk of losses associated with the loans and continue to make provisions for the loans.

The value of loans charged off *(LCO)* is incorporated as it represents realization of losses, while beginning allowances for loan losses *(BegLLA)* indicate the expectation of losses, which may effect adjustments in the current year *LLP* (Dantas et al. 2013). The variables are divided by beginning total assets of the banks to moderate possible effects of heteroskedasticity.

To control for the impact of bank size and bank growth rate, the study include natural logarithm of total assets *(LnASSET)* and the change in total assets *(GROWTH)* of the banks. The annual *GDP* growth rates are included to control for the possible effect of the macro economic environment on *LLP* (Dantas et al. 2013; Ozili 2017). In most prior studies, *BegNPL*, *Δ NPL*, *BegLOAN*, *Δ LOAN*, *LCO*, *LnASSET* show a positive association with *LLP*, while *BegLLA*, *GROWTH* have mixed results. Finally, year dummy variable is included to take into account the possible impact of the economic situations over time.

4 Results

4.1 Descriptive Statistics

Table 1 shows descriptive statistics of the variables. The loan loss provisions *(LLP)* have a mean (median) value of 0.71% (0.63%) of beginning total assets. The average *LLP* accounts for 32.42% of the average *EBTP*. These statistics are higher than those reported for both US and African banks. For example, DeBoskey and Jiang (2012) report the mean (median) value of LLP of 0.21% (0.17%) of beginning total assets, and the proportion of LLP to EBTP of 10.88% for US banks. While Ozili (2017) documents the mean (median) LLP of 0.9% (0.5%) and the average LLP of 28.13% of EBTP for African banks. These figures show that *LLP* is an important expense for Vietnamese banks and is likely to affect the profit reported by the banks in the period. Table 1 shows that non-performing loans have a mean (median) value of 2.34% (1.45%) of total assets, which are also higher than US figures of 0.43% (0.32%) (DeBoskey and Jiang 2012), but lower than African statistics of 7.90% (4.94%) (Ozili 2017).

With regard to the choice of auditors, 82.86% of the observations are audited by a Big 4 firm. Percentage of banks audited by Big 4 firms are similar to some prior studies such as Becker et al. (1998) 82.79%, Antle et al. (2006) 81.9%; and higher than some other studies such as Kanagaretnam et al. (2010) 73.9%, Alzoubi (2018) 72.6%.

The mean (median) of auditor tenure is 0.39 (0). The median value of 0 indicates that there are more banks with auditor tenure from one to three years than banks with tenure from four to five years. This suggests that Vietnamese banks have a

Table 1 Descriptive statistics of the variables

Variable*	Mean	Median	SD	1st quartile	3rd quartile
LLP	0.0071	0.0063	0.005	0.0036	0.0093
EBTP	0.0219	0.0201	0.0127	0.0133	0.0277
BIG4	0.8286	1	0. 3777	1	1
TENURE	0.3905	0	0.4890	0	1
BegNPL	0.0234	0.0145	0.0247	0.0080	0.0281
Δ NPL	0.0081	0.0041	0.0183	−0.0001	0.0101
BegLOAN	0.5389	0.5484	0.1297	0.4403	0.6461
Δ LOAN	0.1446	0.1172	0.1280	0.0717	0.1754
LCO	0.0041	0.0024	0.0050	0.0006	0.0059
BegLLA	0.0082	0.0075	0.0046	0.0049	0.0107
LnASSET	18.199	18.308	1.2296	17.131	19.059
GROWTH	0.2572	0.1831	0.2906	0.0928	0.3462
GDP	6.01	6.10	0.5276	5.42	6.42
N	189				

*Variable definitions: *LLP*: loan loss provisions divided by beginning total assets; *EBTP*: earnings before tax and provisions, divided by beginning total assets; *BIG4*: a dummy variable that equals 1 if the auditor is a Big 4 firm and 0 otherwise; *TENURE*: Dummy variable that equals 1 if the auditor tenure is more than three years and 0 if the auditor tenure is from one to three years; *BegNPL*: beginning of the year non-performing loans divided by beginning total assets; *Δ NPL*: the change in non-performing loans divided by beginning total assets; *BegLOAN*: beginning of the year total loans divided by beginning total assets; *Δ LOAN*: the change in total loans outstanding divided by beginning total assets; *LCO*: net loan write off divided by beginning total assets; *BegLLA*: beginning of the year loan loss allowance divided by beginning total assets; *LnASSET*: Natural logarithm of total assets at the end of the year (in millions Vietnamese Dong); *GROWTH*: the growth of total assets from the beginning to the end of the year; *GDP*: rate of change in real Gross Domestic Product in the year. The initial sample is 210 bank-year observations (21 banks in 10 years), however the actual sample is reduced to 189 observations because of the missing comparative figures for the first year

greater tendency to change auditors rather than maintaining a long-term auditor-client relationship.

4.2 Correlations Among Variables

Table 2 shows the correlation coefficients for the variables used in the models. The relationship between *LLP* and *EBTP* is positive and significant at 1% significant level, which is consistent with the expectation that the banks use *LLP* to smooth earnings. *LLP* is positively correlated with beginning *NPL*, beginning loans, loans charge-off, beginning loan loss allowance, natural logarithm of total assets, and *GDP*. *LLP* has a positive but not significant relationship with Big 4 auditors. The relationship between *LLP* and auditor tenure is negative and insignificant.

Table 2 Correlations among variables

	LLP	EBTP	BIG4	TENURE	ΔNPL	BegNPL	ΔLOAN	BegLOAN	LCO	BegLLA	LnASSET	GROWTH	GDP
LLP	1												
EBTP	0.375***	1											
BIG4	0.036	−0.025	1										
TENURE	−0.080	0.036	−0.076	1									
ΔNPL	0.019	−0.215***	−0.064	0.032	1								
BegNPL	0.224***	−0.301***	−0.202***	−0.170**	0.055	1							
ΔLOAN	−0.041	0.378***	0.038	0.003	0.059	−0.126*	1						
BegLOAN	0.331***	0.095	−0.219***	0.010	−0.026	0.324***	0.005	1					
LCO	0.775***	0.317***	0.043	−0.048	−0.110	0.067	−0.237***	0.257***	1				
BegLLA	0.409***	−0.076	−0.043	−0.133*	−0.090	0.517***	−0.224***	0.571***	0.398***	1			
LnASSET	0.214***	−0.089	0.539***	0.063	−0.053	−0.004	−0.140*	0.11	0.287***	0.300***	1		
GROWTH	−0.205***	0.455***	−0.016	0.050	−0.034	−0.225***	0.780***	−0.054	−0.281***	−0.329***	−0.225***	1	
GDP	0.172**	−0.030	0.015	−0.044	0.052	0.379***	−0.065	0.260***	0.049	0.295***	0.239***	−0.066	1

Notes This table reports correlations among the variables used in the model to estimate the moderating impact of auditor size and auditor tenure on income smoothing through loan loss provisions. The definitions for the variables are given in footnotes of Table 1. ***, **, * indicate significance at 1%, 5% and 10% levels (two-tailed), respectively

All the variables are included in the regression as the correlations are lower than 0.8. The variance inflation factor (VIF) tests are also performed for the models and no multicollinearity problems are identified (results unreported for brevity) with the maximum VIF value of 6.84.

4.3 Regression Results

Table 3 reports the results of regression models. The dependent variable is the loan loss provisions *(LLP)*. The coefficients of interest are those on *EBTP* and the coefficient of the interaction terms *BIG4.EBTP* and *TENURE.EBTP*. Model (1) shows the regression results without auditor size and auditor tenure and the interactions variables, model (2) and model (3) show the results with the interaction terms of auditor size and auditor tenure separately.

To mitigate the impact of the possible serial and cross-sectional correlations in the residuals, the study uses OLS regressions with clustered robust errors for all models (Pertersen 2009). t-statistics and p-value based on clustered standard errors are reported for all models.

Model (1) shows that the coefficient on *EBTP* is positive and significant at conventional levels. This is indicative of Vietnamese banks smoothing their earnings through *LLP*.

Regarding the interaction terms, the coefficient on the interaction variable *BIG4.EBTP* in model (2) is positive and insignificant. In addition, the coefficient on *BIG4* is negative and insignificant. These results therefore do not seem to support hypothesis H1 that the presence of a Big 4 auditor mitigate the positive relationship between *LLP* and *EBTP*.

The coefficient on the interaction term *TENURE.EBTP* is negative and significant at 10% level. In addition, the coefficient on *TENURE* is positive and insignificant. This seems to indicate that longer auditor tenure reduces the positive relationship between *LLP* and *EBTP*.

Regarding the control variables, the change in loans and loan charge-offs are positive and significantly related to *LLP* in all models. The coefficients on the change in non-performing loans, beginning non-performing loans, beginning loans and beginning loan loss allowances are positive as expected and in line with prior research such as Dantas et al. (2013), DeBoskey and Jiang (2012). Bank size and *GDP* growth rates do not have a significant relationship with *LLP*, while banks' growth has a negative and significant association with *LLP* in all models. This result indicates that a growing bank recorded less *LLP*, which is in contrary to prior research such as DeBoskey and Jiang (2012).

Table 3 The impact of auditor size and auditor tenure on banks' income smoothing

Variable	Model (1)	Model (2)	Model (3)
	Coefficient (t-value)	Coefficient (t-value)	Coefficient (t-value)
Intercept	−0.0018 (−0.34)	0.0027 (0.50)	−0.0003 (−0.06)
EBTP	0.1382*** (3.59)	0.0943 (1.61)	0.1764*** (3.92)
BIG4		−0.0004 (−0.19)	
BIG4.EBTP		0.0695 (1.06)	
TENURE			0.0016 (1.45)
TENURE.EBTP			−0.0838* (−1.74)
ΔNPL	0.0222 (1.50)	0.0225 (1.72)	0.0226 (1.59)
BegNPL	0.0324 (1.58)	0.0331* (1.67)	0.0311 (1.43)
Δ LOAN	0.0092*** (3.43)	0.0091*** (3.48)	0.0089*** (3.13)
BegLOAN	0.0018 (1.3)	0.0034* (2.04)	0.0027 (1.65)
LCO	0.5942*** (7.04)	0.5766*** (7.37)	0.5727*** (6.84)
BegLLA	0.0204 (0.27)	0.0201 (0.28)	0.0382 (0.54)
LnASSET	−0.0000 (−0.02)	−0.0003 (0.90)	−0.0001(−0.4)
GROWTH	−0.0041*** (−3.95)	−0.0045*** (−4.56)	−0.0042*** (−4.04)
GDP	0.0002 (0.38)	−0.0001 (−0.22)	0.0001 (0.1)
YEARDUMMIES	Yes	Yes	Yes
N	189	189	189
Adjusted R2	73.99	74.81	74.98
F-value	129.10***	207.96***	172.89***
Root MSE	0.00268	0.00265	0.00264
Maximum VIF	3.89	6.84	5.65

Notes This table reports the results of the regression based on equation (1) and (2). The definitions for the variables are given in footnotes of Table 1. The models are estimated with year dummies, the results of which are not tabulated. All models are estimated using bank-level clustering to produce robust standard errors

In selecting the appropriate regression models for panel data, the study employs F test, Hausman test and Breusch and Pagan Lagrangian multiplier test. The test results (not reported for brevity) show that OLS regression is appropriate for the study. The variance inflation factor (VIF) test is performed, which indicates no multicollinearity problem.

***, **, * indicate significance at 1%, 5%, 10% levels (two-tailed), respectively

4.4 Bayes Factor Analysis

To take into consideration the possibility that the use of p-values exaggerates the strength of evidence against the null hypothesis (Goodman 1999), the study employs Bayes factor to interpret the results of the empirical tests. Table 4 represent the minimum Bayes factors for the main variables of interest: *EBTP*, *BIG4.EBTP* and *TENURE.EBTP*.

Table 4 show that the minimum Bayes factor for *EBTP* is 0.009, which means the probability of banks not smoothing income gets as much as only 1% support.

Table 4 Minimum Bayes factors and the effect of the evidence on the probability of the null hypothesis

Variable	p-value (Z Score)	Minimum Bayes factors	Decrease in the probability of the Null Hypothesis, %	
			From	To no less than
EBTP (From model (1))	0.002 (3.08)	0.009 (1/114) *	90	7.4**
			50	0.9
			25	0.3
BIG4.EBTP (From model (2))	0.300 (1.04)	0.58 (1/1.7)	90	84
			50	37
			16	10
TENURE.EBTP (From model (3))	0.097 (1.66)	0.25 (1/4)	90	69
			50	20
			30	10

Notes This table reports the minimum Bayes factors and the effect of the evidence on the probability of the null hypothesis. Calculations were performed as follows:

*Minimum Bayes factor $= e^{(-z^2/2)}$ where z is the number of standard errors from the null effect

**A probability (*Prob*) of 90% is equivalent to an odds of 9, calculated as $Prob/(1 - Prob)$

$Posterior odds = Bayes factor \cdot prior odds$; thus, $(1/114) \cdot 9 = 0.08$

$Probability = odds/(1 + odds)$; thus, $0.08/1.08 = 0.074$

This shows strong evidence that banks in the study sample smooth their earnings in the period. The result is in line with expectation, due to the nature of *LLP* that is highly subjective to management judgment. The result is also consistent with prior literature that provide evidence of banks' smoothing income in Vietnam such as Dao (2017); and other countries such as DeBoskey and Jiang (2012), Fonseca and González (2008), Ozili (2017), Wahlen (1994).

The interaction term between *BIG4* and *EBTP* has a minimum Bayes factor of 0.58, which indicates that the null hypothesis receive 58% as much support as the best supported hypothesis. The evidence from the study is therefore insufficient for us to conclude that Big 4 auditors have an impact on income smoothing practice of the banks, given the contradicting findings documented in previous research about the impact of big audit firms. The result is consistent with evidence documented from Europe and Africa such as Alves (2013), Antle et al. (2006), Maijoor and Vanstraelen (2006), Ozili (2017).

With regard to the interaction term between *TENURE* and *EBTP*, the minimum Beyes factor is 0.25, indicating that the alternative hypothesis is 4 times as likely as the null hypothesis, based on the observed data. In order to have a final null hypothesis probability of 10%, the prior probability of the null hypothesis must be less than 30%. From a literature review of the topic, it is noted that a majority of prior research supports the hypothesis of the auditor tenure's effect on clients' reporting quality, especially when auditor tenure is around 5 years. Some typical examples include Johnson et al. (2002) with p-value less than 1%, Alzoubi (2018) p-value less than 1%, Chi and Huang (2005) p-value less than 5%, Dantas and Medeiros (2015) p-value less than 5%.

Given the strong support from these prior research, although the evidence for the alternative hypothesis in this study is not strong (minimum Bayes factor is 0.25), it is moderately convincing that longer auditor tenure has a mitigating impact on income smoothing practice of the banks, thus supporting H2.

5 Discussions

The result shows that Big 4 auditors do not have a significant impact on reducing income smoothing through loan loss provisions by Vietnamese banks. The result is not in line with the common belief that big audit firms constrain earnings management, and is in contrast with the evidence documented in a previous research by Kanagaretnam et al. (2010) for a sample of banks from 29 countries. However, it is very similar to evidence documented by Ozili (2017) for a sample of banks from 19 African economies, which shows an insignificant positive relationship between the interaction term and loan loss provisions. It is also consistent with a number of studies that find no difference between big audit firms and other firms in relation to limiting earnings management in non-financial service sectors in European countries such as Bédard et al. (2004), Davidson et al. (2005), Maijoor and Vanstraelen (2006), Piot and Janin (2007). Some researchers comment that evidence of the significant impact of big audit firms on earnings management is more popular in Anglo-Saxon culture than in other countries (Maijoor and Vanstraelen 2006) and big audit firms' conservatism is larger in countries with strong investor protection, a higher litigation risk against auditors and common law tradition (Francis and Wang 2008).

Vietnam is an emerging country with its legal system being developed. The audit profession is young; regulations on legal responsibilities of auditors to third parties are still general. Even though a number of fraud cases in public companies' financial statements were discovered lately, no major litigations against auditors have been noted in the past years (Le 2012, State Audit of Vietnam 2009). According to Piot and Janin (2007), a low litigation risk might increase auditors' tolerance to opportunistic accounting practice.

Audit quality is not only dependent on audit firms but is also affected by regulators and standard setters (Kitiwong 2019). According to World Bank (2016), the supervision and enforcement of compliance with accounting, auditing and ethical standards in Vietnam by the government authorities is still inadequate. This could be another explanation for a weak impact of big audit firms. Ozili (2017) also suggests that weak bank supervision and legal enforcement in Africa could result in less motivation for big audit firms to deliver superior audit quality for banks in the region.

In addition to inadequate enforcement of accounting and auditing standards, shortcomings of the accounting regimes could be another cause for the lack of big auditor impact. Vietnamese accounting standards, which were issued in the period 2001–2005, have not been updated with the recent changes in the International Financial

Reporting Standards. Some important standards, such as financial instruments, fair value accounting, and impairment, have not been issued. In addition to the accounting standards, Vietnamese banks are required to apply accounting regulations set out by the State Bank of Vietnam. World Bank (2016) argues that this practice may negatively affect the reliability and comparability of accounting information, since there may be differences between the accounting standards and other specific regulations in some instances.

Regarding auditor tenure, the regression result shows that longer auditor tenure is effective in constraining banks' earnings management. The result is consistent with evidence documented in prior research such as Alzoubi (2018), Chi and Huang (2005), Johnson et al. (2002), Myers et al. (2003). An auditor may not have sufficient knowledge about a new client. When auditor tenure increases, accumulated knowledge about the client allows the auditor to better assess audit risk and perform the audit. Therefore long auditor tenure may improve the competence of the auditor. Loan loss provision is a high-risk item in a bank's financial statements, due to high level of subjectivity involved. Therefore understanding of the banks' activities and the characteristics of its loan portfolios that is achieved by a long relationship is very important to auditors in assessing estimates and judgments made by management.

6 Conclusions

This paper examines the impact of auditor size and auditor tenure on income smoothing through loan loss provisions in Vietnamese commercial banks. Based on a sample of 21 Vietnamese banks in the period 2008–2017, the findings show that Vietnamese banks smooth their earnings through loan loss provisions. However, Big 4 auditors do not show a significant impact on reducing this income smoothing, while there is moderate evidence that long auditor tenure has a mitigating effect on this kind of earnings management.

The results of the study are consistent with prior empirical evidence that big audit firms do not always make a difference from other audit firms in relation to constraining earnings management, especially in the context of a country with a low risk of litigation against auditors, insufficient enforcement of observance with accounting and auditing standards, and limitations of the accounting regime. Whereas the role of auditor tenure is in line with prior literature that finds long auditor tenure reducing earnings management practice.

The study contributes to the current limited literature of the impact of audit quality dimensions on earnings smoothing in the banking industry in the context of a developing country. The study results have important implications for the regulators and standard setters in improving the national audit environment in Vietnam. Measures should be taken to ensure a greater observance of the accounting and auditing standards. Cases of auditor failures should be closely investigated and harsh penalties taken if necessary to improve audit quality. In addition, Vietnamese accounting standards should be updated and take priority over other specific accounting guidance in order to ensure comparability and reliability of financial statements. Shareholders should take a more active role in corporate governance, including the appointment of auditors.

The study has certain limitation. The study assumes that big audit firms (or non-big firms) are identical in terms of audit quality. In fact, audit firms within the same group may not have homogenous audit quality, and even different national divisions of an audit firm have different audit quality. Therefore, future studies may investigate audit quality among the audit firms within a big firm group.

References

Alves, S. (2013). The impact of audit committee existence and external audit on earnings management: evidence from Portugal. *Journal of Financial Reporting and Accounting, 11*(2), 143–165.

Alzoubi, E. (2018). Audit quality, debt financing, and earnings management: evidence from Jordan. *Journal of International Accounting, Auditing and Taxation, 30*, 69–84.

Antle, R., Gordon, E., Narayanamoorthy, G., & Zhou, L. (2006). The joint determination of audit fees, non-audit fees, and abnormal accruals. *Review of Quantitative Finance and Accounting, 27*(3), 235–266.

Bédard, J., Chtourou, S. M., & Courteau, L. (2004). The effect of audit committee expertise, independence, and activity on aggressive earnings management. *Auditing: A Journal of Practice and Theory, 23*(2), 13–35.

Beaver, W. H., & Engel, E. E. (1996). Discretionary behavior with respect to allowance for loan losses and the behavior of securities prices. *Journal of Accounting and Economics, 21*(1–2), 177–206.

Becker, C., DeFond, M., Jiambalvo, J., & Subramanyam, K. (1998). The effect of audit quality on earnings management. *Contemporary Accounting Research, 15*(Spring), 1–24.

Carcello, J. V., & Nagy, A. L. (2004). Audit firm tenure and fraudulent financial reporting. *Auditing: A Journal of Practice and Theory, 23*(2), 317–334.

Chi, W., & Huang, H. (2005). Discretionary accruals, audit-firm tenure and auditor tenure: an empirical case in Taiwan. *Journal of Comtemporary Accounting and Economics, 11*(1), 65–92.

Craswell, A. T., Francis, J. R., & Taylor, S. L. (1995). Auditor brand name reputations and industry specializations. *Journal of Accounting and Economics, 20*(December), 297–322.

Dantas, J. A., & Medeiros, O. R. (2015). Quality determinants of independent audits of banks. *Revista Contabilidade et Financas, 26*(67), 43–56.

Dantas, J. A., Medeiros, O. R., & Lustosa, P. R. B. (2013). The role of economic variables and credit portfolio attributes for estimating discretionary loan loss provisions in Brazilian banks. *Brazilian Business Review, 10*(4), 65–90.

Dao, N. G. (2017). *Nghien cuu chat luong thong tin loi nhuan cong bo cua cac ngan hang thuong mai Viet Nam*. Hanoi: National Economics University. http://sdh.neu.edu.vn/nghien-cuu-sinh-dao-nam-giang-bao-ve-luan-an-tien-si-226865.html. Accessed 1 March 2019.

Davidson, R., Goodwin-Stewart, J., & Kent, P. (2005). Internal governance structures and earnings management. *Accounting and Finance, 45*(2), 241–267.

DeBoskey, D. G., & Jiang, W. (2012). Earnings management and auditor specialization in the post-sox era: an examination of the banking industry. *Journal of Banking and Finance, 36*(2), 613–623.

Dechow, P., & Skinner, D. (2000). Earnings management: reconciling the views of academics, practitioners, and regulators. *Accounting Horizons, 14*(2), 235–250.

Fonseca, A., & González, F. (2008). Cross-country determinants of bank income smoothing by managing loan-loss provisions. *Journal of Banking and Finance, 32*(2), 217–233.

Francis, J. R., Maydew, E. L., & Sparks, H. C. (1999). The role of Big 6 auditors in the credible reporting of accruals. *Auditing: A Journal of Practice and Theory, 18*(2), 17–34.

Francis, J. R., & Wang, D. (2008). The joint effect of investor protection and Big 4 audits on earnings quality around the world. *Contemporary Accounting Research, 25*(1), 157–191.

González-Díaz, B., Garíca-Fernández, R., & López-Dáaz, A. (2015). Auditor tenure and audit quality in Spanish state-owned foundations. *Revista de Contabilidad -Spanish Accounting Review, 18*(2), 115–126.

Goodman, S. N. (1999). Toward evidence-based medical statistics. 2: The Bayes factor. *Annals of Internal Medicine, 130*(12), 1005–1013.

Jeong, S. W., & Rho, J. (2004). Big Six auditors and audit quality: The Korean evidence. *The International Journal of Accounting, 39*(2), 175–196.

Johnson, V. E., Khurana, I. K., & Reynolds, J. K. (2002). Audit-firm tenure and the quality of financial reports. *Contemporary Accounting Research, 19*(4), 637–660.

Kanagaretnam, K., Lim, C. Y., & Lobo, G. J. (2010). Auditor reputation and earnings management: international evidence from the banking industry. *Journal of Banking and Finance, 34*(10), 2318–2327.

Kanagaretnam, K., Lobo, G. J., & Mathiew, R. (2003). Managerial incentives for income smoothing through bank loan loss provision. *Review of Quantitative Finance and Accounting, 20*(1), 63–80.

Kitiwong, W. (2014). *Earnings management and audit quality: evidence from Southeast Asia.* University of York. http://etheses.whiterose.ac.uk/7007/. Accessed 1 March 2019.

Le, T. T. H. (2012). Trach nhiem phap ly cua kiem toan vien. *Tap chi ngan hang, 7,* 49–51.

Lim, C. Y., & Tan, H. T. (2010). Does auditor tenure improve audit quality? Moderating effects of industry specialization and fee dependence. *Contemporary Accounting Research, 27*(3), 923–957.

Lin, J. W., Li, J. F., & Yang, J. S. (2006). The effect of audit committee performance on earnings quality. *Managerial Auditing Journal, 21*(9), 921–933.

Maijoor, S. J., & Vanstraelen, A. (2006). Earnings management within Europe: the effect of member state audit environment, audit firm quality and international capital markets. *Accounting and Business Research, 36*(1), 33–52.

Myers, J. N., Myers, L. A., & Omer, T. C. (2003). Exploring the term of the audit-client relationship and the quality of earnings: a case for mandatory auditor rotation? *The Accounting Review, 78*(3), 779–799.

Ozili, P. K. (2017). Bank earnings smoothing, audit quality and procyclicality in Africa: the case of loan loss provisions. *Review of Accounting and Finance, 16*(2), 142–61.

Pertersen, M. (2009). Estimating standard errors in finance panel data sets: comparing approaches. *Review of Financial Studies, 22*(1), 435–480.

Piot, C., & Janin, R. (2007). External auditors, audit committees and earnings management in France. *European Accounting Review, 16*(2), 429–454.

State Audit of Vietnam. (2009). 15 nam kiem toan doc lap o Vietnam. State Audit of Vietnam. https://www.sav.gov.vn/Pages/chi-tiet-tin.aspx?ItemID=32987&l=TinTucSuKien. Accessed 1 March 2019.

Tendeloo, B., & Vanstraelen, V. (2008). Earnings management and audit quality in Europe: evidence from the private client segment market. *European Accounting Review, 17*(3), 20–38.

Wahlen, J. M. (1994). The nature of information in commercial bank loan loss disclosures. *The Accounting Review,* 455–478.

World Bank. (2016). *Vietnam report on the observance of standards and codes: accounting and auditing module.* Hanoi: Hong Duc Publishing House.

The Threshold for the Efficient Scale of Vietnamese Commercial Banks: A Study Based on Bayesian Model

Le Ha Diem Chi, Ha Van Dung, and Nguyen Thi Minh Chau

Abstract The study is conducted to determine the relationship between total assets and return on assets (ROA) and to find the optimal size of commercial banks in Vietnam. Based on the data collected from 31 Vietnamese commercial banks in the period 2007–2017, the results show that there is an inverted U-shaped relationship between total assets and ROA of commercial banks in Vietnam. The current total assets of commercial banks are still much below the optimal size. Therefore, the policies of the Central Bank and bank's owners should focus on increasing bank's capital as well as bank's financial capacity.

Keywords Total assets · Threshold · Inverted U-shaped relationship · ROA · Optimal size

1 Introduction

Commercial banks play an important role in Vietnames financial system. Commercial banks ensure the capital circulation in the economy when the growth of capital market is still decent (Truong Van Phuoc 2017). After experiencing rapid growth during the period 2007–2010, Vietnamese commercial banks have faced difficulties such as low capital adequacy ratio, rising non-performing loans, low liquidity ratio and low operating efficiency, which force commercial banks restructure. In addition, the competition climbs up in the financial market when there is the foundation of foreign commercial banks. Many commercial banks have joined the wave of merging

L. H. Diem Chi (✉) · H. Van Dung · N. T. M. Chau
Banking University of Ho Chi Minh City, 56 Hoang Dieu 2, Thu Duc, Ho Chi Minh City, Vietnam
e-mail: chilhd@buh.edu.vn

H. Van Dung
e-mail: dunghv@buh.edu.vn

N. T. M. Chau
e-mail: chauntm@buh.edu.vn

© The Editor(s) (if applicable) and The Author(s), under exclusive license
to Springer Nature Switzerland AG 2021
N. Ngoc Thach et al. (eds.), *Data Science for Financial Econometrics*,
Studies in Computational Intelligence 898,
https://doi.org/10.1007/978-3-030-48853-6_36

and acquisitions in order to form larger banks in order to enhance their competition capability and operating efficiency.

Many empirical studies of banking efficiency suggest that the relationship between bank size and bank efficiency is inconsistent and is highly dependent on the characteristics and the operating environment of banks. The studies of Arif et al. (2013) and Onuonga (2014) show that there is a positive relationship between total assets and bank profitability, while researches conducted by Aladwan (2015) and Le Thanh Tam et al. (2017) discover that total assets and bank profitibilty are negatively related. Meanwhile, other studies found no relationship between bank size and profitability such as that of Parvin et al. (2019) and Dawood (2014). Particularly, Yao et al. (2018) and Goddard et al. (2018) found that both linear and non-linear inverted U-shaped relationship existed between total assets and profitability of Pakistan banks.

In conclusion, the results of the relationship between bank size and profitability are unconcluded. Furthermore, there are still no studies that determine the optimal size for commercial banks. So, this research is conducted to clarify the relationship between bank size and profitability of Vietnamese commercial banks and determine the optimal size of Vietnamese commercial banks if there exists the inverted U-shaped relationship between bank size and profitability. Furthermore, this study gives some suggestions to improve the banks' operating efficiency in the future.

2 Theoretical Framework and Literature Review

Commercial bank profitability is always a concerned topic for researchers. Bank profits depend on the bank-specific characteristics including total assets, outstanding credit, equity, deposits, credit quality or even external factors such as economic growth, inflation, the market competition, etc. If only considering the relationship between size and profit, there are different theories. According to the theory of economy of scale, the bigger the bank, the more profits it can generate because bigger banks allow them to cut their average costs. These banks can also manage to reduce costs by reducing risks through the diversification of products and services (Regehr and Sengupta 2016; Mester 2010). Nevertheless, if the bank is too big, its costs will be high and thus the bank size will no longer be an advantage resulting less profits. For small banks, according to Berger et al. (2005), they take advantage of their local customer networks better than large banks. Consequently, small banks can obtain exclusive and useful information to set better contract terms and prices. This will make up for the disadvantage of economy of scale.

There have been many empirical researches emphasizing determinants of bank profitability. Each research has a different scope of study and methodolody. These studies give inconsistent results about the determinants of banking profitability, including the relationship between total assets and profitability.

Yao et al. (2018) study the impact of bank-specifics, industry-specifics and macroeconomic factors on the profitability of 28 commercial banks in Parkistan in the period 2007–2016. By using the GMM method, the research shows that there is an inverted U-shaped relationship between the bank size and profitability. This means that a rise in the bank profits leads to an increase in the bank assets to a certain threshold. After that, when the total assets increase further, their profits will be decreased. On the contrary, credit quality, growth of banking sector, inflation, and industrial concentration is inversely proportional to bank profitability.

Samad (2015) uses secondary data collected from 42 commercial banks in Bangladesh to investigate determinants of bank profitability. The results show that bank size and microeconomic variables did not impact bank profitability. Variables such as equity ratio, credit quality, operating efficiency and operating expenses have impacts on bank profitability. The findings on the relationship between total assets and profitability of commercial banks in Bangladesh were also consistent with that of Parvin et al. (2019).

Ozili (2017) examines determinants of African bank profitability. Using static and dynamic panel estimation techniques and a sample of 200 listed and non-listed banks across Africa during the period 2004–2013, the study shows that bank size, total capital and loan loss provisions significantly affect bank profitability. Particularly, bank size had a negative impact on profitability of listed banks but not of non-listed banks.

Le Thanh Tam et al. (2017) investigate factors affecting commercial bank profitability in the period 2007–2013. The bank size was determined based on the total assets. The bank profitability was measured through profitability indicators including return on assets (ROA), return on equity (ROE) and net interest margin. By using FEM and REM models, the study showe that bank size has a negative relationship with bank profitability.

In analyzing the relationship between bank size and profitability, Regehr and Sengupta (2016) use bank-specifics and market variables. Defining natural logarithm of total assets as SIZE to reflect the bank size, the results revealed that the SIZE positively affects bank profitability while the SIZE-squared variable negatively affects bank profitability. The findings show that increasing size could help small banks make significant profits but could have a negative impact on large banks. Along with other variable analyses, the article suggests banks could achieve higher profitability if they are better positioned.

3 Methodology

We employ Bayesian regression model to investigate the relationship between total assets and bank profitability in Vietnam. In addition, the paper also identifies whether there exists the optimal size of commercial banks. The Bayesian models are characterized with an adaptive Metropolish-Hastings algorith. The bayesian model uses priors for model parameters and estimates parameters using MCMC by drawing sample which are drawned from the corresponding posterior model. Many likelihood models such as univariate normal linear, multivariate linear, nonlinear, generalized linear, and multiple-equations linear models and prior distribution are given. Model parameters are randomly estimated with posterior distribution formed by combining prior information of parameters with evidence from observed data. Prior information is used to calculate probability distributions parameters while prior distributions do not depend on the observed data. In order to assure the model efficiency, we check the MCMC convergence, MCMC sampling efficiency, and test for interval hypothesis.

4 Research Model

Return on assets (ROA) is used as dependent variable reflecting the bank performance while total assets are independent variables. Some control variables are selected from bank-specific characteristics such as equity, loans, deposits, and liquidity. CR3 represents the banking industry concentration in this model. Other factors including inflation rate, GDP growth and money supply are defined as macroeconomic variables.

$$ROA_{it} = \beta_0 + \beta_1 SIZE + \beta_2 SIZE + \beta_3 ETA + \beta_4 NTA + \beta_5 DEP$$
$$+ \beta_6 LIQ + \beta_7 CR3 + \beta_8 INF + \beta_9 GDP + \beta_{10} M2 + \varepsilon_{it}$$

In particular,

- **ROA_{it}** is the return on assets of bank i in year t which is calculated by dividing net income by average total assets.
- **SIZE** represents the bank total assets and is measured by the natural logarithm of total assets. Iannotta et al. (2007), Flamini et al. (2009), Alper and Anbar (2011), Muda et al. (2013), Al-Jafari and Alchami (2014), Menicucci and Paolucci (2016) … determined that total assets affect bank profitability in their studies. SIZE2 is the square of the SIZE variable.
- **Bank-specific chacracteristics as control variables**:

    ETA is the equity-to-total assets ratio. ETA variable used in this study is used in the study of Athanasoglou et al. (2006), and Flamini et al. (2009).
    NTA is calculated by dividing loans by total assets. NTA variable is used by Athanasoglou et al. (2006), Iannotta et al. (2007), Lee and Hsieh (2013), Ayaydin and Karakaya (2014), Menicucci and Paolucci (2015), and Jabra et al. (2017).

- DEP is the deposit-to-total assets ratio. Researcher using DEP variable in their model were Iannotta et al. (2007), Flamini et al. (2009), and Menicucci and Paolucci (2015).
- LIQ is measured by dividing liquid assets which include cash, deposits and trading securities by total assets. Liquidity has a positive impact on bank profitability is proven in the studies of Ayaydin and Karakaya (2014), Al-Jafari and Alchami (2014), and Jabra et al. (2017).

- **CR3** is measured by dividing total assets of the three largest commercial banks by total assets of commercial banks in the model (Athanasoglou et al. 2006).
- **Macroeconomic determinants as control variables**:

  INF is calculated by using consumer price index (CPI) every year. Inflation has an influence on bank profitability (Athanasoglou et al. 2006; Flamini et al. 2009; Naceur and Kandil 2009; Lee and Hsieh 2013).

  GDP is real gross domestic product growth. The impact of gross domestic product growth on bank profitability can be found in Albertazzi and Gambacorta (2009), Flamini et al. (2009), Lee and Hsieh (2013), and Jabra et al. (2017).

  Money supply M2 growth is calculated by the growth of total means of payments every year in the country. The State Bank monetary policy will have a direct impact on bank net interest income through the usage of required reserved and/or discount and re-discount interest rate.

5 Research Data

Research data was secondary data collected from audited financial statements of 31 Vietnamese commercial banks from 2007 to 2017. The panel data was unbalanced due to the different established dates of the banks, the mergers and acquisitions or the banks which were placed under special control by the State Bank of Vietnam (SBV) (Table 1).

Based on the samples from 2007 to 2017, the average profitability ratio was at 1.02%. The lowest ROA ratio was at 0.01% in 2012 and 2016. LietVietPostBank had the highest ROA ratio, with 5.95% in 2008. Figure 1 showed that during the

Table 1 Descriptive statistics of the variables

Variable	Observations	Average	Standard deviation	Minimum value	Maximum value
ROA	320	0.0102	0.0082	0.0001	0.0595
SIZE	320	17.78	1.31	14.07	20.82
ETA	320	0.1125	0.0611	0.0370	0.4624
NTA	320	0.5149	0.1281	0.1138	0.7680
DEP	320	0.7448	0.0823	0.4299	0.8975
LIQ	320	0.2313	0.1106	0.0486	0.5747
CR3	320	0.4497	0.0492	0.4010	0.5716
INF	320	0.0831	0.0644	0.0060	0.2310
GDP	320	0.0609	0.0059	0.0520	0.0710
M2	320	0.2118	0.0900	0.0927	0.4610

Source Authors' calculation

Fig. 1 ROA of commercial banks. *Source* Authors' calculation

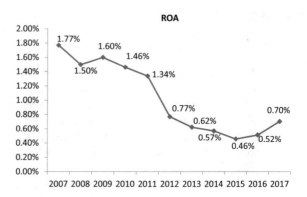

period 2007–2011, the average ROA ratio of the banks was above 1.53%. However, the average ROA ratio only accounted for 0.61% from 2011 to 2017.

The average total assets were VND 52,673,867 million in the period 2007–2017. In 2017, Joint Stock Commercial Bank for Investment and Development of Vietnam had the total asset of VND 1,104,324,861 million. Vietnam Public Joint Stock Bank was the smallest bank, with a total asset of VND 1.295.231 million in 2007. Figure 2 showing the continued rise in assets illustrated total assets of commercial banks from 2007 to 2017.

For bank-specifics, the average equity ratio was for 11.25% while the average ratio of loans over total assets was 51.49%. Figure 3 shows the fluctuation in these variables between 2007 and 2017.

In the period 2007–2017, the average concentration ratio in banking sector was 44.94%, and the inflation rate, the average GDP growth and M2 money supply from 2007 to 2017 are presented in Fig. 4.

Fig. 2 Total assets of commercial banks. *Source* Authors' calculation

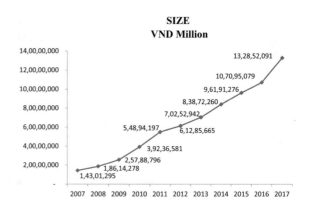

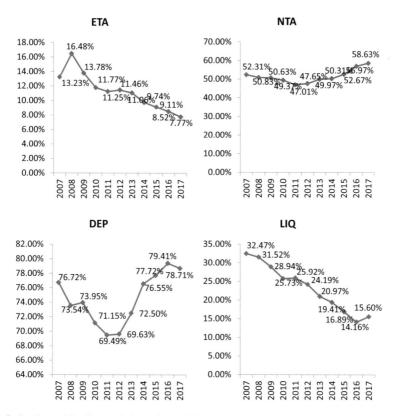

Fig. 3 Bank-specific characteristics as internal determinants. *Source* Authors' calculation

6 Resutls After Using Bayesian Regression Model

Since the study has yet had any information about prior probability distribution and previous studies did not mention most efficient prior probability distribution, the uniform prior probability distribution will be employed. To test the efficiency of coeficients, Trace, Histogram, Autocorrelation, and Density plots will be used. Estimated results are presented in Table 2.

In Figs. 5 and 6, the trace diagram shows the convergence of the SIZE and SIZE2 coefficients while the histogram demonstrates the asymptotic posterior distribution and the normal distribution of the two the coefficients. The autocorrelation chart shows that SIZE and SIZE2 have low autocorrelation coefficients, which do not affect the regression results after lag 40. The Density plot shows that prior distribution, posterior distribution and maximum likelihood distribution are approximately asymptotic. The trace, histogram, autocorrelation and density plots all show that the coefficients of SIZE and SIZE2 are reliable. In addition, the results from efficiency testing (Table 2) are also more than 1% (0.179 and 0.111). Based on the estimation results, it can conclude that total assets (SIZE) has a positive relationship with bank

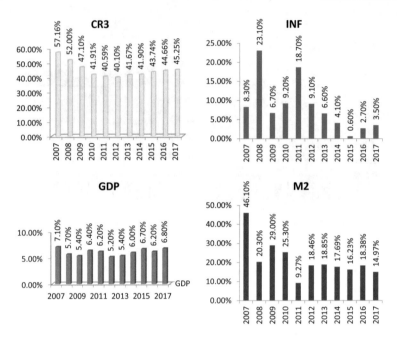

Fig. 4 Macroeconomic variables as external determinants. *Source* Authors' calculation

profitability (ROA) in Vietnam in the period 2007–2017 and that the square of total assets (SIZE2) is negatively related to bank profitability. These two results suggest that there is an inverted U-shaped relationship between total essets and bank profitability and thus there is an existence of an optimal size. This evidence also supports the results of Yao et al. (2018). For Vietnames commercial bank system, total assets have a positive relationship with bank profitability when the total assets are smaller than the optimal size. The commercial banks will decrease their profitability when bank total assets exceed this optimal size.

As the Mean of SIZE and SIZE2 are 0.001948 and –0.000045, respectively, the study finds that the optimal size for Vietnam commercial banks is about VND 3,064,191 billions. Currently, the average total assets of Vietnam commercial banks are VND 132,852 billions (in 2017), which are still far from the optimal size. The results give commercial banks more reasons to raise capital in order to ensure the equity capital following Basel II's capital adequacy ratio (CAR).

Besides the optimal size, this study finds that bank-specifics such as equity, loans and liquidity did positively affect profitability. Raising capital has a negative impact on bank profitability.

Concentration ratio negatively affects bank profitability. The efficiency ratio of macroeconomic variables such inflation, GDP and M2 is larger than 1%. This implies that macroeconomic variables have a positive impact on bank profitability. According to previous studies, inflation could have a positive or negative impact on bank profitability (Perry 1992). Perry (1992), and Said and Tumin (2011) argued that, if banks

Table 2 Bayesian estimation results

Efficiency					Min = 0.0281		
					Avg = 0.2402		
					Max = 0.6108		
Dependent varibles: ROA	Mean	Eficiency	Std.Dev	MCSE	Median	Equal - tailed [95% Cred. Interval]	
SIZE	0.001948	0.1793	0.0140361	0.000105	0.00198	−0.0255726	0.029404
SIZE2	−0.000045	0.1110	0.0003942	0.000004	−0.00005	−0.0008154	0.0007278
ETA	0.048364	0.1091	0.0226045	0.000216	0.04834	0.0039863	0.0926460
NTA	0.019156	0.0928	0.0098009	0.000102	0.01917	−0.0000565	0.038458
DEP	−0.028424	0.2095	0.0108346	0.000075	−0.02843	−0.049590	−0.007144
LIQ	0.021867	0.2173	0.0104918	0.000071	0.02188	0.0011923	0.0423799
CR3	−0.020380	0.5988	0.0238911	0.000098	−0.02040	−0.0671994	0.0264349
INF	0.027407	0.0284	0.0173224	0.000325	0.02740	−0.0066119	0.0614456
GDP	0.234570	0.6108	0.1181338	0.000478	0.23451	0.0023881	0.4648719
M2	0.028413	0.0282	0.0162721	0.000307	0.02842	−0.0034752	0.0605389
_cons	−0.022709	0.2052	0.1331529	0.000930	−0.02271	−0.2837176	0.2383755
Firm: U0:sigma2	0.0007443	0.1599	0.0002063	0.000002	0.0007107	0.0004443	0.0012386
e.roa sigma2	0.0000999	0.5724	0.000009	0.000000	0.0000995	0.0000846	0.0001179

Source Authors' calculation

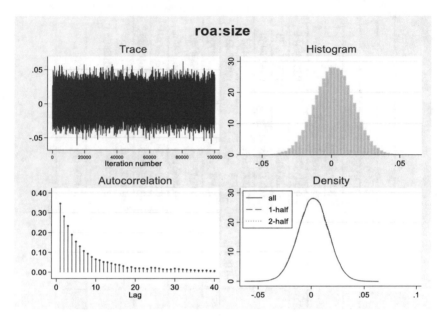

Fig. 5 Testing the SIZE coefficient's efficiency. *Source* Authors' calculation

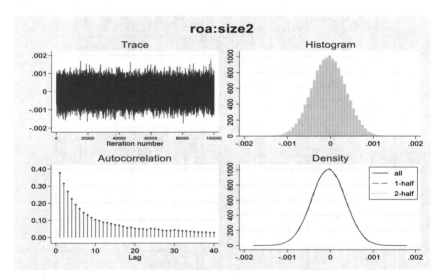

Fig. 6 Testing the SIZE2 coefficient's efficiency. *Source* Authors' calculation

could, in advance, forecast inflation rate, they could adjust their interest rates so that the increase in revenue would be greater than that of costs. If doing so, banks will enhance their operating efficiencies and vice versa. In Vietnam, inflation rate is always announced by the government agencies. Basing on this, commercial banks will issue their interest rates to get the target of profitability. The growth of GDP and M2 also affect bank efficiency. The results are consistent with other studies (Albertazzi and Gambacorta 2009; Flamini et al. 2009; Lee and Hsieh 2013).

7 Conclusion

The study employs Bayesian model and the data from 31 Vietnamese commercial banks in the period 2007–2017. The results indicate that there is an inverted U-shaped relationship exists between total assets and bank profitability and there exists an optimal size for commercial banks. This optimal size is nearly 3 times as much as that of present largest bank. The results lead to the similar implications to SBV policies. It requires commercial banks to raise capital and enhances financial capacity in order to ensure the bank equity capital following Basel II's capital adequacy ratio (CAR).

As the date of applying Basel II is approaching, commercial banks have come up with many solutions to raise their capital such as attracting capital from foreign investors, issuing new shares, merging, etc. Raising capital, increasing size and financial capacity will enhance the bank efficiencies at the same time.

References

Albertazzi, U., & Gambacorta, L. (2009). Bank profitability and the business cycle. *Journal of Financial Stability, 5*(4), 393–409.

Al-Jafari, M. K., & Alchami, M. (2014). Determinants of bank profitability: Evidence from Syria. *Journal of Applied Finance & Banking, 4*(1), 17–45.

Alper, D., & Anba, A. (2011). Bank specific and macroeconomic determinants of commercial bank profitability: empirical evidence from Turkey. *Business and Economics Research Journal, 2*(2), 139–152.

Arif, M., Khan, M. M., & Iqbal, M. (2013). Impact of bank size on profitability: Evidance from Pakistan. *International Journal of Applied Research, 2,* 98–109.

Athanasoglou, P., Delis, M., & Staikouras, C. (2006). Determinants of bank profitability in the South Eastern European region. *MPRA Paper*, 10274.

Ayanda, A. M., Christopher, E. I., & Mudashiru, M. A. (2013). Determinants of banks profitability in a developing economy: Evidence from Nigerian banking Industry. *Interdisciplinary Journal of Contemporary Research in Business, 4,* 155–181.

Ayaydin, H., & Karakaya, A. (2014). The effect of bank capital on profitability and risk in Turkish banking. *International Journal of Business and Social Science, 5*(1).

Berger, A. N., Miller, N. M., Petersen, M. A., Rajan, R. G., & Stein, J. C. (2005). Does function follow organizational form? Evidence from the lending practices of large and small banks. *Journal of Financial Economics, 76*(2), 237–269.

Dawood, U. (2014). Factors impacting profitability of commercial banks in Pakistan for the period of (2009–2012). *International Journal of Scientific and Research Publications, 4,* 1–7.

Flamini, V., Schumacher, M. L., & McDonald, M. C. A. (2009). The determinants of commercial bank profitability in Sub-Saharan Africa. *International Monetary Fund,* 9–15.

Goddard, J., Molyneux, P., & Wilson, P. (2004). The profitability of European banks: A cross sectional and dynamic panel analysis. *The Manchester School, 72,* 363–381.

Iannotta, G., Nocera, G., & Sironi, A. (2007). Ownership structure, risk and performance in the European banking industry. *Journal of Banking & Finance, 31*(7), 2127–2149.

Jabra, W. B., Mighri, Z., & Mansouri, F. (2017). Bank capital, profitability and risk in BRICS banking industry. *Global Business and Economics Review, 19*(1), 89–119.

Tam, L. T., Trang, P. X., & Hanh, L. N. (2017). Determinants of bank profitability: The case of commercial banks listed on the Vietnam's stock exchange. *Journal of Business Sciences, 1*(2), 1–12.

Lee, C. C., & Hsieh, M. F. (2013). The impact of bank capital on profitability and risk in Asian banking. *Journal of International Money and Finance, 32,* 251–281.

Menicucci, E., & Paolucci, G. (2016). The determinants of bank profitability: Empirical evidence from European banking sector. *Journal of Financial Reporting and Accounting, 14*(1), 86–115.

Mester, L. J. (2010). *Scale economies in banking and financial regulatory reform* (pp. 10–13). Federal Reserve Bank of Minneapolis: The Region.

Mohammad Suleiman Aladwan. (2015). The impact of bank size on profitability: An empirical study on listed Jordanian commercial banks. *European Scientific Journal, 11*(34), 217–236.

Muda, M., Shaharuddin, A., & Embaya, A. (2013). Comparative analysis of profitability determinants of domestic and foreign Islamic banks in Malaysia. *International Journal of Economics and Financial Issues, 3*(3), 559–569.

Naceur, S. B., & Kandil, M. (2009). The impact of capital requirements on banks' cost of intermediation and performance: The case of Egypt. *Journal of Economics and Business, 61*(1), 70–89.

Onuonga, S. M. (2014). The analysis of profitability of Kenya's top six commercial banks: Internal factor analysis. *American International Journal of Social Science, 3,* 94–103.

Parvin, S., Chowdhury, A. N. M. M. H., Siddiqua, A., & Ferdous, J. (2019). Effect of liquidity and bank size on the profitability of commercial banks in Bangladesh. *Asian Business Review, 9*(1), 7–10.

Perry, P. (1992). Do banks gain or lose from inflation. *Journal of Retail Banking, 14*(2), 25–40.

Peterson, K. O. (2017). Bank profitability and capital regulation: Evidence from Listed and Non-Listed Banks in Africa. *Journal of African Business, 18*(2), 143–168.

Regehr, K., & Sengupta, R. (2016). *Has the relationship between bank size and profitability changed?* (pp. 49–72). QII: Economic Review.

Said, R. M., & Tumin, M. H. (2011). Performance and financial ratios of commercial banks in Malaysia and China. *International Review of Business Research Papers, 7*(2), 157–169.

Samad, A. (2015). Determinants bank profitability: Empirical evidence from Bangladesh commercial banks. *International Journal of Financial Research, 6,* 173–179.

Van Phuoc, T. (2017). The role of the Vietnamese financial system for economic growth in the period of 2016–2020. *Vietnam Social Sciences, 9,* 12–20.

Yao, H., Haris, M., & Tariq, G. (2018). Profitability determinants of financial institutions: Evidence from banks in Pakistan. *International Journal of Financial Studies, 6*(53), 1–28.

Impact of Outreach on Operational Self-Sufficiency and Profit of Microfinance Institutions in Vietnam

Thuy T. Dang, Quynh H. Vu, and Nguyen Trung Hau

Abstract Paper focuses on the impact of outreach on operational self-sufficiency (OSS) and profit of microfinance institutions (MFIs) in Vietnam in the period 2011–2016. The data collected through statistic of Vietnamese MFIs and the information in the MIX market—one of credible website that is visited by many scholars in the world. Unlike previous studies, which often carried out using the traditional frequentist method, this research based on Bayesian approach, which, in many cases, is thought to be more reliable and effective statistical inference than p-value hypothesis tesing (Briggs WM, Hung TN. Asian J Econ Banking 3(2), 2019). Using Bayesian approach, the paper's results show that outreach has positive impact on OSS and profit of MFIs in Vietnam. Thus, the authors concentrate on the solutions, which aim at increasing the assessment of MFIs in the future.

Keywords Microfinance · Microfinance institutions · Operational self-sufficiency

T. T. Dang (✉)
Vietnam Institute for Indian and Southwest Asian Studies, Vietnam Academy of Social Sciences, 1 Lieu Giai St., Ba Dinh Dist., Hanoi, Vietnam
e-mail: thuy0183@gmail.com

N. T. Hau
State Bank of Vietnam, 49 Ly Thai To St., Hoan Kiem Dist., Hanoi, Vietnam
e-mail: hau.nguyentrung@sbv.org.vn

Q. H. Vu
Research Institute for Banking Academy of Vietnam, 12 Chua Boc St., Dong Da Dist., Hanoi, Vietnam
e-mail: huongquynh90@gmail.com

© The Editor(s) (if applicable) and The Author(s), under exclusive license to Springer Nature Switzerland AG 2021
N. Ngoc Thach et al. (eds.), *Data Science for Financial Econometrics*, Studies in Computational Intelligence 898,
https://doi.org/10.1007/978-3-030-48853-6_37

543

1 Introduction

The birth of microfinance in Europe dates back to tremendous increases in poverty since the sixteenth century in the form of informal lending. The first known micro loan owed its origin to small villages where families and groups of people live in community (Seibe 2003). Microfinance had undergone a complete transformation when numerous international organizations breaking into this new market. As we all know microfinance flourished in Bangladesh in the 1970s. Muhammad Yunus— Professor of Economics at Chittagong University drew the idea of microcredit within his research on poor people at his own homeland when they had been suffering famine in 1974. He decided to set up an organization to help poor people in his country and named Grameen Bank. Henceforth, modern microfinance was officially born (Ledgerwood 1999; Morduch and Haley 2002; Khandker 2003).

The proponents of microfinance claim that financial inclusion can help to substantially reduce poverty (Littlefield et al. 2003; Dunford 2006). Financial inclusion may contribute to a long-lasting increase in income by means of a rise in investments in income generating activities and to an available diversity of sources of income; it may contribute to an accumulation of assets; it may smooth consumption; it may reduce the vulnerability due to illness, drought and crop failures, and it may contribute to better education, health and housing of the borrower. Additionally, financial inclusion may contribute to an improvement of the social and economic situation of women. Finally, microfinance may have positive spillover effects such that its impact surpasses the economic and social improvement of the borrower. The positive assessment of the contribution microfinance can make to reducing poverty has assured numerous governments, non-governmental organizations (NGOs), and individuals to attempt at supporting MFIs and their activities.

Microfinance has also faced criticism. They argure that microfinance does not reach the poorest of the poor (Scully 2004), or that the poorest are consciously eliminated from microfinance programs (Simanowitz and Walter 2002). Firstly, the extreme poor often choose not to take part in in microfinance programs due to the lack of confidence or the fear of risky loans (Ciravegna 2005). The poorest of the poor, the so-called core poor, are generally too risk averse to borrow for investment in the future. They will, therefore, benefit only to a very limited extent from microfinance schemes. Secondly, the core poor are often not accepted in-group lending programs by other group members because they are seen as a bad credit risk (Hulme and Mosley 1996; Marr 2004). Thirdly, staff members of MFIs may prefer excluding the core poor since lending to them is seen as extremely risky (Hulme and Mosley 1996). Fourthly, the way microfinance programs are organized and set up may lead to the exclusion of the core poor (Kirkpatrick and Maimbo 2002; Mosley 2001). The statistics also reflect that many poor customers are unable to access to financial services so addressing the issue is necessary. Thus, how to make MFIs services increasingly popular with poor customers is of great concern to governments, MFIs and many researchers.

Accompanied with the above-mentioned concern, recently, there is a shift from subsidizing MFIs to the sustainability and self-sufficiency of these institutions. Particularly, achieving operational self-sufficiency (OSS), the ability of MFIs to cover all costs through the profits earned by providing financial services to customers without receiving any support (Kimando et al. 2012), is considered as a main factor mentioned by economists as a long-term solution for MFIs (Daher and Le Saout 2013; Quayes 2015). However, there is a debate saying that concentrating on OSS may also go with higher the cost of lending to the poor. According to Hermes et al. (2008), because lending money to the poor—especially the very poor and/or the rural poor—can be very costly, the goal of the outreach and self-suficiency may be conflicting. Thus, a concern raised up in policy circles, questioning that is there a trade-off between outreach and sustainability. Whereas the so-called welfarist view emphasizes the significance of outreach and warns about the too concentration on MFIs' sustainability, the institutionalist view claims that MFI need to be sustainable and their sustainability is the necessary condition for them to realize their social mission.

Recently, there have been numerous studies exploring the factors affecting bank profits, however, most of which use traditional regression method. Briggs and Hung (2019) argued that the current P-value-based hypothesis testing approach has been obsolete since this method appears ineffective in interpreting statistical results in many cases. Therefore, the implementation of forecasting faces many difficulties. In addition, according to Thach, Anh and An (2019), in a model which includes many explanatory variables, there are more than just one model capable of interpreting the same set of variables. Thus, if only one model is selected, it might cause the model to become unfitted. To overcome this drawback, Bayesian statistics consider all possible models (the model space) and then summarize the statistical results based on the weights of the models in model space. Therefore, this study conducted aims to evaluate impact of outreach on operational self-sufficiency and profit of microfinance institutions in Vietnam using Bayes approach.

The remainder of the study is organized as follows. Section 2 discusses the literature on outreach, OSS, profit of MFIs, as well as the nexus between outreach and OSS, profit. Section 3 provides details about data and model specification. Section 4 provides detailed explanation of empirical results and Sect. 5 concludes.

2 Review of Related Literature

2.1 Research on Outreach, OSS and Profit

MFIs outreach is a recognized goal from social and business perspective. When question of sustainability of MFIs raise, it is observed that only few percent of the MFIs are sustainable to run operation without subsidies (Hulme and Mosley 1996). Outreach and impact are integral in nature in attaining microfinance sustainability.

The perception cannot cover in general since in some cases, outreach and sustainability are competitive and sustainability pre-conditioned on the reduction or removal of subsidy on microfinance. Sustainability of microfinance is thus turning into more complicated and doubtful issue from different position and is among the one of the important essential principles of Consultative Group to Assist Poor (CGAP).

Broadly, sustainability of microfinance implies perpetuationof the program. Within microfinance, sustainability can be viewed at many levels-institutional (group and individual) and in term of organizational, managerial and financial aspects. However, financial sustainability of microfinance institutions has become the demanding point of focus in microfinance analysis. In defining sustainability of microfinance, Woller et al. (1999) used the definition given by Brinkerhoff, which encompassed sustainability could as the *"ability of a program to produce outputs that are valued sufficiently by beneficiaries and other stakeholders that the program receives enough resources and inputs to continue production"*. Pollinger etal. (2007) defined sustainability as the ability to cover annual budgets including grants, donations, and other fundraising. Acharya and Acharya (2006) considered view of Sharma and Nepal (1997) where sustainability indicates excess of operating income over operating cost. The concept is from the banker's viewpoint and it includes both financial viability and institutional viability of MFIs. Overall sustainability is not an end in itself. It is just a means to the end of improving the lot of the poor (Schreiner 1997).

Recent literature in microfinance industry, followed by crisis in several countries, directs to the factors that govern sustainability of microfinance institutions. Hermes et al. (2008), carried out an in-depth analysis of the tradeoff between self-sufficiency and depth of outreach and observed a shift from subsidizing MFIs to a focus on financial sustainability and efficiency of the institutions. They found that outreach is negatively relation to efficiency of MFIs. Clearly, MFIs that have lower average loan balance, which is a quota of the depth of outreach, are also less efficient. Ayayi and Sene (2010) identified the factors determining the sustainability of microfinance institutions. They found that a high quality credit portfolio is the most deciding component of financial sustainability of MFIs, followed by the operation of adequate interest rates and effective management to control personnel expenses. Further, they observed that the applicant outreach of microfinance programs and the age of MFIs, whose coefficients are positive and statistically significant, have a lesser effect on financial sustainability of MFIs. Kar (2011) found that expansion in leverage raises profit efficiency in MFIs while cost efficiency deteriorates with decrease in leverage. Further, he observed serious impact of leverage on depth of outreach.

In the argument between outreach versus efficiency, Hashemi and Rosenberg (2006), and Weiss and Montgomery (2005) highlight the importance of outreach as the fundamental mission of MFIs, while Christen (2001) and Isern and Porteous (2005) emphasize the status of efficiency and sustainability in securing long term viability of MFIs. Morduch (2005) argues that outreach and sustainability can be compatible under certain conditions.

2.2 Research on the Impact of Outreach to OSS

Outreach of MFIs generally indicates to both breadth of outreach and depth of outreach (Zeller and Meyer 2002; Yaron et al. 1998). The width of outreach measures the number of credit loans or the total number of borrowers over a fixed period of time. The depth of outreach is determined by the credit accessibility of the poor. It means the poorer people have accessed to the loan, the greater outreach is. Characteristics of MFIs customers in many countries globally in general and in Vietnam in particular are often women (IFAD 2009). With the fast expansion of microcredit, breadth of outreach has also increased not only at the industry level but also at the individual MFI level. Hence, depth of outreach has attracted more attention from all quarters that are anxious about the comprehensive social outreach of microfinance, including policy makers. From the point of providing poor people with access to credit, breadth of outreach can be seen as measuring the quantity of microcredit while depth of outreach measures the quality of microcredit. The most broadly used measure for depth of outreach is average loan balance per borrower. Despite of not a perfect measure of poverty level, it is an excellent proxy for depth of outreach as there is a strong positive correlation between income level and the size of loans. Particually, the poorer the borrower is, the smaller the size of the loan. Hisako (2009) uses average loan size per borrower to measure depth of outreach.

Martínez-González (2008) also capitalized that, most MFIs have been more efficient in pursuing sustainability but fall short of achieving the breath of outreach to the targeted poor people in the country. However, Kinde (2012) identifies factors affecting financial sustainability of MFIs in Ethiopia. The study found that microfinance breadth of outreach, depth of outreach, dependency ratio, and cost per borrower significantly affect the financial sustainability of MFIs in Ethiopia. However, the microfinance capital structure and staff productivity have an insignificant impact on financial sustainability of MFIs in Ethiopia during the period of study. Moreover, other studies also indicate that there is a positive relationship between MFI efficiency and domestic financial development (Hermes et al. 2009).

Although, sustainability of microfinance expresses various levels of sustainability, it is financial sustainability of the MFIs that considered for this paper. Financial sustainability indicates that revenue from the microfinance services should be greater than the charge of providing services. Therefore, self-sufficiency is an implication for the financial sustainability of the MFIs. As the microfinance industry matures, the definition of the self-sufficiency has commenced to slender (Ledgerwood 1999) and currently sustainability refers only two levels of sustainability by the most of the people joining in this industry. These are Operational Self Sufficiency (OSS) and Financial Self Sufficiency (FSS). OSS shows whether or not revenue has been earned to cover all the MFI's direct costs, excluding the cost of capital but including actual financing costs. Thus formula for calculating OSS is: [Operating Income/(Operating Expenses + provision for loan losses]. FSS on the other hand portray the actual financial health of MFIs. Thus, FSS includes cost of capital (adjusted) apart from the components in OSS. On the other hand, Pollinger et al. (2007) refers self-sufficiency as to organizations that can survive and add to their asset base wholly on the basis of income derived from their lending and related operations.

Hermes et al. (2011) used stochastic frontier analysis to examine whether there is a trade-off between outreach to the poor and efficiency of microfinance institutions (MFIs). Findings resolved convincing evidence that outreach is negatively related to efficiency of MFIs. More specifically, the observation found that MFIs that have a lower average loan balance (a measure of the depth of outreach) are also less efficient. Moreover, those MFIs that have more borrowers as clients (again a measure of the depth of outreach) are less efficient. These results remain robustly significant after having added a number of control variables. On the other hand, in 2011, Hudon and Traca (2011) used an original database of rating agencies; this study gave empirical evidence on the impact of subsidy intensity on the efficiency of MFIs. Findings resolved that subsidies have had a positive impact on efficiency, in the sense that MFIs that received subsidies are more efficient than those that do not. However, it also found that subsidization beyond a certain threshold renders the marginal effect on efficiency negative. Moreover, marginal cut on subsidy intensity would increase their efficiency. Furthermore, Hartarska et al. (2013) evaluated the efficiency of microfinance institutions (MFIs) using a structural approach which also captures institutions' outreach and sustainability objectives. The study estimated economies of scale and input price elasticities for lending-only and deposit-mobilizing MFIs using a large sample of high-quality panel data. The outcomes confirm that improvements in efficiency can come from the growth or consolidations of MFIs, as they find substantial increasing returns to scale for all but profitability-focused deposit-mobilizing MFIs. It also supported the existence of a trade-off between outreach and sustainability.

OSS represents the relationship between operating income and total operating costs. This indicator is used to assess whether or not MFIs revenue is earned to cover all the operating expenses such as wages, supplies, loan losses and other administrative costs. OSS = Financial income/(Financial expenses + Provisions for loss of capital + Operating costs) $\times$ 100%. OSS = 100% indicates the breakeven point of MFIs in which operating income is equal to the total cost. Newly established, immature MFIs can take a few years to reach this breakeven point. A MFI will be considered operationally sustainable if OSS is >100%, which means that all MFI's revenue earned can cover all operating expenses. However, international practice shows that in order to achieve long-term sustainability, OSS should be equal to or greater than 120% (IFAD 2000).

2.3   Research on Profits of Microfinance Institutions

Along with OSS rate, some other studies have used profitability as a proxy for sustainability because they say that profit is a signal of sustainability (Jaramillo Millson 2013). In the author's opinion, profitability is the ability that MFIs generate profits on assets (ROA) and on equity (ROE). Therefore, in this study, the author uses additional indicators of net income in Return On Assets (ROA) and Return on equity

(ROE) to reflect the level of sustainability in MFIs. ROA measures profitability on the average total assets of MFIs. The higher the ratio, the greater the profitability of a MFI over 1 dong of assets, however, if the rate is too high, MFI is likely to face the potential risks, because it heavily invests highly risky assets. According to international practices, ROA > 2% and ROE ≥ 15% prove the sustainability of microfinance institutions (IFAD 2000).

Gonzales and Rosenberg (2006) analyze the link between outreach, profitability and poverty handling data reported to Microcredit Summit Campaign (MSC) and the MIX market platformcommonly. In the Mix Market data, the interaction between average loan size and profitability (measured by return on assets) is very feeble and slop of the curve is low. Applying the percentage of the poorest clients and operational self-sufficiency from the MSC dataset, they find that the correlation between these variables and the slope for the relationship are still very weak and low. They conclude that there may be relatively little conflict between improving sustainability and reaching poorer clients and it can find individual MFIs who are strongly profitable while serving the very poorer clients.

It is hard to reach a coherent conclusion on this issue as the microfinance sector is stillderiving. Besides, more varied institutions with mixed clientele join the sector, making it harder to conclude. Recent factual studies start to divide the total sample dataset into subsamples based on certain features that may affect the sustainability-outreach relationship.

Cull et al. (2007) give clue that MFIs can control depth of outreach and remain profitable but they observe a tradeoff between profitability and outreach to the poorest. Navajas et al. (2000) reveals that while MFIs provide loans to many house-holds that are adjacent the poverty line, most of these poor people are also the richest amongst the poor. Schreiner (2002) argues how enhanced breadth, length and scope of outreach can compensate for reduced depth of outreach. Using a modified poverty outreach index, Paxton and Cuevas (2002) rebuffs the notion of any trade-off between financial sustainability and outreach.

The study uses nine indicators as proxies for sustainability and profitability of MFIs. The indicators include OSS, ROA, yield on gross loan, operating expenses to asset ratio, financial revenue to gross loan portfolio ratio, gross loan to asset ratio, debt to equity ratio, cost per borrower ratio and borrowers per staff ratio. The study adopts the outreach framework by subdividing outreach to the poor into six variables of the breadth of outreach, depth of outreach, length of outreach, scope of outreach and the worth of outreach (Schreiner 2002; USAID 2006). The breadth of outreach is the size or scale of MFIs, depth of outreach is the value the society attach to the net gain of given client, length of outreach is the sustainability of supply of microfinance services, scope of outreach refers to the number of distinct product and services offered to clients, cost of outreach is sum of price cost and transaction cost and worth of outreach refers to the value of products and services consumed and the client's willingness to pay (USAID 2006).

According to USAID (2006), ROA implies how well an MFI is managing its assets to optimize its profitability. The ratio includes not only the return on the portfolio, but also all other income generated from investments and other operating activities.

If an institution's ROA is fairlystable, this ratio can be used to estimate earnings in future periods. Unlike ROE, this ratio measures profitability regardless of the institution's underlying funding structure; it does not discriminate against MFIs that are funded mainly through equity. Thus, ROA is a good measurement to compare commercial and noncommercial MFIs. Because non-commercial MFIs with low debt/equity ratios have low financial expenses and pay fewer taxes, they can often achieve higher ROA than their commercial counterparts. ROA should be positive. MFIs have achieved unusually high ROA in recent years. A positive interaction exists between this ratio and Portfolio to Assets; the ratio is higher for institutions that maintain a large percentage of the assets in the Gross Loan Portfolio (GLP).

In a for-profit MFI, ROE is the most decisive profitability indicator; it calculates an MFI's ability to reward its shareholders' investment, build its equity base through retained earnings, and promote additional equity investment. For a non-profit MFI, ROE indicates its capacity to build equity through retained earnings, and increased equity enables the MFI to leverage more financing to grow its portfolio. By excluding donations and non-operating revenues, this ratio demonstrates an institution's ability to achieve income from its core financial service activity. Some mature MFIs have achieved remarkably high ROE exceeding those of banks. Young organizations may take several years to achieve this, and even a mature MFI's ROE can be temporarily depressed due to unplanned events (such as natural disasters) or planned undertakings (such as expansion). ROE tends to fluctuate more than ROA. Monthly measurements of ROE can be deceiving because many MFI expenditures may not be recorded until the end of the fiscal year. Managers should look for a positive trend over several years and a ratio similar or better than competitors. Since the market becomes saturated and competition increases, ROE may plateau. MFIs that are financed solely through equity donations will find this ratio less meaningful because donors rarely base their future investment decisions on ROE. ROE is, however, a good indicator of how well the MFI has used retained earnings and donor money to become sustainable.

In summary, almost all previous studies used traditional statistical methods for analysis, therefore leading to unreliable results (Wasserstein et al. 2019). By applying Bayesian estimation method, our study contributes to the growing body of literature regarding the assessment mechanism of the impact of outreach on operational self-sufficiency and profit of microfinance institutions in emerging market economies.

3 Data and Model Specification

3.1 Research Data

This study employs panel data covering 28 microfinance institutions in Vietnam over the period of 2011–2016. The data collected through statistics of Vietnamese MFIs and the MIX market. Based on the literature review, we propose three research models as follows:

$$\begin{aligned}
\text{OSS}_{it} = {} & \beta_0 + \beta_1 \text{Age}_{it} + \beta_2 \text{Office}_{it} + \beta_3 \text{Borrower}_{it} + \beta_4 \text{Aveloanpebor}_{it} \\
& + \beta_5 \text{Loangrowth}_{it} + \beta_6 \text{Depositratio}_{it} + \beta_7 \text{Operexpratio}_{it} \\
& + \beta_8 \text{Par}_{it} + \beta_9 \text{formal}_{it} + \mu_{it}
\end{aligned} \tag{1}$$

$$\begin{aligned}
\text{ROA}_{it} = {} & \beta_0 + \beta_1 \text{Age}_{it} + \beta_2 \text{Office}_{it} + \beta_3 \text{Borrower}_{it} + \beta_4 \text{Aveloanpebor}_{it} \\
& + \beta_5 \text{Loangrowth}_{it} + \beta_6 \text{Depositratio}_{it} + \beta_7 \text{Operexpratio}_{it} \\
& + \beta_8 \text{Par}_{it} + \beta_9 \text{formal}_{it} + \mu_{it}
\end{aligned} \tag{2}$$

$$\begin{aligned}
\text{ROE}_{it} = {} & \beta_0 + \beta_1 \text{Age}_{it} + \beta_2 \text{Office}_{it} + \beta_3 \text{Borrower}_{it} + \beta_4 \text{Aveloanpebor}_{it} \\
& + \beta_5 \text{Loangrowth}_{it} + \beta_6 \text{Depositratio} + \beta_7 \text{Operexpratio}_{it} \\
& + \beta_8 \text{Par}_{it} + \beta_9 \text{formal}_{it} + \mu_{it}
\end{aligned} \tag{3}$$

where OSS_{it} operational self-sufficiency level of MFI i at year t. Similarly, ROA_{it}, ROE_{it} represents profits of MFI i at year t. Following the literature as Bogan et al. (2007), Kipesha and Zhang (2013), Kyereboah-Coleman and Oesi (2008), we use OSS, ROA and ROE as proxies for MFIs; independent variables used for modelling consist of age, size, borrower, office, average loan per borrower, loan growth, deposit ratio, operating expense ratio, par >30 and formal. According to our review of the literature, all the independent variables as listed above may influence OSS, ROA and ROE.

Particularly, in the context of Vietnam, when MFIs are mainly in the start-up and self-sustaining stages of their lifecycle, it is proper to use OSS as a reliable source of data to determine the self-sufficiency of MFIs. In addition, ROA (the dependent variable) represents the level of effectiveness in the business portfolio of MFIs. As IMFs are monetary business, in which credit activities are the core activities, through the profitability of the loan portfolio, MFIs can develop microfinance activities and expand their customer networks through increasing capital from business activities. The dependent variable ROE shows the management's ability to generate income from the equity available of MFIs.

MFIs from loan interest are primarily therefore the list of loans is the most important asset. Quality items reflect overdue loans and future income decisions as well as the organization's ability to increase access and service to existing customers. The most important indicator for assessing portfolio quality is measured by par >30 days.

The MFIs' income mainly from loan interest, therefore the loans are the most important asset. Quality list reflects overdue loans and future income decisions as well as the organization's ability to increase access and service to existing customers. The most important indicator to measure the portion of the loan portfolio is par30.

Not only affected by the level of access, OSS, ROA and ROE of MFIs are also affected by the characteristics of the organization (age, target borrower), effective (average loan per borrower, loan growth, deposit ratio, operating expense ratio), risk level (PAR > 30) and classification of MFI (formal).

The GLP is used to measure the size of an institution; it can increase either due to an increase in the number of borrowers without any change in the average loan per borrower, or with an increase in the average loan size with no increase in the number of borrowers. In the first case, the level of outreach would remain unchanged while the level of outreach would decline in the latter case. As a result, either the size of the firm will have no effect on outreach or it will have a negative impact on outreach. Since the primary goal of an MFI is to provide loans to the very poor, a larger level of equity would allow it to achieve a greater level of outreach, especially for the nonprofit MFIs. Following the same thread of logic, debt to equity ratio can also be expected have a positive effect on outreach. The higher cost of disbursing and administering smaller loans are amongst the key reasons that prevent formal financial institutions from being able to provide microcredit to the poor. Therefore, a higher total operation ratio would indicate a smaller average loan size and better outreach. Following our logic for total operation ratio, one would expect the cost per borrower to be positively related to outreach. However, group loans are smaller per borrower in comparison to other loans offered by MFIs. Consequently, the cost per borrower for the smallest loans may be lower than larger loans. This may result in a positive correlation between loan size and cost per borrower. Otherwise, one would expect outreach to be positively correlated with average loan per borrower.

3.2 Data and Sample Overview

The research sample is Vietnamese MFIs listed in the MFIs directory by the Vietnam Microfinance Working Group as well as financial data on the MIX market. However, due to the continuous fluctuation in the number of organizations joining and withdrawing from Vietnam's microfinance sector during the period of 2011–2016, in this study, the authors mainly focused on data of organizations that have yearly operational data for 6 years. Particularly, 28 organizations (accounting for about 92% of the microfinance market share) include 24 semi-formal MFIs and 04 formal MFIs, which are licensed to operate under the Law on Credit Institutions 2010 by the State Bank.

With 28 organizations selected as the observed variable and included in the analysis model. Table 1 provides a full description of the research data of each observed variable by data such as OSS, age, borrower, average loan per borrower, loan growth, deposit ratio, operating expense ratio, PAR 30 and formal (model 1) to assess the impact of outreach to OSS of MFIs and ROA with similar variables (model 2) and ROE with similar variables (model 3) to evaluate the impact of outreach on MFIs' profits.

Through the descriptive statistics in table above, the average of MFI's OSS is 139.67%. It suggests that MFIs can generally offset operating costs from revenue generated from their operations. This result is completely appropriate to the statistics of the research on microfince in almost developing countries (Marakkath 2014; Mwangi et al. 2015) and in Vietnam (Ngo 2012).

Table 1 Summary statistics

	Obs.	Mean	Std. Dev.	Min	Max
OSS	151	139.6742	38.83433	30.85	290.08
Age	150	9.36	6.166896	1	26
Offices	151	6.807947	11.04096	1	60
Borrower	151	8.746378	1.392162	5.220356	12.64416
Aveloanper-r	151	228.5352	110.8832	43.83	555.93
Loangrowth	147	32.07463	58.80253	−36.34	532.45
Depositratio	151	0.2680066	0.2145276	0	0.878
Operexpratio	146	13.82637	12.16418	1.64	146
Par	126	0.4146032	0.9850447	0	7.17
Formal	151	0.1456954	0.3539746	0	1

Source The authors

The number of offices as well as age of MFIs approved by Tchuigoua (2015) and Khachatryan, Hartarska, and Grigoryan (2017) concludes that if MFIs have many offices as well as organizations longer time of operation in microfinance sector, it would be easier for organizations to reach customers and the mobilizing capital and credit due to the expansion of the network.

The number of customers has a positive correlation with OSS of MFIs. Because of MFIs operate based on fixed costs and variable costs, thus a rise in the number of customers reduce the average fixed cost. For other variables such as the average loan per borrower is expected to be positive relation to OSS of MFIs. When a customer takes out a loan (even only a small amount), the institution also loses a cost namely the transaction costs (Hermes et al. 2009).

The growth rate of total outstanding loans and deposits are assumed to have a positive correlation with the sustainability of MFIs, which means the growth rate of total outstanding loans indicates an increase in the number of customers as well as loans. As mentioned above, the more customers and loans are, the more sustainable MFIs will be. Besides, deposits/total outstanding loans are indicators that compare with the amount of deposits with total outstanding loans. Typically, MFIs will be sustainable when the amount deposited is higher than the amount lent. Whether or not institutions are sustainable depends largely on loans as well as subsidies to customers (Hermes et al. 2011). According to the research of Cull et al. (2009), Muriu (2011) and Mwangi, Muturi, and Ombuki (2015), all authors consider deposits as a low-cost source of capital and the increase in the proportion of deposit will help reduce operational costs and increase profits, thus it creates a positive impact on OSS of MFIs.

There are two variables in OSS, including operating expense ratio and Par30, which are assumed to have opposite implications for the sustainability of MFIs (Mazlan 2014; Gibson 2012; Arnone et al. 2012; Rahman and Mazlan 2014). The operating expense ratio of MFIs which are different from other financial institutions will be considered low and effective, in other words, these low rate will make OSS rise. The more effective the MFIs are, themore sustainable are, the more sustainable are. For Par30, the ratio of loan portfolio is expected not to be available. PAR30 provides the researcher an indication of the quality of loans in the MFIs portfolio. The correlation between PAR30 and sustainability is negative, thus the higher this ratio is, the less sustainable the MFI is likely to be (Gershwin Long and Marwa 2015; Tehulu 2013; Marakkath 2014).

Two variables used to evaluate the institution's profit include ROA and ROE. MFIs can achieve OSS, but in order to achieve higher levels of sustainability and increase outreach, it is necessary to take into account two indicators of ROA and ROE. OSS and ROA, ROE have a negative impact on outreach to the poor. MFI that increase its focus on sustainability saves self few wealthier clients since most of the poor cannot pay the market price of the services. Similarly, Kipesha and Zhang (2013) examined the presence of tradeoffs between sustainability, profitability and outreach using a panel data of 47 MFIs for four years of 2008 to 2011 from Mix market. Data using unbalanced panel regression analysis model. Using Welfarists approach the study found the presence of negative tradeoffs between profitability and outreach to the poor and did not show presence of tradeoffs between financial sustainability and outreach measures.

3.3 Methodology

The previous studies were conducted according to the traditional method frequentist. However, according to Briggs and Hung (2019), p-value testing does not provide a good measurement tool for evidence related to models or hypotheses. Briggs and Hung (2019) also claimed that p-value hypothesis test has become obsolete because this method is not effective in interpreting statistical results in many cases. Therefore, the implementation of forecasts faces many difficulties. Beside, Thach et al. (2019) show that a model has many independent variables, there are more than one model, which are capable of interpreting variables. Therefore, if we just have one model, there is a high risk that the model selected is not suitable. The Bayesian approach can overcome this drawback by considering all possible models in the model space and then summarizing of statistical results based on the weights of the models in the model space. Because of those reasons, we will conduct this research follow Bayesian approach. We use Bayesian linear regression with Metropolis–Hastings (MH) algorithm and Gibbs sampling method.

The MH is the algorithm that simulates samples from a probability distribution by using of the full joint density function $\pi(.)$ and proposal distributions for each of the variables of interest. The process of MH algorithm was summarized by Yildirim (2012) as follows.

This algorithm is begun by initializing the sample value for each random variable $y^{(0)} \sim q(y)$. The main loop of this Algorithm include three components: Frist, create a candidate (or a proposal) sample $y^{(propsal)}$ from the proposal distribution

$q(y^{(i)}|y^{(i-1)})$; Second, calculate the acceptance probability by using the acceptance function $\beta(y^{(propsal)}|y^{(i-1)})$ based on the the the proposal distribution and the full joint density. Third, decide to accept or reject a proposal by relying on the acceptance probability; the acceptance function in the form: $\gamma(y^{(i-1)}|y^{(i-1)}) = \min\left\{1, \frac{\pi(a^{(x)})}{\pi(a^{(x-1)})}\right\}$. If a candidate number were smaller than γ we would accept that value, otherwise we would reject it.

Gibbs sampling is a special case of MH sampling.

We assume that we have n sample of vector $y = (y_1, y_2, ..., y_n)$ from a jont probability p $(y_1, y_2..., y_n)$. Denote the *ith* sample by $y^{(i)} = \left(y_1^{(i)}, ..., y_n^{(i)}\right)$. We process as follow:

(1) We begin with some initial value $y^{(i)}$. We have a vector $y^{(i)} = \left(y_1^{(i)}, y_2^{(i)} ..., y_n^{(i)}\right)$.

(2) Name this next sample is $y^{(i+1)}$. We consider this status $y^{(i)} = \left(y_1^{(i)}, y_2^{(i)} ..., y_n^{(i)}\right)$. We sample n dimension.

We sample $y_1^{i+1} \sim p\left(y_1|y_1^{(i)}, y_2^{(i)} ..., y_n^{(i)}\right)$. New status is updated $y = (y_1^{i+1}, y_2^i ..., y_n^i)$.

We sample $x_2^{i+1} \sim p\left(x_2|x_1^{(i+1)}, x_3^{(i)} ..., x_n^{(i)}\right)$. This status is update $y^{(i)} = \left(y_1^{(i+1)}, y_2^{(i+1)}, y_3^{(i)}, ..., y_n^{(i)}\right)$

(3) Repeat the above step n times.

4 Empiricial Results

4.1 Regression Analysis Results

In this paper, we conducted Bayesian regression approach and Gibbs sampling method. Two important criteria for measuring the effectiveness of MCMC (Markov chain Monte-Carlo) are the chain acceptance rate and the degree of autocorrelation. Convergent diagnostics are carried out to ensure that Bayes reasoning based on MCMC pattern is reasonable. To test the convergence of MCMC by parameter estimates, we can use visual diagnostics. The graphical diagnosis mainly consists of a trace plot, an autocorrelation plot, a Histogram and a Kernel Density plot covered with estimated density using the first and second half of the MCMC sample. Converged diagnostics for variables OSS, ROE and ROA are shown in the following figure.

As Fig. 1, it is seen that all parameter estimates in the graphs are relatively reasonable: the trace graphs and the correlation graphs show low autocorrelation; the shape of the chart is unimodal. In addition, all six graphs show a good mix, whereby the correlation coefficient fluctuates around 0.02. This shows the suitability of distributed simulation density. The autocorrelation plot reflects all delays within the efficiency limits. Histogram and Density plots both show the simulation of the shape of the

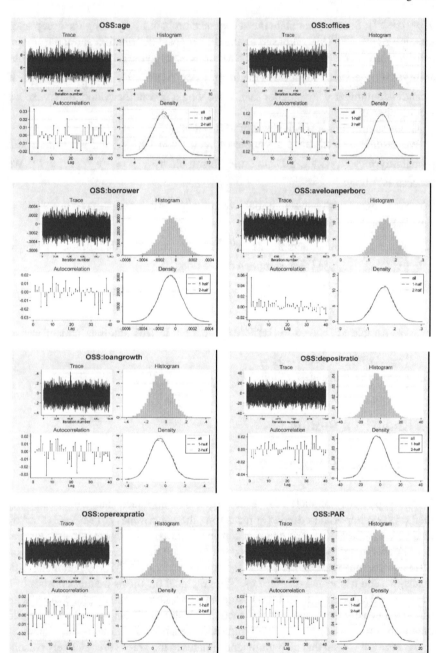

Fig. 1 Convergence standard for OSS

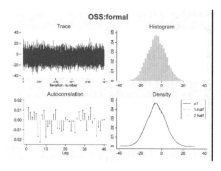
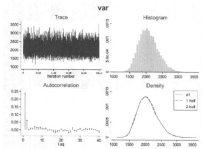

Fig. 1 (continued)

normal distribution of parameters. MCMC series graph of var also shows the reasonable. The coefficients representing the degree of autocorrelation are also just below 0.02. Histogram chart with density function.

Effective sample size is considered a necessary tool for assessing convergence level in addition to visual convergence diagnosis of MCMC series. Accordingly, an ESS value compared to an MCMC sample size, which is greater than 1%, is acceptable. In this study, the ESS value is close to the MCMC sample size, ie 10,000, thus this result is acceptable.

The convergent diagnostic results for ROA and ROE variables have similar results. The model parameter estimates of these two variables are relatively reasonable. The autocorrelation coefficient is below 2%. Histograms and Density plots both show the shape of the normal distribution of parameters.

Previous studies have been conducted on a frequency basis, so we do not have a priori information for this study. However, because the sample in this study is large with a total of 151, the prior information does not affect much accuracy of the model. Therefore, in this study, we use the usual a priori distribution Normal (1, 100) for the regression coefficients and Invgamma (2,5 2,5) for the variance of the error.

4.2 Impact of Outreach on OSS

Table 2 shows that the results of the models are similar in age, offices, borrowers, depositratio and formal of MFIs have a positive impact on OSS. In contrast, factors such as avaloanperborrower, loangrowth, operexratio, par have the opposite effect on OSS.

According to the research results, the positive impact of age of the microfinance institution suggests that the greater year the institution operates, the higher operational self-sufficiency they can achieve. This result is similar to the conclusions made by Tchuigoua (2015) and Khachatryan et al. (2017). Particularly, they assert that the longer the organization operates in the field of microfinance, the more reputation they earn expanding customer network in both mobilization and credit easily.

Table 2 Summary of regression results

Regressant InQ	OSS			ROA			ROE		
	Mean	Sta. Dev	Equal-tailed [95%Cred. Interval]		Mean	Sta. Dev	Equal-tailed [95%Cred. Interval]		Mean	Sta. Dev	Equal-tailed [95%Cred. Interval]	
Age	0.1418	0.3109	−0.4587	0.7575	0.1797	0.2431	−0.2949	0.6605	0.1418	0.3109	−0.4587	0.7576
Offices	0.1559	0.1793	−0.51127	0.1946	−0.0367	0.1409	−0.3130	0.2362	−0.1559	0.1796	−0.5113	0.1946
Borrowers	0.0001	0.00001	−0.0001	0.0001	0.0001	0.0001	−0.0001	0.0001	0.0001	0.0001	−0.0001	−0.0001
Aveloanperborrower	−0.0402	0.01322	0.0139	0.6622	0.0098	0.0104	−0.0106	0.0303	0.0402	0.0132	0.0139	0.0663
Loangrowth	−0.0048	0.03173	−0.1102	0.1275	−0.2947	0.0244	−0.0777	0.0190	−0.0484	0.03173	−0.1102	0.0127
Depositratio	4.8232	5.8605	−6.8002	16.3409	1.2101	0.4830	−8.0092	10.8092	4.8232	5.8606	−6.8002	16.3409
Operexration	−0.1041	0.1047	−0.3112	0.0992	−0.0642	0.0808	−0.2240	0.0946	−0.1041	0.1047	−0.3112	0.0992
PAR	−7.3828	1.2362	−9.8464	−4.9424	−6.7826	0.9757	−8.7103	−4.8455	−7.3828	1.2362	−9.8464	−4.9424
Formal	2.6417	4.9478	−12.3607	7.0174	4.0860	4.0650	−12.0538	3.9792	2.6417	4.9478	−12.3607	7.0174
_cons	2.6417	4.1761	−6.1131	10.3179	3.2841	3.3709	−3.3380	9.9929	2.1482	4.1762	−6.1132	10.3179
Var	179.5588	24.3518	138.2683	233.7433	106.4891	14.4177	81.9579	138.2468	179.5588	24.3518	138.2683	233.7433

Source The authors

Besides, these two studies also indicate that the number of MFIs' offices in the study results is positive, indicating that MFIs having more offices are more likely to achieve the self-sufficiency of MFIs. They have confirmed that if MFIs have various offices as well as instititutions operating longer in the microfinance sector, it means that it is easier for instititutions to reach customers due to expanding the network and the mobilization of capital and credits.

The number of borrowers as well as the deposit ratio has the same relationship with OSS in accordance with the research direction of Hermes, Lensink, and Meesters (2009). This is perfectly reasonable. As the number of borrowers increases, the value of loans per borrower increases, which will help MFIs increase profits and towards OSS. Besides, when microfinance institutions attract a large amount of customers' deposits, the organization will not entirely depend on support grant, investments of other organizations achieving operational self-sufficiency. Typically, MFIs will be sustainable when the amount deposited is higher than the amount lent. According to the research of Cull et al. (2009), Muriu (2011) and Mwangi et al. (2015), all authors consider deposits as a low-cost source of capital and the increase in the proportion of deposit will help reduce operational costs and increase profits, thus it creates a positive impact on OSS of MFIs.

The formal factor also affects the sustainability of microfinance institutions in Vietnam. According to the results of this study, this factor is in the same direction as the OSS. It means the four formal MFIs are often more self-sufficient. This result is entirely reasonable because according to Vietnam's Law on Credit Institutions 2010, these 4 institutions must fully comply with the principles as an formal financial institution, subject to the constraints, management of The State Bank of Vietnam, so the development, expansion as well as targeting the customer segment, increasing loans, calling for deposits must fully comply with the regulations of the State Bank and direct regulatory agencies.

Also according to the study of this group of authors, the loan growth has the same relationship with OSS eventually. Furthermore, the growth rate of re-loans proves that strong MFIs will develop lending activities, improve scale advantages as well as improve their risk and cost management capabilities gradually (Tehulu 2013; Tchuigoua 2015). According to the research of Hartarska Shen and Mersland (2013), the loan growth increase also positively affected the OSS of MFIs in Vietnam.

However, our study shows that loan growth has a negative impact on OSS. Our argument present that when microfinance institutions only concentrate on increasing loans too quickly, it may pose an increase in risk. In the past, many microfinance institutions in Vietnam carried out evaluation and approval before lending simply, quickly and inaccurately due to the purpose of lending as much as possible. However, these institutions encountered bad debts from customers and thus, there was reduction in their OSS.

Operating expense ratio and PAR30 have negative values according to the results of the model, indicating that these two indices have negative effects on OSS or in other words, the higher these two values, the lower ability to reach self-sufficiency. This result supports the conclusions drawn by Gibson (2012), Arnone et al. (2012) and Rahman and Mazlan (2014). Specifically, the researchers agreed that the increase in operating expense ratio would impair the performance of the institutions and therefore, the ability to increase profits and achieve OSS would be negatively affected. The impact of the PAR30 to OSS of Vietnam MFIs is negative, which is similar to the conclusions of Tehulu (2013) and Marakkath (2014).

4.3   Impact of Outreach on Profit

We also found negative correlation between ROA with number of offices, loan growth, deposit ratio, operating expense ratio and formal. The results on ROA were also supported by the positive correlation results between ROA and age, number of borrowers, average loan per borrower, operating expense ratio and par. Tương tự, ROE has negative correlation with offices, loan growth, operating expense ratio and par. The results on ROE also supported by the positive correlation results between ROE to age, number of borrowers and average loan per borrower, deposit ratio and formal.

If MFIs want to expand their operations through opening more offices, transaction offices will reduce profits of MFIs (reduction in ROA, ROE). In comparision with other financial institutions in Vietnam, expansions more offices in various localities and villages will help reach more customers. However, the characteristics of almost customers of MFIs are poverty and remote areas. In order to attract this customer segment, instead of opening more offices and trading offices, the solution is that the credit officers penetrate to the remote areas and help customers carry out transactions. This is considered an effective method to reach the right customer segment and increase profits for MFIs.

The increasing number of customers as well as the average loan per borrower will increase the profit of MFIs (increase in ROA, ROE), which is perfectly reasonable, when the MFIs are popular among individual lenders, they can increase profits. In addition, the average loan per borrower also represents customer trust in the institutions. Microfinance sector is considered a specific industry having its customers suffering low education levels and often build the trust in institutions through their relatives and friends. As the loan and customer base increase, it is also the time for MFIs to profit and develop sustainably. The emergence of microfinance aims to improve economic conditions for poor customers, bringing opportunities to develop, improve their lives by providing financial services. It is based on the fact that the poor often have less connection to formal financial services due to a number of barriers, especially no collateral when accessing to formal credit services. In addition to the direct goal of providing opportunities for the poor to access to their financial and non-financial services directly, MFIs also aim to provide long-term goals and help their customers be able to sustainable access to formal financial services. Therefore, microfinance is considered as a development tool of the poor, not merely a financial service and it is aimed at the following specific development objectives (i) creating jobs, increasing incomes for the poors; (ii) building capacity for target groups, especially women, (iii) reducing vulnerability for the poor when facing difficulties, unexpected risks and (iv) helping the poor to develop sustainably.

Although, loan growth and operating expense ratio are also considered as two important factors in the development of MFIs. The deposit rate has opposite correlation to ROA and positive correlation with ROE indicating that in the current development period, the MFIs not only seek to increase their profits simply through each deposit mobilization activity but also aim at many other derivative activities to increase profits fast such as calling for investment in capital contribution, business ... Besides, the model results indicate that PAR30 has opposite correlation with ROA, ROE. This is completely understandable, if the MFIs want to increase profit, the Par30 must always be low. This result is completely consistent with the model's expectations and the research of Vanroose and D'Espallier (2009), 2016; and is explained by economic theory of scale and risk. Accordingly, when risks decrease and thus the profits increase. This also implies that in order to increase the maximum profit for MFI, MFIs need to minimize their external risk investments.

Besides, the age and formal factors also have a positive impact on the ROA and ROE of microfinance institutions. Specifically, microfinance institutions will develop through four stages: start-up, self-reliance, financial sustainability and profit. During stage 1, the main sources of funding for microfinance institutions are subsidies, soft loans and compulsory savings (Calvin 2001) because these microfinance institutions are not reputable enough to access to other funding sources. Moreover, due to limited financial capacity, microfinance institutions are also unable to afford other higher-cost sources. However, funding from public funds may make microfinance institutions less motivated to monitor debt collection (Ngo 2012), while private funds mainly help microfinance institutions achieve social goals (Westley 2006). Therefore, step by step, microfinance institutions tend to replace their capital sources from non-profit organizations to commercial enterprises as microfinance institutions grow to the next stages (Hoque et al. 2011). As these institutions grow to stage 2 and 3 in which they have developed in terms of operations, they will be eligible to meet the requirements of loans from credit institutions. In addition, achieving financial self-sufficiency will help microfinance institutions increase their ability to mobilize capital. In stage 4, NGOs and microfinance organizations officially achieve stable after-tax profit, retained earnings and equity. During this stage, the formal organizations will focus on mobilizing voluntary savings, interbank loans, issuing bonds and stocks. These factors will ensure revenue and sustainable development of microfinance institutions. Therefore, the growth of a microfinance institution is crucial to increasing its profitability. The formal factor for ROA and ROE is as significant as for OSS. When a microfinance institution is subject to legal constraints, as well as operational regulations, it will help the development of institution is a step in right direction, gradually towards autonomy, sustainability and profitability in each business activity.

5 Conclusions

Results from paper show that the important indictors of outreach as number of borrowers, operating expense ratio, par, loan growth, etc.. impact on OSS and profit of MFIs. Therefore, the study will focus on providing solutions to help MFIs to attract more customers, reduce operating costs and control par, specifically:

First, in order to attract customers to borrow from and deposit at MFIs, MFIs need to actively diversify capital mobilization products. The products and services of MFIs in Vietnam are not really diversified, so it is necessary to design more new products suitable for different customers. To accomplish this, MFIs need to conduct a survey on the demand and saving capacity of the microfinance market, thereby helping to build and design appropriate products. In addition, MFIs should support, protect customer rights and increase their reputation, transparent their basic information, such as: interest rates, contract terms, financial statements... so as to ensure all customer loan terms are known and explained clearly. The rights and obligations of depositors and borrowers are clearly reflected in the regulations and publicly posted. Implementing periodic independent auditing of financial reports to increase the transparency of the institution, thereby the prestige of the MFI is built and consolidated. It is the transparency and clarity in lending activities that consolidate customers' faith and long-term commitment to MFIs.

Second, MFIs need to increase mobilization of savings with different, diversifying forms of capital mobilization with flexible interest rates. In addition, MFIs need to search for and utilize relatively cheap mobilized sources such as donors' capital, development investors, entrusted capital of credit institutions. This is the most important basis for MFIs to reduce lending rates and increase their ability to operate sustainably. In addition, MFIs should strengthen market research, innovation and application of new products and services, such as: diversifying the way of repaying and paying loan interest, savings mobilization to meet many different needs; developing a number of services such as telephone or internet transfer, insurance agents, collection agents, etc. to meet the increasing financial needs of poor and low-income households and micro-enterprises. In fact, no MFIs can be distributed across all districts, towns and hamlets, the fragmented, uneven distribution in Vietnam has limited the ability to provide general financial services and deposit and deposit services from residents. The solution is that MFIs can use the network of infrastructure and information technology infrastructure to be able to develop automated transaction points with the aim to effectively manage and save operating costs.

Finally, MFIs support customers to access credit loans widely and fairly, minimizing the asymmetry between borrowers and lenders. Better credit quality allows MFIs to assess their risks accurately and improve the quality of customer portfolios. When MFIs assess the total outstanding loans of borrowers, it is possible to estimate the repayment ability of borrowers. Since then MFIs can improve profitability, reduce operating costs, increase margin profit, ensure sufficient capital and meet backup requirements.

References

Acharya, Y. P. (2006). Sustainability of MFI from small farmer's perspective: A case study of rural Nepal. *Institutional Review of Business Research Papers, 2*(2), 117–126.

Mazlan, A. R. (2014). Determinants of operational efficiency of microfinance institutions in Bangladesh. *Article in Asian Social Science.*

Arnone, M., Pellegrini, C. B., Messa, A., & Sironi, E. (2012). Microfinance institutions in Africa, Asia, and Latin America: an empirical analysis of operational efficiency, institutional context and costs. *International Journal of Economic Policy in Emerging Economies, 5*(3), 255–271.

Kar, A. K. (2011).Does capital and financing structure have any relevance to the performance of microfinance institutions? *International Review of Applied Economics.*

Ayayi, G. A., & Sene, M. (2010). What drives microfinance institution's financial sustainability. *The Journal of Developing Areas, 44,* 303–324.

Bogan, V. L., Johnson, W., & Mhlanga, N. (2007). Does capital structure affect the financial sustainability of Microfinance Institutions.

Briggs, W. M., & Hung T. N.: Clarifying ASA's view on P-values in hypothesis testing. *Asian Journal of Economics and Banking, 3*(2).

Calvin, B. (2001). Framework for financing sources. In *An introduction to key issues in microfinance, on microfinance network website.*

Christen, R. P. (2001). Commercialization and mission drift: the transformation of microfinance in Latin America. Vol. Occasional Paper No. 5. Washington, DC: CGAP.

Ciravegna, D. (2005). The role of microcredit in modern economy: The case of Italy.

Cull, R., Asli Demirgüç-Kunt, A., & Morduch, J. (2009). Microfinance meets the marke. *Journal of Economic Perspectives, 23*(1), 167–192.

Cull, R., Demirguc-Kunt, A., & Morduch, J. (2007). Financial performance and outreach: A global analysis of leading microbanks. *Economic Journal, 117,* F107–F133.

Daher, L., & Le Saout, E. (2013). Microfinance and financial performance. *Strategic Change, 22*(1–2), 31–45.

Dunford, C. (2006). *Evidence of microfinance's contribution to achieving the millennium development goals.* USA Davis, CA: Freedom from Hunger.

El-Maksoud, S. A. (2016). *Performance of microfinance institutions.* Cardiff School of Management: A doctor of philosophy.

Long, G., & Marwa, N. (2015). Determinants of financial sustainability of microfinance institutions in Ghana. *Journal of Economics and Behavioral Studies, 7*(4), 71–81. (ISSN: 2220-6140).

Gibson, A. B. (2006). Determinants of operational sustainability of microfinance institutions in Kenya. Unpublished MBA thesis, School of Business, University of Nairobi.

Gonzalez, A., & Rosenberg, R. (2006). The state of microfinance: Outreach, profitability and poverty. Paper presented at the World Bank Conference on Access to Finance, Washington, DC.

Seibe, H.D. (2003). History matters in microfinance: Small enterprise development. *An International Journal of Microfinance and Business Development, 14*(2), 10–12.

Hartarska, V., Shen, X., & Mersland, R. (2013). Scale economies and input price elasticities in microfinance institutions. *Journal of Banking and Finance, 37*(1), 118–131.

Hashemi, S., & Rosenberg, R. (2006). Graduating the poor into microfinance: Linking safety nets and financial services. Vol. Focus Note No. 34. Washington, DC: CGAP.

Hermes, N., Lensink R., & Meesters, A. (2008). Outreach and efficiency of microfinance institutions. SSRN Working Paper Series.

Hermes, N., Lensink, R., & Meesters, A. (2009). Financial development and the efficiency of microfinance institutions.

Hermes, N., Lensink, R., & Meesters, A. (2011). Outreach and Efficiency of Microfinance Institutions. *World Development, 39*(6), 938–948.

Hoque, M., Chishty, M., & Halloway, R. (2011). Commercialization and changes in capital structure in microfinance institutions: An innovation or wrong turn? *Managerial Finance, 37*(5), 414–425.

Hudon, M., & Traca, D. (2011). On the efficiency effects of subsidies in microfinance: An empirical inquiry. *World Development, 39*(6), 966–973.

Hulme, D., & Mosley, P. (1996). *Finance Against Poverty* (Vol. 1). New York: Routledge.

IAFD. (2009). Gender and rural microfinance: Reaching and empowering women. Guide for practitioners.

IAFD. (2000). IAFD rural finance policy. In: *Executive board*, Sixty-Ninth Session, Rome, 3–4.

Yildirim, I. (2012). Bayesian inference: Metropolis-hastings sampling. In: *Department of brain and cognitive sciences University of Rochester*. NY 14627: Rochester.

Isern, J., & Porteous, D. (2005). Commercial banks and microfinance: Evolving models of success. Focus Note No. 28. Washington, DC: CGAP.

Jaramillo, M. (2013). *ELLA policy brief: Latin American innovation in microfinance technology*. Lima, Peru: ELLA Practical Action Consulting.

Hisako, K. (2009). Competition and wide outreach of microfinance institutions. Munich Personal RePEc Archive. Kobe University.

Khachatryan, K., Hartarska, V., & Grigoryan, A. (2017). Performance and capital structure of microfinance institutions in Eastern Europe and Central Asia. *Eastern European Economics, 55*(5), 395–419.

Khandker, S. R. (2003). Micro-finance and poverty: Evidence using panel data from Bangladesh. Policy research working paper, no. 2945.

Kimando, L., Kihoro, J. M., Njogu, G. W. (2012). Factors influencing the sustainability of micro-finance institutions in Murang'a municipality. *International Journal of Business and Commerce, 1*(10), 21–45.

Kinde, B. A. (2012). Financial sustainability of microfinance institutions in Ethiopia. *European Journal of Business and Management, 4*(15), 1–10.

Kipesha, E. F, & Zhang, X. (2013). Sustainability, profitability and outreach tradeoffs: Evidences from microfinance institutions in East Africa. *European Journal of Business and Management, 5*(8), 136–148.

Kirkpatrick, C and M Maimbo.: The Implications of the Evolving Microfinance Agenda for Regulatory and Supervisory Policy. Development Policy Review, Vol 20, No 3, pp 293–304 (2002)

Kyereboah-Coleman, A., & Osei, K. (2008). Outreach and profitability of microfinance institutions: The role of governance" . *Journal of Economic Studies, 35*(3), 236–248.

Ledgerwood, J. (1999). *Microfinance handbook: An institutional and financial perspective, Washington*. D.C: World Bank.

Littlefield, E., Morduch, J., & Hashemi, S. (2003). Is microfinance an effective strategy to reach the Millennium Development Goals? *Focus Note, 24*(2003), 1–11.

Marakkath, N. (2014). *Sustainability of Indian microfinance institutions*. DE: Springer Verlag.

Marr, A. A. (2004). Challenge to the Orthodoxy Concerning Microfinance and Poverty Reduction. *Journal of Microfinance, 5*(2), 1–35.

Martínez-González, A.: Technical Efficiency Of Microfinance Institutions: Evidence From Mexico. The Ohio State University (2008)

Morduch, J. & Haley, B. (2002). Analysis of the effects of microfinance on poverty reduction. NYU Wagner Working Paper No. 1014.

Morduch, J. (2005). Smart subsidies for sustainable microfinance: Finance for the Poor. *ADB Quarterly Newsletter of the Focal Point for Microfinance, 6*, 1–7.

Mosley, P. (2001). Microfinance and Poverty in Bolivia. *Journal of Development Studies, 37*(4), 101–132.

Muriu P. (2011). Microfinance Profitability: Does financing choice matter? Birmingham Business School University of Birmingham.

Mwangi. M, Muturi.W & Ombuki. C.: The effects of deposit to asset ratio on the financial sustainability of deposit taking micro finance institutions in Kenya. *International Journal of Economics, Commerce and Management, 3*(8), 504–511.

Navajas, S., Schreiner, M., Meyer, R. L., et al. (2000). Microcredit and the poorest of the poor: Theory and evidence from Bolivia. *World Development, 28,* 333–346.

Ngo, T. V. (2012). Capital structure and microfinance performance: A cross country analysis and case study of Vietnam. PhD Thesis.

Thach, N. N., Anh, L. H., & An, P. T. H. (2019). The effects of public expenditure on economic growth in asia countries: A bayesian model averaging approach. *Asian Journal of Economics and Banking, 3*(1), 126–149.

Paxton, J. and Cuevas, C. E.: Outreach and sustainability of member-based rural financial interme-diaries. In Zeller, M. & Meyer, R. L. (Eds), *The triangle of microfinance: Financial sustainability, outreach and impact* (pp. 135–51). Johns Hopkins University Press for the International Food Policy Research Institute, Baltimore, MD.

Pollinger, J. J., Outhwaite, J., & Guzmán, H. C. (2007). The Question of Sustainability for Microfinance Institutions. *Journal of Small Business Management, 45*(1), 23–41.

Quayes, S. (2015). Outreach and performance of microfinance institutions: A panel analysis. *Applied Economics, 47*(18), 1909–1925.

Rahman, M. A., & Mazlan, A. R. (2014). Determinants of operational efficiency of microfinance institutions in Bangladesh. *Asian Social Science, 10*(22), 322.

Schreiner, M. (2002). Aspects of outreach: A framework for discussion of the social benefits of microfinance. *Journal of International Development, 14,* 591–603.

Schreiner, M. J. (1997). A Framework for the analysis of the performance and sustainability of subsidized microfinance organizations with application to bancosol of Bolivia and Grameen Bank of Bangladesh. Unpublished Doctoral Dissertation, Graduate School of The Ohio State University.

Scully, N. (2004). *Microcredit no panacea for poor women, Washington.* DC: Global Development Research Centre.

Sharma, S. R., & Nepal, V. (1997). Strengthening of credit institutions. In Programs for Rural Poverty Alleviation in Nepal, United Nations, Economic and Social Council (ECOSOC) for Asia and Pacific, Bangkok, Thailand.

Simanowitz, A., & Walter, A. (2002). Ensuring impact: Reaching the poorest while building finan-cially selfsufficient institutions and showing improvement in the lives of the poorest women and their families. In D. Harris (Ed.), *Pathways out of poverty, connecticut*, USA: Kumarian Press.

Tchuigoua, H. T. (2015). Capital structure of microfinance institutions. *Journal of Financial Services Research, 47*(3), 313–340.

Tehulu, T. A. (2013). Determinants of financial sustainability of microfinance institutions in East Africa. *European Journal of Business and Management, 5*(17), 152–158.

USAID. (2006). Evaluating microfinance institutions social performance: A measurement tool. Micro Report No. 35. USA: USAID.

Vanroose, A., & D'Espallier, B. (2009). Microfinance and Financial Sector Development. CEB Working Paper No. 09/040.

Wasserstein, R. L., Schirm, A. L., & Lazar, N. A. (2019). Moving to a World Beyond "p<0.05". *The American Statistician, 73*(1), 1–19.

Weiss, J., & Montgomery, H. (2005). Great expectations: Microfinance and poverty reduction in Asia and Latin America. *Oxford Development Studies, 33,* 391–416.

Westley, G. D. (2006). *Strategies and structures for commercial banks in microfinance.* Inter-American Development Bank.

Woller, G. M., Dunford, C., & Warner, W. (1999). Where to Microfinance. *International Journal of Economic Development, 1*(1), 29–64.

Yaron, J., Benjamin M., & Charitonenko, S. (1998). *Promoting efficient rural financial intermedi-ation* (Vol. 13). The World Bank Research Observer.

Zeller, M., & Meyer, R. L. (Eds). (2002). The triangle of rural finance: Financial sustainability, outreach, and impact. Johns Hopkins UniversityPress in collaboration with the International Food Policy Research Institute (IFPRI), Baltimore and London.

Credit Growth Determinants—The Case of Vietnamese commercial Banks

Trang T. T. Nguyen and Trung V. Dang

Abstract The study empirically identifies the determinants which explain credit growth of 27 commercial banks of Vietnam for the period from 2008 to 2018 by applying the general least squares (GLS) method and Bayesian approach. The results indicates that two group of determinants, included the micro and macro factors impact significantly to the credit growth. The results show that liquidity and profitability of bank and have the positive association with credit growth, the other factors, named non-performing loan, bank size, CPI and FDI have the negative association with credit growth.

Keywords Bayesian approach · Credit growth · Commercial banks · Panel · Vietnam

1 Introduction

Credit growth is an important prerequisite for businesses as well as individuals and household to expand production and consumption, increasing production and then promoting the whole economy. Supervision and regulators care about credit growth because it affects the efficiency of monetary policy and partly reflect the commercial bank's strength and strategy. For period from 2008 to 2012, Vietnam exhibits the slowdown in the credit growth from 25.4 to 14.31%. However, from 2012 to 2016, the banking credit growth was comparatively high, from 8.9% in 2012 to 18.7% in 2016, following the continuously decrease till the year of 2018. In 2018, the credit growth rate is 14% overall, lower that 18% target set up by State Bank of Vietnam.[1]

---

[1] State bank of Vietnam, annual report

T. T. T. Nguyen (✉) · T. V. Dang
Banking Faculty, Banking Academy of Vietnam, Hanoi, Vietnam
e-mail: trangntt@hvnh.edu.vn

© The Editor(s) (if applicable) and The Author(s), under exclusive license
to Springer Nature Switzerland AG 2021
N. Ngoc Thach et al. (eds.), *Data Science for Financial Econometrics*,
Studies in Computational Intelligence 898,
https://doi.org/10.1007/978-3-030-48853-6_38

567

In Vietnam, State bank of Vietnam set up the credit growth target for whole banking system as well as for each group of banks. Based on the credit growth limit, commercial banks in Vietnam could set up their own credit growth ratio target which is most suitable with their banking activities. There for, a deeper understanding of the determinants of credit growth will help commercial bank to build and maintain more accurate credit growth targets as well as adjust their credit policies.

The main objective of the present study is to identify the determinants of credit growth in commercial banks of Vietnam. Section 2 presents the review of the existing literature, followed by methodology and data in Sect. 3. The empirical results and discussion are presented in Sect. 4 and the last section is conclusion.

2   Literature Review

There are many factors proved influencing the credit growth of commercial banks.

For macro factors, the stronger economic condition with higher GDP growth rate and low inflation rate leads to higher credit growth in banking sectors. The study of Imran and Nishat (2013) studied the commercial banks of Pakistan from 1971 to 2010 and concluded that economic growth has a significant positive impact on credit growth in Pakistan. Pouw and Kakes (2013) studied the macro-economic factor influence the credit growth of banks in 28 countries from 1980 to 2009. The study applied the FEM model and founded that GDP has a cyclical and positive impact on the credit growth of banks. The author has found a positive effect on long-term and short-term interest rates that has a negative impact on lending operations. Guo and Stepanyan (2011) examine the changes credit growth across a range of emerging market economies in 38 countries, from the 1st quarter of 2001 to the 2nd quarter of 2010. The results show that real inflation has actually slowed down the rate of private credit. Better infrastructure growth leads to a higher demand of credit, when the deposit interest rate higher, the monetary policy is more tight and the credit market is lessen. Besides, study of Deniz Igan and Zhibo Tan (2015) about 33 countries from the period 1980–201 concluded that net FDI inflows are not significantly connected with credit developments except for corporate sector credit with only marginally significant.

For micro factors, Carlson et al. (2011) conducted a research study on the influences of capital to credit growth from 2001 to 2011 using Fixed Effects model. The authors recognize that the impact of the equity ratio will change from period to period: The ratio of equity has a great impact on credit market during a period of financial crisis (2008–2010) rather than non-financial crisis (from 2001 to 2007 and 2011). When deeply analyzing the credit market components, the real estate and commercial real estate market will be more sensitive with the equity ratio than other sectors. Laidroo (2014) conducted a study of credit growth of 247 commercial banks in 11 countries in Central and Eastern Europe, with the period from 2004 to 2012. The factors that affect the bank's credit growth are bank size, credit risk and credit risk. In the crisis period, the liquidity rate has a strong positive relationship with the development credit activities.

There are many researchers studied the impacts of both macro and micro factors to credit growth in commercial banks. Chen and Wu (2014) studied about the credit growth of more than 900 commercial banks in 24 countries in emerging markets. The study concludes that commercial banks tend to supply more credit in the case of higher GDP rate, higher capital, higher profitability and higher liquidity. Besides, study of Tamirisa and Igan (2008) about the credit growth of 217 commercial banks in Central and Eastern Europe countries from 1995 to 2004 and found that weaker banks keep the higher credit growth rate, and higher GDP growth and lower interest rate have a positive and significant impact to the credit supply. More efficient management with better profitability and liquidity could promote credit growth but bank size remains the negative association with credit growth.

In the case of Vietnam, there are some papers studying about credit growth. Chu Khanh (2012) estimated the impact of monetary policy to credit growth in banking system as a whole, rather than credit growth of difference commercial banks in Vietnam. Trung and Chung (2018) assess factors affecting the credit growth of 21 commercial banks in Vietnam from 2008 to 2015 as the indicator of finance stability in Vietnam. The research conducted that monetary policy instruments (including required reserve ratio, discount rate) have stronger affect than macro prudential policy instruments (including capital ratio, liquidity ratio, credit to deposit ratio). With richer time-series and cross-section data set, the study aims to estimate the factors effecting the credit growth in commercial banks in Vietnam.

3 Methodology and Data

This study applies the following research model:

$$CG_{it} = \beta_o + \beta_1 * LIQ_{it} + \beta_2 * SIZE_{it} + \beta_3 * ROA_{it} + \beta_4 * NPL_{it} \\ + \beta_5 * CAP_{it} + \beta_6 * GDP_{it} + \beta_7 * CPI_{it} + \beta_8 * FDI_{it} + u_{it}$$

where:

CG   Credit growth
GDP Gross domestic product
LIQ Liquidity ratio
CPI Customer Price Index
SIZE Bank size
FDI Foreign direct investment
ROA Return on asset
NPL Non-performing loan
CAP Bank equity capital

Table 1 Description and expected signs of variables

No.	Variables	Formula	Expectation	Previous study
Dependence variable				
1	CG	$\frac{\text{Credit year}t - \text{Credit year}(t-1)}{\text{Credit year}(t-1)}$		
Explanatory variables				
2	LIQ	$\frac{\text{Short term asset}}{\text{Total asset}}$	+	Laidroo (2014), Tamirisa and Igan (2008), Thong (2018)
3	SIZE	Ln (total asset)	±	Laidroo (2014), Thong (2018), Tamirisa and Igan (2008)
4	ROA	$\frac{\text{Net income}}{\text{Total asset}}$	+	Aydin (2008), Chen and Wu (2014), Tamirisa and Igan (2008)
5	NPL	$\frac{\text{Non performing loan}}{\text{Total loan}}$	−	Mark (2011), Thong (2018)
6	CAP	$\frac{\text{Equity capital}}{\text{Total aset}}$	±	Tracey (2011), Carlson et al. (2011), Chen and Wu (2014), Thong (2018)
7	GDP	Gross Domestic Product (GDP) growth	+	Aydin (2008), Trung and Chung (2018)
8	FDI	Ln (FDI)	+	
9	CPI	Consumer product index	−	Imran and Nishat (2013), Thong (2018), Trung and Chung (2018)

The description of variables and the expected signs are detailed in Table 1.

Sources for micro data come from official annual reports of commercial banks and sources for macro data come from World Bank (WB), Asian Development Bank (ADB) and International Monetary Fund (IMF). The data set is strong balance panel data with 297 observations of 27 commercial banks in Vietnam for the period of 11 years from 2008 to 2018.

4 Empirical Results and Discussion

4.1 Empirical Results

The descriptive table in Appendix 1 describes mean, standard deviation, min and max of variables. The multicollinearity between the variables is examines by correlation matrix of the variables (Appendix 2) and VIF indicators (Appendix 3). The results shows that none off the correlation coefficient between pairs of variables higher than 0.8 and the highest VIF indicators is lower than 10. Therefore, there is no multicollinearity in this model.

For the panel data, the research employs Pooled OLS, fixed effect model (FEM) and Random effects model (REM). Pooled OLS estimation is OLS technique run on Panel data so all individually specific effects of each commercial bank are completely ignored. FEM and REM care about specific effects of each commercial bank. The difference between FEM and REM is that in REM there is no correlation between random element of each bank and explanatory variables but FEM. The results of Pooled OLS, FEM and REM regression are detailed in Appendixes 4, 5 and 6.

Next, the series of test is applied to choose which model is the best among OLS, FEM and REM models. F-test is applied to choose the better model between pooled OLS and FEM. The results from Appendix 7 show that FEM is better model with 5% significantly. Then, Breusch—Pagan LM test applied to choose the better model between pooled OLS and REM. The results from Appendix 8 show that REM is better model with 1% significantly. Then, Hausman test is applied to choose the better model between REM and FEM. The results from Appendix 9 shows that REM is more appropriate model.

After choosing the most appropriate model, Modified Wald test is applied to test the heteroscedasticity and Wooldridge test is applied to test the autocorrelation to make sure about reliable results. Appendixes 10 and 11 show that the model exhibits both heteroscedasticity and autocorrelation. According to Wooldridge (2002), general least squares method (GLS) can be applied to solve this problem (Appendix 12). Finally, Bayesian approach is applied to promote the reliable results (Appendix 13). The results of GLS model and Bayesian approach is showed in Table 2.

Table 2 Regression results from GLS method and Bayesian approach

Variables	Coefficients	Strength of evidence
LIQ	0.2496461	Moderate
SIZE	−5.204684	Strong to very strong
ROA	3.462474	Moderate
NPL	−2.145125	Moderate to strong
CAP	−0.8170769	Moderate
GDP	3.211486	Weak
CPI	−1.26621	Strong to very strong
FDI	−27.9261	Strong to very strong
c	657.9848	Strong to very strong
Prob > chi^2	0.0000	

5   Discussion

5.1   For Micro Determinants

Similarity to Laidroo (2014), Tamirisa and Igan (2008) liquidity ratios exhibits a positive association with credit growth. This due to the fact that banks with high liquidity assets, which are in excess of capital, will have the goal of maximizing profits by boardening the credit activities. In line with the previous literature reviews of Laidroo (2014), Tamirisa and Igan (2008) bank size exhibits the negative association with credit growth. This could related to more strict lending condition of bigger banks compared with smaller banks. The result shows the positive association between profitability and credit growth of commercial banks in Vietnam. This is consistent with previous research findings of Aydin (2008); Chen and Wu (2014); Tamirisa and Igan (2008). When banks have high profitability, it means that banks have used assets and capital effectively. This creates the foundation and motivation for banks to promote their credit and lending activities.

In contrast, NPL and capital have a negative association with credit growth. Poor performance of commercial banks in Vietnam is associated with lower subsequent credit growth. This is in line with the study of Mark (2011) and Thong (2018). When the NPL ratio increased, forcing banks to find solutions to collect NPL and make more strict regulations when lending. Besides, if the capital increases, credit growth will decrease, as the conclusion of Mark (2011); Carlson et al. (2011); Chen and Wu (2014), Thong (2018).

5.2   For the Macro Determinants

Differ with other research proving positive relationship between GDP and credit growh, such as Aydin (2008), Guo and Stepanyan (2011), Chen and Wu (2014), Lane and McQuade (2013), Tamirisa and Igan (2008), Trung and Chung (2018), the results show that there is no strong evidence about relationship between GDP and credit growh in commercial banks in Vietnam. Other factors, CPI and FDI are significantly affecting credit growth of commercial banks in Vietnam. Banks tend to remain lower credit growth when inflation goes up, as the conclusion of Guo and Stepanyan (2011), Thong (2018). When inflation increases, the State Bank of Vietnam will use tight monetary policy to increase interest rates, which will reduce lending activities off commercial banks.

For FDI determinant, there is strong negative relationship between FDI and credit growth. The reason is that FDI enterprises often borrow from foreign banks because they have relationships with parent companies of FDI enterprises. Besides, there are many FDI enterprises cooperate with foreign funds and other foreign enterprise to jointly invest, so these enterprises are also gradually decrease credit sources from domestic commercial banks.

6 Conclusion

The study examined factors that affect the credit growth of 27 commercial banks in Vietnam form 2008 to 2018. GLS method and Bayesian approach are applied to eliminate the heteroscedasticity and autocorrelation in the model and to make sure about the reliable of results. The macro determinants include CPI and FDI and the micro determinants include bank size, bank capital, liquidity and non-performing loan. While liquidity and profitability of bank have the positive association with credit growth, the other factors, named non-performing loan, bank size, CPI and FDI have the negative association with credit growth. By considering the research results, commercial banks could apply to estimate their suitable credit growth and the supervisors could apply to understand more about the banking activities.

There is some limitation of this study. First, the study just focus on the commercial banks as whole, rather than deferent groups such as, foreign and domestic banks, or Big 4 banks and other jointed banks as well as compare the lending behavior between groups. Second, the recommendation and policy support could be better if the commercial banks supply the credit growth for difference sectors, such as, enterprise and individual or difference industries.

Appendix 1: Descriptive Table

Variable	Max	Min	Mean	Std. Dev.	Obs
Dependence variable					
CG	165	−24.59	27.56949	25.49642	297
Explanatory variables					
LIQ	61.09695	4.519005	20.4825	10.50298	297
SIZE	20.99561	14.69872	18.23943	1.29984	297
ROA	11.90369	−8.676205	0.9291272	1.214542	297
NPL	11.4	0.28	2.327572	1.490063	297
CAP	46.24462	3.23	10.28161	5.947945	297
GDP	7.08	5.23	6.149091	0.6257937	297
CPI	19.87	0.6	7.783636	5.926508	297
FDI	19.90165	18.98628	19.38866	0.2959772	297
M2	22.9532	21.20701	22.17008	0.5396759	297

Appendix 2: Correlation Matrix of the Variables

	CG	LIQ	SIZE	ROA	NPL	CAP	GDP	CPI	FDI	M2
CG	1.0000									
LIQ	0.3117	1.0000								
SIZE	-0.2794	-0.2786	1.0000							
ROA	0.2108	0.2209	-0.1516	1.0000						
NPL	-0.1541	-0.0311	-0.0846	-0.0943	1.0000					
CAP	0.1489	0.2042	-0.7185	0.4138	0.1129	1.0000				
GDP	-0.1789	-0.2218	0.1954	-0.0701	-0.2282	-0.1473	1.0000			
CPI	0.0491	0.4077	-0.3312	0.2319	0.1078	0.3130	-0.1245	1.0000		
FDI	-0.3146	-0.4539	0.4197	-0.2349	-0.1826	-0.3392	0.6376	-0.6313	1.0000	
M2	-0.3188	-0.4708	0.4524	-0.2782	-0.1210	-0.3778	0.4738	-0.7540	0.9576	1.0000

Appendix 3: VIF Indicators

```
. estat vif
```

Variable	VIF	1/VIF
FDI	3.92	0.254896
CAP	2.66	0.376264
SIZE	2.47	0.404556
GDP	2.22	0.449667
CPI	2.21	0.452938
ROA	1.40	0.714911
LIQ	1.35	0.738230
NPL	1.12	0.894457
Mean VIF	2.17	

Appendix 4: Pooled OLS Regression Results

```
. regress CG LIQ SIZE ROA NPL CAP GDP CPI FDI
```

Source	SS	df	MS		Number of obs	=	297
					F(8, 288)	=	14.05
Model	54002.4252	8	6750.30315		Prob > F	=	0.0000
Residual	138417.559	288	480.616524		R-squared	=	0.2806
					Adj R-squared	=	0.2607
Total	192419.984	296	650.067514		Root MSE	=	21.923

CG	Coef.	Std. Err.	t	P>|t|	[95% Conf. Interval]	
LIQ	.4651821	.1412034	3.29	0.001	.1872606	.7431037
SIZE	-5.696721	1.541253	-3.70	0.000	-8.73027	-2.663172
ROA	3.514198	1.240838	2.83	0.005	1.071938	5.956458
NPL	-2.640975	.9042096	-2.92	0.004	-4.420672	-.8612779
CAP	-.7482602	.3492529	-2.14	0.033	-1.435672	-.0608484
GDP	4.996292	3.036525	1.65	0.101	-.9803026	10.97289
CPI	-1.601757	.3194738	-5.01	0.000	-2.230557	-.9729576
FDI	-40.23261	8.52734	-4.72	0.000	-57.01642	-23.4488
_cons	894.3228	152.3463	5.87	0.000	594.4695	1194.176

Appendix 5: FEM Regression Results

```
. xtreg CG LIQ SIZE ROA NPL CAP GDP CPI FDI, fe

Fixed-effects (within) regression              Number of obs    =        297
Group variable: ID                             Number of groups =         27

R-sq:                                          Obs per group:
    within  = 0.2731                                        min =         11
    between = 0.2921                                        avg =       11.0
    overall = 0.2535                                        max =         11

                                               F(8,262)         =      12.30
corr(u_i, Xb)  = -0.3307                        Prob > F         =     0.0000
```

CG	Coef.	Std. Err.	t	P>|t|	[95% Conf.	Interval]
LIQ	.4110654	.1907957	2.15	0.032	.0353772	.7867536
SIZE	-11.32931	4.510345	-2.51	0.013	-20.21045	-2.448167
ROA	3.923857	1.34908	2.91	0.004	1.267437	6.580276
NPL	-2.49714	.9728796	-2.57	0.011	-4.412798	-.5814821
CAP	-1.171637	.4188858	-2.80	0.006	-1.996449	-.3468263
GDP	4.185616	3.052881	1.37	0.172	-1.82569	10.19692
CPI	-1.620291	.318836	-5.08	0.000	-2.248098	-.9924838
FDI	-32.22354	11.32602	-2.85	0.005	-54.52515	-9.921927
_cons	851.648	164.8842	5.17	0.000	526.9812	1176.315

```
sigma_u |  9.7649191
sigma_e |  21.343849
    rho |  .17308277   (fraction of variance due to u_i)

F test that all u_i=0: F(26, 262) = 1.61                    Prob > F = 0.0344
```

Appendix 6: FEM Regression Results

```
. xtreg CG LIQ SIZE ROA NPL CAP GDP CPI FDI, re

Random-effects GLS regression              Number of obs    =        297
Group variable: ID                         Number of groups =         27

R-sq:                                      Obs per group:
     within  = 0.2693                                       min =        11
     between = 0.3454                                       avg =      11.0
     overall = 0.2802                                       max =        11

                                           Wald chi2(8)     =     109.62
corr(u_i, X)    = 0 (assumed)              Prob > chi2      =     0.0000
```

CG	Coef.	Std. Err.	z	P>|z|	[95% Conf. Interval]	
LIQ	.4375773	.154008	2.84	0.004	.1357273	.7394273
SIZE	-6.264206	1.829882	-3.42	0.001	-9.85071	-2.677702
ROA	3.669118	1.257605	2.92	0.004	1.204257	6.133978
NPL	-2.587569	.9139756	-2.83	0.005	-4.378928	-.7962096
CAP	-.8525115	.3608483	-2.36	0.018	-1.559761	-.1452618
GDP	4.955448	2.952288	1.68	0.093	-.8309309	10.74183
CPI	-1.590668	.3109055	-5.12	0.000	-2.200032	-.9813046
FDI	-39.94834	8.48239	-4.71	0.000	-56.57352	-23.32316
_cons	900.6956	149.4474	6.03	0.000	607.7841	1193.607

sigma_u	5.8898372	
sigma_e	21.343849	
rho	.07076017	(fraction of variance due to u_i)

Appendix 7: F Test for Choosing FEM or OLS

F-stat	Prob	Result
1.61	0.0344	FEM

With P-value = 0.0344 < 0.05 (5% significant) so FEM is better model with 5% significantly

Appendix 8: Breusch—Pagan LM Test for Choosing REM or OLS

```
. xttest0

Breusch and Pagan Lagrangian multiplier test for random effects

        CG[ID,t] = Xb + u[ID] + e[ID,t]

        Estimated results:
                          |       Var     sd = sqrt(Var)
                 ---------+-------------------------------
                      CG  |    650.0675       25.49642
                       e  |    455.5599       21.34385
                       u  |    34.69018       5.889837

        Test:   Var(u) = 0
                            chibar2(01) =       2.98
                       Prob > chibar2 =      0.0422
```

Appendix 9: Hausman Test for Choosing FEM or REM

```
. hausman fe re
```

	—— Coefficients ——			
	(b)	(B)	(b-B)	sqrt(diag(V_b-V_B))
	fe	re	Difference	S.E.
LIQ	.4110654	.4375773	-.0265119	.1126257
SIZE	-11.32931	-6.264206	-5.0651	4.122468
ROA	3.923857	3.669118	.2547391	.4883106
NPL	-2.49714	-2.587569	.0904287	.3333818
CAP	-1.171637	-.8525115	-.3191259	.2127294
GDP	4.185616	4.955448	-.7698321	.7772242
CPI	-1.620291	-1.590668	-.0296229	.0706695
FDI	-32.22354	-39.94834	7.724798	7.505187

```
                        b = consistent under Ho and Ha; obtained from xtreg
           B = inconsistent under Ha, efficient under Ho; obtained from xtreg

    Test:  Ho:  difference in coefficients not systematic

              chi2(8) = (b-B)'[(V_b-V_B)^(-1)](b-B)
                      =         2.43
              Prob>chi2 =      0.9649
```

Appendix 10: Modified Wald Test for Heteroskedasticity

```
. xttest3

Modified Wald test for groupwise heteroskedasticity
in fixed effect regression model

H0: sigma(i)^2 = sigma^2 for all i

chi2 (27)   =     2302.28
Prob>chi2 =       0.0000
```

Appendix 11: Wooldridge Test for Autocorrelation

```
. xtserial CG LIQ SIZE ROA NPL CAP GDP CPI FDI

Wooldridge test for autocorrelation in panel data
H0: no first-order autocorrelation
    F(  1,      26) =     12.363
           Prob > F =      0.0016
```

Appendix 12: GLS Regression Results

```
. xtgls CG LIQ SIZE ROA NPL CAP GDP CPI FDI, panels(h) corr(ar1)

Cross-sectional time-series FGLS regression

Coefficients:  generalized least squares
Panels:        heteroskedastic
Correlation:   common AR(1) coefficient for all panels   (0.2327)

Estimated covariances       =        27        Number of obs      =        297
Estimated autocorrelations  =         1        Number of groups   =         27
Estimated coefficients      =         9        Time periods       =         11
                                               Wald chi2(8)       =     112.07
                                               Prob > chi2        =     0.0000
```

CG	Coef.	Std. Err.	z	P>|z|	[95% Conf. Interval]	
LIQ	.2496461	.1166438	2.14	0.032	.0210285	.4782637
SIZE	-5.204684	1.246871	-4.17	0.000	-7.648506	-2.760862
ROA	3.462474	1.353693	2.56	0.011	.8092836	6.115665
NPL	-2.145125	.6256336	-3.43	0.001	-3.371345	-.9189059
CAP	-.8170769	.3397383	-2.41	0.016	-1.482952	-.151202
GDP	3.211486	1.895105	1.69	0.090	-.5028512	6.925823
CPI	-1.26621	.1980366	-6.39	0.000	-1.654355	-.8780658
FDI	-27.9261	5.858663	-4.77	0.000	-39.40887	-16.44333
_cons	657.9848	104.8436	6.28	0.000	452.4951	863.4745

Appendix 13: Regression Results from GLS Method and Bayesian Approach

	p-Value (Z score)	Minimum Bayens factor	Decrease in probability of the null hypothesis (%)		Strength of evidence
			From	To no less than	
LIQ	**0.032** (1.85)	**0.17** (1/6)	90 50 25	60.00[a] 14.29 5.21	Moderate
SIZE	**0.000** (3.29)	**0.0045** (1/224)	90 50 25	3.9 0.4 0.1	Strong to very strong
ROA	**0.011** (2.290)	**0.07** (1/14)	90 50 25	39.13 6.67 2.30	Moderate

(continued)

(continued)

	p-Value (Z score)	Minimum Bayens factor	Decrease in probability of the null hypothesis (%)		Strength of evidence
			From	To no less than	
NPL	**0.001** (3.09)	**0.008** (1/118)	90 50 25	7.09 0.84 0.28	Moderate to strong
CAP	**0.016** (2.14)	**0.1** (1/10)	90 50 25	47.37 9.09 3.19	Moderate
GDP	**0.090** (1.34)	**0.5** (1/2)	90 50 25	81.82 33.33 14.16	Weak
CPI	**0.000** (3.29)	**0.0045** (1/224)	90 50 25	3.9 0.4 0.1	Strong to very strong
FDI	**0.000** (3.29)	**0.0045** (1/224)	90 50 25	3.9 0.4 0.1	Strong to very strong
_cons	**0.000** (3.29)	**0.0045** (1/224)	90 50 25	3.9 0.4 0.1	Strong to very strong

[a]Calculations were performed as follows

A probability (Prob) of 90% is equivalent to an odds of 9, calculated as Prob/(1 − Prob)

Posterior odds = Bayes factor × prior odds; thus (1/6) × 9 = 1.5

Probability = odds/(1 + odds); thus 1.5/2.5

References

Aydin, B. (2008). Banking Structure and Credit Growth in Central and Eastern European Countries (September 2008). IMF Working Papers, Vol., pp. 1–44, 2008. Retrieved from SSRN: https://ssrn.com/abstract=1278426.

Carlson, M. A., Warusawitharana, M., & Shan, H. (2011). Capital ratios and bank lending: A matched bank approach. *Journal of Financial Intermediation, 22*(4).

Chen, G., & Wu, Y. (2014). Bank ownership and credit growth in emerging markets during and after the 2008–09. Financial crisis—A cross-regional comparison. No. 14/171, IMF Working Papers.

Chu Khanh, L. (2012). Discussion about the impact of monetary policy to credit growth in Vietnam. *Banking Journal*. Number 13, 7/2012.

Guo, K., & Stepanyan, V. (2011). Determinants of Bank Credit in Emerging Market Economies. IMF working paper WP/11/51.

Igan, D., & Tan, Z. (2015). Capital inflows, credit growth, and financial systems. IMF Working Papers WP/15/193.

Imran, K., & Nishat, M. (2013). Determinants of Bank Credit in Pakistan: A Supply Side Approach. Proceedings of 2nd International Conference on Business Management (ISBN: 978-969-9368-06-6).

Laidroo, L. (2014). Lending Growth and Cyclicality in Central and Eastern European Banks. TUTECON Working Paper No. WP-2014/4.

Lane, P. R., & McQuade, P. (2013). Domestic credit growth and international capital flows. ECB Working Paper, No. 1566, European Central Bank (ECB), Frankfurt a. M.

Mark, T. (2011). The impact of non-performing loans on loan growth: an econometric case study of Jamaica and Trinidad and Tobago.

Pouw, L., & Kakes, J. (2013). What drives bank earnings? Evidence for 28 banking sectors. *Applied Economics Letters, 20*(11), 1062–1066.

Tamirisa, N. T., & Igan, D. O. (2008). Are weak banks leading credit booms? Evidence from emerging Europe (September 2008). IMF Working Papers (pp. 1–21). Retrieved from SSRN: https://ssrn.com/abstract=1278430.

Thong, D. M. (2018). Factors affecting the credit growth in jointed-stock banks of Vietnam. Master thesis. The university of economy in Ho Chi Minh city.

Trung, N. D., & Chung, N. H. (2018). Impact of monetary policy and macro prudential policy to finance stability in Vietnam- Perspective through credit growth. *Journal of banking technology,* (1 and 2), 59–73.

Wooldridge, J. (2002). *Econometric analysis of cross section and panel data.* Cambridge: MIT press.

Measuring Dependence in China-United States Trade War: A Dynamic Copula Approach for BRICV and US Stock Markets

Worrawat Saijai, Woraphon Yamaka, and Paravee Maneejuk

Abstract The phenomena of trade war between China and United States (US) leads us to examine the spillover effects of US stock market volatility on the BRICV stock markets (Brazil, Russia, India, China, and Vietnam). Thus, the dynamic correlations between US and each BRICV stock market, is measured using the flexible dynamic conditional correlations based bivariate GARCH-with-jumps model. The result of both classical bivariate GARCH(1,1) model and bivariate GARCH(1,1)-with-jumps model show that all stock returns have high volatility persistence with the value higher than 0.95. Moreover, the result of DCC-Copula part shows a dynamic correlations between US and each stock in BRICV. We find that the dynamic correlations for all pairs are similar and are not constant. We also find that US stock market has a positive correlations with BRICV stocks between 2012 and 2019. When, we compare the correlations between pre and post trade war in 2018, we observe that bivariate copula between US-China, US-Vietnam and US-Brazil seems to be affected by the trade war as there exhibit a large drop of the correlations after 2018.

Keywords GARCH · DCC-GARCH · Copula GARCH · Trade war

1 Introduction

Trade war is an economic conflict which leads to the raising of economic barrier(s) between the countries. The key tools for raising the barrier are either tariffs or non-tariffs to reduce trade deficit and provide good environment to domestic industries

W. Saijai · W. Yamaka (✉) · P. Maneejuk
Center of Excellence in Econometrics Faculty of Economics, Chiang Mai University,
Chiang Mai 50200, Thailand
e-mail: woraphon.econ@gmail.com

W. Saijai
e-mail: worrawat13@hotmail.com

P. Maneejuk
e-mail: mparavee@gmail.com

© The Editor(s) (if applicable) and The Author(s), under exclusive license
to Springer Nature Switzerland AG 2021
N. Ngoc Thach et al. (eds.), *Data Science for Financial Econometrics*,
Studies in Computational Intelligence 898,
https://doi.org/10.1007/978-3-030-48853-6_39

to be stronger in market competitions. Recently, China-United States (US) trade war has been playing a significant role in the world economy and it contributes many effects to the world economy. Both countries increase their protective level against each other which aims to decrease/increase trade deficit/surplus after the one country raise tariff and/or non-tariff policies to, this situation keep continue when countries employ policies to response each other.

Theoretically, US's imposition of additional and higher tariffs and its implementation of qualitative measures could encourage more investment in the country from both domestic investors and foreign countries. Thus, this could lead to the increase in commodity, stock, and real estate prices in the USA. Furthermore, the economic boom could result in higher capital investment, more spending and development activities. However, these trade barriers may render both beneficial and non-beneficial effects to other countries, especially China. China, which is the large exporter to USA, has suffered from the new barriers and its exports to the US went down abruptly.

When one country declares some unilateral policies, the directly affected countries may employ some policies in response. China uses retaliatory trade barrier as a response, for example, by lowering the bean importing quota from US in order to show its power in the trade war. From the idea of trade policy, when a big economy start doing trade barrier or increasing import tariff, other countries may response by using trade policy and more countries follow this way to protect themselves. That's why trade war from two countries may widespread to the whole world and could lead to world economic slowdown. BRICV (Brazil, Russia, India, and China, and Vietnam) economies are not an exception. They are very important in the world trade with remarkable potential in the future. In the case of BRIC economies (Brazil, Russia, India, and China), they are the fastest growing market economies nowadays due to their investment opportunity, favourable market environment, and abundant natural resources. These countries have a high potential to be leaders in the world economy. Among the world's emerging economies, one with very high potential is Vietnam in Southeast Asia which has exhibited a high and rapid economic growth in recent years because of its escalation of trade openness. Vietnam is believed to soon catch up with the BRIC economies and some advanced economies and thereby reaching the same level of development. Therefore, BRICV economies can be viewed as a good representative to study the impacts of trade war.

The effect of trade war could strengthen US economy because of the growing investment funds from both domestic and foreign sources. However, when the economy is growing, it rises the inflation, the expected returns of assets, or stock and commodity prices in US could become higher, thus attracting more investment from overseas. Investors always seek for higher return, indicating there could be a relationship between US stock price and foreign stock markets including BRICV. Fundamentally, each country in BRICV has different structure of trade and investment relationship with the US and we believe that the relationship structure after the trade war should have changed; the positive relationship may turn to be negative, the strong correlation may turn to be weak. The change in correlation can arise from the financial contagion.

In the literature, financial contagion is the situation in which different markets have co-movement of exchange rates, stock prices, sovereign spreads, and capital flows usually occurs with negative shock, the shock which temporaliry deviate index value from its structure, and then neutralize to normal. Dornbusch et al. (2001). Longin and Solnik (1995) found the contagion effect between advanced markets and emerging markets during 1960–1990, and the correlations of financial indexes in those countries was higher during the high volatility period. Samarakoon (2011) also revealed the contagion effect between US market and emerging markets, except Latin America. Forbes and Rigobon (2002); Corsetti et al. (2005); and Yilmaz (2010) confirmed the existence of the contagion effect of financial market of East Asian Crisis which the crisis in a country in the group is the precursor of the crisis in another country in East Asian. Finally, Chiang et al. (2013) found the contagion effects of the US sup-prime crisis on the stocks of BRICV countries, whereas Russia and Vietnam were found to have a substantial effect, the effect is higher than other countries in the group.

As mentioned above, the contagion effects have occurred in several cases. The US economy is one of the large economic systems which produce a large contagion effect to many countries. Its so called sub-prime mortgage crisis, which eventually led to the global financial crisis, had generated spillover effects to both emerging and developed countries.

In this study, we investigate the contagion effect before and after the trade war between China-US. However, study on trade war effect is quite a new challenge because the war started in the middle of 2018 and has continued until today. To the best of our knowledge, we investigates the contagion effects between the US and the BRICV stock markets under the context of the US-China trade war. Therefore, we attempt to fill the gap of the literature by measuring the dynamic contagion effects using the dynamic conditional copula based GARCH of Jondeau and Rockinger (2006). However, we aim to replace the classical GARCH model with GARCH-with-jumps model in order to model marginal distributions as it has an ability to measure the sudden movement, the extreme case, of returns.

This paper adds several important economic and econometric contributions to the literature: First, it employs a GARCH-with-jumps model to estimate the marginal distribution of the stock returns. We also compare the conditional variance obtained from classical GARCH and GARCH-with-jumps in order to capture the abnormal extreme events through Poisson distributed jump Duangin et al. (2018). Second, this current paper compares the contagion effects of US stock on BRICV stocks for both pre-trade war and post-trade war periods to provide a more complete knowledge of the impact of trade war on stock markets. Third, several GARCH-with-jumps distributions and dynamic conditional copulas are introduced in this empirical study.

This paper is organized as follows: Sect. 2 gives a brief overview of various models for estimating GARCH-with-jumps and dynamic copulas. Section 3 explains the data used in this study. Section 4 discusses empirical analysis. Concluding remarks are presented in Sect. 4.

2 Methodology

2.1 Structure of GARCH Models with Jumps

The GARCH models are applied to estinate the volatility of variables and they are
very useful in financial time series. The GARCH typed model has an ability to remove
autocorrelation and heteroscedasticity that often appear in financial time series data.
However, a distinctive feature of the modern financial series is the presence of jump
dynamics of financial data, Therefore, GARCH-with-jumps model was introduced
by Jorion (1988). This model contains one additional error term to capture the effect
of jumps in volatility. GARCH-with-jumps innovation can be shown as vector in
Eq. (1)

$$r_{i,t} = \mu_{i,t} + \varepsilon_{i,1t} + \varepsilon_{i,2t}, \tag{1}$$

where $r_{i,t}$ is return of stock i, $\mu_{i,t}$ is the mean of return or expected return of $r_{i,t}$,
the volatility of the stock return can be separated into 2 different terms which help
to capture the normal innovation ($\varepsilon_{i,1t}$) and unusual innovation (jump effect) ($\varepsilon_{i,2t}$).
The normal component is represented by

$$\varepsilon_{i,1t} = \sqrt{h_{i,t}} z_{i,t} \tag{2}$$

where $h_{i,t}$ is the conditional variance of return i in time t which is generated by using
GARCH(1,1) model, $z_{i,t}$ is the standardized residuals with the property of being
independent and identically distributed (i.i.d.), $E(z_{i,t}) = 0$ and $E(z_{i,t}z_{i,t}^T) = I$. The
GARCH(1,1) specification can be written as

$$h_{i,t} = \omega_{i,0} + \alpha_{i,1}\varepsilon_{i,t-1}^2 + \beta_{i,1}h_{i,t-1}^2 \tag{3}$$

The expected effects of normal innovation is zero, $E(\varepsilon_{1,t}|I_{t-1}) = 0$. Moreover, all
parameters are positive, thus $\alpha_0, \alpha_1, \beta_1 > 0$. The second component is the unusual
innovation or jump in volatility ($\varepsilon_{i,2t}$). It is also assumed that the effects of unusual
innovation are sum up to be zero, thus $E(\varepsilon_{i,2t}|I_{t-1}) = 0$. This unusual innovation
can be computed by

$$\varepsilon_{i,2t} = J_{i,t} - E(J_{i,t}|I_{i,t-1}). \tag{4}$$

$$\varepsilon_{i,2t} = \sum_{k=1}^{n_{i,t}} Y_{i,t,k} - \theta_i \lambda_{i,t}. \tag{5}$$

where $Y_i \sim N(\theta_i, \delta_i^2)$, θ is the size of jumps, and $\lambda_{i,t}$ is the intensity of jumps. n_t is the number of jump which can be random from Poisson distribution ($n_t \sim Rpoisson(\lambda_t)$) with conditional jump intensity $\lambda_{i,t}$,

$$\lambda_{i,t} = a_i + b_i(\lambda_{i,t-1}) + c_i \xi_{i,t-1} \tag{6}$$

where

$$\zeta_{i,t-1} = E(n_{i,t-1}|I_{i,t-1}) - \lambda_{i,t-1} = \sum_{j=0}^{\infty} j P(n_{i,t-1} = j|I_{i,t-1}) - \lambda_{i,t-1}. \tag{7}$$

$$= \frac{f(\varepsilon_{i,t}|n_{i,t-1} = j, I_{i,t-2}, \Theta_i) P(n_{i,t-1} = j)|I_{i,t-2})}{\sum_{j=0}^{\infty} j (f(\varepsilon_{i,t}|I_{i,t-2}, \Theta_i))} \tag{8}$$

$$= \frac{\frac{exp(-\tilde{\lambda}_{i,t-1})\tilde{\lambda}_{i,j,t-1}}{j!} \frac{1}{\sqrt{2\pi(\bar{\sigma}_{i,t-1}^2 + j\delta_i^2)}} exp\left(-\frac{(\varepsilon_{i,t-1} + \theta_i\lambda_{i,t-1} - \theta_i j)^2}{2(\bar{\sigma}_{i,t-1}^2 + j\delta_i^2)}\right)}{\sum_{j=0}^{\infty} j (f(\varepsilon_{i,t-1}|I_{i,t-2}, \Theta_i))} \tag{9}$$

2.2 DCC-Copula GARCH

The Dynamic Conditional Covariance Model (DCC model) is introduced by Engle (2002) to investigate the time-varying correlations between n asset prices. In the financial time series, we can apply the DCC to study correlations of asset returns. Typically, this model is constructed based on the multivariate normal distribution; thus GARCH equaiton is generally assumed to follow normal distribution as well. However, the normality assumption of GARCH model is often rejected by many empirical studies (Pastpipatkul et al. 2015, 2016). Thus, in this study, we consider the DCC copula approach of Jondeau and Rockinger (2006) to join the stock returns. This model is more flexible than the traditional DCC of Engle (2002) as it can join any distribution of the stock returns. In DCC model, the matrix of conditonal covariance is

$$H_t = D_t R_t D_t \tag{10}$$

where D_t is a diagonal matrix of time-varying conditional standard deviation of the disturbance ε_{1t}, which is obtained from GARCH(1,1)-with-jumps model. In this study, we aim to find the conditional corrrelation between US stock return and each stock return in BRICV. Hence, the bivariate DCC Copula GARCH model is employed. Then, we can write matrix D_t as follows:

$$D_t = \begin{bmatrix} \sqrt{h_{1,t}} & 0 \\ 0 & \sqrt{h_{2,t}} \end{bmatrix}. \tag{11}$$

R_t is correlation matrix, which contains time-varying conditional correlations of the standardized disturbances $z_{i,t} = \varepsilon_{i,t} D_t^{-1}$. R_t has symmetric positive semi-definite matrix, and has the value of one along the diagonal.

$$R_t = \begin{bmatrix} 1 & \rho_{12,t} \\ \rho_{22,t} & 1 \end{bmatrix}, \tag{12}$$

We can also show R_t as

$$R_t = Q_t^{*-1} Q_t Q_t^{*-1} \tag{13}$$

and Q can be shown as

$$Q_t = (1 - a - b)\bar{Q} + a\varepsilon_{t-1}\varepsilon'_{t-1} + bQ_{t-1}, \tag{14}$$

where $\bar{Q}$ is the unconditional covariance matrix which contains standardized errors $z_{i,t}$. This also guarantees the positive definite property of H_t.

2.3 Estimation

Engle (2002) suggested estimating GARCH and DCC parts separately, by starting with GARCH model to obtain the standardized residuals z_t and the conditional variance h_t. Then we can estimate time-varying correlations from DCC approach in the second step. In this study we also follow the two-step estimation, thus, the stock returns are modelled by the GARCH(1,1)-with-jumps. Note that four distributions are considered in this study, namely normal, student-t, skewed student-t and skewed normal distributions. We then employ the Gaussian copula to join the standardized residuals. To estimate all unknown parameters in both GARCH(1,1)-with-jumps and DCC copula parts, the maximum likelihood estimator (MLE) is employed. Note that the bivariate Copula based GARCH-with-jumps is used, hence the likelihood function can be shown as the function of joint density by the following equation.

$$f_{1,2}(z_{1,t}, z_{2,t}) = \frac{\partial^2 F_{1,2}(z_{1,t}, z_{2,t})}{\partial z_{1,t}, \partial z_{2,t}} \tag{15}$$

$$= \frac{\partial^2 C(F_1(z_{1,t}, z_{2,t}))}{\partial z_{1,t}, \partial z_{2,t}}$$

$$= \frac{\partial^2 C(F_1(z_{1,t}), F_1(z_{2,t}))}{\partial u_{1,t}, \partial u_{2,t}} \cdot \frac{\partial F_1(z_{1,t})}{\partial z_{1,t}} \cdot \frac{\partial F_2(z_{2,t})}{\partial z_{2,t}}$$

$$= c_{R_t}(u_{1,t}, u_{2,t}) \cdot f_1(z_{1,t}) \cdot f_2(z_{2,t})$$

Thus, we can construct the likelihood function as

$$L(r_{1t}, r_{2t} \mid R_t, \Psi) = \prod_{i=1}^{T} c_{R_t}(u_{1,t}, u_{2,t}) \cdot f_1(z_{1,t}) \cdot f_2(z_{2,t}) \tag{16}$$

$$c_{R_t}(u_{1,t}, u_{2,t}) = (\sqrt{R_t})^{-1} \left(\frac{1}{2} \left(\Phi_1^{-1}(u_1), \Phi_2^{-1}(u_2) \right) \cdot (R_t - I) \cdot \begin{pmatrix} \Phi_1^{-1}(u_1) \\ \Phi_2^{-1}(u_2) \end{pmatrix} \right) \tag{17}$$

$$f_i(z_{i,t}) = \frac{exp(-\tilde{\lambda}_{i,t-1})\tilde{\lambda}_{i,j,t-1}}{j!} \frac{1}{\sqrt{2\pi(\tilde{\sigma}_{i,t-1}^2 + j\delta_i^2)}} exp\left(-\frac{(\varepsilon_{i,t-1} + \theta_i\lambda_{i,t-1} - \theta_i j)^2}{2(\tilde{\sigma}_{i,t-1}^2 + j\delta_i^2)} \right) \tag{18}$$

where Φ_1^{-1} is the quantile function, $\Psi = (\mu, \omega, \alpha, \beta, \delta, \theta, \lambda, a, b)$ is the vector of estimated parameters of DCC-Copula GARCH model. Note that we can maximize only the copula density $c_{R_t}(u_{1,t}, u_{2,t})$ and fix $f_i(z_{i,t})$ to be constant. $f_i(z_{i,t})$ is obtained by maximizing the density function of GARCH-with-jumps model.

3 Data

In this study, we investigate the dynamic correlations between stock price of US and BRICV. Thus, Down Jones Industrial average (DJIA), Brazil Bovespa 50 (BVSP), Russia RTS index (RTSI), India Sensex Index (BSE 30), China Shanghai Composite index (SSEC), and Vietnam Ho Chi Minh index (VNI) are collected. The data are obtained from Investing.com covering the period from January 2, 2009 to July 27, 2019. The daily stock price returns are presented in Table 1. In this study, we employ Minimum Bayes factor (MBF) instead of p-value to check the significant result. If $1 <$ MBF $< 1/3$, $1/3 <$ MBF $< 1 = 10$, $1 = 10 <$ MBF $< 1/30$, $1/30 <$ MBF $< 1/100$, $1/100 <$ MBF $< 1/300$ and MBF $< 1/300$, it can be considered as, weak evidence, moderate evidence, substantial evidence, strong evidence, very strong evidence and decisive evidence, respectively, to reject the null hypothesis. The data exhibit the negative

Table 1 Descriptive statistics of stock returns

	BVSP	RTSI	BSE	SSEC	VNI	DJIA
Mean	0.0002	−0.0001	0.0003	0.0001	0.0001	0.0004
Median	0.0002	0.0003	0.0005	0.0004	0.0008	0.0005
Maximum	0.0624	0.1324	0.0502	0.0603	0.0392	0.0486
Minimum	−0.0916	−0.1325	−0.0478	−0.0887	−0.0696	−0.05706
Std. Dev.	0.0140	0.01779	0.0099	0.0139	0.0112	0.0087
Skewness	−0.1683	−0.3975	−0.1549	−0.8179	−0.5896	−0.4999
Kurtosis	5.1021	10.0761	4.7268	9.0683	5.8107	7.3702
Jarque-Bera	388.61a	4,356.38a	259.4578a	3,298.326a	796.7063a	1,797.185a
Unit root test (ADF)	−45.4068a	−42.4667a	−46.3891a	−48.8681a	−28.05426a	−48.1265a
Observations	1818	1825	1784	1820	1770	1817

Note "a" denotes strongly support to reject the null hypothesis: the variable is not normal distributed, and the alternative hypothesis: the variable is normal distributed, according to Minimum Bayes Factor (MBF) (see, Maneejuk and Yamaka 2020)

skewness and high kurtosis indicating that our data are not normally distributed. We thus employ the Jarque-Bera test and the result shows that our data are not normally distributed. In addition, the Augmented Dickey Fuller (unit root) test points out that stock returns are stationary.

3.1 Estimation Results

Prior to interpreting our volatility and correlation results, we firstly compare the performance of the bivariate GARCH model with jumps and without jumps. Note that four distributions, namely normal distribution, skew-normal distribution, student's t

Table 2 AIC estimates for GARCH(1,1) with and without jumps

		DJIA	BVSP	RTSI	BSEC	SSEC	VNI
Norm	AIC-Jump	−0.4546	−1.7471	−1.9818	−0.9001	−1.3991	−1.0970
	AIC	−0.5167	**−1.7737**	**−2.0736**	−0.9198	**−1.4933**	−1.1648
Snorm	AIC-Jump	−0.4535	−1.7460	−1.9807	−0.8990	−1.3980	−1.0959
	AIC	−0.5051	−1.7722	−2.0634	−0.9144	−1.4875	−1.1398
Std	AIC-Jump	**−0.5170**	−1.7468	−2.0729	**−0.9990**	−1.3980	**−1.1794**
	AIC	−0.4550	−1.7498	−1.9767	−0.8978	−1.3933	−1.0980
Sstd	AIC-Jump	−0.4524	−1.7661	−2.0758	−0.8978	−1.3968	−1.1783
	AIC	−0.4539	−1.7486	−1.9754	−0.8951	−1.3918	−1.0886

Note Norm, Snorm, Std, and Sstd are referred to normal distribution, skew-normal distribution, student's t distribution, and skew-student's t distribution, respectively

Table 3 The result of the best fit GARCH(1,1) model

Indexes	ω	α	β	μ
BVSP	0.0114a	0.0449a	0.9238a	0.0224a
RTSI	0.0097a	0.0660a	0.9182a	0.0124
BSE	0.0063a	0.0835a	0.8776a	0.0319a
SSEC	0.0014a	0.0562a	0.9417a	0.0101
VNI	0.0073a	0.1160a	0.8556a	0.0310a
DJIA	0.0086a	0.1882a	0.7464a	0.0345a

Note "a" denotes strongly support to reject the null hypothesis: the coefficient is not statistically significant, and the alternative hypothesis: the coefficient is statistically significant, according to Minimum Bayes Factor (MBF)

distribution, and skew-student's t distribution are considered for bivariate GARCH models. Table 2 shows that GARCH(1,1)-with-jumps under student's t distribution show the lowest AIC for DJIA (from the pair of DJIA and BVSP), BSEC, and VNI while classical GARCH(1,1) under normal distribution of BVSP, RTSI, and SSEC show the lowest AIC. This indicates that there exists an unusual error or jumps in the volatilities of AIC for DJIA, RTSI. To give a clear picture, we plot the volatilities obtained from GARCH(1,1) (blue line) and GARCH(1,1)-with-jumps (red line). We can observe a lot of peaks of the volatilities of DJIA, BSEC, and VNI along the sample period. (see Fig. 1). Additionally, we observe that the volatility estimates obtained from GARCH(1,1) are lower than those from GARCH(1,1)-with-jumps, except for RTSI case. Moreover, the graphs of RTSI, SSEC, and VNI show heavy jumps on the returns as their volatilities are greater than 1.

From Tables 3 and 4, we show the estimated coefficients of our two models. In these two tables, we show only the best GARCH(1,1) and GARCH(1,1) with jumps specifications that provide the lowest AIC. (see, Table 2). We find that $\alpha + \beta$ is more than 0.95, implying a high volatility persistence in our stock returns. According to Table 4, the jump size effect is considered by θ and δ. The results show a negative value of θ. It indicates that the decrease in price is related to the bad shock. We find a negative value in the cases of RTSI, BSE, SSEC, and DJIA. However, the jumps in BVSP and VNI are related to the good shock ($\theta > 0$). However, standard deviation of jumps, δ, is very high in SSEC, BSE, and, DJIA (0.97, 0.83, and 0.70). The frequency of jumps λ is very high in SSEC and DJIA.

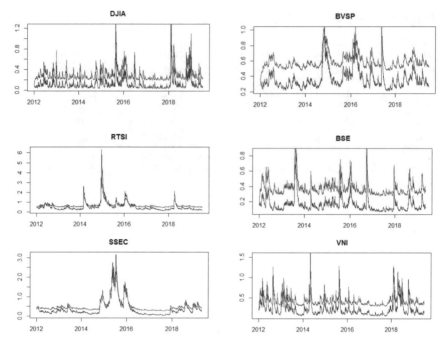

Fig. 1 GARCH(1,1) (blue) and GARCH(1,1)-with-jumps (red)

Table 4 The result of the best fit GARCH(1,1)-with-jumps

Indexes	ω	α_1	β	δ	θ	λ	μ
BVSP	0.0305	0.0336	0.9284	0.0000	0.0004	0.0633	0.0224
RTSI	0.0376	0.0373	0.9187	0.0000	−0.0012	0.0541	0.0136
BSE	0.0380	0.1083	0.8548	0.7007	−0.1542	0.2723	0.0283
SSEC	0.0161	0.0416	0.9388	0.9721	−0.0170	0.8685	0.0120
VNI	0.0463	0.1203	0.8407	0.0000	0.0159	0.1665	0.0317
DJIA	0.0372	0.2076	0.8014	0.8324	−0.1258	0.5373	0.0272

Note All parameters are strongly support to reject the null hypothesis: the coefficient is not statistically significant, and the alternative hypothesis: the coefficient is statistically significant, according to Minimum Bayes Factor (MBF)

To understand the effects of the trade war in 2018, we can compare the dynamic correlations of each index and DJIA by investigating the correlations between US stock return and BRICV stock returns in two sample periods, say pre-2018 and post-2018. We estimate the DCC-Copula based on the best GARCH(1,1)-with-jumps specifications. The estimated results are illustrated in Table 5. We can measure the correlations persistence (higher correlation intensity is persistent) by investigating a+b and the results show that BVSP shows the highest correlation persistence while SSEC shows the lowest persistence.

Table 5 Estimated parameters of DCC-Copula part (Eq. 18)

Coefficients	US-BVSP	US-RTSI	US-BSEC	US-SSEC	US-VNI
a	0.0542a	0.0428a	0.1031a	0.0956a	0.1038a
b	0.8380a	0.8368a	0.7744a	0.7495a	0.7504a
a+b	0.8922	0.8796	0.8775	0.8451	0.8542

Note "a" denotes strongly support to reject the null hypothesis: the coefficient is not statistically significant, and the alternative hypothesis: the coefficient is statistically significant, according to Minimum Bayes Factor (MBF)

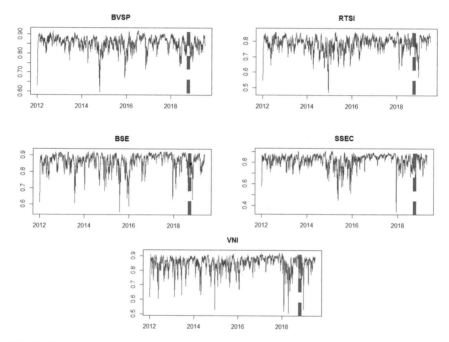

Fig. 2 Dynamic conditional correlations student-t copula: DJIA-BRICV stock markets: The vertical red dash line presents the starting of trade war

We also plot the dynamic conditional correlations in Fig. 2. The result shows that the characteristics of the dynamic correlations of all pairs are not much different. We also observe a positive correlation along 2009–2019 for all pairs. In the period of pre-2018, the volatility of stock returns of all indexes dropped in some periods, especially in 2015. In the period of post-2018, it obviously shows that the dynamic correlations of DJIA-BSE, DJI-SSEC and DJIA-VNI pairs dropped after 2018. This indicates that the trade war between US-China would contribute to a large drop of the correlations between US stock market and three stock markets of BRICV, namely Vietnam, Brazil, and China.

4 Conclusion

In this paper, we estimate the degree of dependence across US stock market and BRICV stock markets before and after China-United States Trade War. The degree of dependence is modelled by DCC-Copula GARCH-with-jumps model. Two-step estimation is employed to estimate this model. First, we estimate the standardized residuals of our six stock returns using GARCH(1,1)-with-jumps. In this step, GARCH(1,1)-with-jumps is based on four distributions, namely normal, student's t, skew-student's t, and skew-normal. We also compare the model with the classical GARCH(1,1) model using the AIC. In the second step, the obtained standardized residuals from GARCH(1,1)-with-jumps under the best distribution are further joined by the DCC-Copula based on Gaussian distribution.

The findings in the first step provide the evidence that GARCH(1,1)-with-jumps under student's t distribution show the lowest AIC for DJIA, BSEC, and VNI while classical GARCH(1,1) under normal distribution of BVSP, RTSI, and SSEC show the lowest AIC. The estimated GARCH parameters also show the high unconditional volatility of the stock returns. In the second step, the dynamic conditional correlations is provided. We find similar dynamic correlations between US stock market and each stock market of BRICV. However, when we compare the dynamic correlations between pre and post trade war, we find that the correlations between US-China, US-Vietnam and US-Brazil has clear decreasing trends after the start of trade war in 2018. Therefore, the investors in China, Vietnam, and Brazil should be careful of what the US government will do during the trade war with China.

References

Cheung, W., Fung, S., & Tsai, S. C. (2010). Global capital market interdependence and spillover effect of credit risk: Evidence from the 2007–2009 global financial crisis. *Applied Financial Economics, 20*(1 & 2), 85–103.

Chiang, S., Chen, H., & Lin, C. (2013). The spillover effects of the sub-prime mortgage crisis and optimum asset allocation in the BRICV stock markets. *Global Finance Journal, 24*, 30–43.

Corsetti, G., Pericoli, M., & Sbracia, M. (2005). Some contagion, some interdependence: More pitfalls in tests of financial contagion. *Journal of International Money and Finance, 24*, 1177–1199.

Dornbusch, R., Park, Y., & Claessens, S. (2001). Contagion: How it spreads and how it can be stopped? In S. Claessens & K. Forbes (Eds.), *International financial contagion* (pp. 19–42). New York: Kluwer Academic Publishers.

Duangin, S., Yamaka, W., Sirisrisakulchai, J., & Sriboonchitta, S. (2018). Volatility jump detection in Thailand stock market. In *International symposium on integrated uncertainty in knowledge modelling and decision making* (pp. 445–456).

Engle, R. (2002). Dynamic Conditional Correlation: A simple class of multivariate generalized autoregressive conditional heteroskecasticity models. *Journal of Business & Economic Statistics, 20*(3), 339–350.

Forbes, K., & Rigobon, R. (2002). No contagion, only interdependence: Measuring stock markets comovements. *The Journal of Finance, 57*, 2223–2261.

Jondeau, E., & Rockinger, M. (2006). The copula-garch model of conditional dependencies: An international stock market application. *Journal of international money and finance, 25*(5), 827–853.

Jorion, P. (1988). On jump processes in the foreign exchange and stock markets. *The Review of Financial Studies, 1*(4), 427–445.

Longin, F., & Solnik, B. (1995). Is the correlation in international equity returns constant: 1960–1990. *Journal of International Money and Finance, 14*, 3–26.

Maneejuk, P., & Yamaka, W. (2020). Significance test for linear regression: how to test without P-values? *Journal of Applied Statistics, 47*, 1–19. https://doi.org/10.1080/02664763.2020.1748180.

Pastpipatkul, P., Yamaka, W., & Sriboonchitta, S. (2015). Co-movement and dependency between New York stock exchange, London stock exchange, Tokyo stock exchange, oil price, and gold price. In *International symposium on integrated uncertainty in knowledge modelling and decision making* (pp. 362–373).

Pastpipatkul, P., Maneejuk, P., & Sriboonchitta, S. (2016). The best copula modeling of dependence structure among gold, oil prices, and US currency. *International symposium on integrated uncertainty in knowledge modelling and decision making* (pp. 493–507).

Samarakoon, L. (2011). Stock market interdependence, contagion, and the U.S. financial crisis: The case of emerging and frontier markets. *Journal of International Financial Markets Institutions and Money, 21*.

Yilmaz, K. (2010). Return and volatility spillovers among the East Asian equity markets. *Journal of Asian Economics, 21*, 304–313.

The Effect of Governance Characteristics on Corporate Performance: An Empirical Bayesian Analysis for Vietnamese Publicly Listed Companies

Anh D. Pham, Anh T. P. Hoang, and Minh T. H. Le

Abstract This study extends the literature by conducting a Bayesian inference analysis of the relationship between governance characteristics and financial performance in Vietnamese publicly listed firms over a six-year span. The empirical results reveal that two main components of corporate governance, namely board gender diversity and blockholder ownership, tend to foster firm performance. Besides, no new evidence was found concerning the role of other governance characteristics such as board size, CEO duality (i.e. a concurrent position in a company that combines the duties of the CEO and the board chair) and non-executive director representation. From the above findings, we recommend Vietnamese firms to attach greater importance to corporate governance characteristics as a fundamental driver for better financial outcomes in the short run and sustainable development in the long run. Our results are confirmed by extensive robustness tests.

Keywords Corporate governance · Firm performance · Board of directors · Blockholder ownership · Bayesian analysis

A. D. Pham (✉)
Vietnam Banking Academy, 12 Chua Boc St., Dong Da Dist., Hanoi, Vietnam
e-mail: anhpd@hvnh.edu.vn

A. T. P. Hoang · M. T. H. Le
University of Economics Ho Chi Minh City, 196 Tran Quang Khai St., Dist. 1, Ho Chi Minh City, Vietnam
e-mail: anhtcdn@ueh.edu.vn

M. T. H. Le
e-mail: minhtcdn@ueh.edu.vn

N. Ngoc Thach et al. (eds.), *Data Science for Financial Econometrics*,
Studies in Computational Intelligence 898,
https://doi.org/10.1007/978-3-030-48853-6_40

1 Introduction

Corporate governance has been an issue of interest to an enormous number of academics worldwide, particularly after the collapse of major global corporations and international banks, such as WorldCom and Commerce Bank, due to weaknesses in corporate governance. The question arises as to whether, and by what means, corporate governance structures affect firm performance. Among the typical measures of corporate governance, board characteristics and ownership structure could be considered to be of the highest importance.

So far, studies on the impact of board characteristics and ownership structure on firms' financial performance have taken into account factors including board size, CEO duality, female representation on the board, the presence of independent directors, or blockholder ownership. For instance, Campbell and Mínguez-Vera (2008) and Pham and Hoang (2019) found a significantly positive correlation between the number of female directors on the board and firm performance, while others reported a negative relationship (e.g., Adams and Ferreira 2009) or found no evidence for such a relationship. Likewise, empirical works associated with other aspects have yielded mixed results.

This paper aims to justify the importance of governance characteristics in boosting business performance of firms listed on the Ho Chi Minh City Stock Exchange (HOSE). Unlike most prior studies for Vietnam that have utilised traditional performance metrics based on book values, for example return on total assets (ROA) as per Vo and Phan (2013), or return on equity (ROE) as per Doan and Le (2014), we present new perspectives on the governance-performance nexus through the application of a market-based indicator, namely Tobin's Q, as a proxy for performance. A remarkable advantage of Tobin's Q is that it helps to predict the financial outcomes of a firm with high reliability (since this measure is reflected in the market value of a firm's shares), implying a market assessment of the potential profitability of the firm.

Another noteworthy contribution is that, in this study, we conduct Bayesian estimation and testing for the corporate governance—financial performance relationship in the context of emerging market economies, which is unprecedented in the previous literature. This approach could improve on traditional p-value estimations in several ways. Specifically, instead of working backward by calculating the probability of our data if the null hypothesis were true under the p-value testing mechanism which might lead to substantial misinterpretation and errant conclusions, Bayesian statistics allow us to work forward by calculating the probability of our hypothesis given the available data. This methodology provides a purely mathematical means of incorporating our "prior probabilities" from previous study data to generate new "posterior probabilities" (Buchinsky and Chadha 2017).

The remainder of the paper is structured as follows. Section 2 reviews the theoretical background and empirical literature on the effectiveness of governance characteristics. Section 3 describes data and econometric approach used in the study. Main results are discussed in Sect. 4, while Sect. 5 delivers concluding remarks.

2 Literature Review

2.1 Theoretical Background

2.1.1 Agency Theory

Agency theory is the grounding theoretical perspective in corporate governance studies. As indicated by Daily et al. (2003), the dominance of agency theory in governance research could be explained by two reasons. First, it is just a simple theory, in which large corporations are reduced to two participants—managers and shareholders—and the interests of each are assumed to be both clear and consistent. Second, agency theory holds that both groups of participants tend to be self-interested instead of willing to sacrifice individual interests for the interests of others. While shareholders expect managers to act in the best interests of the business, managers might not necessarily make decisions for the goal of shareholder wealth maximisation; instead, they may act in their own self-interest. This may lead to the reality that managers act for self-interest, not for the sake of the owner. As the issue of conflicts of interest is highly likely to occur in joint stock companies, it might create "agency costs". Thus, a key problem posed by agency theory is how to guarantee the interests of a company's owners while reducing agency costs. Hillman and Dalziel (2003) argue that the board of directors is the key to the reconciliation of benefits between shareholders and managers. Accordingly, among the most urgent measures in today's corporate governance is devising an effective board structure.

2.1.2 Resource Dependence Theory

Unlike agency theory, which concerns issues between ownership and management, the focus of resource dependence theory is on the association of enterprises with their external environment. Encompassing various different resources such as labour, equipment, raw materials and information, the external environment plays an important role in the decision-making process in an organisation. Therefore, the board of directors acts as a bridge between the enterprise and the external environment, thus reducing the uncertainty in operations from external and non-controllable factors. According to Gabrielsson and Huse (2004), resource dependence theory appears to be useful for the analysis of board functions and actions.

Eisenhardt (1989) proposed that agency theory only explains part of the "big picture" of a business. In addition, this theory seems unable to sufficiently mirror the reality of corporate governance in all contexts, analysed by differences in corporate characteristics in each country (Young et al. 2008). Based on similar arguments, Hillman and Dalziel (2003) and Nicholson and Kiel (2007) suggest that agency theory should be supplemented by resource dependence theory in the corporate governance literature.

2.2   Empirical Evidence

2.2.1   Impact of Board Diversity on Firm Performance

There are numerous studies in the literature on the role of women in strengthening firm performance. Empirical results seem inconsistent with regard to the relationship between board gender diversity and business performance. Some studies have found a positive association between board diversity and the performance of firms, while others have reached the conclusion that there is a negative link, or even no link.

Erhardt et al. (2003) conducted a study on the relationship between gender diversity in the boardroom and the performance of 127 major corporations in the US over the period of 1993–1998. By employing two dependent variables, namely ROA and return on investment (ROI) to measure firm performance, and the percentage of female directors on the board to represent board diversity, the research results reveal that the proportion of female directors on the board appears to be positively correlated with both financial performance indicators, viz. ROA and ROI. This proves that board diversity has a positive impact on a firm's financial performance.

Campbell and Mínguez-Vera (2008) studied the relationship between gender diversity on the board and the performance of 68 Spanish companies between 1995 and 2000, employing the fixed-effect model and the two-stage least squares (2SLS) approach to control endogenous problems. The board diversity variable was measured by the percentage of female directors on the board, the Shannon index and the Blau index. Business performance was proxied by Tobin's Q ratio. The research findings confirm that board diversity positively affects firm performance and the causal effects seem negligible.

Most recently, in a study of 170 non-financial listed companies in Vietnam over the period from 2010 to 2015, Pham and Hoang (2019) also confirmed that gender diversity, measured by the proportion and number of female directors on the board, exerts a significantly positive influence on firm performance. Such effects are primarily derived from women directors' executive power and management skills rather than their independence status.

In the opposite direction, based on the dataset of major corporations in the US between 1996 and 2003, Adams and Ferreira (2009) found that gender diversity on the board tends to strengthen monitoring functions, yet the empirical results pointed to a negative correlation between the percentage of female directors on the board and Tobin's Q index.

Likewise, in a study on 248 enterprises in Norway over the period of 2001–2009, Ahern and Dittmar (2012) concluded that as the proportion of female directors on the board rises by 10%, firms' financial performance, characterised by the Tobin's Q index, reduces by 12.4%. Though there exist both positive and negative dimensions, Rose (2007) found no evidence of the impact of board gender diversity on the performance (measured by Tobin's Q) of Danish companies.

In addition, Farrell and Hersch (2005) contend that women tend to be appointed to work for firms with higher performance. Specifically, based on a sample of 300

Fortune-500 companies during the 1990s, these authors concluded that businesses with a high level of ROA tend to appoint female directors to the board. If that is the case, board diversity should be treated as an endogenous variable in studies of the relationship between gender diversity and firm performance. There has been much debate in recent research such as Adams and Ferreira (2009) that gender diversity might only be an endogenous problem, implying that ignorance of the endogenous nature of such relationship may lead to unreliable estimates.

2.2.2 Impact of Board Size on Firm Performance

The positive impact of board size on firm performance has been explored in numerous studies. For instance, Beiner et al. (2006) investigated the impact of corporate governance on firm value, based on a dataset of 109 businesses in Switzerland. They affirmed a positive relationship between board size and firm value (measured by Tobin's Q index). This study also suggests that a large board would be beneficial to the management activities due to the complexity of the business environment as well as the diversity of corporate culture.

Meanwhile, other economic scholars have admitted a negative association between board size and business performance. Employing a large sample of 452 major industrial enterprises in the US between 1984 and 1991 and Tobin's Q index as a measure of firm value, Yermack (1996) points out that the size of the board is negatively correlated with the performance of firms, since an increase in the size of the board would cause far more agency costs and difficulties in reaching uniform decisions. In addition, on investigating the effect of board size on firm value (measured by Tobin's Q) in Singapore and Malaysia, Mak and Kusnadi (2005) found an inverse relationship between the number of directors on the board and business value. These findings seem to be in line with those in some other markets, such as the US (Yermack 1996). Such inverse correlation between the size of the board and the performance of firms can be generalised for different corporate governance systems.

Besides the positive and negative tendencies, Schultz et al. (2010), when examining the relationship between governance characteristics and business performance of firms (measured by the ASX 200 index) during the period from 2000 to 2007, indicated a statistically insignificant correlation between board size and firm performance after correcting for endogeneity issues.

As for the case of Vietnam, the study of Vo and Phan (2013) on 77 enterprises listed on the Ho Chi Minh City Stock Exchange over the 2006–2011 period admits that there exists an inverse correlation between the size of the board and firm value; in other words, the more directors sitting in the boardroom, the worse the firm's value becomes.

2.2.3   Impact of Non-Executive Directors on Firm Performance

According to the agency theory, a perfect board should have a higher proportion of non-executive members who are believed to produce outstanding performance thanks to their independence from supervisory activities. Fama and Jensen's (1983) study showed that non-executive directors have more motivation to protect the interests of shareholders, because of the importance of protecting the firm's reputation, as well as their reputation on the external labour market. Nicholson and Kiel (2007) argue that if the monitoring of the board results in high corporate performance, there would be less opportunities for managers to pursue self-interest at shareholders' costs, meaning that shareholders' benefits could be guaranteed. Therefore, the agency theory suggests that a higher proportion of non-executive directors would lead to better monitoring by the board.

Furthermore, the above consideration is consistent with the view of the resource dependence theory. Daily et al. (2003) argue that non-executive directors provide access to important resources in accordance with business requirements, and therefore, a higher proportion of non-executive directors could contribute positively to business performance improvement.

Bhagat and Bolton (2008) conducted a study on the relationship between corporate governance and business performance using two different measures. The correlation between non-executive directors and firm performance was found to be negative in the case of performance measured by ROA, yet insignificant in the case of Tobin's Q.

In addition, Kiel and Nicholson (2003) investigated the relationship between board structure and the performance of 348 listed companies in Australia. They demonstrated that the number of non-executive directors on the board showed no correlation with business performance measured by ROA. However, the study found a positive correlation in the case of firm performance measured using Tobin's Q index. Meanwhile, Hermalin and Weisbach (1991) argued that board structure has no impact on business performance; however, during the research process, these authors recognised that firm performance is mainly driven by managerial experience, but not by the proportion of non-executive board directors.

2.2.4   Impact of CEO Duality on Firm Performance

Empirical research on the relationship between CEO duality and business performance yields conflicting results.

Some scholars have pointed out that this relationship tends to be positive. Specifically, Donaldson and Davis (1991) observed 321 companies in the US and confirmed that CEO duality helps to improve business performance. Accordingly, the benefits for shareholders would increase when there is CEO duality, compared to the separation of the board chair and CEO (average increase of 14.5% as measured by ROE). Meanwhile, in the East Asian context, Haniffa and Hudaib (2006) have shown a significant negative relationship between duality and business performance (measured by ROA), implying that the separation of the positions of board chair and CEO could lead to better performance for firms. However, the shortcoming of the research conducted by Haniffa and Hudaib (2006) lies in not considering the endogeneity problems linked with corporate governance characteristics, which leads to less reliable estimates. It is argued that a high concentration of managerial function and monitoring function in a group of major shareholders (including members who are both board directors and senior executive managers) may pose serious challenges in terms of protecting the interests of other minority shareholders and maintaining an effective monitoring function. In other words, such a board leadership structure may facilitate self-interest behaviour among majority shareholders, which in turn may reduce firm performance, as predicted by the agency theory.

Despite conflicting results regarding the relationship between duality and business performance, there still remains consensus between policy makers, investors and shareholders that managerial duties should be separated from control decisions. In other words, a board chair should not act as the CEO of the company (non-CEO duality). In European countries, over 84% of companies distinguish between the chairman of the board and the CEO (Heidrick and Struggles 2007). In Vietnam, in accordance with Clause 3, Article 10 of Circular No. 121/2012/TT-BTC regulating corporate governance applicable to Vietnamese public companies: "*The chairman of the board of management must not concurrently hold the position of chief executive officer (or general director), unless it is annually approved at the annual general meeting of shareholders*".

2.2.5 Impact of Blockholder Ownership on Firm Performance

Agency theory suggests that concentration of ownership is one of the important mechanisms for monitoring managerial behaviour. The concentrated ownership by shareholders (such as institutional and individual investors, and blockholders) helps to mitigate agency problems arising from the separation of ownership and control (Shleifer and Vishny 1986). Hence, it is argued that the larger the proportion of shares held by blockholders, the stronger the power they will have to make management work for their benefits. Furthermore, holding a large proportion of the company assets provides institutional investors and/or blockholders with incentives to monitor managerial behaviour (Haniffa and Hudaib 2006). Although blockholder ownership is regarded as a mechanism to reduce the conflict between shareholders and management, it may be a potential source of conflict of interest between minority and majority shareholders.

However, the empirical evidence regarding the relationship between concentrated ownership (blockholder ownership is used as a proxy) and firm financial performance is unclear and inconclusive across different markets and also within the same market. For example, some studies have found no statistically significant relationship between ownership concentration and firm performance. The study of Demsetz and Lehn (1985) empirically tested 511 large enterprises in the US, with the observation of different forms in the firms' ownership structure, including ownership by individual investors, ownership by institutional investors and ownership by top five shareholders. The research results indicate that there is no link between ownership structure and business performance. In addition, a positive relationship has been found by others—in one example, Xu and Wang (1999) conducted a study of 300 Chinese listed enterprises during the period from 1993 to 1995, and found a positive correlation between centralised ownership structure and the profitability of an enterprise.

3 Data and Methodology

3.1 Sample and Data

As recommended in previous research, finance companies and banks were excluded from our sample since their liquidity and governance could be affected by different regulatory factors (see, e.g., Mak and Kusnadi 2005; Schultz et al. 2010; Nguyen et al. 2015). Therefore, our final sample only consists of 152 enterprises listed on HOSE. The study period spans six years, from 2011 to 2016. All data were collected from annual reports, management reports and board of direcrors' resolutions of sampled companies published on the Viet Stock's website (*finance.vietstock.vn*). Data on market capitalization (market value of firm's equity) and stock held by the 10 largest shareholders (*blocktop10*) were provided exclusively by Tai Viet Corporation.

3.2 Description of Variables

- **Dependent variable: Firm performance**
  In line with previous studies (e.g. Coles et al. 2012), this study incorporates Tobin's Q as a proxy for business performance. Tobin's Q could be worked out as follows.

$$Tobin's\ Q = \frac{Market\ value\ of\ firm's\ stock\ +\ Book\ value\ of\ debt}{Book\ value\ of\ total\ assets}$$

- **Explanatory variables: Governance characteristics**

  Explanatory variables in this study encompass:
  – The percentage of female directors on the board (Female), representing board diversity;
  – The percentage of non-executive directors on the board[1] (Nonexe);
  – The percentage of independent directors on the board (Indep);

---

[1] According to Clause 2, Article 2 of Circular No. 121/2012/TT-BTC, a non-executive director of the board of management is defined as a member of the board who is not the director (general director),

– CEO duality (Dual), a dummy variable, taking the value of 1 if the board chair is also CEO, and 0 otherwise;

– Board size (Bsize), indicating the total number of directors on the board;

– The percentage of ordinary shares held by shareholders with at least 5% holding to the total number of ordinary shares of a company (Block);

– The percentage of ordinary shares held by the ten largest shareholders to the total number of ordinary shares of a company (Blocktop10).

- **Control variables**

Control variables used for the regression model include: (i) firm size (Fsize), measured by taking the natural logarithm of the book value of total assets; (ii) firm age (Fage), indicating the number of years from the time the company first appeared on the HOSE; and (iii) leverage (Lev), measured as the ratio of the company's debt to its total assets.

3.3 Model Specification

Derived from Wintoki, Linck, and Netter (2012) research, we construct the baseline equation illustrating the impact of governance characteristics on firm performance as follows.

$$ln\,Q_{it} = \alpha + \beta Governance + \delta Control + \eta_i + \varepsilon_{it} \tag{1}$$

where:

- **Q**: Tobin's Q, denoting firm performance (dependent variable). To mitigate the possible effects of outliers, the natural logarithm of this index is taken.
- **Governance**: Corporate governance variables, which include the percentage of female directors on board (Female), the percentage of non-executive directors on board (Nonexe), CEO duality (Dual), board size (Bsize), the percentage of shares held by block-holders (Block).
- **Control**: Control variables, which consist of firm age (Fage), firm size (Fsize) and corporate leverage (Lev).

Theoretically, to estimate longitudinal data, either pooled ordinary least squares, fixed effect or random effect model could be adopted. Yet, as pointed out by Wintoki et al. (2012), endogeneity concerns might exist when considering governance characteristics. Endogeneity issues might arise from two main sources: unobservable characteristics across enterprises and simultaneity. In addition, as corporates' current performance and governance characteristics are influenced by their past financial performance, the relationship between corporate governance and firm performance is

---

deputy director (deputy general director), chief accountant or any other manager appointed by the board of management.

deemed dynamic in nature. To overcome such problems, system generalized method of moments (SGMM) estimator proposed by Blundell and Bond (1998) might have been a viable option.

A serious concern here is that, according to the American Statistical Association (ASA), the use of p-values in the aforementioned modelling approaches as a passport for arrving at an assertation in science could lead to unsettling distortion of the scientific process, thereby providing unreliable results (Amrhein et al. 2017). As remarked by Nguyen (2019), the use of p-values for conducting experimental tests should be put to rest once for all. The situation is not alike to a debate (in computer science) some time ago on "How to quantify uncertainty? As an additive or as a non additive measure?" where there was things such as"In defense of probability". There should be no "in defense of p-values". Rather, as in several areas such as physics, we should look for how to live without p-values, even this culture has been around with us for over a century, and, as a quote, "we have taught our students the wrong thing too well". Hence, the quest for more reliable alternatives to p-values in testing is essential.

One possibility may be to follow example of natural sciences, where models are compared and justified by their capacity to correctly predict, e.g., when we divide the available data into the training set and testing set, use only training set to determine parameters of the model and then check its quality by checking how well this model works on examples from the testing set. Yet, this ideal approach may require new experiments and innovative ideas, which seem mission impossible in a short time. Under the conditions of limited resources, Bayesian parameter estimation could be the optimum solution for this study. With this approach, the expert information—formally represented by the corresponding prior distribution—is used to make conclusions more reliable. Besides, the fairly straightforward transformation from p-value-based explanation into a more reliable Bayesian one is also a definite advantage when adopting such a new analysis framework.

4   Results

4.1   Descriptive Statistics Analysis

Table 1 reports summary statistics for the main variables used in this study. As can be seen, the mean of Tobin's Q is 1.04, which is higher than that of studies in other countries, such as Nguyen, Locke, Reddy (2014) for Singaporean market at 0.82. This proves that sampled companies have relatively high performance, since the mean of Tobin's Q ratio in excess of one implies that the business is running at a profit, under which stimulating investment is necessary. In terms of board diversity, the mean percentage of female directors on board stands at an approximate 16%, which is far higher than that of the Asian region (6%), as per Sussmuth-Dyckerhoff et al. (2012), Singapore (6.9%) and China (8.5%), according to Catalyst's statistics

Table 1 Summary statistics

Variable	Obs	Mean	Median	Std. Dev	Min	Max
Tobin's Q	912	1.04	0.94	0.45	0.34	5.83
The percentage of female directors (Female)	912	15.98	16.67	16.39	0.00	80.00
The percentage of non-executive directors (Nonexe)	912	62.24	60.00	16.71	0.00	100.00
The percentage of independent directors (Indep)	912	15.25	16.67	16.40	0.00	80.00
CEO duality (Dual)	912	0.31	0.00	0.46	0.00	1.00
Board size (Bsize)	912	5.75	5.00	1.21	4.00	11.00
Blockholder ownership (Block)	912	50.70	51.53	17.70	10.49	97.07
Blockholder ownership top 10 (Blocktop10)	912	56.84	58.50	16.89	20.00	98.00
Firm age (Fage)	912	5.84	6.00	2.76	1.00	15.00
Firm size (Fsize)	912	27.82	27.68	1.23	25.57	32.61
Leverage (Lev)	912	46.79	49.54	20.41	0.26	87.41

Source The Authors

(2012). Subsequently, on considering the independence of the board, on average, about 62% of board directors are non-executive and 15% are independent directors. Aside from that, as regards duality, about 31% of the chairpersons concurrently hold the CEO positions.

As for board size, the mean number of directors on board is five. Finally, in respect of the concentration of ownership, the mean value of the percentage of shares held by shareholders owning at least 5% of the common stock (Block) is about 51%, while 57% is the percentage of shares held by ten largest shareholders. Clearly, the concentration of share ownership in Vietnamese firms seems rather high.

4.2 Results and Discussion

Table 2 summarises the Bayesian estimation results based on Eq. (1). To decompose the role of the governance aspects in shaping firm performance, the baseline specification is split into two sub-models, specifically:

- Model (1) tests the effects of governance characteristics on corporate performance, in which explanatory variables encompass: the percentage of female directors on the board (Female), the percentage of non-executive directors (Nonexe), CEO duality (Dual), board size (Bsize) and blockholder ownership (Block);
- Model (2) re-estimates the baseline specification by replacing 'Nonexe' with 'Indep' (denoting the independence of the board), 'Block' with 'Blocktop10'

Table 2 Bayesian estimation results

Regressant: lnQ	Model (1)			Model (2)		
	Mean	Std. Dev	Equal-tailed (95% Credible Interval)		Mean	Std. Dev	Equal-tailed (95% Credible Interval)	
Female	0.522*	0.067	0.382	0.636	0.639*	0.073	0.501	0.793
Indep					0.033	0.086	−0.133	0.208
Nonexe	0.076	0.079	−0.079	0.235				
Dual	0.059	0.029	−0.004	0.114	0.056	0.031	−0.010	0.111
lnBsize	0.034	0.070	−0.106	0.176	0.081	0.068	−0.046	0.219
Blocktop10					0.597*	0.080	0.446	0.761
Block	0.495*	0.076	0.342	0.495				
Fage	−0.035	0.025	−0.084	0.013	−0.037	0.025	−0.083	0.013
Fsize	0.101*	0.006	0.091	0.113	0.101*	0.004	0.094	0.109
Lev	−0.457	0.058	−0.573	−0.348	−0.478	0.070	−0.611	−0.343
Constant	−1.962	0.037	−2.033	−1.889	−2.102	0.114	−2.322	−1.883

Note *Significance at the 5% level
Source The Authors

(a proxy for concentrated ownership structure) to check the robustness of our findings to alternative proxies for governance structures.

As can be seen from the results of models (1) and (2) (see Table 2), the coefficients of 'Indep' (a proxy representing board independence) and 'Blocktop10' (a proxy for ownership concentration) are analogous to those of 'Nonexe' and 'Block' in terms of sign and magnitude. It should be noted that no evidence was found related to the influence of either non-executive directors (Nonexe) and independent directors (Indep) on business performance. On the other hand, the coefficients on the remaining corporate governance variables appear essentially unchanged. Therefore, it is concluded that the robustness conditions for our Bayesian estimation persist to alternative proxies for corporate governance structures.

First, board gender diversity, i.e. the presence of female directors on the board, is found to be positively correlated with firm performance. This result favours the resource dependence theory, which claims that firms assemble benefits through three channels: advice and counselling, legitimacy and communication (Pfeffer and Salancik 2003). A gender-diverse board could help to reinforce these three channels. For instance, businesses may appoint female entrepreneurs to their board to sustain relationships with their female trade partners and consumers. Some firms regard their female leaders as fresh inspiration and connections with their female workers. Others, meanwhile, desire to incorporate female views into the key decisions of the board. Hence, gender diversity on the board helps to strengthen the board's reputation and the quality of their decisions, thereby benefiting businesses as a whole.

Second, the study recognises a positive yet statistically insignificant correlation between the size of the board and the financial outcomes of firms. This result is in line with the resource dependence theory, assuming that firms with a larger board size tend to foster their linkages with external resources, as well as deriving extra benefits thanks to the board directors' knowledge and experience. Alternatively, our finding could also be compared to the viewpoint of Beiner et al. (2006), that a large board could benefit the management of business outcomes a great deal owing to improvements in the quality of support and counselling activities, as well as the formation of corporate cultural diversity.

Third, our results also reveal that the presence of non-executive directors on the board has no impact on firm performance. According to Bhagat and Black (2002), there is no evidence that enterprises with numerous non-executive directors on the board deliver better performance than those without. These authors indicated that the performance of firms does not rely on the number of non-executive directors on the board, but that in reality, every single enterprise has its distinct non-executive board structure depending on the size and growth of its business. Furthermore, we found no evidence that CEO duality has an impact on business performance, conforming with the research of Mak and Kusnadi (2005) for the Singapore market.

Fourth, the concentration of ownership by blockholders (represented by 'block' variables) shows a positive correlation with firm performance at the 5% significance level. Under the agency theory, ownership concentration is among the most important mechanisms for monitoring management behaviour, as it helps to alleviate agency concerns arising from the separation between ownership and control decisions. Therefore, the larger the proportion of shares owned by blockholders, the stronger the power they will have to push management to serve their best interests (Shleifer and Vishny 1986). This result also coincides with the research of Xu and Wang (1999), revealing that the higher the ownership proportion of blockholders, the more likely it is that firm performance will be enhanced.

Finally, regarding control variables, whilst financial leverage (Lev) and firm age (Fage) show no significant impact, firm size (Fsize) is found to have a positive correlation with firms' financial performance at the 5% level of significance.

5   Concluding Remarks

This study empirically evaluates the relationship between governance characteristics (characterised by Female, Nonexe, Dual, lnBsize and Block) and corporate performance (measured by Tobin's Q) on a sample consisting of 152 non-financial enterprises listed on HOSE during the period of 2011–2016. By applying the Bayesian parameter estimation approach and then replacing necessary variables, namely 'Nonexe' and 'Block' with 'Indep' and 'Blocktop10', respectively, for robustness checking, we find that governance characteristics, viz. board gender diversity and blockholder ownership, exert a positive influence on firm performance. These findings tend to favour the propositions that: (i) gender diversity on the board helps to fortify corporate governance through a more severe monitoring and supervision of management activities (Adams and Ferreira 2009); (ii) the likelihood of errors when making strategic decisions in firms would be alleviated since the presence of female directors in board meetings helps improve the quality of discussions on complicated issues (Huse and Grethe Solberg 2006; Kravitz 2003; Pham and Hoang 2019); and (iii) the concentration of blockholder ownership might help alleviate the agency concerns arising from the separation between ownership and control decisions. Thus, an increase in the proportion of stock held by blockholders would bring greater motivation for them to monitor managers' performance towards their best interests. It is recommended from our findings that enterprises attach great importance to corporate governance characteristics as a fundamental requirement, thereby creating fresh momentum for better financial outcomes in the short run as well as sustainable development goals in the long run.

References

Adams, R. B., & Ferreira, D. (2009). Women in the boardroom and their impact on governance and performance. *Journal of Financial Economics, 94*(2), 291–309.

Ahern, K. R., & Dittmar, A. K. (2012). The changing of the boards: The impact on firm valuation of mandated female board representation. *The Quarterly Journal of Economics, 127*(1), 137–197.

Amrhein, V., Korner-Nievergelt, F., & Roth, T. (2017). The earth is flat (p > 0.05): Significance thresholds and the crisis of unreplicable research. *Peer J, 5*, e3544.

Beiner, S., Drobetz, W., Schmid, M. M., & Zimmermann, H. (2006). An integrated framework of corporate governance and firm valuation. *European Financial Management, 12*(2), 249–283.

Bhagat, S., & Black, B. (2002). The non-correlation between board independence and long term firm performance. *Journal of Corporation Law, 27*, 231–274.

Bhagat, S., & Bolton, B. (2008). Corporate governance and firm performance. *Journal of Corporate Finance, 14*(3), 257–273.

Blundell, R., & Bond, S. (1998). Initial conditions and moment restrictions in dynamic panel data models. *Journal of Econometrics, 87*(1), 115–143.

Buchinsky, F. J., & Chadha, N. K. (2017). To P or not to P: Backing Bayesian statistics. *Otolaryngology-Head and Neck Surgery, 157*(6), 915–918.

Campbell, K., & Mínguez-Vera, A. (2008). Gender diversity in the boardroom and firm financial performance. *Journal of Business Ethics, 83*(3), 435–451.

Coles, J. L., Lemmon, M. L., & Meschke, J. F. (2012). Structural models and endogeneity in corporate finance: The link between managerial ownership and corporate performance. *Journal of Financial Economics, 103*(1), 149–168.

Daily, C. M., Dalton, D. R., & Cannella, A. A., Jr. (2003). Corporate governance: Decades of dialogue and data. *Academy of Management Review, 28*(3), 371–382.

Demsetz, H., & Lehn, K. (1985). The structure of corporate ownership: Causes and consequences. *Journal of Political Economy, 93*(6), 1155–1177.

Doan, N. P., & Le, V. T. (2014). The impact of corporate governance on equitized firm's performance in Vietnam. *Journal of Economic and Development, 203*(May), 56–63.

Donaldson, L., & Davis, J. H. (1991). Stewardship theory or agency theory: CEO governance and shareholder returns. *Australian Journal of Management, 16*(1), 49–64.

Eisenhardt, K. M. (1989). Agency theory: An assessment and review. *Academy of Management Review, 14*(1), 57–74.

Erhardt, N. L., Werbel, J. D., & Shrader, C. B. (2003). Board of director diversity and firm financial performance. *Corporate Governance: An International Review, 11*(2), 102–111.

Fama, E. F., & Jensen, M. C. (1983). Separation of ownership and control. *The Journal of Law and Economics, 26*(2), 301–325.

Farrell, K. A., & Hersch, P. L. (2005). Additions to corporate boards: The effect of gender. *Journal of Corporate Finance, 11*(1–2), 85–106.

Gabrielsson, J., & Huse, M. (2004). Context, behavior, and evolution: Challenges in research on boards and governance. *International Studies of Management & Organization, 34*(2), 11–36.

Haniffa, R., & Hudaib, M. (2006). Corporate governance structure and performance of Malaysian listed companies. *Journal of Business Finance & Accounting, 33*(7–8), 1034–1062.

Heidrick and Struggles (2007). 10th Annual Board Effectiveness Study, 2006–2007, USC Center for Effective Organizations.

Hermalin, B. E., & Weisbach, M. S. (1991). The effects of board composition and direct incentives on firm performance. Financial management, 101–112.

Hillman, A. J., & Dalziel, T. (2003). Boards of directors and firm performance: Integrating agency and resource dependence perspectives. *Academy of Management Review, 28*(3), 383–396.

Huse, M., & Grethe Solberg, A. (2006). Gender-related boardroom dynamics: How Scandinavian women make and can make contributions on corporate boards. *Women in Management Review, 21*(2), 113–130.

Kiel, G. C., & Nicholson, G. J. (2003). Board composition and corporate performance: How the Australian experience informs contrasting theories of corporate governance. *Corporate Governance: An International Review, 11*(3), 189–205.

Kravitz, D. A. (2003). More women in the workplace: Is there a payoff in firm performance? *Academy of Management Perspectives, 17*(3), 148–149.

Mak, Y. T., & Kusnadi, Y. (2005). Size really matters: Further evidence on the negative relationship between board size and firm value. *Pacific-Basin Finance Journal, 13*(3), 301–318.

Nguyen, H. T. (2019). How to test without p-values? *Thailand Statistician, 17*(2), i–x.

Nguyen, T., Locke, S., & Reddy, K. (2014). A dynamic estimation of governance structures and financial performance for Singaporean companies. *Economic Modelling, 40*, 1–11.

Nguyen, T., Locke, S., & Reddy, K. (2015). Does boardroom gender diversity matter? Evidence from a transitional economy. *International Review of Economics & Finance, 37*, 184–202.

Nicholson, G. J., & Kiel, G. C. (2007). Can directors impact performance? A case based test of three theories of corporate governance. *Corporate Governance: An International Review, 15*(4), 585–608.

Pfeffer, J., & Salancik, G. R. (2003). The external control of organizations: A resource dependence perspective. Stanford University Press.

Pham, A. D., & Hoang, A. T. P. (2019). Does female representation on board improve firm performance? A case study of non-financial corporations in Vietnam. In: V. Kreinovich, N. Thach, N. Trung, D. Van Thanh (Eds), *Beyond traditional probabilistic methods in economics* (Vol. 809). ECONVN 2019. Studies in Computational Intelligence. Cham: Springer.

Rose, C. (2007). Does female board representation influence firm performance? The Danish evidence. *Corporate Governance: An International Review, 15*(2), 404–413.

Schultz, E. L., Tan, D. T., & Walsh, K. D. (2010). Endogeneity and the corporate governance-performance relation. *Australian Journal of Management, 35*(2), 145–163.

Shleifer, A., & Vishny, R. W. (1986). Large shareholders and corporate control. *Journal of political economy, 94*(3, Part 1), 461–488.

Sussmuth-Dyckerhoff, C., Wang, J., & Chen, J. (2012). *Women matter: An Asian perspective.* Washington, DC, USA: McKinsey & Co.

Vo, H. D., & Phan, B. G. T. (2013). Corporate governance and firm performance: Evidence from Vietnamese listed companies. *Journal of Economic Development., 275*(September), 1–15.

Wintoki, M. B., Linck, J. S., & Netter, J. M. (2012). Endogeneity and the dynamics of internal corporate governance. *Journal of Financial Economics, 105*(3), 581–606.

Xu, X., & Wang, Y. (1999). Ownership structure and corporate governance in Chinese stock companies. *China Economic Review, 10*(1), 75–98.

Yermack, D. (1996). Higher market valuation of companies with a small board of directors. *Journal of Financial Economics, 40*(2), 185–211.

Young, M. N., Peng, M. W., Ahlstrom, D., Bruton, G. D., & Jiang, Y. (2008). Corporate governance in emerging economies: A review of the principal–principal perspective. *Journal of Management Studies, 45*(1), 196–220.

Applying LSTM to Predict Firm Performance Based on Annual Reports: An Empirical Study from the Vietnam Stock Market

Tri D. B. Le, Man M. Ngo, Long K. Tran, and Vu N. Duong

Abstract Forecasting the firm performance based on financial information has been an interesting and challenging problem for analysts and managers. The results will help managers set a proper strategy to improve their firm performance and achieve goals. Therefore, there is a wide range of quantitative forecasting techniques that have been developed to deal with this problem. Among them, Long Short-Term Memory (LSTM) is a specific recurrent neural network (RNN) architecture that was designed to model temporal sequences and their long-range dependencies. In this study, we used Return on Assets (ROA) as a proxy to measure firm performance. Based on the financial items from financial reports of all firms listed on the Vietnam stock market from 2007 to 2017, we deployed the LSTM technique to predict the firms' ROA and compare the results to other previous methods. We found that LSTM networks to outperform memory-free classification methods, i.e., random forest (RAF), and logistic regression classifier (LOG) in detecting the negative ROA.

Keywords Firm performance forecast · Long short-term memory · Neural networks

T. D. B. Le
State Capital and Investment Corporation, Ho Chi Minh City, Vietnam
e-mail: ledinhbuutri@yahoo.com

M. M. Ngo (✉) · V. N. Duong
John von Neumann Institute, Ho Chi Minh City, Vietnam
e-mail: man.ngo@jvn.edu.vn

V. N. Duong
e-mail: vu.duong@jvn.edu.vn

L. K. Tran
Banking University of HCMC, Ho Chi Minh City, Vietnam
e-mail: longtk@buh.edu.vn

© The Editor(s) (if applicable) and The Author(s), under exclusive license
to Springer Nature Switzerland AG 2021
N. Ngoc Thach et al. (eds.), *Data Science for Financial Econometrics*,
Studies in Computational Intelligence 898,
https://doi.org/10.1007/978-3-030-48853-6_41

1 Introduction

The firm's performance forecasting plays a vital role in strategic planning and business decision-making. Firms' managers are now concerned about the utilisation of their assets to increase the performance of the company. To keep the competitiveness of a firm, managers are in pressure to enhance the efficiency of assets, and they must always keep an eye on company performance indicators. Therefore, forecasting performance is not only the goal but also the responsibility of the managers.

Different studies used different indicators to measure the performance of the company and most of the studies focus on profitability as proxy to evaluate the performance. Many indicators have been used, namely return on assets (ROA), return on equity (ROE), return on sales (ROS), and return on investment (ROI). Among them, ROA is the most popular measures used in previous studies (Maria Teresa 2010; Gul et al. 2011; Issah and Antwi 2017; Sorana 2015). ROA is defined as earnings before interest and taxes divided by total assets. It also shows how efficiently a company uses its assets to generate income. Performance efficiency is better measured by ROA because ROA eliminates the effect of leverage when a firm uses debt financing. In this study, we also use ROA as a proxy to measure the performance of firms.

Along with the development of data mining algorithms, data scientists have attempted to build predictive models that produce more accurate results. The following is a brief introduction of some algorithms widely applied in finance in the past ten years, namely support vector machines, decision trees, random forests and artificial neural networks.

The Support Vector Machine (SVM) has been developed since the middle of the 60s and has widely used in the field of forecasting the firm performance recently. Ding et al. (2008) used SVM to forecast the financial distress of high-tech Chinese listed manufacturing companies. They confirmed that SVM models outperform conventional statistical methods and back-propagation neural network. In Taiwan, (Yuan 2012) also used the Support Vector Regression (SVR) to forecast the sales volume of manufacturers. He introduced the new method, GA-SVR model, which is the combination between the Genetic Algorithms (GA) and the SVM. In his method, GA was used to optimise free parameters of SVM. GA-SVR improved the forecasting accuracy than artificial neural network and other traditional models. Zhang et al. (2015) combined Principal Component Analysis (PCA) and SVM to build the model for predicting the profitability of listed construction Chinese firms from 2001 to 2012. Their model also improved the protability forecasting result significantly with the accuracy exceeded 80% on average in 2003–2012. Up until now, SVM and its combinations are still good choices for classification and forecasting problems.

Another data mining algorithms are commonly used in finance are Decision Tree and Random Forest. Delen et al. (2013) applied the decision tree method to discover the potential relationships between the rm performance and nancial ratios. They used four popular decision tree algorithms, including CHAID, C5.0, QUEST and CRT, and made a comparative analysis to determine the most appropriate model. The results showed that the CHAID and C5.0 decision tree algorithms produced the best prediction accuracy. Recently, Zhou et al. (2017) also used the decision tree method combined with the improved filter feature selection method to predict the listing status of Chineses listed firms. They found that the decision tree C5.0 achieves the highest F-measure values, implying that it is the best method among decision tree algorithms. In Asia, Daisuke et al. (2017) applied weighted random forest to predict firm performance measures and compare the result with the score assigned by the credit agency. The research was conducted on a large sample of 1,700,000 Japanese firms from 2006 to 2014. Their proposed model showed high performance in forecasting of exit, sales growth, and profit growth.

In the same development process with the above two methods, Artificial Neural Network (ANN) techniques have emerged as a better predictive method with high accuracy result. Ince and Trafalis (Huseyin and Theodore 2008) used ANN and SVM to predict stock prices in the US market from 2000 to 2006. Their sample includes 1500 firms, in which 1300 were used for the training set and 200 were used for the testing set. Another critical research is the study of Geng et al. (2015) who would like to find the best appropriate model to predict the financial distress of 107 Chinese firms listed on the Shanghai Stock Exchange and Shenzhen Stock Exchange from 2001 to 2008. They used various models based on statistical probability theory, consisting of Neural network, C5.0 Decision Tree, CR Tree, QUEST, CHAID, logistic regression, decision list, Bayes net, discriminant analysis, and SVM. The results indicated that the neural network is more accurate than other classifiers. Besides, financial indicators, such as net profit margin of total assets, return on total assets, earnings per share, and cash flow, are the main contributing factors in the prediction of deterioration in profitability. Recenty, Lee et al. (2017) used a Restricted Boltzmann Machine (RBM) as the main component of a Deep Belief Network (DBN). After pre-training using the RBM, the proposed model was ne-tuned using a backpropagation algorithm to predict the corporate performance of 22 bio-pharmaceutical companies listed on the US stock market. They found that the prediction ability of the RBM-DBN was superior to that of the models constructed using the DBN, SVR, and FNN algorithms. Furthermore, the prediction accuracy of time-series data can be improved by using relatively recent data when ne-tuning the pre-trained network with a RBM.

In finance, Long Short-Term Memory (LSTM) is still rather new with little research confirming its validation. Most studies that applied LSTM mainly focus on forecasting the stock price. To name a few, Roondiwala et al. (2017) applied LSTM to predict the stock returns of the National Stock Exchange of India (NIFTY 50). Hiransha et al. (2018) applied four types of deep learning algorithms, including Multilayer Perceptron (MLP), Recurrent Neural Networks (RNN), LSTM and Convolutional Neural Network (CNN) for predicting the price of stocks listed on National Stock Exchange (NSE) of India and New York Stock Exchange (NYSE). CNN was confirmed outperforming the other models. Recently, Kim and Kim (2019) propose a model called the feature fusion long short-term memory-convolutional neural network (LSTM-CNN) to forecast stock prices. This method learned to predict stock prices from the combination of various features, which represents the stock characteristics and the stock chart images. They discovered that a candlestick chart is the most appropriate stock chart image to use to forecast stock prices. Thus, this study suggested that prediction error can be efficiently reduced by using a combination of temporal and image features rather than using these features separately.

In Vietnam, machine learning in finance is a new field, and there is no research applying LSMT to forecast the firms' performance. Can the LSTM be applied to predict the performance of Vietnamese companies? The main object of this study is to validate the LSTM model in forecasting the firm performance on the Vietnam stock market. Another object is to confront the performance of the LSTM and other previous models, using the same dataset from the Vietnam stock market in the period 2007 to 2017.

The rest of this article is organised as follows. Section 2 describes the methodology used in this paper, including the data, the LSTM method and the metrics. Section 3 shows the empirical results and discusses the relevant findings. The last section concludes the main findings, indicates implications and limitations.

2   Methodology

2.1   Data

The sample consists of all non-financial firms listed on the HOSE and HNX for which the data is available from 2007 to 2017. We use accounting information constituents from the FiinPro database. The number of companies has changed over the years, starting from 583 companies in 2007 and increasing to 1438 companies in 2017. The predicted variable—firm's performance—are measured through the status of ROA growth. It is a dummy variable, which takes the value of 1 if the growth rate is negative and takes the value of zero in other cases. Predictor variables are accounting items or financial ratios collected from the companies' financial statements. Most of the sections comply with the regulations on accounting information disclosure in accordance with the law of Vietnam. We attempt to use a large number of predictor variables to exploit the predictive power of data mining algorithms for big data. In particular, financial information of current year t is used to predict status of ROA of the next year $t + 1$.

2.2   Long Short Term Memory

In theory, classic recurrent neural networks (RNNs) have been used to describe the long term relationship in the input sequences. However, in practice, RNNs could face the computational problem. For examples, because of the computations involved in the process, which use finite-precision numbers, when using back-propagation, the gradients could tend to zero (vanishing gradient problem) or tend to infinity (exploring gradient problem). To deal with these above gradient problems, Hochreiter and

Fig. 1 The structure of
LSTM memory cell
following Graves and Olah
(Alex Graves 2013)

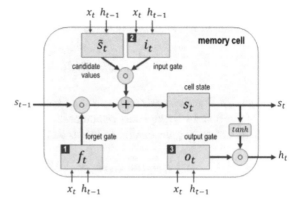

Schmidhuber (1997) have introduced the new method - RNNs using LSTM. Compare to standard feedforward neural networks, LSTM has a feedback connection, and it can process not only single data points but also entire sequences of data. Therefore, LSTM networks worked well in classifying, processing and making predictions based on time series data.

LSTM networks consist of an input layer, one or more hidden layers, and an output layer. The difference of LSTM compared to other RNNs methods is the memory, contained in hidden layers (memory cells). To maintain and adjust the cell states, each of the memory cells has three gates: input gate (i_t), output gate (o_t) and forget gate (f_t). Each gate is responsible for a specific function in the process. The input gate defines which information is added to the cell state, while the forget gate determine which information is removed from the cell state. The last gate, the output gates keep the information used as output from the cell state. Figure 1 illustrates the three components of the LSTM memory cell.

The following notations are used in Fig. 1 to describe the process.

- x_t is the input vector at timestep t
- $W_{f,x}, W_{f,h}, W_{\bar{s},x}, W_{\bar{s},h}, W_{i,x}, W_{i,h}, W_{o,x}, W_{o,h}$ are weight matrices.
- $b_f, b_{\bar{s}}, b_i, b_o$ are bias vectors.
- f_t, i_t, o_t are vectors for the activation values of the respective gates.
- $s_t, \bar{s}_t$ are vectors for the cell states and candidate values.
- h_t is a vector for the output of the LSTM layer.

The followings are four steps of the LSTM process.

Step 1: Computing the activation values f_t

The activation value f_t of the forget gates at time step t are computed based on the current input x_t, the outputs h_{t-1} of the memory cells at the previous time-step $(t-1)$, and the bias terms b_f of the forget gates:

$$f_t = sigmoid°(W_{f,x}x_t + W_{f,h}h_{t-1} + b_f),$$

where $sigmoid^{\circ}$ denote the element-wise application of the $sigmoid$ function to a vector or matrix. The $sigmoid$ function is defined as:

$$x \in \mathbf{R}, \quad sigmoid(x) = \frac{1}{1 + e^{-x}}.$$

The $sigmoid$ function scales all activation values into the range between 0 (completely forget) and 1 (completely remember).

Step 2: Computing the candidate values and the input gate values

$$\bar{s}_t = tanh^{\circ}(W_{\bar{s},x}x_t + W_{\bar{s},h}h_{t-1} + b_{\bar{s}}),$$

$$i_t = sigmod^{\circ}(W_{i,x}x_t + W_{i,h}h_{t-1} + b_i),$$

where $tanh^{\circ}$ denote the elementwise application of the $tanh$ function to a vector or matrix. The $tanh$ function is defined as:

$$x \in \mathbf{R}, \quad tanh(x) = \frac{e^x - e^{-x}}{e^x + e^{-x}}.$$

Step 3: Computing the new cell states

The LSTM layer determines which information should be added to the network's cell states s_t. The new cell states st are calculated based on the results of the previous two steps with ∘ denoting the Hadamard product (element-wise product):

$$s_t = f_t \circ s_{t-1} + i_t \circ \bar{s}_t$$

Step 4: Deriving the output

The output h_t of the memory cells is derived as denoted in the following two equations:

$$o_t = sigmoid^{\circ}(W_{o,x}x_t + W_{o,h}h_{t-1} + b_o)$$

$$h_t = o_t \circ tanh^{\circ}(s_t)$$

During training, and similar to traditional feed-forward networks, the weights and bias terms are adjusted in such a way that they minimize the loss of the specified objective function across the training samples. Since we are dealing with a classification problem, we use cross-entropy as the objective function.

2.3 Benchmark Models

In this study, we choose random forest and logistic regression for benchmarking the LSTM. We briefly outline the way we calibrate the benchmarking methods.

Random forest: Random forest algorithm was suggested by Ho (1995) and was expanded by Breiman (2001). It is an "ensemble learning" technique consisting of the bootstrapping of a large number of decision trees, resulting in a reduction of variance compared to the single decision trees. We used the Decrease Gini Impurity as the splitting criterion for each decision tree. The final classification is performed as a majority vote.

Logistic regression: Logistic regression is a form of regression which is used to predict probabilities of the presence of a specific object based on a set of attributes in any kind (continuous, discrete, or categorical).

Random forest and logistic regression serve as a baseline so that we can derive the added value of LSTM networks and make a comparative analysis with other standard classifications.

2.4 List of Performance Measures

The performance of models used in binary (two-groups) is often measured by using a confusion matrix (Table 1). A confusion matrix contains valuable information about the actual and predicted classifications created by the classification model (1998). For purposes of this study, we used well-known performance measures such as overall accuracy, AUC (Area Under ROC Curve), Precision, Recall and F-measure. All of these measures were used to evaluate each model in our study, after which the models were compared on the basis of the proposed performance measurements.

Overall Accuracy (AC): Accuracy is defined as the percentage of records that are correctly predicted by the model. It is also defined as being the ratio of correctly predicted cases to the total number of cases.

$$Accuracy = \frac{TP + TN}{TP + TN + FP + FN}$$

Precision: Precision is defined as the ratio of the number of True Positives (correctly predicted cases) to the sum of the True Positive and the False Positive.

Recall: Recall is also known as the Sensitivity or True Positive rate. It is defined as the ratio of the True Positive (the number of correctly predicted cases) to the sum of the True Positive and the False Negative.

Table 1 Confusion matrix for financial performance of firms

Model		Actual	
		Positive ROA(0)	Negative ROA(1)
Predicted	Positive ROA(0)	True positive	False negative
	Negative ROA(1)	False positive	True negative

Specificity: This is also known as the True Negative Rate (TN). It is defined as the ratio of the number of the True Negative to the sum of the True Negative and the False Positive.

F-Measure: F-measures take the harmonic mean of the Precision and Recall Performance measures. Therefore, it takes into consideration both the Precision and the Recall Performance as being important measurement tools for these calculations.

$$F_{measure} = \frac{2 \times Precision \times Recall}{Precision + Recall}$$

3   Empirical Results

Tables 2 and 3 show the average confusion matrix of LSTM and random regression models, respectively. Overall, the LSTM model has a strong forecast performance with a 62.15% accuracy rate, a 74.99% recall rate and 39.03% specificity rate, a 68.91% precision rate.

From Table 4, the LSTM demonstrated lower accuracy rates than random forest with 0.62% and 0.67%, respectively. However, LSTM has higher recall rate compare to random forest, suggesting that our proposed model can predict label 1 case better than random forest.

Table 2 Confusion matrix of LSTM model

Model		Actual	
		Positive ROA(0)	Negative ROA(1)
Predicted	Positive ROA(0)	1898	856
	Negative ROA(1)	633	548

Table 3 Confusion matrix of random forest model

Model		Actual	
		Positive ROA(0)	Negative ROA(1)
Predicted	Positive ROA(0)	2287	1027
	Negative ROA(1)	244	377

Table 4 Comparative analysis of LSTM and random forest

Metric	LSTM	Random forest
Accuracy	0.6216	0.6770
Recall	0.3903	0.2685
Specificity	0.7499	0.9036
Precision	0.4640	0.6071

Table 5 Comparative analysis of LSTM, random forest and logistic regression

Trials	Metrics	LSTM	Random Forest	Logistic Regression
1	F-measure	0.4166	0.3635	0.2194
	AUC	0.5847	0.6672	0.6432
2	F-measure	0.4173	0.3774	0.2485
	AUC	0.5967	0.6695	0.6557
3	F-measure	0.4208	0.3862	0.2265
	AUC	0.5886	0.6604	0.6441
4	F-measure	0.4537	0.3515	0.2028
	AUC	0.6541	0.6769	0.6330
5	F-measure	0.4125	0.3820	0.2085
	AUC	0.5821	0.6725	0.6498

As shown in Table 5, after five trials, the F-measure of the proposed LSTM model is always higher than that in random forest and logistic regression models. Note that in scikit-learn package, F-measure is calculated based on the forecast of negative-ROA firms (label 1). On the other hand, the AUC of LSTM is lower than other methods. The reason could be that our sample is unbalance, i.e, the number of positive-ROA firms (label 0) outweighs the number of negative-ROA firms (77% compared to 33%). In such a case, the F-measure is more significant than the AUC.

Because no research is close to perfection, we believe that some limitations exist and these could pave the way for further research over time. Firstly, the Vietnam stock market is still small and relatively young compared to other developed countries' stock markets. Therefore, the time horizon (120 months) in this research is still short. This problem will be gradually lessened by further research in the future when the market becomes more mature. Furthermore, most of the predictor variables in our sample are financial items extracted from financial statements which do not include external macro variables, such as economic growth, inflation rate or interest rate. However, in the long term, we believe that these factors will be reflected in the firms' financial ratios.

4 Conclusion

This study aims to forecast the Return on Assets of all Vietnamese firms listed on the Hanoi and Ho Chi Minh Stock Exchange based on financial information. The LSTM algorithm is used as the primary prediction tool, and the result is compared to other common models, such as random forest and logistic regression. The result showed that LSTM outperforms than other methods in forecasting negative ROA firms, through higher F-measure and recall rate. The research also lays the foundation for further studies of the using of complex neural networks as an alternative method to traditional forecasting methods in finance.

References

Breiman, L. (2001). Random forests. *Machine Learning*, *45*(1), 5–32.

Daisuke, M., Yuhei M., & Christian, P. (2017) Forecasting firm performance with machine learning: Evidence from Japanese firm-level data. *Research Institute of Economy, Trade and Industry (RIETI)*. Retrieved from https://www.rieti.go.jp/jp/publications/dp/17e068.pdf.

Delen, D., Kuzey, C., & Uyar, A. (2013). Measuring firm performance using financial ratios: A decision tree approach. *Expert Systems with Applications*, *40*(10), 3970–3983.

Ding, Y., Song, X., & Zen, Y. (2008). Forecasting financial condition of Chinese listed companies based on support vector machine. *Expert Systems with Applications*, *34*(4), 3081–3089.

Geng, R., Bose, I., & Chen, X. (2015). Prediction of financial distress: An empirical study of listed Chinese companies using data mining. *European Journal of Operational Research*, *241*(1), 236–247.

Graves, A. (2013). Generating sequences with recurrent neural networks. arXiv preprint arXiv:1308.0850.

Gul, S., Irshad, F., & Zaman, K. (2011). Factors affecting bank profitability in Pakistan. *Romanian Economic Journal,14*(39).

Hiransha, M., Ab Gopalakrishnan, E., Menon, V. K., & Soman, K. P. (2018). NSE stock market prediction using deep-learning models. *Procedia Computer Science,132*, 1351–1362.

Hochreiter, S., & Schmidhuber, J. (1997). Long short-term memory. *Neural Computation*, *9*(8), 1735–1780.

Huseyin, I., & Theodore, B. T. (2008). Short term forecasting with support vector machines and application to stock price prediction. *International Journal of General Systems,37*(6), 677–687.

Issah, M., & Antwi, S. (2017). Role of macroeconomic variables on firms' performance: Evidence from the UK. *Cogent Economics & Finance,5*(1), 1405581.

Kim, T., & Kim, H. Y. (2019). Forecasting stock prices with a feature fusion LSTM-CNN model using different representations of the same data. *PloS One,14*(2), e0212320.

Lee, J., Jang, D., & Park, S. (2017). Deep learning-based corporate performance prediction model considering technical capability. *Sustainability,9*(6), 899.

Maria Teresa Bosch-Badia. (2010). Connecting productivity to return on assets through financial statements: Extending the Dupont method. *International Journal of Accounting & Information Management*, *18*(2), 92–104.

Provost, F., & Kohavi, R. (1998). Guest editors' introduction: On applied research in machine learning. *Machine Learning*, *30*(2), 127–132.

Roondiwala, M., Patel, H., & Varma, S. (2017). Predicting stock prices using LSTM. *International Journal of Science and Research (IJSR)*, *6*(4), 1754–1756.

Sorana, V. (2015). Determinants of return on assets in Romania: A principal component analysis. *Timisoara Journal of Economics and Business*, *8*(s1), 32–47.

Yuan, F.-C. (2012). Parameters optimization using genetic algorithms in support vector regression for sales volume forecasting. *Applied Mathematics*, *3*(10), 1480.

Zhang, H., Yang, F., Li, Y., & Li, H. (2015). Predicting profitability of listed construction companies based on principal component analysis and support vector machine—Evidence from China. *Automation in Construction,53*, 22–28.

Zhou, L., Si, Y.-W., & Fujita, H. (2017). Predicting the listing statuses of Chinese-listed companies using decision trees combined with an improved filter feature selection method. *Knowledge-Based Systems,128*, 93–101.

Does Capital Affect Bank Risk in Vietnam: A Bayesian Approach

Van Dung Ha, Tran Xuan Linh Nguyen, Thu Hong Nhung Ta, and Manh Hung Nguyen

Abstract In this study, the effects of capital on bank risk in Vietnam are investigated over the period 2007–2017. Loan losses reserves/total assets (LLR) and non-performing loans/total loan (NPL) are considered as the proxies for bank risk. We also analyzed the effect of macroeconomic factors on the risk of bank besides capital that is measured by ratio of equity to total assets and the value of equity. Based on the Bayesian approach and with a dataset of annual reports of 30 banks, this paper reveals that bank capital and bank risk have a negative relationship. In addition, macroeconomic factors also have a relatively significant influence on the risk of bank.

Keywords Bank capital · Bayesian approach · Bank risk · Loan · Reserve · Assets

1 Introduction

Recent financial crises have emphasized the role of bank capital in coping with the spillover effects of shocks, especially from the perspective of competitiveness and survival of banks. In response to crises, capital safety regulations have been initiated globally in order to ensure a sound and stable banking system. Vietnam is also a part of this trend. The influence of financial crises on commercial banks of Vietnam has proved the importance of maintaining the stability and soundness of the

V. D. Ha (✉) · T. H. N. Ta · M. H. Nguyen
Banking University of Ho Chi Minh city, Ho Chi Minh City, Vietnam
e-mail: dunghv@buh.edu.vn

T. H. N. Ta
e-mail: nhungtth@buh.edu.vn

M. H. Nguyen
e-mail: hungnm@buh.edu.vn

T. X. L. Nguyen
Ho Chi Minh City Industry and Trade Colledge, Ho Chi Minh City, Vietnam
e-mail: xuanlinh86@gmail.com

N. Ngoc Thach et al. (eds.), *Data Science for Financial Econometrics*,
Studies in Computational Intelligence 898,
https://doi.org/10.1007/978-3-030-48853-6_42

banking system. Therefore, the identification of these factors, especially the role of capital on risks in banking activities, attracted the attention of many researchers but the results of the research are found to be non-consistent. In addition, these studies are primarily following the traditional frequentist econometric method. However, according to Briggs and Hung (2019), the p-value techniques testing is no longer appropriate in many cases because it inaccurately interprets statistical hypotheses and predictions. Therefore, this study was conducted for assessing the impact of capital and some macroeconomic factors on the risks of commercial banks in Vietnam. Based on research data of 30 domestic commercial banks in the period 2007–2017, the results of this study show that the size of equity and the ratio of equity to total assets are negatively related to the risk of banks. In addition, other macroeconomic factors, such as economic growth, M2 money supply, and inflation, also show an impact on bank risks.

2 Literature Review

With regard to the effect of capital on bank stability and risk, different economic theories have reported different forecasts. One such theory is the "regulatory hypothesis," which is often used to analyze the relationship between risks and capital of banks. This theory proposes that banks with low capital face regulations must increase their capital by reducing dividends to shareholders. Anginer and Demirgüç-Kunt (2014) reported that the goal of banks must be to gain high capital ratio in order to resist the income shocks and ensure adequate financial capacity in order to meet the withdrawal of deposits and other agreements of customers. They also proposed that higher capital buffers help bank owners to be more cautious and skillful in making their investment decisions. Accordingly, the policy "more skin in the game" helps to improve monitoring and controlling bank risks. However, the higher capital ratio reduces the pressure on bank liabilities and the risk of acquiring government bailouts (Demirgüç-Kunt et al. 2013). Many empirical studies supported this view. Jacques and Nigro (1997) have proved that high risk-based capital measures can reduce bank risks. Similarly, Aggarwal and Jacques (1998) used data of Federal Deposit Insurance Corporation (FDIC) of commercial banks from 1990 to 1993 and reported that banks tend to maintain capital ratios above the required reserves for preventing disruptions from unexpected serious circumstances. Furthermore, Editz et al. (1998) analyzed the relationship between regulation and bank stability using research data from commercial banks of UK, which showed that the provisions on reserve requirement exhibit a positive impact on the stability and soundness of the banking system and do not distort the lending ability of commercial banks. Moreover, Berger and Bouwman (1999) reported that capital shows a positive impact on the survival of small banks. Tan and Floros (2013) with data from commercial banks of China and Anginer and Demirgüç-Kunt (2014) with data from commercial banks of 48 countries reported an inverse relationship between measures of capital and risks of bank.

However, other theories do not follow "regulatory hypothesis," and assert that unqualified banks tend to accept excessive risks in order to maximize the shareholder value at the expense of depositors. In fact, managers can take advantage of deposit insurance schemes in order to carry out risky activities because the money of depositors is guaranteed which investments do not pay off (Lee et al. 2015). Demirgüç-Kunt and Kane (2002) supported the moral hazard hypothesis and suggests a negative relationship between capital and risk. When analyzing risks and capital of banks, the same pattern that is often applied is that the too-big-to-fail hypothesis has created ethical risk behaviors that lead to excessive risk activities taking into consideration both deposit insurance and government bailouts. Based on empirical studies, Koehn and Santomero (1980) have shown that a higher capital ratio increases variance in the total risk of the banking sector. Blum (1999) applied a dynamic framework and concluded that an increase in capital will eventually lead to increased risk. He explained that if raising capital to meet future standards is too costly, currently the only solution for banks is to increase the risk of the portfolio with the expectation that high returns will be received in order to meet the minimum capital requirements in the future. Previously, Kahane (1997), Koehn and Santomero (1980), and Kim and Santomero (1988) also supported this hypothesis. They claimed that banks could respond to regulatory actions that force them to increase their capital by increasing asset risk. While analyzing the relationship between capital size and risks of large banks in Europe from 1999 to 2004, Iannotta et al. (2007) observed a positive correlation between capital and loan losses.

From the above analyses, it can be concluded that the studies on the relationship between capital and risk of banks have demonstrated heterogeneous results. In addition, our review indicated that there is no specific study to investigate the impact of capital on the risks of the banking system of Vietnam. Therefore, an analysis of the effects of capital on risks in banking operations is conducted in this study. In addition, the effects of macroeconomic factors on the risks of the banking system of Vietnam are also considered in this study.

3   Research Method

Literature reviews indicate that the research on the relationship between capital and risk is carried out by using the frequentist approach, that is, by using the p-value to test the statistical hypothesis. However, Briggs and Hung (2019) reported that the interpretation of results and forecasting by this method are not accurate in many cases. The complicated algorithm of Bayes is relatively easier to handle with the development of computer science. Therefore, Bayesian econometric is being widely used in social research because of its effectiveness and flexibility besides the quantum approach (Nguyen et al. 2019). One advantage of the Bayesian approach is that the results do not depend adequately on the sample size. Therefore, the impact of capital on bank risk by using the Bayesian approach is assessed in this study. Bayesian statistics in contrast with traditional statistics are based on the assumption that all model

Table 1 Summary of the variables

Variable		Formula	Notation
Dependent	Loan loss reserve	Loan losses reserves/total assets	LLR
	Non-performing loans	Non-performing loans/total loan	NPL
Independent	Capital rate	Equity/total assets	CAP
	Equity	Log (equity)	EQT
	Inflation rate		INF
	GDP growth rate		GDP
	M2 growth		M2

parameters are random numbers, along with Bayes rules, which suggest combining prior information. It is evident from the fact that data collection is used to create the posterior distribution of model parameters (Table 1).

$$Posterior \propto Likelihood \times Prior$$

Based on the presented literature, the following research model is proposed:

$$\text{Model 1}: \ \text{LLR} = \alpha_0 + \alpha_1\text{CAP} + \alpha_2\text{EQT} + \alpha_3\text{INF} + \alpha_4\text{GDP} + \alpha_5\text{M2} + \varepsilon_1$$

$$\text{Model 2}: \ \text{NPL} = \beta_0 + \beta_1\text{CAP} + \beta_2\text{EQT} + \beta_3\text{INF} + \beta_4\text{GDP} + \beta_5\text{M2} + \varepsilon_2$$

where

Data of LLR, NPL, CAP, and EQT are collected from the financial statement of 30 domestic commercial banks of Vietnam for the period 2007–2017. Similarly, INF and GDP are obtained from the World Economic Outlook Database (IMF, 2019) and M2 is adapted from the report of the State Bank of Viet Nam.

We do not have prior information about the parameters of the model because the relevant studies were conducted according to traditional methods. In addition, the prior information does not have a large impact on the accuracy of the model because the sample size is relatively large. Therefore, we formulated three regression simulation models in this study and then selected the most suitable one among these three models. Prior information for the simulators is categorized as follows: completely noninformative, mildly informative, and more strongly informative priors. A simulation model is performed with a prior flat that is completely noninformative prior to the coefficients of the model. This prior information is widely applied in the Bayesian statistics when the studies have absolutely no prior information about the coefficients. The g-prior of Zellner is a relatively weak prior that is used for simulations with mild prior information. The standard distribution prior is applied

Table 2 The results of OLS regression

Source	SS	df	MS			
Model	.001094875	5	.000218975	Number of obs	=	273
Residual	.008301254	267	.000031091	F(5, 267)	=	7.04
				Prob > F	=	0.0000
				R-squared	=	0.1165
Total	.009396128	272	.000034545	Adj R-squared	=	0.1000
				Root MSE	=	.00558

Source The authors' calculation

to the third simulation. After regression with prior information as mentioned above, the Bayes factor is used to select the appropriate model.

Simulation 1 (for model 1) is expressed as follows:

Likelihood model:

$$LLR \sim N\left(\mu, \sigma^2\right)$$

Prior distributions:

$$\alpha_i \sim 1 \text{ (flat)}$$

$$\sigma^2 \sim \text{Jeffreys}$$

where μ is the average normal distribution of LLR, α is the vector of the coefficients (with i = 1, 2, 3, 4, 5), and δ^2 is the variance of the error term.

Simulation 2 is expressed as follows:

Likelihood model:

$$LLR \sim N\left(\mu, \sigma^2\right)$$

Prior distributions:

$$\alpha|\sigma^2 \sim \text{Zellner's } g \left(\text{dimension, df, prior mean, } \sigma^2\right)$$

$$\sigma^2 \sim \text{Invgamma}\left(\frac{v_0}{2}, \frac{v_0\sigma_0^2}{2}\right)$$

where v_0 is the prior degree of freedom—(df) and σ_0^2 is the residual of MS. We get the The value values of dimensions,df ,prior mean, σ_0^2, and v_0 obtained from the Ordinary ordinary least squares least-squares (OLS) regression results (Table 2).

Therefore, we have the following prior distributions:

$$\alpha|\sigma^2 \sim \text{zellnersg } (6, 272, .031, .002, -.012, .12, .001, -.03)$$

$$\sigma^2 \sim \text{Invgamma } (136, .004)$$

Likelihood model:

$$LLR \sim N\left(\mu, \sigma^2\right)$$

Prior distributions:

$$\alpha \sim N\,(1,\,100)$$

$$\sigma^2 \sim \text{Invgamma} \ (2.5, 2.5)$$

Similarly, simulations for model 2 are implemented.

4 Empirical Results

4.1 LLR

After performing the regression analysis, the Bayesian factor test is conducted (Table 3). Table 4 demonstrates the results.

In the Bayesian analysis, the model with the highest Log (BF) is selected. In addition, it is possible to consider the addition of Log (ML) and DIC. According to the analysis results, simulation 2 shows the highest Log (BF). This simulation also shows the largest Log (ML) and the smallest DIC. Therefore, the prior information of simulation 2 is considered the most suitable for model 1.

We used convergence diagnostics in order to prove that the Bayesian inference is valid. In this study, the diagnosis is performed visually by using the trace plot, autocorrelation plot, histogram, and kernel density plot. The results show that all parameter estimates illustrated in the diagrams are relatively reasonable. The trace plot and the autocorrelation plot show good mixing, i.e., low autocorrelation. Moreover, this diagram also demonstrates that all the latency is within the efficiency limit. In addition, the histograms (Fig. 1) match the density. The shape of the diagram is unimodal. Furthermore, histogram and density plots demonstrate the simulation of the standard distribution of parameters.

In addition, the visual diagnostics for the convergence of the MCMC chains also shows the rationality. The coefficient exhibiting the degree of autocorrelation also fluctuates at less than 0.02. The histogram also matches the density function.

According to regression results presented in Table 2, the higher the ratio of equity/total assets and the size of equity the higher the loan loss reserves/total assets rate. Specifically, the regression results of simulation 2 that is selected show that CAP increased by 1%, LLR increased by 0.02%, and Ln EQT increased by 1, and LLR increased by 0.018. This means that banks tend to increase lending when they have a high share of equity ratio and their equity size. Therefore, the reserve fund as compared to the total assets is also increased in order to control the operational risks. In addition, macroeconomic factors also have an impact on LLR. The regression results demonstrate that when GDP increases by 1%, LLR increases by 0.09%, M2 money supply increases by 1%, and LLR increases by 0.0004%, whereas an increase in inflation by 1% reduces LLR by 0.009%.

Table 3 Summary of regression results of model 1 with different prior information

	Mean	Std. Dev.	MCSE	Median	Equal-tailed [95% Cred. Interval]	
Simulation 1						
CAP	0.023131	0.005838	0.000289	0.023106	0.0110436	0.0347972
EQT	0.001767	0.000414	0.00002	0.001751	0.0009661	0.0025578
GDP	0.07775	0.050372	0.00511	0.079127	−0.020336	0.1721052
M2	−0.00033	0.004014	0.000248	−0.00052	−0.0083505	0.0077123
INF	−0.00954	0.005195	0.000194	−0.00964	−0.0194417	0.0007562
_cons	−0.02825	0.008054	0.000426	−0.02839	−0.0434699	−0.0123076
var	2.82E–05	2.17e	6	4.7e	8	0.0000281
Simulation 2						
CAP	0.023569	0.006243	0.000285	0.023422	0.0113935	0.0354219
EQT	0.001827	0.000407	0.000026	0.001827	0.0010636	0.0026241
GDP	0.09755	0.050015	0.005188	0.09571	0.0025182	0.1969749
M2	0.000419	0.003803	0.00015	0.000206	−0.0068507	0.0083107
INF	−0.00861	0.005079	0.000301	−0.00859	−0.0186238	0.0011435
_cons	−0.03063	0.008012	0.000771	−0.03049	−0.0464631	−0.014836
var	2.98E–05	1.73e	6	3.7e	8	0.0000297
Simulation 3						
CAP	0.005974	0.138171	0.009317	0.006239	−0.2657924	0.2555133
EQT	0.000751	0.009222	0.000665	0.000487	−0.0183064	0.0185061
GDP	0.055294	1.252078	0.077198	0.065536	−2.428667	2.508151
M2	0.004729	0.090186	0.004819	0.007455	−0.1665701	0.1815896
INF	0.007121	0.120938	0.007071	0.008676	−0.2313408	0.2421076
_cons	−0.01133	0.181193	0.013363	−0.01754	−0.3649928	0.3518753
var	0.015691	0.00125	0.000028	0.015651	0.0133995	0.0183095

Source The authors' calculation

Table 4 Analysis results of the Bayesian factor

	IC	Log (ML)	Log BF
Simulation 1	−2461.44	1206.35	1011.4
Simulation 2	−2462.99	1206.41	1011.46
Simulation 3	−737.7352	194.95	

Source The authors' calculation

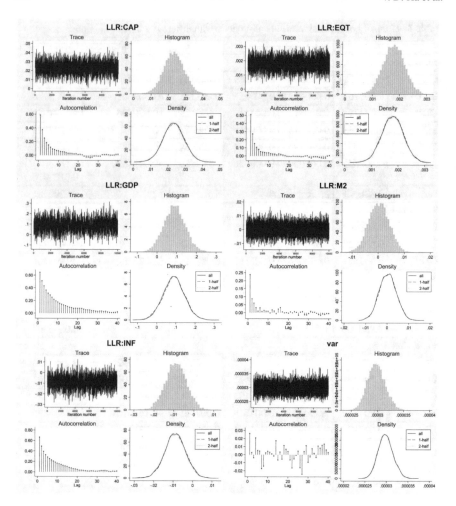

Fig. 1 Convergence test for model 1

4.2 *NPL*

In this paper, model 2 shows the relationship between NPL and capital and macrofactors. Model 2 is also analyzed with three simulations similar to as discussed above:

Simulation 4:

$$\beta_i \sim 1 \text{ (flat)}$$

$$\sigma^2 \sim \text{Jeffreys}$$

Simulation 5:

$$\beta|\sigma^2 \sim \text{Zellner's g } (6, 272, -0.008, -0.001, -0.017, -0.744, -0.03)$$

$$\sigma^2 \sim \text{Invgamma } (136, 0.048)$$

Simulation 6:
$$\beta \sim N (1, 100)$$

$$\sigma^2 \sim \text{Invgamma } (2.5, 2.5)$$

According to analysis results, simulation with prior information flat for coefficients and Jeffery for variance are found to be most suitable.

Convergence diagnostic results also show that the parameter estimates in the model are reasonable, variance fluctuates around 0.02, and thus the Bayesian inference is considered to be valid.

According to the regression results from 2007 to 2017, the higher the scale of equity and capital ratio the lower the value of NPL. Specifically, the ratio of the equity/total capital increased by 1%, the value of NPL decreased by only 0.0077%, the Ln of the equity increased by 1, and the NPL decreased by only 0.00087. This indicates that banks with an advantage in equity size and the ratio of equity/total assets tend to select a safe lending portfolio (Tables 5 and 6). Therefore, they have lower bad debt ratio/total loan. However, the impact of capital on bad debt is lower as

Table 5 Analysis results of the Bayesian factor

	DIC	Log (ML)	Log BF
Simulation 4	−1739.889	852.1541	659.2266
Simulation 5	−1731.661	829.0314	639.1039
Simulation 6	−728.2858	192.9274	

Source The authors' calculation

Table 6 Regression result of model 2

		Mean	Std. Dev.	MCSE	Median	Equal-tailed [95% Cred. Interval]	
NPL							
	CAP	-.0077119	.0175376	-.001738	-.0075312	-.0407883	.0271615
	EQT	-.00087	.0011937	.000093	-.0008661	-.0032192	.0015374
	GDP	-.7061874	.1588397	.041914	-.7108461	-.990547	-.3693992
	M2	-.0312049	.0115406	.000486	-.030861	-.0539744	-.0089238
	INF	-.0158642	.0149511	.001075	-.0156798	-.0441282	.0139563
	_cons	-.089717	.0228693	.004191	-.0907699	-.044688	-.1322454
var		.0002633	.0000207	4.3e-07	.0002622	.000226	.0003076
Acceptance rate		.3903					
Efficiency: min		.001436					
	avg	.0483					
	max	.2312					

Source The authors' calculation

compared to that of macroeconomic factors. Consequently, if GDP increases by 1%, the ratio of bad debt/total debt decreases to 0.7%, which indicates that enterprises operate effectively when the economy is prospering, and as a result their creditworthiness is better. In addition, money supply also helps in reducing the bad debt ratio. Therefore, when the money supply increases, interest rates in the economy decrease and businesses access capital more easily, and thus banks operate more efficiently, their financial capability becomes better, and bad debts reduce. Interestingly, inflation also helps in reducing bad debts. The fact that the State Bank of Viet Nam (SBV) introduces monetary tightening policies to reduce inflation when the increase in inflation proves the above results. Hence, banks reduce their lending and thus provide credit to customers with good financial capacity, and consequently increase the quality of the loan.

5 Conclusion

In this study, the impact of capital on bank risks in Vietnam is evaluated. According to the above results, banks have the advantage of the size of equity. The higher the ratio of equity/total assets, the lower the ratio of bad debt/total debt and the higher the loan losses reserves/total assets ratio, and thus it is safe for banks to operate. In addition to capital factor, the impact of macroeconomic factors on banking activity is also assessed in this paper and it is observed that the impact of these factors is relatively large. When the economy prospers and the economic growth is high, SBV increases the money supply to stimulate economic growth, the operation of enterprises becomes more favorable, their creditworthiness becomes better, and consequently the risk of the banking system is reduced. In this study, inflation is considered to be an interesting finding as it improves the NPL ratio of the banking system. The fact that inflation has restricted the supply of credit, so banks tend to select customers with financial capacity, good repayment ability, and thus bad debt is reduced proves the above results.

According to the results of the above analysis, capital has an impact on bank activities, but its role is not critical for the risk of banking operations. However, the economic situation shows a significant impact on the risks of banking operations. Therefore, banks should rely on the economic situation in order to devise operational strategies for ensuring operational safety.

References

Aggarwal, R. K., & Jacques, K. (1998). Assessing the impact of prompt corrective action on bank capital and risk. *Economic Policy Review, 4*(3), 23–32.
Anginer, D., Demirgüç-Kunt, A., & Zhu, M. (2014). How does bank competition affect systemic stability? *Journal of Financial Intermediation, 23*(1), 1–26.

Beck, T., Demirgüç-Kunt, A., and Merrouche, O. (2013). Islamic vs. conventional banking: Business model, efficiency and stability. *Journal of Banking & Finance, 37*(2), 433–447.

Berger, A. N., & Bouwman, C. H. S. (1999). How does capital affect bank performance during financial crises? *Journal of Financial Economics, 109*(1), 146–176.

Blum, J. (1999). Do capital adequacy requirements reduce risks in banking? *Journal of Banking Finance, 23*(5), 755–771.

Briggs, W. M., and Hung T. Nguyen (2019). Clarifying ASA's view on P-values in hypothesis testing. *Asian Journal of Economics and Banking, 3*(2).

Lee, C., Ning, S., & Lee, C. (2015). How does bank capital affect bank profitability and risk? Evidence from China's WTO accession. *China & World Economy, 23*(4), 19–39.

Demirgüç-Kunt, A., & Kane, E. (2002). Deposit insurance around the world: Where does it work? *Journal of Economic Perspectives, 16*(2), 175–95.

Editz, T., Michael, I. and Perraudin, W. (1998). The impact of capital requirements on U.K. bank behaviour. *Reserve Bank of New York Policy Review, 4*(3), 15–22.

Iannotta, G., Nocera, G., & Sironi, A. (2007). Ownership structure, risk and performance in the European banking industry. *Journal of Banking & Finance, 31*(7), 2127–2149.

Jacques, K., & Nigro, P. (1997). Risk-based capital, portfolio risk and bank capital: A simultaneous equations approach. *Journal of Economics and Business, 49*(6), 533–547.

Kahane, Y. (1997). Capital adequacy and the regulation of financial intermediaries. *Journal of Banking and Finance, 1*(2), 207–18.

Kim, D., & Santomero, A. M. (1988). Risk in banking and capital regulation. *Journal of Finance, 43*(5), 1219–33.

Koehn, M., & Santomero, A. (1980). Regulation of bank capital and portfolio risk. *Journal of Finance, 35*(5), 1235–45.

Nguyen, H. T., Sriboonchitta, S., Thach, N. N., & (2019). On Quantum Probability Calculus for Modeling Economic Decisions. In: Kreinovich V., Sriboonchitta S. (eds) Structural Changes and their Econometric Modeling. TES,. (2019). *Studies in computational intelligence, 808* (pp. 18–34). Cham: Springer.

Tan, Y., & Floros, F. (2013). Risk, capital and efficiency in Chinese banking. *International Financial Markets, Institutions, and Money, 26*, 378–393.

Printed in the United States
by Baker & Taylor Publisher Services